Proceedings of International Conference on Advanced Materials in Engineering Sciences – ICAMES2023

N. V. R. Naidu, Chief Editor

Dr. N. V. R. Naidu is currently serving as Principal at M S Ramaiah Institute of Technology, Bangalore, India. With over 42 years of experience in academia and Research. Dr. Naidu academic journey includes B.Tech. in Mechanical Engineering, M.Tech and Ph.D. in Industrial Engineering from Sri Venkateshwara University. He has received Dr. J Mahajan Award (2008–09) from the Indian Institution of Industrial Engineering. Throughout his illustrious career, he has held pivotal roles such as Dean Faculty of Engineering and Executive council Member at Visvesvaraya Technological University, reflecting his dedication to the advancement of academic excellence. He is serving on various academic boards and contributing his expertise to prestigious institutions in the field of Engineering. Dr. Naidu's expertise extends to research, evident in his 140+ published papers in prestigious national and international journals and conferences. He contributes as a reviewer for reputable journals and has chaired sessions at various conferences. Dr. Naidu has authored nine textbooks spanning Management, Total Quality Management, and Operations Research. Dr. Naidu has significantly impacted academic research, guiding seven doctoral candidates in areas like robust design and supply chain networks and he is adjudicator for PhD theses across India. His international engagements in countries like the USA, Japan, and Canada have enhanced industry-institution interactions. He has executed numerous funded programs and projects, advancing academia and industry collaboration.

G. M. Madhu, Editor

Dr. G. M. Madhu is currently working as a Professor at the Department of Chemical Engineering and serves as the IQAC Coordinator at M S Ramaiah Institute of Technology, Bangalore, India. He holds degrees in Chemical Engineering at the undergraduate, postgraduate, and doctoral levels. With twenty-three years of experience in both research and teaching, his expertise lies in nanomaterial synthesis and applications, water purification technologies, and nano composites. Dr. Madhu has contributed significantly to academia, having authored 105 papers in peer-reviewed national and international journals, as well as presenting 65 papers at national and international conferences. His research endeavours have been supported by funded projects from government and non-governmental agencies. Under his mentorship, ten students have successfully completed their PhDs, while two students have attained M.Sc. Engineering degrees. Currently, Dr. Madhu is supervising five students pursuing their PhDs. He had delivered numerous invited talks at both national and international conferences. Dr. Madhu is an esteemed life member of IIChE, IEI, InDA, and IAENG.

Nagaraju Kottam, Co-editor

Dr. Nagaraju Kottam is currently working as an Associate Professor in the Department of Chemistry, Ramaiah Institute of Technology, Bengaluru. He has obtained his PhD in Physical Chemistry from Central College, Bangalore University. His research interests include the design and development of advanced nanomaterials and their composites for wastewater treatment, hydrogen generation, electro-chemical, and biosensors applications. He has to his credit more than 75 research publications in refereed international journals, 06 Patents, 03 book chapters. He has guided 09 Ph.D's for their doctoral degrees and another 02 are on-going in which one is DST-Inspire fellow. He has mentored more than 15 UG/PG projects and the majority have received best project awards. Recently, received as the **best innovative project award of the year 2024** and 2020 from KSTA and KSCST, Bengaluru. He earned some government/institutional-funded projects and received several prestigious awards including the **"Award of Excellence in Teaching" from MSRIT for the year 2018–19** for his contribution to research and teaching. There is on-going institute level DST-FIST program. He had reviewed more than hundred articles in international journals and was awarded the Outstanding Reviewer award from an Elsevier journal and also served as guest editor for a few journals. Dr. Nagaraju was a recipient of the INSA-IAS summer research fellowship in 2014. He was also nominated as an expert member for evaluating Switzerland's government-funded project in 2020.

G. N. Anil Kumar, Co-editor

Dr. G. N. Anil Kumar is working as Assistant Professor in the Department of Physics, Ramaiah Institute of Technology. He has obtained his PhD in chemical crystallography from Bangalore University in 2012. His Research interests include Single crystal and Powder x-ray diffraction studies on organic & inorganic materials, Co-crystal design and *ab-initio* studies of solids. He has published 50 research articles in international peer reviewed journals and completed an external funded project.

Advanced Materials in Engineering Applications

Proceedings of International Conference on Advanced
Materials in Engineering Sciences – ICAMES2023

Edited by
N. V. R. Naidu

G. M. Madhu

Nagaraju Kottam

G. N. Anil Kumar

CRC Press
Taylor & Francis Group
Boca Raton London New York

CRC Press is an imprint of the
Taylor & Francis Group, an **informa** business

First edition published 2024
by CRC Press
4 Park Square, Milton Park, Abingdon, Oxon, OX14 4RN

and by CRC Press
2385 NW Executive Center Drive, Suite 320, Boca Raton FL 33431

CRC Press is an imprint of Informa UK Limited

British Library Cataloguing-in-Publication Data
A catalogue record for this book is available from the British Library
ISBN: 9781032900469 (pbk)
ISBN: 9781003545941 (ebk)

DOI: 10.1201/9781003545941

Typeset in Sabon LT Pro
by HBK Digital

Contents

List of figures

List of tables

1 Formability characteristics of Al 8011 alloy sheets

A. Mohan[1,a], R. B. Uppara[2], P. Jagadeesh[3], B. S. Navaneeth[4], R. K. Pandey[5], and S. Muthukumarasamy[1]

[1]Department of Mechanical Engineering, Vel Tech Rangarajan Dr. Sagunthala R&D Institute of Science and Technology, Chennai, India
[2]Department of Civil Engineering, Srinivasa Ramanujan Institute of Technology, Ananthapur, India
[3]Department of Mechanical Engineering, K.S.R. College of Engineering, Tiruchengode, India
[4]Department of Aeronautical Engineering, Nehru Institute of Technology, Coimbatore, India
[5]Department of Civil Engineering, MATS University, Raipur, India

Abstract

The formability features of sheets made of the alloy Al 8011 are examined experimentally and the results are compared with the numerical ones in this research. Through an axisymmetric finite element simulation of the Erichsen cupping test, formability characteristics were evaluated. The Erichsen cupping test was used to examine the effects of several factors, including friction at the punch-sheet contact and sheet thickness. The nonlinear finite element method is used to calculate the dome height, stress, and strain values for the aluminum sheet, and the results are then compared to the numerical ones. The findings demonstrated that the Al 8011 alloy's formability greatly rises with increasing sheet thickness. The formability is significantly impacted by the lubricant. The application of the finite element technique to forecast the formability of Al 8011 alloy.

Keywords: Formability, Al 8081, Erichsen cupping test, finite element method, tensile strength

1. Introduction

The primary step in the manufacturing of automobiles is sheet forming. Reducing vehicle weight has been seen as the best approach as the automotive industry strives to improve the performance and economy of cars [1]. The body covers for automobiles are often composed of formed sheets that fulfill the purposes of serving as a cover, being strong enough to withstand impact, and maintaining a smooth surface quality [2]. Aluminum rolled sheets that have been formed into the desired profile in accordance with the vehicle design make up the automotive sheets. Sheet metal is plastically deformed throughout the shaping process using a hydraulic punch and die to get the desired pattern [3]. Sheet metal is plastically deformed throughout the shaping process using a hydraulic punch and die to get the desired pattern. Due to the material's shrinking cross-sectional area, the real stress in a forming process is always higher than the engineering stress [4]. Using finite element analysis tools, the deformation behavior of deep drawing and v-Die bending on sheet metal. The most appropriate value for several parameters was found after completing numerous simulation experiments in the deep drawing and v-die bending process [5]. Additionally, an elastic-plastic computational program was constructed to simulate subsequent deep drawing and v-die bending process. The outcomes of simulations and experiments with various process conditions were consistent [6]. With the use of a finite element modeling technique, the formability

[a]mohan@veltech.edu.in

DOI: 10.1201/9781003545941-1

properties of the hydraulic brake booster system in the prediction of deep drawing process parameters such die radius, punch diameter, and friction coefficient [7]. They arrived at the conclusion that the majority of the faults, including necking and thinning, were discovered in the second and third processes by using the Kevlar equation to anticipate the formation of a limit diagram [8]. Instability or defect prediction is important in sheet metal forming processes. Because they reduce part weight and offer a fashionable appearance, the sheets are punched before the forming process [9]. The aperture size and tendon width are the two main factors that determine how easily the punctured sheets can be formed. The failures are measured and simply depicted as a forming limit diagram in the experimental technique [10–12]. This study used an aluminum 8011 alloy sheet as the work material, and it compared the results from simulations and experiments for the features of formability at various thicknesses and lubrication levels [13–15].

2. Materials and Methods

The aluminum alloy 8011, chosen as the work material, has a high level of corrosion resistance in both industrial and saltwater environments. Better properties for weldability and formability can be found in the high-strength alloy 8011 [16–18]. This alloy is frequently used in pressure vessels, containers, welded tubes, irrigation, desalination units, and boiler vessels in the chemical industry. Si: 0.25%, Cu: 0.10%, Mn: 0.10%, Mg: 2.2–2.8%, Zn: 0.1%, Cr: 0.15–0.35%, other small impurities: 0.15%, and the remaining Aluminium make up the chemical makeup of the chosen material [19–21]. Figure 1.1 depicts the optical microstructure of the alloy Al 8011.

2.1. *Tensile Test Method*

The findings of the tensile test provide important knowledge about the behavior of the

material. Understandable material characteristics include ductility, where low ductile materials have a low resistance to fracture [22–23]. As illustrated in Figure 1.2, the tensile test specimens are extracted from three different places on the rolled Al 8011 sheets. As per ASTM standard E8, sheets were rolled in 0°, 45°, and 90° directions to create a tensile test specimen measuring 50 mm gauge in length and 12.5 mm width, as shown in Figure 1.3.

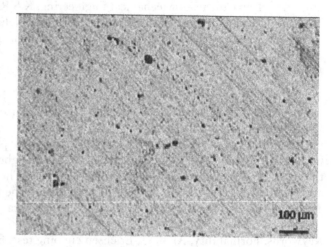

Figure 1.1: Optical microstructure of Al 8011 alloy.

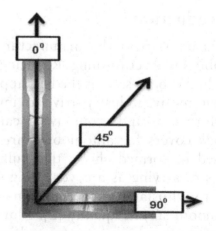

Figure 1.2: Tensile test specimen cut from three rolling directions.

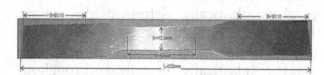

Figure 1.3: The dimensions of the tensile test specimen.

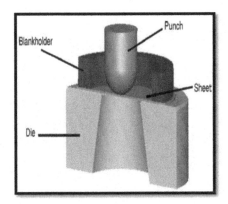

Figure 1.4: 3D design of Erichsen test design layout with major components.

Table 1.1: Process parameter for Erichsen cupping test

SI. NO.	Parameter	Values
1	Cross section of the sheet	85 × 85 mm
2	The initial thickness of the sheet	1 mm, 1.5 mm, 2 mm
3	Punch diameter	20 mm
4	Blank holder inner diameter	33 mm
5	Blank holder force	10 KN
6	Die inner diameter	27 mm
7	The die shoulder's radius of curvature	0.75 mm

The dimensions of the tensile test specimen are shown in Figure 1.3. These are L for overall specimen length, G for gauge length, B for grip section length, and C for grip section width.

The Erichsen cupping test, also known as a stretch forming test, is used to assess a material's properties of stretch formability [24]. Figure 1.4 depicts the Erichsen cupping testing apparatus and the specification. On formability characteristics, the impacts of sheet thickness of 1 mm, 1.5 mm, and 2 mm and frictional state (dry condition, "=0.3," Grease lubricant, "=0.21") between the sheet and punch were investigated.

In this project, a laser engraving equipment was used to create grid markings. The circles for the grid marking were drawn using the coral draw program, and the laser engraving machine used this drawing as input [16]. The specimens, which were square sheets measuring 85 mm × 85 mm, were cut from sheets that were 1, 1.5, and 2 mm thick. The circles were drawn on the sheets after the paste had been applied to them and they had been placed on the machine. The process settings for the Erichsen cupping test were set to the specified values shown in Table 1.1.

The Erichsen test experimental test process is described in detail. Slice the specimen on both sides using common grease. The punch top is set up for the test such that the micrometer reads zero and is in contact with

the plane [25]. The punch is initially squeezed gradually at a constant pace of 0.1 mm/sec. From the specimen's surface to the spot on the specimen's back side where the first crack appears, the Erichsen number is displayed in mm. When a crack in the sheet is noticed in the back mirror, the punch that is being pressed at the set speed is halted. Figure 1.5 depicts the specimen that was tested. Given that grid strain analysis is costly and time-consuming, as suggested by Keeler and Goodwin, fewer tests were conducted.

This was accomplished by etching before the specimen was formed circular designs on it. The circles are transformed into ellipses at the conclusion of the forming process, which corresponds to the minor and major strains caused in the specimen. The FLD plot, as depicted in Figure 1.6, which has major

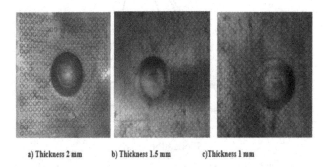

a) Thickness 2 mm   b) Thickness 1.5 mm   c) Thickness 1 mm

Figure 1.5 (a, b, c): Stretched specimen for various thicknesses.

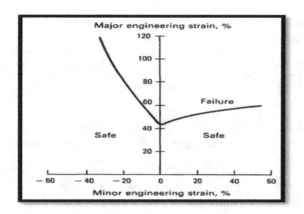

Figure 1.6: Forming Limit Diagram.

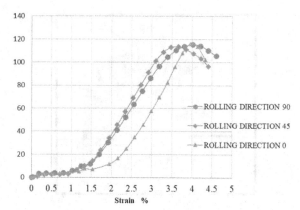

Figure 1.8: Stress vs. strain curve for the tensile test results.

and minor strains, predicts when the metal will fail. Forming limit diagrams illustrates the limiting strains that sheet metals can bear over a broad range of major-to-minor strain ratios.

3. Results and Discussion

Figure 1.7 depicts the Al8011 alloy's load extension curve for three distinct rolling directions. According to the findings, elongation is greater in the 0° rolling direction than in the 45° and 90° rolling directions. The elongation behavior for the 0° rolled aluminum sheets demonstrates their great ductility. This stress vs strain curve was also used to forecast the qualities of the material, as illustrated in Figure 1.8.

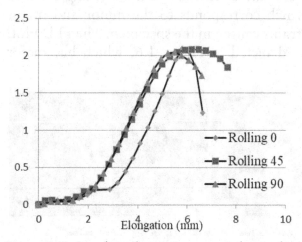

Figure 1.7: Load vs elongation curve shows the results of the tensile test.

Tensile tests were performed on the specimens for the rolling directions of 0°, 45°, and 90° at 30, 40, and 50 speed conditions.

Uniaxial tension testing yields values for a variety of material parameters relevant to formability, including tensile strength (114.66 MPa), uniform elongation (4.91%), young's modulus (69 GPa), poisson's ratio (0.3), yield strength (95 MPa), and tangent modulus (67 GPa). Figure 1.9(a) depicts a stretched specimen of 1 mm thickness creating a limit diagram under dry conditions. This number indicated a major strain value of 88% and a minor strain value of 4.5 percent. The failure zone and safe zone regions are depicted in the forming limit diagram. The ranges of the major and minor strain levels are 2.5% to 85% and 3.5% to 88%, respectively. There is an indication of a fracture zone when the major strain value exceeds 60%. Figure 1.9(b) depicts a stretched specimen with a thickness of 1 mm that creates a grease condition limit diagram. This figure represented the expected major strain of 7.9% and the expected minor strain of 4.5%. The range of minor strain levels is 4.5% to 55%, whereas the range of major strain values is 7.5% to 79%. In this figure, a fracture zone is indicated by the primary strain value of 65%.

A 1.5 mm thick stretched specimen developing a limit diagram under dry conditions is shown in Figure 1.9(c). This result indicated a major strain value of 88% and a minor strain value of 4.5%.

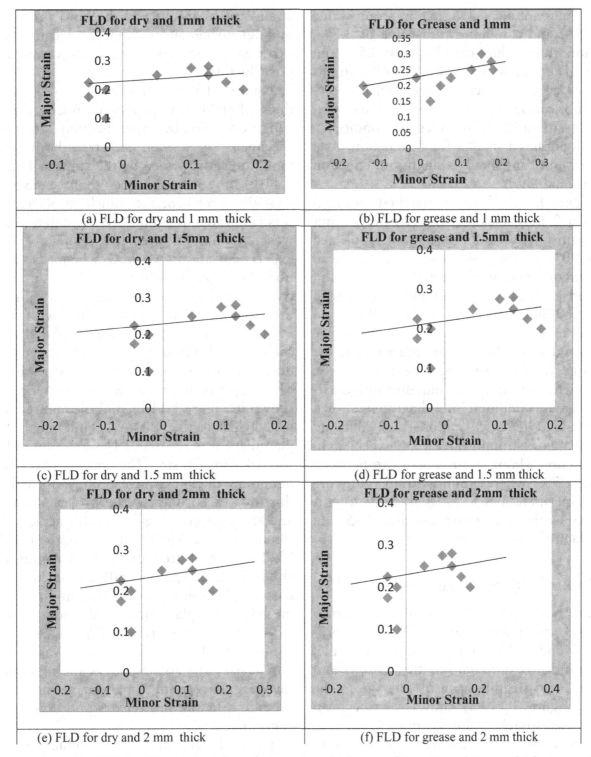

Figure 1.9 (a–f): FLD for 8011 alloy sheet for varying thickness of 1, 1.5, and 2 mm thickness and under dry and grease conditions.

There is variation in both the major and minor strain values, which range from 2.5% to 69% for the minor strain and from 3.5% to 88% for the main strain. If the primary strain value in this figure is greater than 60%, it indicates a fracture zone. A 1.5 mm thick stretched specimen creating a grease condition limit diagram is shown in Figure 1.9(d).

This figure indicated a major strain value of 78% and a minor strain value of 3.5%. Minor strain values vary between 3.5% and 55%. The primary strain value of 60% in this figure denotes a fracture zone. When applying lubrication, the FLD displays fewer failure sites than it would under dry conditions, which implies a smaller fracture region. As a result, lubricant shows a significant effect on the formability.

Figure 1.9(e) depicts a stretched specimen with a 2 mm thickness that creates a limit diagram in dry conditions. It was projected that 90% of the major strain and 2.5% of the minor strain would occur. The major and minor strain values exhibit fluctuations, with corresponding ranges of 3.5% to 90% and 2.5% to 65%. Major strain values above 65% and above 40% in this figure denote the fracture zone and necking zone, respectively. A stretched specimen of 2 mm thickness creating a grease condition limit diagram is shown in Figure 1.9(f). 6.5% minor strain value and 77% major strain value were expected in this figure. Major strain values are between 6 and 77%, whereas minor strain values are between 4.5% and 55%. A fracture zone is shown by the primary strain value of 65% in this figure.

3.1. *Finite Element Analysis Results*

Figure 1.10(a) shows the simulated outcome for a 1 mm thick Al 8011 alloy sheet produced under dry circumstances with FLD. According to the contour, the tension and strain build steadily from the die rim to the pole. Figure 1.11 (a) shows the specimen that was stretched during the experiment. In this situation, the failure happens close to the pole. 8.9 mm were taken as the punch displacement measurement. The Erichsen index was 8.32 mm in size. Results from the experiment and FEM showed a stronger link. Figure 1.12(a, b) depicts the outline of the major and minor strains in an aluminum sheet (1 mm thick) under dry conditions. The

strain-stress contour and the FEM projected findings further suggest that this location has experienced both the largest stress and the least thickness.

Figure 1.10(b) shows the FLD of a 1 mm thick Al 8011 sheet produced under greasing conditions and the experimentally stretched specimen. The tension and strain both grow monotonically as seen by the contour Figure 1.11(b) shows the specimen that was experimentally stretched; the punch displacement was calibrated at 9 mm. The Erichsen index was 8.56 mm in size. The findings of the experiment and FEM show a better correlation. Additionally, it was noted that in comparison to dry conditions, the stress value in the area around the pole was the lowest.

Figure 1.10(c) depicts the simulated outcome for a 1.5 mm thick Al 8011 sheet produced under dry circumstances with FLD. The experimentally stretched material is shown in Figure 1.11(c), and the failure happened close to the pole. A 11 mm punch displacement was recorded. The Erichsen index was 10.7 mm in size. The greatest stress and minimum thickness have both happened in this area, according to the projected strain and stress contour.Figure 1.10(d) depicts the simulated outcome for a 1.5 mm thick Al 8011 sheet produced under grease conditions and FLD. The experimentally stretched specimen is shown in Figure 1.11(d), and the failure has taken place close to the pole. It was determined that the punch displacement was 11.60 mm. The Erichsen index was 10.34 mm in size. The greatest stress and minimum thickness have both happened in this area, according to the projected strain and stress contour.

Figure 1.10(e) depicts the simulated outcome for a 2 mm thick Al 8011 sheet produced under dry circumstances and FLD. The illustration shows that the strain increases toward the pole as the tension increases monotonically from the die rim to the pole. The experimentally stretched specimen is shown in Figure 1.11(e), and the failure has

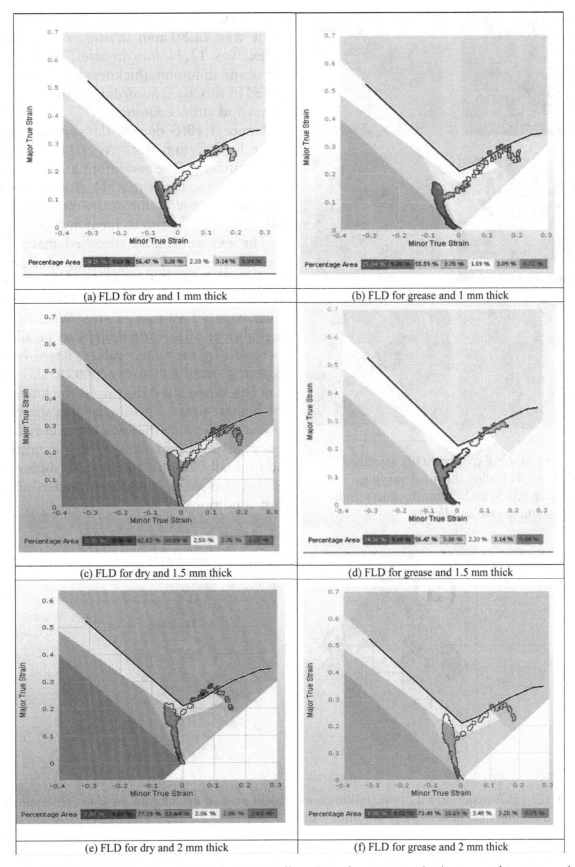

(a) FLD for dry and 1 mm thick

(b) FLD for grease and 1 mm thick

(c) FLD for dry and 1.5 mm thick

(d) FLD for grease and 1.5 mm thick

(e) FLD for dry and 2 mm thick

(f) FLD for grease and 2 mm thick

Figure 1.10 (a–f): The simulated FLD for 8011 alloy sheet for varying thicknesses of 1, 1.5, and 2 mm thickness and under dry and grease conditions.

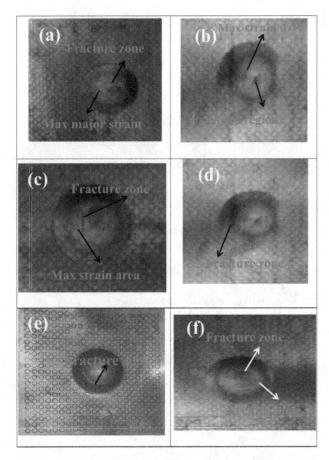

Figure 1.11 (a–f): Experimentally stretched specimen for 011 alloy sheet for varying thickness of 1, 1.5, and 2 mm thickness and under dry and grease conditions.

taken place close to the pole. Punch displacement was 12.80 mm in size. The Erichsen index was 12.34 mm in size. The greatest stress and minimum thickness have both happened in this area, according to the projected strain and stress contour.

Figure 1.10(f) depicts the simulated outcome for a 2 mm thick Al 8011 sheet produced under grease conditions and FLD. It is clear from the figure that the strain increases toward the pole as the stress monotonically rises from the die rim to the pole.

The experimentally stretched material is shown in Figure 1.11(f), and the failure has happened close to the pole. A 13 mm punch displacement was recorded. The Erichsen index was 12.49 mm in size. When applying lubricant, the formability can be silently increased by the stated values. Based on the expected strain and stress contour, this location has experienced the lowest thickness and the highest stress. This image also shows the lowest stress value that was experienced near the pole in comparison to dry conditions. Additionally, the fracture zone is diminished as sheet thickness is increased under lubrication conditions.

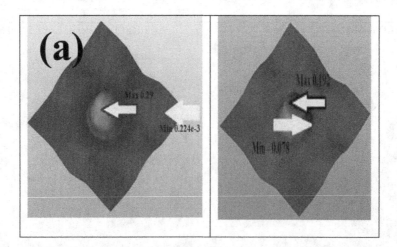

Figure 1.12: (a) A major strain of one millimeter in thickness and dry aluminum sheet; (b) a minor strain of the same thickness and dry aluminum sheet.

4. Conclusion

The formability of aluminum alloy sheet manufactured from alloy 8011 has been investigated in this study using the axisymmetric elastic-plastic finite element method. For the Erichsen test, calculations have been made and compared to experimental findings. The analysis above led to the following findings. Excellent agreement between the strain distribution and limit dome height and the experimental data is evident. With increasing sheet thickness, aluminum 8011 alloys become more formable. In this investigation, the impact of lubricants demonstrated a striking improvement in formability properties.

Acknowledgements

The authors express their sincere gratitude to the Department of Mechanical Engineering, Vel Tech Rangarajan Dr. Sagunthala R&D Institute of Science and Technology, Chennai, for providing the research facilities and necessary support throughout the entire phase of this research work.

References

[1] Anastasiou K. S. (2002). Optimization of the aluminum die-casting process based on the Taguchi method. *J. Eng. Manuf.*, 216, 969–977.

[2] Baik, S. C., Han, H. N., Lee, S. H., Oh, K. H., and Lee, D. N. (1997). Plastic behavior of perforated sheets under biaxial stress state, *Int. J. Mech. Sci.*, 39, 781–793.

[3] Baik, S. C., Han, H. N., Lee, S. H., Oh, K. H., and Lee, D. N. (2000). Plastic behavior of perforated sheets with slot-type holes under biaxial stress state, *Int. J. Mech. Sci.*, 42, 523–536.

[4] Baik, S. C., Oh, K. H., and Lee, D. N. (1995). Forming limit diagram of perforated sheet, *Scr. Metall. Mater.*, 33, 1201–1207.

[5] Chen, F. K. (1993). Analysis of plastic deformation for sheet metals with circular perforations, *J. Mater. Process. Technol.*, 37, 175–188.

[6] Garc, C., Celentano, D., Flores, F., Ponthot, J.-P., and Oliva, O. (2006), Numerical modeling and experimental validation of steel deep drawing processes, *J. Mater. Process. Technol.* 172, 461–471

[7] Tari, D. G., Worswick, M. J., Mckinley, J. and Bagheriasl, R. (2010), AZ31 magnesium deep drawing experiments and finite element simulation, *Int. J. Mater. Form.*, 3, 159–162.

[8] Yasar, M. and Kadi, I. (2007). High velocity forming of aluminum cylindrical cups-experiments and numerical simulations. *J. Mater. Sci. Technol*, 23(2), 230–236.

[9] Lademo, O. G., Pedersen, K. O., Berstad, T., Furu, T., and Hopperstad, O. S. (2008), An experimental and numerical study on the formability of textured AlZnMg alloys, *Eur. J. Mech. A/Solids*, 27, 116–140.

[10] Huang, Y. M., Tsai, Y. W., and Li, C. L. (2008). Analysis of forming limits in metal forming processes. *J. Mater. Process. Technol.*, 201(1–3), 385–389.

[11] Vijayakumar, M. D., et al. (2020). Experimental investigation on single point incremental forming of IS513Cr3 using response surface method, *Mat. Today Proc.*, 21, 902–907.

[12] Banu, M., Madhavan, V. R. B., Manickam, D., and Devarajan, C. (2021). Experimental investigation on stacking sequence of kevlar and natural fibres/epoxy polymer composites, Polimeros: Ciencia e Tecnologia, 31, 1–9.

[13] Murali, B., Ramnath, B. M. V., Rajamani, D., Nasr, E. A., Astarita, A., and Mohamed, H. (2022). Experimental investigations on dry sliding wear behaviour of kevlar and natural fiber-reinforced hybrid composites through an RSM–GRA hybrid approach, *Materials*, 15, 1–16.

[14] Murali, B., Yogesh, P., Karthickeyan, N. K., and Chandramohan, D. (2022). Multi-potency of botanicals (Flax, Hemp and Jute Fibers) as composite materials and their medicinal properties: a review, Mater. *Today Proc.*, 62, 1839–1843.

[15] Karthik, K. and Manimaran, A. (2020). Wear behaviour of ceramic particle reinforced hybrid polymer matrix composites. *Int. J. Ambient Energy*, 41, 1608–1612.

[16] Ganesh, R., Karthik, K., Manimaran, A., and Saleem, M. (2017). Vibration damping characteristics of cantilever beam using piezoelectric actuator. *Int J Mech Eng Technol*, 8(6), 212–221.

[17] Karthik, K., Ganesh, R., and Ramesh, T. (2017). Experimental investigation of hybrid polymer matrix composite for free vibration test. *Int. J. Mech. Eng. Technol.*, 8, 910–918.

[18] Ahmed, I., Renish, R. R., Karthik, K., and Karthik, M. (2017). Experimental investigation of polymer matrix composite for heat distortion temperature test. *Int. J. Mech. Eng. Technol*, 8(8), 520–528.

[19] Murali, B. and Nagarani, J. (2013, April). Design and fabrication of construction helmet by using hybrid composite material. In *2013 International Conference on Energy Efficient Technologies for Sustainability* (pp. 145–147). IEEE.

[20] Karthik, K., Prakash, J. U., Binoj, J. S., Mansingh, B. B. (2022). Effect of stacking sequence and silicon carbide nanoparticles on properties of carbon/glass/Kevlar fiber reinforced hybrid polymer composites, polymer composites, *Polym. Compos.*, 43, 6096–6105. https://doi.org/10.1002/pc.26912.

[21] Fayaz, H., Karthik, K., Christiyan, K. G., Arun Kumar, M., Sivakumar, A., Kaliappan, S., Mohamed, M., Subbiah, R., and Yishak, S. (2022). An investigation on the activation energy and thermal degradation of biocomposites of Jute/Bagasse/Coir/Nano TiO2/epoxy-reinforced polyaramid fibers. *J. Nanomater.*, 2022, 5. https://doi.org/10.1155/2022/3758212.

[22] Karthik, K., Rajamani, D., Venkatesan, E.P., Shajahan, M.I., Rajhi, A.A., Aabid, A., Baig, M. and Saleh, B. (2023). Experimental investigation of the mechanical properties of carbon/basalt/SiC nanoparticle/polyester hybrid composite materials. Crystals, 13, 415.

[23] Ramesh, V., Karthik, K., Cep, R. and Elangovan, M. (2023). Influence of stacking sequence on mechanical properties of basalt/ramie biodegradable hybrid polymer composites. *Polymers*, 15, 985.

[24] Jebarose Juliyana, S., Udaya Prakash, J., Čep, R. and Karthik, K. (2023). Multi-objective optimization of machining parameters for drilling LM5/ZrO2 composites using grey relational analysis. Materials, 16, 3615.

[25] Siva Kumar, M., Rajamani, D., El-Sherbeeny, A.M., Balasubramanian, E., Karthik, K., Hussein, H.M.A. and Astarita, A. (2022). Intelligent modeling and multi-response optimization of AWJC on fiber intermetallic laminates through a hybrid ANFIS-Salp swarm algorithm. *Materials*, 15, 7216.

2   Minds and materials in flight: The dynamic duo of smart materials and machine learning in aerospace evolution

Abhay Bhandarkar[1,a], Nagaraju Kottam[2,b], and D. Vishwachetan[3,c]

[1]Computer Science and Engineering Ramaiah Institute of Technology, Bangalore, India
[2]Department of Chemistry Ramaiah Institute of Technology, Bangalore, India
[3]Computer Science and Engineering Ramaiah Institute of Technology, Bangalore, India

Abstract

The development of smart materials and nanotechnology have led to a paradigm shift in the aerospace industry which have led to a transformation in the design and development of next generation of aircrafts. Equipped with the remarkable ability to intelligently respond to external stimuli, they have unlocked numerous possibilities to aerodynamics, structural health monitoring (SHM) and safety. Also, materials such as shape memory alloys, piezoelectric materials, carbon fibre reinforced polymers, and giant magnetostrictive materials have proven to have diverse applications in Aerospace. With production of materials using nanotechnology, materials have evolved to show unique properties and also exhibit self-healing capabilities. The structures that are developed using Nanotechnology have also offered remarkable control over aerodynamic flow, drag reduction and protection against corrosion, erosion and extreme temperatures. Moreover, the application of advanced Machine Learning algorithms and Generative AI into these materials has led to optimising manufacturing processes redefining the precision and performance of these materials by enabling advanced features such as design and selection of the best components and materials for a particular Aerospace application. It also has shown advanced safety features such as predicting failures due to fatigue and damage of aircrafts during flight and fractures in structures. This paper reviews the application of smart materials in aviation and the numerous possibilities of integrating AI into these materials.

Keywords: Carbon fibre-reinforced polymers (CFRPs), giant magnetostrictive materials (GMs), piezoelectric materials, shape memory alloys (SMA), structural health monitoring (SHM)

1. Introduction

Within the vast domain of aerospace engineering, the trajectory which initiated from the Wright brothers' rudimentary wooden materials used in the manufacture of their aircraft to the contemporary elegance of metallic marvels signifies a profound evolution in materials. In the early days, aviation relied on the simplicity of wood and fabric, serving as the foundational elements propelling us into the epoch of flight. Subsequently, the era of metals unfolded, with aluminium and steel assuming prominence, imparting robustness and dependability to the fundamental structures of aircraft [1]. In the present generation, aerospace materials transcend their conventional static nature, emerging as intelligent, dynamic entities reminiscent of science fiction. Envision an aircraft effecting self-repairs mid-flight or wings harnessing power from ambient vibrations – achievements facilitated by materials demonstrating adaptability and responsiveness to their surroundings. Nature has played an instructive role and has contributed significantly to this journey.

[a]abhaybhandarkar@gmail.com, [b]nagaraju@msrit.edu, [c]vishwachetan@msrit.edu

DOI: 10.1201/9781003545941-2

Engineers have drawn inspiration from the intricate brilliance of honeycomb structures and the mathematical elegance of Fibonacci sequences inherent in the natural world. These principles serve as guiding frameworks for crafting materials that not only exhibits strength but also possesses a remarkable lightweight quality.

The integration of artificial intelligence (AI) and materials, further advanced by the advent of Generative AI, revolutionises the conventional design paradigm. This AI technique, endowed with the capacity to autonomously design and optimise components, stands as a transformative force like a designer in an Engineering team but also with unparalleled precision. The exploration of structural possibilities reaches unprecedented heights under the creative guidance of this technological marvel. Simultaneously, machine learning (ML) enables materials to exhibit heightened capabilities such as the capability to self-monitor their structural integrity, anticipate impending maintenance requirements, and dynamically adapt to evolving environmental conditions. This automation enhances better maintainance and safety features of aircraft to predict the failure and fractures in its structures even those which are unseen to the human eye. As this exploration examines the confluenced interplay between smart materials and machine learning in aerospace engineering, we aim to uncover the structural intricacies concealed within materials to comprehending the transformative impact of AI on the design processes and their application which has led the aviation sector to promote sustainability and innovation.

2. Smart Materials in Aviation

2.1. *Shape Memory Alloys*

Shape memory alloys are metals that can return to a predetermined shape after being heated to a critical transformation temperature and can also endure large deformations

that can be recovered after unloading (pseudoelasticity). Their use is based on a specific martensitic transition (thermoelastic) in certain metal alloys. They are well-known in the aircraft sector for their usage as actuators in smart wings, propulsion systems, and smart switches [2]. Nitinol (Ni-Ti) is the most often used SMA owing to its high superelasticity and sensitivity to composition and manufacturing heat gradients.

2.1.1. Applications of SMAs in Wings of Aircraft

Wing morphing is carried out using SMA wire actuators in planes which enables the smooth changing of wing shape thereby preventing aerodynamic losses [3]. Morphed wings are known to generate smoother wing surfaces which in turn increase the agility and efficiency of the aircraft. The morphing wing has a quadrilateral frame fixed with spars and ribs. When the temperature of the SMA wires rises to the actuation temperature, there is shrinkage in the actuating wire and the quadrilateral frame is deformed by the actuator as shown in Figure 2.1 wherein the upper skin goes back and the edge of the wing shifts down like a flap.

2.1.2. Application of SMAs in Space

SMAs also have space applications which solve problems of actuation and vibration attenuation during the launch of spacecrafts during zero atmosphere and microgravity conditions [4]. Microvibrations due to a spaceborne cryocooler during an in orbit operations has known to effect the image quality of satellites. [5] has demonstrated a solution to this problem using a pseudoelastic SMA mesh washer that ensures adequate isolation of vibration in launches as well as those due to the cryocooler on orbit as shown in Figure 2.2.

Figure 2.3 shows the diverse applications of SMAs in various structures of an

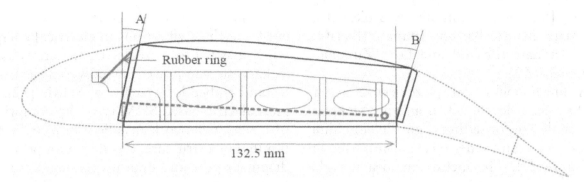

Figure 2.1: Flap morphing wing mechanism using SMA wire actuators showing deflection in the trail end of skin.

Source: [3].

aircraft such as for wing morphing, landing gear release mechanism, engine, fuselage and other components.

2.2. *Piezoelectric Materials*

A ferroelectric class of materials, piezoelectric materials have diverse applications in smart health monitoring (SHM), damage detection, sensors and harvesters in the aviation industry [7]. It has gained prominence in the recent times in the effort to invent electric aircrafts through an electromechanical system which is known for its fast response and high-power

Figure 2.2: Pseudoelastic SMA mesh washer.

Source: [5].

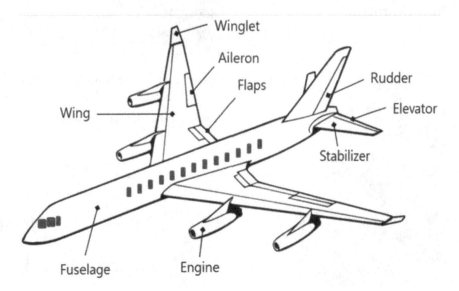

Figure 2.3: Application of SMAs in Aircraft structures.

Source: [6].

density [8]. Some commonly used piezoelectric materials are Barium Titanate (BaTiO), Lead Titanate (PbTiO) and Lead Zirconate Titanate (PZT).

A linear inchworm piezoelectric actuator has been developed in another stepped-motion design for setting engine inlet guiding vanes via a crank slider mechanism as shown in Figure 2.4. The inchworm idea may be employed to provide rotational motion to directly drive the unison ring, hence minimising the gearbox mechanism from the preceding linear actuator.

Piezos find applications in the defense sectors too as Lockheed Martin and NextGen built large-scale change UAVs under DARPA's MAS programme. Lockheed Martin's folding wing concept has hinges powered by large-throw actuators as shown in Figure 2.5 that folds the wing from a flat, loiter-efficient configuration to a much smaller, sweeping wing in 30 seconds [9]. Flexible silicone skins reinforced with a metal mesh cover the hinge locations, boosting durability by 150%. The group investigated EAP actuators as a source of motive force. During the initial operation of an EAP actuator, a tiny fold on the wing's inboard leading edge is created, which deploys against the fuselage due to the square wind current along the fold.

PZTs are used to convert mechanical energy such as vibrations to electrical energy which has led to piezoelectric materials used as harvesters to power low powered electronics and embedded systems of SHMs [12]. A piezoelectric-based SHM typically comprises of a sensor network installed atop or incorporated in a structure, together with portable diagnosis gear and data processing software [10]. Figure 2.6 shows a typical SHM diagram using piezoelectric transducers.

A novel method of SHMs have been developed with the innovation of a lightweight diagnostic film with actuators and sensors. Embedded using multipath wiring using inkjet printing of silver nanoparticle conductive wires onto a polyimide film Kapton with DuraAct transducers as PZT sensors, it delivers a flexible film with a systematic, accurate

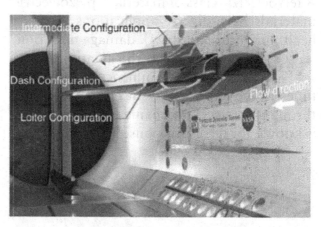

Figure 2.5: Wing folding design of Lockheed Martin using hinge-driven mechanism.

Source: (Bashir and Rajendran 2019).

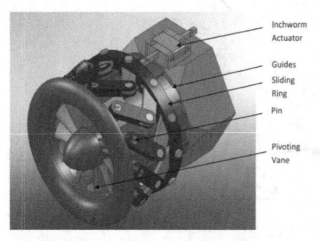

Figure 2.4: Linear inchworm piezoelectric actuator.

Source: [7].

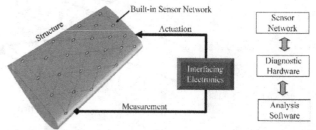

Figure 2.6: Typical structure of piezoelectric transducer based SHM.

Source: [10].

and uniform sensor placement. A thermoplastic film whose melting point is 150C was used to bond Kapton film to a composite surface. This solves the general challenges faced by SHMs such as in terms of cost, installation and precision [11].

2.3. Carbon Fiber Reinforced Polymers and Glass Fiber Reinforced Polymers

CFRPs are composite materials that have diverse applications in the field of Aerospace Engineering given their properties such as high strength, elasticity and lightweighted nature. Composite materials made up of polymer matrix reinforced with Fiber like Fiber reinforced plastics (FRPs) are reinforced with carbon fibers and CFRP polymer composites have provided self-healing abilities for components like coatings, actuators, and antennae and have demonstrated higher specific strength and better corrosion and fatigue resistance and have also been employed into some of the engine

parts [13]. When Boeing launched their 787 series of aircrafts which were the successors to the 767 series, the use of CFRP increased dramatically with the percentage of aluminium dropping substantially. Table 2.1 shows the comparison between the materials used in Boeing 767 and Boeing 787 aircrafts.

After the success of implementing CFRPs to the structure of Boeing 787 Dreamliner aircrafts Airbus A380 followed the same and included CFRPs into their flight components. Figure 2.7 shows the applications of thermoplastics and CFRPs in Airbus A380.

Table 2.1: Comparison of use of materials in structures of Boeing 767 and Boeing 787 aircrafts.

Structures	Boeing 767	Boeing 787
Airframe	Aluminium	CFRP
Main Wings	Aluminium	CFRP
Tail Wing	Aluminium	CFRP
Flaps	CFRP	CFRP

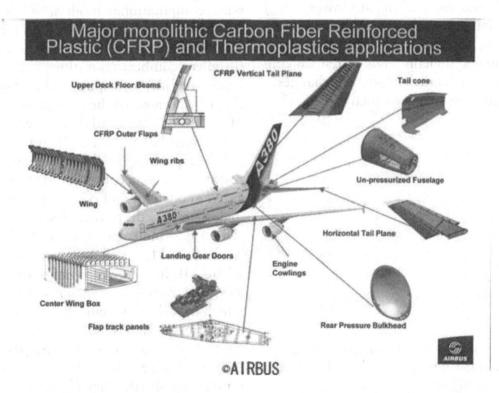

Figure 2.7: Applications of CFRPs and Thermoplastics in Airbus A380 flight structures.
Source: ©AIRBUS.

CFRP composites integrated with piezo-electric ceramics have proven beneficial for real-time SHM. Impact sensors were created by combining a GFRP composite laminate with a combination of epoxy resin and piezo-electric powder [14]. They are also known to have a variety of uses in aircraft wings, fuselages, seats, floor panels, and tail.

2.4. *Magnetostrictive Materials*

Magnetostrictive materials are advanced materials that convert mechanical energy into magnetic energy and also have a greater energy density compared to piezoelectric materials. However, due to hysteresis and complex physical properties, their applications are limited. Magnetostrictive materials may be represented as a collection of magnetic domains whose orientation is determined by the interplay of magnetic and mechanical energy [15]. Magnetostrictive materials, in layman's words, exhibit stress and strain when subjected to a magnetic field. When no magnetic field is supplied in Figure 2.8, the magnetic domain's orientation is random, resulting in a lower energy and more stable overall state. When an external magnetic field is supplied, the orientation of the magnetic domain tends to remain constant while the overall displacement changes, a process known as magnetostrictive [16].

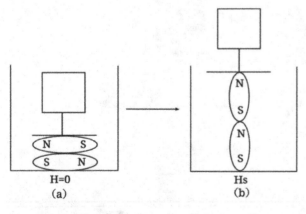

Figure 2.8: Showing Magnetostrictive effect. (a) Without applied magnetic field and (b) With external magnetic field.

Source: [16].

Magnetostrictive patch sensors (MPT) with an electromagnetic coupling mechanism may be more effective in gathering single-mode, non-dispersive guided waves in ultrasonic Guide Wave Monitoring applications. [17] studied the use of magnetostrictive sensors to generate a non-dispersive shear horizontal wave mode (SH0) to track the deterioration of non-ferromagnetic plate structures and demonstrated the effectiveness of magnetostrictive sensors after optimising the design using commercial software, ANSYS. 2019 [18] MPT may be suitable for non-destructive testing of thin-walled structures.

2.5. *Machine Learning in Aviation*

With the advent of machines into the aerospace sector, they have been a driving force in applications such as fluid mechanics, aerodynamics, combustion, and aircraft structural evaluation, all of which strive to create a more sustainable future in aviation. Historically, combustion research has dealt with vast amounts of data using experimental and large-scale computational models. While exascale computers will allow for more realistic conditions, direct numerical simulation (DNS) of turbulent combustion is already an important study area for developing and assessing models [19]. Because of the enormous complexity of the linked issues and their inherent nonlinearities in turbulent flows, ML approaches have a lot of promise in fluid mechanics. Future research will concentrate on developing innovative modelling and control procedures that make use of deep reinforcement learning's (DRL) enormous potential [20]. Large-eddy simulation (LES) is another simulation domain where ML has shown promise. In contrast to Reynolds-Averaged Navier-Stokes (RANS), which models all turbulent scales using Reynolds stresses LES simulates just a subset of (larger) scales and employs low-pass filtering. A subgrid-scale model (SGS) is then used to represent scales that have larger wavenumbers (i.e., with smaller wavelengths) [21]. To create such SGS models, numerous deep-learning-based approaches are available [22].

With the rapid growth of data science, data-driven aerodynamic models are emerging to compensate for the lack of accuracy in classical aerodynamic models based on empirical principles. These models are more precise than theoretical models and can provide information about the physical understanding of the problems. They also open up opportunities for applying semi-empirical models to domains such as new flow control, optimisation, aeroelasticity, flow reconstruction, and flight dynamics [23].

Design optimisation is another critical topic in the aviation field, where the use of machine learning has lately gained prominence. Aerodynamic design is concerned with turbine engine blades, aeroplane wings, unmanned aerial vehicles (UAVs), airfoil geometry, flying circumstances, and so on. [24] released a work on the use of machine learning for design optimisation, which included design space, constraint modelling, objective modelling, and optimisation convergence rate. Using deep learning models such as GAN, realistic aerodynamic shapes were created. Applications have expanded from two-dimensional airfoils to three-dimensional wings. Deep learning geometric filtering algorithms are now being trained to detect complex abnormalities in order to improve sophisticated aerodynamic form design.

A strategy for identifying and characterising damages in a steel bridge joint model was implemented [25]. The method made use of electrical impedance's sensitivity to damage, combining it with the neural network's capabilities for quantified damage evaluation. There were two major milestones in the implementation process. To begin, the electrical impedance's susceptibility to damage was used for the first detection by placing the body in the vicinity of PZT sensors. Following that, neural network technology was used, especially a neural network with a sigmoid interpolation function and 10

neurons in the hidden layer. The implementation was carried out using Matlab's Neural Network Toolbox, with the backpropagation technique used for training. The proposed approach was successfully tested on a quarter-size bridge segment as well as a space truss structure which has shown its practical application in real-world scenarios. Given is a flowchart on damage detection in Figure 2.8 using the proposed impedance method along with neural networks.

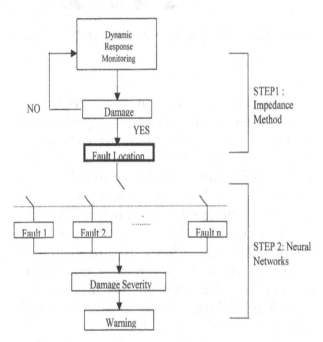

Figure 2.9: Flow diagram of Damage detection.
Source: [25].

2.6. Engineered Nanomaterials and Polymers for Aerospace Applications

Engineered nanoparticles enhance the structural and non-structural components of nearly all aircraft systems. Examples include weight reduction, increased flexibility while maintaining structural integrity, multicomponent tracking with redundant backup sensors, efficient power production, storage, and transmission, improved radiation protection for electronic devices and people, and

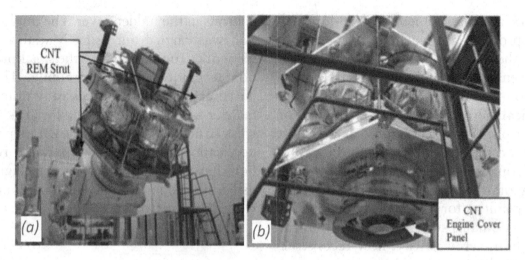

Figure 2.10: Phenolic impregnated carbon ablator (Nano-PICA).
Source: [26].

long-duration exploration life-support systems [26]. Nanomaterials are being employed in thermal protection materials to improve thermal stability and mechanical strength across a wide range of aerothermal flow conditions. Research is also being conducted to add multifunctionality to some task-specific material systems, like as thermal protection systems (TPSs), which are used to shelter the spacecraft from the severe temperatures of re-entry into the atmosphere while simultaneously providing radiation protection which is shown in the Figure 2.9.

Carbon nanotechnology discoveries allow us to capitalise on its multifunctional qualities, such as the inclusion of structural, thermal, and electrical capabilities in material composites. The combination of CNT sheet-based composites and cutting-edge manufacturing technologies saves money and time by removing labor-intensive surface preparation processes for composite parts. Flat laminates, sandwich panels, and proto-flight tubes with a carbon nanotube outer ply on a composite substrate were created utilising typical composite manufacturing techniques. Microstructural examination, non-destructive inspection, mechanical property testing, and EMI/ESD tests were conducted on the test materials. NASA's Juno spacecraft

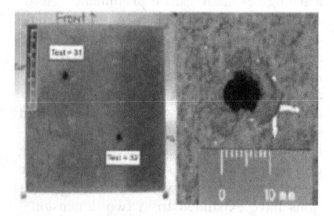

Figure 2.11: (a) CNT composite struts supporting the Juno spacecraft (b) CNT Engine cover panel on Juno spacecraft.
Source: [28].

flew CNT-based REM support struts and an engine cover panel (Figure 2.11). CNT features including electrical and thermal conductivity, as well as electromagnetic shielding, benefited the Juno mission. [27,28].

The water droplets in clouds cause the building up of ice within the engines and on the surfaces of aircrafts. This leads to hazards such as increased drag on the aircraft and raises the stalling speed which thereby reduces the thrust that can effect the propellers efficiency. A nano inspired surface coating $La_2Zr_2O_7$ has been investigated to solve this [31,32]. This was confirmed by measuring the contact angle between the nano-coated surface and ice formed by super-cooled liquid in mid-air during flight, which surpassed 150°. Electrochemical experiments on modified LZ-NPs-carbon paste electrodes show good redox potentials as measured by Cyclic Voltammetry (CV) and Electrochemical Impedance Spectroscopy (EIS) at various scan speeds. The EIS Nyquist plot revealed the increased resistance of the constructed LZ-NPs-carbon electrode. Over the last two decades, much research has been undertaken on the addition of carbon-based elements to epoxy matrices, such as carbon nanotubes (CNTs) and carbon black, in order to improve the composite's electrical, mechanical, and thermal properties. Despite their ability to improve the electrical properties of epoxy resin, these carbon compounds have various limitations that make them unsuitable for usage in a variety of applications. Carbon nanotubes, for example, remain expensive due to low productivity [33–35], whereas carbon black (CB) is cheaper than CNT, but effective electrical conduction of carbon black/polymer composites requires a relatively high concentration of carbon black [36]. Graphene sheets, on the other hand, may be produced at a lower cost through mass production than CNTs, and they can also provide enough electrical conductivity to polymer composites at low concentrations when compared with carbon black. As a result, graphene nanoplatelets (GNPs) have emerged as a potential nanofiller for polymer matrices [37].

Radar is a commonly used tool for detecting and tracking aeroplanes. Although radar is an essential instrument in aviation traffic control, it presents a dilemma in offensive military operations in the defence sector as they would require the aircraft to assault their objective and then flee without being discovered. Radar includes the emission of radio waves into the atmosphere, which are subsequently reflected back to a receiving antenna by the aircraft. Radar-absorbing material (RAM) is a polymer-based substance applied to the surfaces of stealth military aircraft such as the F-22 Raptor (Figure 2.12) [29] and F-35 Lightning II (Figure 2.13) [30] to reduce radar cross-section and hence make them more difficult to detect by radar. These materials are also utilised to develop stealth versions of tactical unmanned aerial vehicles, such as the Boeing X-45. RAM is applied to the skin's surface or to places with a high radar reflectivity, such as edges. RAM works by having the aircraft absorb electromagnetic wave energy in order to weaken the reflected signal. RAMs, along with other stealth technologies such as concealed engines and planar designs, are used to make military aircraft harder to detect.

Figure 2.12: F22 Raptor.

Source: Wikipedia.org Lockheed Martin F22 Raptor.

Figure 2.13: F35 Lightning II.

Source: Wikipedia.org Lockheed Martin F22 Raptor.

The bulk of RAMs are composed of ferromagnetic particles placed in a polymer matrix with a high dielectric constant. Iron ball paint, which consists of small metal-coated spheres suspended in an epoxy-based paint, is one of the most common RAMs. Ferrite or carbonyl iron is utilised to coat the spheres. Another type of RAM is a neoprene sheet containing carbon black particles or ferrite. This material was used on early versions of the F-117A Nighthawk, which works similarly to iron ball paint by converting radar signals to heat. Some US Air Force stealth aircraft currently feature radar-absorbent coatings made of ferrofluidic and nonmagnetic materials [38].

3. Conclusion

With continued advancement of smart, nano and polymer composites, there will always be the need for a better material with self-healing, light and smart characteristics. With application of GANs, we must aim to automate the generation of structural components of aircrafts which can be designed based on initial datasets and functionality inputs. The goal should always be to promote sustainable flying and promote the conjunction of materials and AI and make Green Flying a reality.

Acknowledgements

I extend my heartfelt gratitude to our esteemed professors, Dr. Nagaraju Kottam and Vishwachetan D, who are co-authors of this paper, for their invaluable guidance and support throughout the research process. Their expertise and mentorship have not only enriched our understanding of aerospace materials and artificial intelligence but also inspired me to explore new frontiers in this dynamic field. I am profoundly grateful for the opportunity they provided us to delve into this fascinating realm and for their unwavering encouragement every step of the way. Their dedication to fostering academic excellence has been a driving force behind our collective research endeavours. We thank the International Conference on Advanced Materials and Engineering Sciences for giving us the best platform to showcase and present our paper. We also thank Ramaiah Institute of Technology Bengaluru for providing a brilliant foundation for the conduction of this conference.

References

[1] Pritchard, J. L. (1954). The work of the wright brothers for aviation. *RSA J.*, 102(4916), 112–128. http://www.jstor.org/stable/41365639

[2] Dadbakhsh, S., Speirs, M., Van Humbeeck, J., and J.- Kruth, P. (2016). Laser additive manufacturing of bulk and porous shape-memory NiTi alloys: from processes to potential biomedical applications. *MRS Bull*, 41(10), 765–774. http://dx.doi.org/10.1557/mrs.2016.209

[3] Kang, W. -R., Kim, E. -H., Jeong, M. -S., Lee, I., and Ahn, S. -M. (2012). Morphing wing mechanism using an SMA wire actuator. *IJASS*, 13(1), 58–63. http://dx.doi.org/10.5139/ijass.2012.13.1.58

[4] Costanza, G., and Tata, M. E. (2020). Shape memory alloys for aerospace, recent developments, and new applications: a short review. *Mater.*, 13(8), 1856. http://dx.doi.org/10.3390/ma13081856

[5] Kwon, S. -C., Jeon, S. -H., and Oh. H. -U. (2015). performance evaluation of space-borne cryocooler micro-Vibration isolation system employing pseudoelastic SMA mesh washer. *Cryogenics*, 67, 19–27. http://dx.doi.org/10.1016/j.cryogenics.2015.01.002

[6] Sohn, J. W., Ruth, J. S., Yuk, D. -G., and Choi, S. -B. (2023). Application of shape memory alloy actuators to vibration and motion control of structural systems: a review. *Appl. Sci.*, 13(2), 995. http://dx.doi.org/10.3390/app13020995.

[7] Vo, T. V. K., Lubecki, T. M., Chow, W. T., Gupta, A., and Li, K. H. H. (2021). Large-scale piezoelectric-based systems for more electric aircraft applications. *Micromachines*, 12, 140. http://dx.doi.org/10.3390/mi12020140.

[8] Viswanathan, V., and Knapp., B. M. (2019). Potential for electric aircraft. *Nat. Sustain.,* 2(2), 88–89. http://dx.doi.org/10.1038/s41893-019-0233-2.

[9] Bashir, M., & Rajendran, P. (2019). Recent Trends in Piezoelectric Smart Materials and Its Actuators for Morphing Aircraft Development. International Review of Mechanical Engineering (IREME), 13(2), 117. https://doi.org/10.15866/ireme.v13i2.15538

[10] Qing, X., Li, W., Wang, Y., and Sun. H. (2019). Piezoelectric transducer-based structural health monitoring for aircraft applications. *Sensors,* 19(3), 545. http://dx.doi.org/10.3390/s19030545.

[11] Bekas, D., Sharif-Khodaei, Z., and Aliabadi, M. H. (2018). An innovative diagnostic film for structural health monitoring of metallic and composite structures. *Sensors,* 18(7), 2084. http://dx.doi.org/10.3390/s18072084.

[12] Safaei, M., Sodano, H. A., and Anton, S. R.(2019). A review of energy harvesting using piezoelectric materials: state-of-the-art a decade later (2008–2018). *SMS,* 28(11), 113001. http://dx.doi.org/10.1088/1361-665x/ab36e4.

[13] Goyal, M., Agarwal, S. N., and Bhatnagar, N. (2022). A review on self-healing polymers for applications in spacecraft and construction of roads. *J. Appl. Polym. Sci.,* 139(37). http://dx.doi.org/10.1002/app.52816.

[14] Komagome, R., Katabira, K., Kurita, H., and Narita. F. (2022). Characteristics of carbon fiber reinforced polymers embedded with magnetostrictive Fe–Co wires at room and high temperatures. *Compos. Technol.,* 228, 109644. http://dx.doi.org/10.1016/j.compscitech.2022.109644.

[15] Deng, Z. and Dapino. M. J. (2018). Review of magnetostrictive materials for structural vibration control. *SMS,* 11, 113001. http://dx.doi.org/10.1088/1361-665x/aadff5.

[16] Wang, W., Xiang, Y., Yu, J., and Yang, L. (2023). Development and prospect of smart materials and structures for aerospace sensing systems and applications. *Sensors,* 23(3), 1545. http://dx.doi.org/10.3390/s23031545.

[17] Zhou, L., Yang, Y., Yuan F. -G. (2012). Design of a magnetostrictive sensor for structural health monitoring of non-ferromagnetic plates. *CORE Reader,* https://core.ac.uk/reader/323313160.

[18] Park, C. I., Seung, H. M., and Kim, Y. Y. (2019). Bi-annular shear-horizontal wave MPT tailored to generate the SH1 mode in a plate. *Ultrasonics,* 99, 105958. http://dx.doi.org/10.1016/j.ultras.2019.105958.

[19] Wick, A., Attili, A., Bisetti, F., and Pitsch, H. (2020). DNS-driven analysis of the flamelet/progress variable model assumptions on soot inception, growth, and oxidation in turbulent flames. *Combust. Flame.,* 214, 437–449. http://dx.doi.org/10.1016/j.combustflame.2020.01.012.

[20] Kurz, M., Offenhäuser, P., and Beck, A. (2023). Deep reinforcement learning for turbulence modeling in large eddy simulations. *IJHT,* 99, 109094. http://dx.doi.org/10.1016/j.ijheatfluidflow.2022.109094.

[21] Le Clainche, S., Ferrer, E., Gibson, S., Cross, E. A. Parente, E., and Vinuesa, R. (2023). Improving aircraft performance using machine learning: a review. *Aerosp. Sci. Technol.,* 138, 108354. http://dx.doi.org/10.1016/j.ast.2023.108354.

[22] Beck, A., Flad, D., and Munz, C.-D.(2019). Deep neural networks for data-driven les closure models. *J. Comput. Phys.,* 398, 108910. http://dx.doi.org/10.1016/j.jcp.2019.108910.

[23] Kou, J. and Zhang, W.(2021). Data-driven modeling for unsteady aerodynamics and aeroelasticity. *Prog. Aerosp. Sci.,* 125, 100725. http://dx.doi.org/10.1016/j.paerosci.2021.100725.

[24] Li, J., Du, X., and Martins, J. R. R. A. (2022). Machine learning in aerodynamic shape optimization. *Prog. Aerosp. Sci.,* 134, 100849. http://dx.doi.org/10.1016/j.paerosci.2022.100849.

[25] Lopes, V., Park, G., Cudney, H. H., and Inman, D. J. (2000). Impedance-based structural health monitoring with artificial neural networks. *J. Intell. Mater. Syst. Struct.* 11(3), 206–214. http://dx.doi.org/10.1106/h0ev-7pwm-qyhw-e7vf.

[26] Arepalli, S. and Moloney, P. (2015). Engineered nanomaterials in aerospace. *MRS Bull.,* 40(10), 804–811. http://dx.doi.org/10.1557/mrs.2015.231.

[27] Rawal, S., Brantley, J., and Karabudak, N.(2013). Development of carbon nanotube-based composite for spacecraft components. 2013 6th International Conference on Recent Advances in Space Technologies (RAST). http://dx.doi.org/10.1109/rast.2013.6581186.

[28] Rawal, S. (2018). Materials and structures technology insertion into spacecraft systems: successes and challenges. *Acta Astronaut.*, 146, 151–160. http://dx.doi.org/10.1016/j.actaastro.2018.02.046.

[29] Anon. (2023). Lockheed Martin F-22 Raptor. *Wikipedia.* https://en.wikipedia.org/wiki/Lockheed_Martin_F-22_Raptor.

[30] Anon. (2023). Lockheed Martin F-35 Lightning II. *Wikipedia.* https://en.wikipedia.org/wiki/Lockheed_Martin_F-35_Lightning_II.

[31] Rudresh, M., Surendra, B. S., Gurushantha, K., Dinamani, M., Basavaraju, N., Deepa, H. A., Mehrotra, R., Golchha, T., Mahesh, A., and Ruhal, A. (2023). A critical exploratory investigation on nano-inspired (La 2 Zr 2 O 7 NPs) surface coatings for aircraft icing mitigation and electrochemical studies. *Aust. J. Mech. Eng.*, 1–12. https://doi.org/10.1080/14484846.2023.2295109

[32] Surendra, B., Gagan, R., Soundarya, G., and Naresh. (2020). Thermal barrier and photocatalytic properties of La2Zr2O7 NPs synthesized by a Neem extract assisted combustion method. *App. Surf. Sci. Adv.*, 1, 100017. https://doi.org/10.1016/j.apsadv.2020.100017

[33] Mallick, P. K. (2007). Fiber-Reinforced Composites: Materials, Manufacturing, and Design. CRC Press, Boca Raton.

[34] Volat, C., Farzaneh, M., Leblond, A. (2005). De-icing/anti-icing techniques for power lines: current methods and future direction. Presented at: 11th International Workshop on Atmospheric Icing of Structures; Montreal, Canada.

[35] Chu, H., Zhang, Z., Liu, Y., and Leng, J. (2014). Self-heating fiber reinforced polymer composite using meso/macropore carbon nanotube paper and its application in deicing. *Carbon*, 66, 154–163. https://doi.org/10.1016/j.carbon.2013.08.053

[36] Gohardani, O. (2011). Impact of erosion testing aspects on current and future flight conditions. *Prog. Aerosp. Sci.*, 47(4), 280–303. https://doi.org/10.1016/j.paerosci.2011.04.001

[37] Brittingham, D. L., Prybyla, S. G., Christy, D. P. (2014). Methods of protecting an aircraft component from ice formation. *US Patent No. 8752279B2.*

[38] Setua, D. K., Mordina, B., Srivastava, A. K., Roy, D., and Eswara Prasad, N. (2020). Carbon nanofibers-reinforced polymer nanocomposites as efficient microwave absorber. fiber-reinforced nanocomposites. *Fund. Appl.*, 395–430. http://dx.doi.org/10.1016/b978-0-12-819904-6.00018-9.

3 A mini review on the production of biodiesel using photocatalysts

M. S. Akansha[a], S. Veluturla, G. Akshay, and V. Vijay Kumar

Department of Chemical Engineering, M. S. Ramaiah Institute of Technology, Bangalore, India

Abstract

In recent years, heightened awareness of the environmental repercussions stemming from fossil fuel usage has spurred intensive exploration into sustainable alternatives, with biodiesel emerging as a frontrunner. While traditional methods of biodiesel production, including both heterogeneous and homogeneous catalytic routes, have demonstrated efficacy, they often entail significant energy consumption and yield only partial recoverability. In contrast, photocatalysis has emerged as a promising avenue for biodiesel synthesis, offering the potential for enhanced energy efficiency and reduced environmental impact. This review delves into the strides made in photocatalytic transesterification, elucidating the underlying mechanisms driving biodiesel production via photocatalysis. It surveys an array of photocatalysts employed in transesterification reactions, highlighting their respective merits and limitations. Additionally, the review addresses the current challenges confronting photocatalytic biodiesel production, such as optimising photocatalyst performance, understanding intricate reaction pathways, and scaling up production for industrial applications. Overcoming these hurdles presents compelling opportunities to unlock the full potential of photocatalysis in biodiesel production, paving the way for a more sustainable and eco-friendly energy landscape.

Keywords: Fossil fuels, biodiesel, photocatalysis, transesterification

1. Introduction

Biodiesel is a very promising alternative fuel source to eventually replace the use of fossil fuels. It is considered to be ecofriendly due to its characteristics such as low emission profile, nonhazardous and biodegradable [1]. Due to the current global energy crisis, the usage of biodiesel as a substitute for fossil fuels is growing in popularity [2]. Biodiesel is often produced by alkaline transesterification of waste cooking oils, vegetable oil, animal fat, and lipid feedstock with short-chain alcohols like methanol or ethanol. This approach has some limitations despite being effective. Since the base-catalyzed process is particularly sensitive to high FFA contents, saponification results from residual oil's high Free Fatty Acid (FFA) content, which lowers biodiesel yields and causes soap to develop

[3,4]. Esterification using an acid catalyst is used to address this problem. However, the enormous amounts of methanol needed to use an acid catalyst results in significant expenses resulting in terms of waste treatment and the essential requirement of stainless steel equipment [5].

A potential energy efficient solution to this problem is utilising solar energy via photocatalysis. Extensive research has examined how well photocatalysis maintains the environment by degrading organic contaminants [6–9], reduction of CO2 [10] and so forth. Additionally, in recent studies for the production of biodiesel, photocatalysts were used in processes including the transesterification of Jatropha curcas crude oil (JCCO) utilising ZnO/SiO_2 and UV radiation [11], the use of Graphitic carbon nitrides in the

[a]akshaygmsrit@gmail.com

DOI: 10.1201/9781003545941-3

transesterification of canola oil [12] and the synthesis of biodiesel using $SrTiO_2/g$-C3N4 photocatalyst [13] which show that photocatalysis in biodiesel production is simple, highly energy efficient and is eco-friendly.

The domain of biodiesel feedstocks and catalysts for its synthesis is a vibrant tapestry of innovation and sustainability. From first-generation biofuels, rooted in food crops like corn and soybeans, to second-generation alternatives derived from non-food sources such as lignocellulose materials, and the futuristic promise of third-generation biofuels sourced from algae, the evolution is marked by efficiency gains and environmental consciousness. Catalysts play a pivotal role, with homogeneous catalysts like sodium hydroxide and potassium hydroxide enabling transesterification reactions, while heterogeneous counterparts such as calcium oxide and zirconium oxide offer improved separation and reusability. The emerging frontier of photocatalysis presents a tantalising prospect, harnessing the power of semiconductors to drive esterification and transesterification reactions under light irradiation.

Photocatalysts are a promising technology for the future of renewable energy and biofuel production because they minimise energy use, harness solar energy, and promote selective reactions, which significantly reduces both energy usage and waste formation [14]. Under favorable circumstances, photocatalysis usually occurs at ambient temperature and atmospheric pressure. By minimising the amount of energy used throughout the biodiesel synthesis process, photocatalysis differs from typical thermocatalytic processes, which frequently call for high temperatures and pressures. This is especially important because it lowers the overall energy requirements for the production process by eliminating the need for energy-intensive heating and pressurisation. Without the use of intermediary energy carriers, photocatalysis allows light energy to be directly converted into chemical energy.

Photocatalysis is a more efficient way to synthesis biodiesel because it avoids energy losses related to the generation and transportation of energy carriers through this direct conversion process.

Photocatalysts exhibit high selectivity in promoting specific chemical reactions, such as the transesterification of triglycerides to produce biodiesel [15]. This high selectivity ensures that the desired products are formed with minimal side reactions, resulting in a purer end product. Reduced side reactions mean fewer unwanted byproducts, leading to lower waste generation during the production process. Photocatalysts can be easily regenerated and reused multiple times without significant loss of activity. This reusability reduces the need for frequent replacement of catalysts, thereby decreasing the generation of waste associated with catalyst disposal. Additionally, the longevity of photocatalysts contributes to the overall sustainability of the biodiesel production process by minimising the consumption of catalyst materials and reducing waste generation over time.

However, employing photocatalysts presents difficulties because most exhibit low electron- hole pair separation efficiency and only exhibit high catalytic efficiency when exposed to ultraviolet radiation. [14,15]. At present, there is a lack of sufficient studies on photocatalysis in biodiesel production, however, there has been good progress made in improving photocatalytic activity and efficiency through doping techniques, morphological changes, chemical modification [16] and heterojunctions [17]. Table 3.1 summarises various catalyst used for biodiesel synthesis.

This mini review aims to summarise the progress made in photocatalytic approaches to biodiesel production.

2. Mechanism of biodiesel production using photocatalysis

The mechanism of (trans)esterification using photocatalysts takes place in four steps,

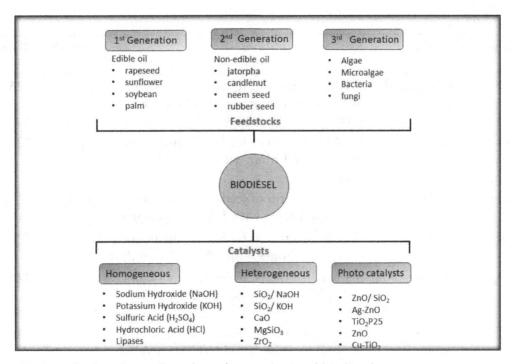

Figure 3.1: Various feedstocks and catalysts for synthesis of biodiesel.

Table 3.1: Depicting the various catalysts employed in the production of biodiesel

Sl no	Catalyst	Feedstock	Reaction conditions	Yield %	References
1	H_2SO_4	Waste tallow (chicken)	1:30 Molar ratio of feedstock: methanol at 50°C for 24 hours with 1.25% w/v% catalyst	99	[18]
2	H_2SO_4	Palm fatty acid	7.2:1 Molar ratio of feedstock: methanol at 70°C for 2 hours with 1.25% w/v% catalyst	99.6	[19]
3	KOH	Sunflower oil	7.2:1 Molar ratio of feedstock: methanol at 70°C for 2 hours with 0.55% w/v% catalyst	96	[20]
4	Sodium methoxide	Jojoba oil-wax	7.5:1 Molar ratio of feedstock: methanol at 60°C for 4 hours with 1% w/v% catalyst	55	[21]
5	KOH	Brassica carinata	7.5:1 Molar ratio of feedstock: methanol at 60°C for 4 hours with 1% w/v% catalyst	98.27	[22]
6	KOH	Canola oil	6:1 Molar ratio of feedstock: methanol at 25°C for 0.33 hours with 0.5% w/v% catalyst	86.1	[25]
7	KOH	Jatropha curcas	6:1 Molar ratio of feedstock: methanol at 30°C for 1 hour with 0.5% w/v% catalyst	92	[26]

(continued)

Table 3.1: continued

Sl no	Catalyst	Feedstock	Reaction conditions	Yield %	References
8	NaOH	Cottonseed oil	6:1 Molar ratio of feedstock: methanol at 55°C for 1 hour with 0.5% w/v% catalyst	77	[25]
9	KOH	Roselle oil	8:1 Molar ratio of feedstock: methanol at 60°C for 1 hour with 1.5% w/v% catalyst	99.4	[26]
10	NaOH	Rubber seed oil	6:1 Molar ratio of feedstock: methanol at 60°C for 1 hour with 1% w/v% catalyst	84.46	[27]
11	NaOH	Mahua oil	6:1 Molar ratio of feedstock: methanol at 60°C for 2 hours with 1% w/v% catalyst	92	[28]
12	H_2SO_4	Rice bran oil	3:1 Molar ratio of feedstock: methanol at 100°C for 8 hours with 2% w/v% catalyst	98	[29]
13	KOH	Waste cooking oil	8:1 Molar ratio of feedstock: methanol at 50°C for 2 hours with 0.75% w/v% catalyst	98	[30]
14	H_2SO_4	Jatropha oil	9:1 Molar ratio of feedstock: methanol at 60°C for 2 hours with 0.5% w/v% catalyst	95	[31]
15	KOH	Karanja oil	6:1 Molar ratio of feedstock: methanol at 65°C for 3 hours with 1% w/v% catalyst	98	[32]

following the Langmuir-Hinshelwood reaction pathway. When the heterogeneous photocatalyst is placed under UV irradiation (stronger than band gap energy of the semiconductor), electron-hole pairs are generated and move to the surface to participate in the chemical reaction. Free radicals transformed from surrounding molecules with the help of free electrons (e⁻) and holes (h⁺) [33]. The photogenerated electrons (e⁻) migrate from the valence band to the conduction band, leading to the formation of holes (h⁺) on the valence band [34,12,13,35]. From the study by Corro et al. [34] the steps can be broken down as follows. The free photogenerated electrons (e⁻) and holes (h⁺) make the surrounding methanol (CH_3OH) and FFA (R-COOH) molecules (from WCO) turn into free radicals. CH_3OH reacts with the holes (h⁺), under irradiation, generating ($CH_3O^•$) radicals [33,35]. Meanwhile, the photogenerated electrons (e⁻) react with the FFAs (R-COOH) in the WCO, generating R-COO radicals [35]. Next, the $CH_3O^•$ radicals attack R-COOH radicals to form an intermediate. The intermediate formed undergoes dehydration and rearrangement to form Fatty acid methyl ester (FAME) or biodiesel. The FAME is then desorbed from the surface of the catalyst and is then separated. The transfer of free fatty acid (FFA) and methanol to the surface of the photocatalyst and the subsequent separation of product from the photocatalyst surface can be accelerated via vigorous stirring. Pd^{2+} doping increases the photocatalytic activity of TiO_2 [36].

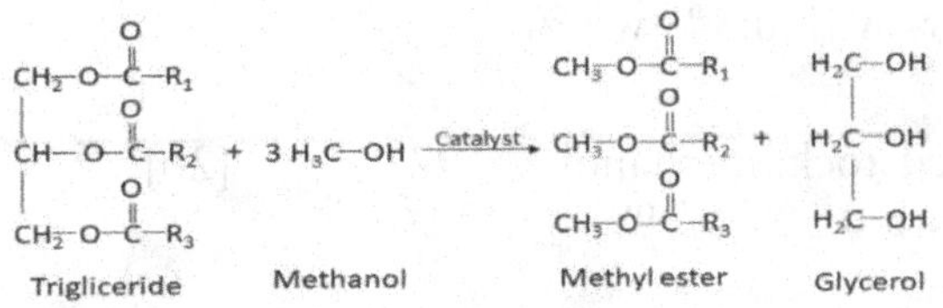

Figure 3.2: Basic chemical reaction for the synthesis of biodiesel.

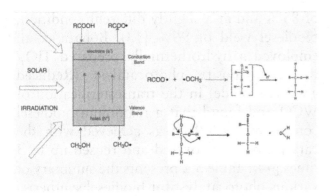

Figure 3.3: A diagrammatic summation of the photocatalytic mechanism of biodiesel production as adapted from Corro et al. [33].

3. Photocatalysts Used for Biodiesel Production

3.1. ZnO based Catalysts

Over the years, many studies have tested and affirmed the efficiency of using a photocatalyzed approach as an alternative for the production of biodiesel from various oils. Among the first was a study utilising a ZnO/SiO2 semiconductor photocatalyst to carry out FFA esterification from JJCO [11]. A very high FFA% conversion of 96% was obtained during a four-hour duration under UV radiation and the ZnO/SiO2 composite catalyst was recovered and recycled 10 times without change in catalytic performance. A future study by Corro et al. used Cr/SiO2 to photocatalyzed biodiesel production from waste cooking oil under solar irradiation [19]. The combination of Cr and SiO2 helped enhance photocatalytic activity by inhibiting aggregation. Another study was conducted in which Ag was loaded on ZnO, and the Ag-ZnO catalyst was used in the transesterification of simarouba oil [37]. The study found that Ag-ZnO showed better photocatalytic activity than undoped ZnO nanoparticles due to Ag acting as a sink to the photogenerated electrons flowing from the ZnO nanostructure. A yield of about 84.5% of biodiesel was obtained by using the doped catalyst.

Along with metal nanoparticle doping, ZnO can be mixed with other metallic oxides,

as done in a particular study by Guo et al. where the photocatalyzed transesterification of waste cooking oil (WCO) using a CuO/ZnO catalyst was carried out [38]. The maximum yield obtained was 93.5% and the catalyst was recycled 6 times with a yield of 81.4%, proving it be stable and have high activity. Another novel photocatalyst had been synthesised in a prior study, mixing metal ions and metallic oxides with ZnO [38]. A La^{3+}/ZnO-TiO_2 catalyst was prepared via sol-gel method and used to catalyze the esterification of the FFA in WCO. A conversion rate of about 87% was retained after five cycles of reuse. The enhanced photocatalytic activity was because doping with La^{3+} improved separation of the photogenerated carriers due to oxygen vacancies being formed. Archna, et al. reported the synthesis of biodiesel with ZnO nanoparticles obtained using Moringa oleifera fuel and obtained a yield of 80.6% [37].

3.2. TiO₂ Based Catalysts

TiO_2 is among the most commonly used photocatalysts. This is due to its large band gap, chemical stability, high surface area, high reusability, corrosion resistance and easy availability [39–42]. TiO_2 nanoparticles have high photocatalytic efficiencies due to their short lateral diffusion lengths and low reflectivity [33,40]. A study compared TiO_2 nanotubes (TNT) with TiO_2 P25 catalyst for the esterification of oleic acid [41,44]. The yield obtained using P25 was 86.0% and 59.3% for TNT. The paper mentions factors such as the large band gap, surface area and TNT crystalline phase as being important to catalytic performance and theorises that the higher number of hydroxyl groups compared to P25 could have contributed to a low ester yield. The main drawback however is that due to the band gap for TiO_2 being large (3.2 eV), it is active only under UV radiation [42].

To improve the photocatalytic activity of TiO_2 under visible light, methods such as doping with metal ions such as Pt, Ag, Au, Fe and Zn [43,44] have been employed, among which noble metals showed the highest photocatalytic activity due to their inert, resistant nature. A paper used Cu impregnated TiO_2 to photocatalyzed the production of biodiesel from palm oil and obtained a yield of almost 91% [47]. Another method involves the mixing TiO_2 with other metal oxides such as, SnO_2, ZnO, MgO and Fe_2O_3 [45–47]. A study showed that ZnO has a similar band gap (3.2 eV) and staggered band position to TiO_2 [48]. Therefore, introducing ZnO inhibits the growth of TiO_2 particles [37], thereby increasing catalytic activity by combining the electron-hole pairs [38].

3.3. *Carbon Nanoparticle Catalysts*

Carbonaceous nanostructure materials have been studied and used as catalysts for a large number of chemical reactions within the manufacturing sector for over fifty years. The various forms of carbonaceous nanoparticles include graphene, carbon nanosheets, and nanotubes [12,13]. Polymeric carbon nitride (PCN) is a highly active photocatalyst which has excellent chemical and physical properties, good photo-corrosion resistance, rich earth abundance and suitable electronic structure [39–41]. Graphite carbon nitrides have certain disadvantages such as having weak absorption of long wavelength light, smaller specific surface area, and a rapid recombination rate of photogenerated carriers, due to which it is normally utilised in a composite form.

A study on graphitic carbon nitrides involved the synthesis of a TiO_2 composite with g-C3N4 to produce biodiesel via transesterification of waste cooking oil [42]. The study concludes that the synthesised catalyst was effective with a yield of 89.5%. The TiO_2/g-C3N4 was also employed for the synthesis of biodiesel from jatropha oil and reported a conversion of 97% at 250C [43]. A different study focused on the pretreatment of Waste frying oil (WFO) using a SrTiO3/g-C3N4 composite to catalyze the esterification of WFO, obtaining an FFA conversion

of 85% and at a slightly different condition, biodiesel yield of 96% [13]. Bohara et al. employed a hydrothermally prepared TiO_2/RGO composite photocatalyst (Reduced graphene oxide) in the transesterification of WCO and found that a maximum biodiesel conversion of 98% was achieved with the catalyst being recovered and reused up to 3 times [44]. Table 3.2 presents the summary of various photocatalyst for biodiesel synthesis.

3.4. *Supplementary Co-catalysts*

In addition to the ones mentioned above, numerous other photocatalyst types have also been used in the synthesis of biodiesel. A metal-free Brønsted acid-functional porphyrin grafted with benzimidazolium-based ionic liquid (BAPBIL) was studied in the catalysis of the esterification of oleic acid under visible light irradiation [45]. The photocatalyst could be recycled up to 5 times without loss in photocatalytic activity and at optimal conditions, yielded a maximum conversion of 96%. Another approach was the surface modification of TiO_2 nanoparticles to create a novel surface functionalised TiO_2/PrSO3H catalyst. The synthesised catalyst had larger mesopores with the pore diameter being larger than that of TiO_2 nanoparticles which was favorable for the esterification and transesterification reaction. The transesterification of used cooking oil (UCO) was carried out, and the yield obtained was 98.3% and the catalyst had was reusable for 4 cycles under optimal conditions [26]. Nanofibers such as SiO2-Cu@Fe2O3 nanofibers have also been tested in the transesterification of waste edible oil, where the study mentioned a biodiesel yield of 98% [46]. Fe2O3 tends to separate rapidly from the porous surface of silica, due to which certain techniques can be used to strengthen the interactions between Fe2O3 and SiO2 [8,41] such as a hydrothermal method using cetyltrimethylammonium bromide as a surfactant. This improves both interfacial adhesion strength and material strength. Even though there have been

Table 3.2: Depicting the various photo catalysts employed for biodiesel production

SI.No	Catalyst	Oil	Reaction Conditions	Yield/ Conversion	Ref
1.	ZnO/SiO2	JJCO	12:1 molar ratio (MR) of methanol/ JJCO, at 20°C for 4 hrs, catalyst/ JJCO mass ratio of 15%	59.3%	[29]
2.	Ag-ZnO	Simarouba Oil	9:1 MR of methanol/oil at 64°C for 2 hrs, 1.5% w/v Ag-ZnO	86.04%	[29]
3.	CuO/ZnO	WCO	9:1 MR of ethanol/WCO at 65°C for 2 hrs, 5wt% catalyst dosage	90.93%	[32]
4.	La^{3+}/ ZnO-TiO$_2$	WCO	12:1 MR of ethanol/WCO at 35°C for 3 hrs, 4 wt% catalyst dosage	96.14%	[20]
5.	TiO$_2$ TNT	Oleic acid	3:1 MR of methanol/oleic acid at 110°C for 48 hrs	59.3%	[26]
6.	TiO$_2$ P25	Oleic acid	3:1 MR of methanol/oleic acid, 15% catalyst	86.04%	[26]
7.	Cu- TiO$_2$	Palm oil	20:1 MR of methanol/oil at 45°C	90.93%	[29]
8.	TiO$_2$/g-C3N4	WCO	9:1 MR of methanol/oil at 60°C for 1 hr, 2% catalyst concentration (TiO2/20%g-C3N4)	89.5%	[42]
9.	SrTiO$_3$/g-C3N4	WFO	12:1 MR of methanol/oil ratio at 65°C for 2 hrs with 600 rpm stirring speed, 1.25 % catalyst dose	96%	[13]
10.	TiO$_2$/RGO	WCO	12:1 MR of methanol/oil at 65°C for 3 hrs, 1.5 wt% catalyst loading	98%	[44]
11.	TiO$_2$/g-C3N4	JatrophaOil	10:1 MR of ethanol/oil at 25°C for 2hrs,4wt% catalyst loading	97%	[43]
12.	BAPBIL	Oleic acid	MR of 5:1 alcohol/oleic acid at room temperature, 20 mg catalyst amount	95%	[45]
13.	TiO$_2$/ PrSO3H	UCO	15:1 MR of methanol/oil at 60°C for 9 hrs, 4.5 wt% catalyst loading	98.3%	[26]
14.	SiO2-Cu@ Fe2O3	WCO	15:1 MR of methanol/oil at 70°C for 7 hrs, 2 wt% catalyst loading.	98%	[46]
15.	ZnO	Aegle marmelos oil	1.5% w/v of ZnO to methanol (9:1) added to the preheated oil and reaction was carried out at 65oC for 2h.	80.6%	[22]
16	SnS$_2$/Ag/ HAp	Oleic acid	8:1 MR of methanol/oil at 70°C for 1 hr, 1 wt% catalyst loading.	98%	[23]
17	Graphene oxide	WCO	1:1 MR of methanol/oil for 270 sceonds, 5 wt% catalyst loading.	96.95%	[24]

significant advances in the study of photocatalyzed biodiesel production, it is clear from the outcomes and constraints that additional research on photocatalysts is necessary to increase their activity and make them economically viable for biodiesel production.

3.5. *Challenges and Limitations*

3.5.1. Photocatalyst Stability and Recyclability

Photocatalysts, typically used in processes like photocatalytic water splitting or pollutant degradation, face degradation over time due to exposure to harsh conditions or photo-corrosion. This degradation can lead to decreased efficiency and the need for frequent replacement [46].

Developing photocatalysts that can maintain their activity over multiple cycles of use is crucial for sustainability. Methods such as immobilisation of catalysts on stable supports or surface modification can enhance recyclability by protecting the active sites from degradation [46].

3.5.2. Effect of Impurities in Feedstocks

Impurities present in biodiesel feedstocks, such as free fatty acids (FFAs) and water, can negatively impact catalyst performance and product quality. FFAs can react with the catalyst to form soaps, reducing catalyst activity and complicating product separation. Water content can also decrease catalyst reactivity and promote unwanted side reactions [47].

To mitigate the effects of impurities, pre-treatment steps such as esterification or dehydration may be employed to remove FFAs and water from the feedstock before biodiesel production. Alternatively, catalysts with higher tolerance to impurities can be developed to maintain performance in the presence of contaminants [47].

3.5.3. Economic Considerations

The overall cost of biodiesel production significantly influences its commercial viability. Factors such as the cost of catalysts, feedstock, energy consumption, and waste disposal contribute to production expenses [48].

To reduce production costs, strategies may include using low-cost feedstocks like waste cooking oils or agricultural residues. Additionally, utilising catalysts derived from biomass or waste materials can lower catalyst expenses. Process optimisation to minimise energy consumption and waste generation also plays a critical role in cost reduction [48].

3.5.4. Scalability Challenges

Scaling up biodiesel production from laboratory-scale to industrial levels presents various challenges, including reactor design, process optimisation, and logistical considerations. Maintaining consistent product quality and maximising efficiency becomes increasingly complex at larger scales [49].

Ensuring a reliable and sustainable supply chain of feedstock, catalysts, and other resources is essential for large-scale biodiesel production. This involves efficient logistics, reliable sourcing of raw materials, and establishing partnerships with suppliers to meet demand while minimising costs and environmental impact [49].

4. Conclusion and Scope

Biodiesel stands as a beacon of hope in the quest to replace fossil fuels, offering a renewable, sustainable alternative to traditional energy sources. The burgeoning field of photocatalysis presents a tantalising prospect for revolutionising biodiesel production, offering a more energy-efficient, easily recoverable, and environmentally friendly alternative to conventional thermal-based heterogeneous catalysis. A comprehensive review of recent studies underscores the immense potential of photocatalytic processes in biodiesel production, showcasing a myriad of promising avenues for improvement and optimisation. Central to this paradigm shift is the exploration of various photocatalysts, each with its unique properties and mechanisms, which promise to propel the efficiency and scalability of biodiesel production. Through meticulous experimentation and analysis, researchers have elucidated the general mechanisms underlying transesterification and esterification processes, shedding light on the intricate steps involved in harnessing solar energy to drive chemical transformations. Notably, the performance of photocatalysts has been substantially enhanced through ingenious strategies such as metal/metal ion doping, synergistic combinations with other metal oxides, incorporation of graphitic carbon nitrides, and implementation of vigorous stirring regimes during reaction cycles. These interventions serve to fine-tune band gaps, optimise light absorption within the visible spectrum, facilitate the efficient separation of photogenerated charge carriers, and mitigate the recombination rate of electrons and holes, thereby bolstering overall catalytic activity. Despite the remarkable progress achieved thus far, the journey towards realising the full potential of photocatalysis in biodiesel production remains rife with challenges and opportunities. Continued research efforts are imperative to deepen our understanding of the underlying mechanisms, unravel the intricacies of photocatalytic pathways, and engineer commercially viable and economically efficient photocatalysts suitable for large-scale deployment. A steadfast commitment to innovation and collaboration holds the key to unlocking the transformative potential of photocatalysis, ushering in an era of greener, more sustainable, and economically viable biodiesel production. By harnessing the power of sunlight to drive chemical reactions, we can forge a path towards a brighter, cleaner, and more prosperous future, where energy independence and environmental stewardship converge harmoniously. Through sustained investment and collective endeavor, we can catalyze the transition towards a more resilient and equitable energy landscape, where the promise of biodiesel as a cornerstone of sustainable development is fully realised.

References

[1] Ahmed, M. H., Byrne, J. A., and Keyes, T. E. (2014). Investigation of the inhibitory effects of TiO2 on the β-amyloid peptide aggregation. *Mater. Sci. Eng. C*, 39, 227–234. https://doi.org/10.1016/j.msec.2014.03.011.

[2] Alves, A. K., Berutti, F. A., Clemens, F. J., Graule, T, and Bergmann, C. P. (2009). Photocatalytic activity of titania fibers obtained by electrospinning. *Mater. Res. Bull.*, 44, 312–317. https://doi.org/10.1016/j.materresbull.2008.06.001.

[3] Ani, I. J., Akpan, U. G., Olutoye, M. A., and Hameed, B. H. (2018). Photocatalytic degradation of pollutants in petroleum refinery wastewater by TiO2- and ZnO-based photocatalysts: recent development. *J. Cleaner Prod.*, 205, 930–954. https://doi.org/10.1016/j.jclepro.2018.08.189.

[4] Aranda, D. A. G., Santos, R. T. P., Tapanes, N. C. O., Ramos, A. L. D., and Antunes, O. A. C. (2008). Acid-catalyzed homogeneous esterification reaction for biodiesel production from palm fatty acids. *Catal. Lett.* 122, 20–25. https://doi.org/10.1007/s10562-007-9318-z.

[5] Babu, V. J., Kumar, M. K., Nair, A. S., Kheng, T. L., Allakhverdiev, S. I., and Ramakrishna, S. (2012). Visible light photocatalytic water splitting for hydrogen production from N-TiO2 rice grain shaped electrospun nanostructures. *Int. J. Hydrogen Energy*, 37, 8897–8904. https://doi.org/10.1016/j.ijhydene.2011.12.015.

[6] Bandara, J., Hadapangoda, C. C., and Jayasekera, W. G. (2004). TiO2/MgO composite photocatalyst: the role of MgO in photoinduced charge carrier separation. *Appl. Catal. B Environ.* 50, 83–88. https://doi.org/10.1016/j.apcatb.2003.12.021.

[7] Barange, S. H., Raut, S. U., Bhansali, K. J., Balinge, K. R., Patle, D. S., and Bhagat, P. R. (2021). Biodiesel production via esterification of oleic acid catalyzed by brønsted acid-functionalized porphyrin grafted with benzimidazolium-based ionic liquid as an efficient photocatalyst. *Biomass Convers. Biorefin.*, 13, 1873–1888. https://doi.org/10.1007/s13399-020-01242-7.

[8] Bhatkhande, D. S, Pangarkar, V. G, and Beenackers, A. A. C. M. (2002). Photocatalytic degradation for environmental applications - a review." *J. Chem. Technol. Biotechnol.*, 77, 102–16. https://doi.org/10.1002/jctb.532.

[9] Borah, M. J., Devi, A., Saikia, R. A., and Deka, D. (2018). Biodiesel production from waste cooking oil catalyzed by in-situ decorated TiO2 on reduced graphene oxide nanocomposite." *Energy* 158, 881–889. https://doi.org/10.1016/j.energy.2018.06.079.

[10] Chen, J., Qiu, F., Xu, W., Cao, S., and Zhu, H. (2015). Recent progress in enhancing photocatalytic efficiency of TiO_2-based materials. *Appl. Catal. A General,* 495, 131–140. https://doi.org/10.1016/j.apcata.2015.02.013.

[11] Choi, W.. 2006. Pure and modified TiO2 photocatalysts and their environmental applications. *Catal. Sur. Asia,* 10, 16–28. https://doi.org/10.1007/s10563-006-9000-2.

[12] Corro, G., Pal, U., and Tellez, N. (2013). Biodiesel production from jatropha curcas crude oil using ZnO/SiO2 photocatalyst for free fatty acids esterification. *Appl. Catal. B Environ.*, 129, 39–47. https://doi.org/10.1016/j.apcatb.2012.09.004.

[13] Corro, G., Sánchez, N., Pal, U., Cebada, S., and Fierro, J. L. G. (2017). Solar-irradiation driven biodiesel production using Cr/SiO2 photocatalyst exploiting cooperative interaction between Cr6+ and Cr3+ moieties. *Appl. Catal. B Environ.* 203, 43–52. https://doi.org/10.1016/j.apcatb.2016.10.005.

[14] Cui, W., Xu, C., Zhang, S., Feng, L., Lü, S., and Qiu, F. (2005). Hydrogen evolution by photocatalysis of methanol vapor over Ti-Beta. *J. Photochem. Photobiol. A Chem.* 175, 89–93. https://doi.org/10.1016/j.jphotochem.2005.04.020.

[15] De, A. and Boxi, S. S. (2020). Application of Cu impregnated TiO2 as a heterogeneous nanocatalyst for the production of biodiesel from palm oil." *Fuel* 265, 117019. https://doi.org/10.1016/j.fuel.2020.117019.

[16] deKrafft, K. E., Wang, C., and Lin, W. (2012). Metal-organic framework templated synthesis of Fe_2O_3/TiO_2 nanocomposite for hydrogen production. *Adv. Mater.* 24, 2014–2018. https://doi.org/10.1002/adma.201200330.

[17] Djurišić, A. B. et al. (2020). Visible-light photocatalysts: prospects and challenges. *APL Mater.* 8, 030903. https://doi.org/10.1063/1.5140497.

[18] Dong, H., Zeng, G., Tang, L., Fan, C., Zhang, C., He, X., and He, Y. (2015). An overview on limitations of TiO2-based particles for photocatalytic degradation of organic pollutants and the corresponding countermeasures. *Water Res.* 79, 128–146. https://doi.org/10.1016/j.watres.2015.04.038.

[19] Encinar, J. M., González, J. F., and Rodríguez-Reinares, A. (2007). Ethanolysis of used frying oil. Biodiesel preparation and characterization." *Fuel Process. Technol.,* 88, 513–522. https://doi.org/10.1016/j.fuproc.2007.01.002.

[20] Gardy, J., Hassanpour, A., Lai, X., Ahmed, M. H., and Rehan, M. (2017). Biodiesel production from used cooking oil using a novel surface functionalised TiO2 nano-catalyst." *Appl. Catal. B Environ.*, 207, 297–310. https://doi.org/10.1016/j.apcatb.2017.01.080.

[21] Gautam, A. et al. (2023). Metal- and ionic liquid-based photocatalysts for biodiesel production: a review. *Environ. Chem. Lett.*, 21, 3105–3126. https://doi.org/10.1007/s10311-023-01637-8.

[22] Ghani, N., Iqbal, J., Sadaf, S., Bhatti, H. N., and Asgher, M. (2021). A facile approach for the synthesis of $SrTiO_3$/G-C_3 N_4 photocatalyst and its efficacy in biodiesel production. *Chem. Select.*, 6, 12082–12093. https://doi.org/10.1002/slct.202101787.

[23] Guo, M., Jiang, W., Chen, C., Qu, S., Lu, J., Yi, W., and Ding, J. (2021). Process optimization of biodiesel production from waste cooking oil by esterification of free fatty acids using La3+/ZnO-TiO2 photocatalyst. *Energy Convers. Manage.*, 229, 113745. https://doi.org/10.1016/j.enconman.2020.113745.

[24] Guo, M., Jiang, W., Ding, J., and Lu, J. (2022). Highly active and recyclable CuO/

ZnO as photocatalyst for transesterification of waste cooking oil to biodiesel and the kinetics. *Fuel* 315, 123254. https://doi.org/10.1016/j.fuel.2022.123254.

[25] Hoffmann, M. R., Martin, S. T., Choi, W., and Bahnemann, D. W. (1995). Environmental applications of semiconductor photocatalysis. *Chem. Rev.* 95, 69–96. https://doi.org/10.1021/cr00033a004.

[26] Huang, J. et al. (2022) Research progress on the photo-driven catalytic production of biodiesel." *Front. Chem.*, 10, 904251. https://doi.org/10.3389/fchem.2022.904251.

[27] Hwu, J. R., Hsu, C.-Y., and Jain, M. L. (2004). Efficient photolytic esterification of carboxylic acids with alcohols in perhalogenated methane. *Tetrahedron Lett.*, 45, 5151–5154. https://doi.org/10.1016/j.tetlet.2004.04.155.

[28] Khan, M., Farah, H., Iqbal, N., Noor, T., Amjad, M. Z. B., and Bukhari, S. S. E. (2021). "A TiO_2 composite with graphitic carbon nitride as a photocatalyst for biodiesel production from waste cooking oil. *RSC Adv.*, 11, 37575–83. https://doi.org/10.1039/D1RA07796A.

[29] Liao, S., Donggen, H., Yu, D., Su, Y., and Yuan, G. (2004). "Preparation and characterization of ZnO/TiO2, SO42–/ZnO/TiO2 photocatalyst and their photocatalysis. *J. Photochem. Photobiol. A Chem.*, 168, 7–13. https://doi.org/10.1016/j.jphotochem.2004.05.010.

[30] Luque, R., Lovett, J. C., Datta, B., Clancy, J., Campelo, J. M., and Romero, A. A. (2010). Biodiesel as feasible petrol fuel replacement: a multidisciplinary overview." *Energy Environ. Sci.* 3, 1706. https://doi.org/10.1039/c0ee00085j.

[31] Manique, M. C., Silva, A. P., Alves, A. K., and Bergmann, C. P. (2016). "Application of hydrothermally produced TiO2 nanotubes in photocatalytic esterification of oleic acid. *Mater. Sci. Eng. B*, 206, 17–21. https://doi.org/10.1016/j.mseb.2016.01.001.

[32] Medeiros, T. V. de, Macina, A., and Naccache, R. (2020). "Graphitic carbon nitrides: efficient heterogeneous catalysts for biodiesel production." *Nano Energy*, 78, 105306. https://doi.org/10.1016/j.nanoen.2020.105306.

[33] Nagaraju, G., Udayabhanu, S., Prashanth, S. A., Shastri, M., Yathish, K. V., Anupama, C., and Rangappa, D. (2017). Electrochemical heavy metal detection, photocatalytic, photoluminescence, biodiesel production and antibacterial activities of Ag–ZnO nanomaterial. *Mater. Res. Bull.* 94, 54–63. https://doi.org/10.1016/j.materresbull.2017.05.043.

[34] Naveed, A. B., Javaid, A., Zia, A., Ishaq, M. T., Amin, M., Farooqi, Z. U. R., and Mahmood, A. (2023). TiO_2/g-C_3 N_4 binary composite as an efficient photocatalyst for biodiesel production from jatropha oil and dye degradation. *ACS Omega*, January, acsomega.2c04841. https://doi.org/10.1021/acsomega.2c04841.

[35] Nemiwal, M., Zhang, T. C., and Kumar, D. (2021). Recent progress in G-C3N4, TiO2 and ZnO based photocatalysts for dye degradation: strategies to improve photocatalytic activity. *Sci. Total Environ.*, 767, 144896. https://doi.org/10.1016/j.scitotenv.2020.144896.

[36] Shah, S. I., Li, W., Huang, C.-P., Jung, O., and Ni, C. (2002). Study of Nd^{3+}, Pd^{2+}, Pt^{4+}, and Fe^{3+} dopant effect on photoreactivity of TiO_2 nanoparticles." *Proc. Natl. Acad. Sci.*, 99, 6482–6486. https://doi.org/10.1073/pnas.052518299.

[37] Singh, J., Kumar, S., Rishikesh, Manna, A. K., and Soni, R. K. (2020). Fabrication of ZnO–TiO2 nanohybrids for rapid sunlight driven photodegradation of textile dyes and antibiotic residue molecules. *Opt. Mater.* 107, 110138. https://doi.org/10.1016/j.optmat.2020.110138.

[38] Song, C., Wang, Z., Yin, Z., Xiao, D., and Ma, D. (2022). Principles and applications of photothermal catalysis." *Chem. Catal.*, 2, 52–83. https://doi.org/10.1016/j.checat.2021.10.005.

[39] Ul Haq, Z., Tahir, K., Aazam, E. S., Almarhoon, Z. M., Al-Kahtani, A. A., Hussain, A. A., Nazir, S., Khan, A. U., Subhan, A., and Rehman, K. U. (2021). Surfactants assisted SiO 2 -Cu@Fe 2 O 3 nanofibers: ultra efficient photocatalyst for photodegradation of organic compounds and transesterification of waste edible oil to biodiesel. *Environ. Technol. Innov.* 23, 101694. https://doi.org/10.1016/j.eti.2021.101694.

[40] Veluturla, S., Archna, N., Subba Rao, D., Hezil, N., Indraja, I. S., and Spoorthi, S. (2018). Catalytic valorization of raw

glycerol derived from biodiesel: a review." *Biofuels* 9, 305–314. https://doi.org/10.108 0/17597269.2016.1266234.

[41] Volokh, M., Peng, G., Barrio, J., and Shalom, M. (2019). Carbon nitride materials for water splitting photoelectrochemical cells. *Angew. Chem. Int. Ed.* 58, 6138–6151. https://doi.org/10.1002/anie.201806514.

[42] Wang, J., Wang, G., Wei, X., Liu, G., and Li, J. (2018). ZnO nanoparticles implanted in TiO2 macrochannels as an effective direct Z-scheme heterojunction photocatalyst for degradation of RhB. *Appl. Surf. Sci.* 456, 666–675. https://doi.org/10.1016/j. apsusc.2018.06.182.

[43] Wang, Z., Hu, X., Liu, Z., Zou, G., Wang, G., and Zhang, K. (2019). Recent developments in polymeric carbon nitride-derived photocatalysts and electrocatalysts for nitrogen fixation. *ACS Catal.* 9, 10260–10278. https://doi.org/10.1021/acscatal.9b03015.

[44] Wei, P., Liu, J., and Li, Z. (2013). Effect of Pt loading and calcination temperature on the photocatalytic hydrogen production activity of TiO2 microspheres. *Ceram. Int.* 39, 5387–5391. https://doi.org/10.1016/j. ceramint.2012.12.045.

[45] Wu, G., Chen, T., Su, W., Zhou, G., Zong, X., Lei, Z., and Li, C. (2008). H2 production with ultra-low CO selectivity via photocatalytic reforming of methanol on Au/TiO2 catalyst." *Int. J. Hyd. Energy* 33, 1243–1251. https://doi.org/10.1016/j.ijhydene.2007. 12.020.

[46] Zhang, K., Wang, L., Sheng, X., Ma, M., Jung, M. S., Kim, W., Lee, H., and Park, J. H. (2016). Tunable bandgap energy and promotion of H_2O_2 oxidation for overall water splitting from carbon nitride nanowire bundles." *Adv. Energy Mater.*, 6, 1502352. https://doi.org/10.1002/aenm.201502352.

[47] Zhang, S., Ye, H., Hua, J., and Tian, H. (2019). Recent advances in dye-sensitized photoelectrochemical cells for water splitting. *Energy Chem.*, 1, 100015. https://doi. org/10.1016/j.enchem.2019.100015.

[48] Zhang, Y., Wang, Q., Wang, K., Liu, Y., Zou, L., Zhou, Y., Liu, M., Qiu, X., Li, W., and Li, J. (2022). Plasmonic Ag-decorated Cu_2O nanowires for boosting photoelectrochemical CO_2 reduction to multi-carbon products. *Chem. Commun.*, 58, 9421–9424. https:// doi.org/10.1039/D2CC03167A.

[49] Zhen, B., Li, H., Jiao, Q., Li, Y., Wu, Q., and Zhang, Y. (2012). $SiW_{12}O_{40}$-based ionic liquid catalysts: catalytic esterification of oleic acid for biodiesel production." *Indus. Eng. Chem. Res.* 51, 10374–10380. https://doi. org/10.1021/ie301453c.

4 Comparative analysis of photoelectric properties on monolayer WSe$_2$ and MoS$_2$ by Zn doping

A. Rushitha[a] and S Dawnee[b]

M S Ramaiah Institute of Technology, Bangalore, India

Abstract

Among various 2D materials, transition metal dichalcogenides (TMD) are one of the promising materials for high performance photovoltaic cells due to atomically thin structures with strong light adsorption characteristic, tunable band gap structures and exceptional optical property. In this work, the effect of Zinc doping on the photoelectric properties of 2D TMD materials like WSe$_2$ and MoS$_2$ is explored. Under zero biasing conditions photocurrent is generated due to the photo galvanic effect (PGE), hence a detailed analysis of the PGE from first principles is presented. The analysis is based on non-equilibrium Green's function density-functional theory. Quantum transport simulations were performed to study the photo galvanic effect on WSe$_2$ and MoS$_2$ monolayer augmenting the optical property by substitution doping. The WSe$_2$ and MoS$_2$ monolayer was vertically illuminated by linearly polarised light to generate photocurrent. Zn substitution doping has evidently increased photocurrent generated and high polarisation sensitivity was obtained. The asymmetry created in the device created by doping enhanced the photocurrent generated. The improved performance obtained from doped structures can be attributed to the impurities introduced in the WSe$_2$ and MoS$_2$ monolayer since the impurity energy bands which cross the fermi energy level serve as pathways for electronic transitions thus improving the photocurrent. This paper provides an insight into enhancement of photocurrent in 2D TMD materials like WSe$_2$ and MoS$_2$ monolayer for potential applications in optoelectronics including photovoltaic cells.

Keywords: 2D material, transition metal dicalcogenides, photo galvanic effect

1. Introduction

With the increase in exploration of new materials for better performance devices, TMDs which belong to a class of 2D materials have gained massive interest from the researchers. The increase in the attention to explore various TMD materials by researchers is due to the exceptional structural, optical and chemical properties [1,2]. 2D materials have potential application in the field of optoelectronics and photovoltaic cells owing to divergent optical properties of the materials [3–6]. Among the various TMD materials, the most explored materials by researchers are WSe$_2$ and MoS$_2$ due to their indirect bandgap, high carrier mobility and their properties which resemble graphene. The unique properties of WSe$_2$ and MoS$_2$ make them a potential candidate for optical applications [7,8]. MoS$_2$ has been characterised with an indirect bandgap of 1.2 eV. Reduction in the thickness of the single layer MoS$_2$ films will change the indirect bandgap to a direct gap which increases the photoluminesce of the single layer MoS$_2$ [9]. The efficient light harvesting property of MoS$_2$ made it a potential candidate for photoelectric applications [10–12]. The exceptional optical efficiency of single layer MoS$_2$ makes it a potential application photovoltaic

[a]ar2002.atla@gmail.com, [b]dawnee@msrit.edu

DOI: 10.1201/9781003545941-4

cells [13–15]. WSe_2 has a wide bandgap which makes it a potential material for optical applications. The bandgap of WSe_2 lies between 1.16 eV to 1.54eV which falls in the similar range of visible light absorption [16]. Similarity of crystal structural and photoelectric properties exist between WSe_2 and MoS_2 [17–19]. Researchers have been working to improve the performance of WSe_2 and MoS_2 in optical applications by inducing various dopant materials. Photoelectric properties are enhanced by doping various materials in WSe_2 structure [20]. A lot of ongoing research in the field of MoS_2 doping in both theory and experimental work for photoelectric applications [21].

In this paper, a monolayer of $2H\text{-}WSe_2$ and $2H\text{-}MoS_2$ structure was modeled and Zn dopant was introduced in the structure. Using the non-equilibrium Green's function density—functional theory, the photocurrent generated under zero and 0.5eV biasing conditions. A comparative analysis was done based on the normalised photocurrent generated with zinc dopant in both the materials. Calculations show that by doping zinc atom in the monolayer of $2H\text{-}WSe_2$ and $2H\text{-}MoS_2$ enhance their photoelectric properties. These results may provide a novel approach for designing experiments and implementing photovoltaic conversion devices and solar cells in future applications.

2. Modeling and Computational Method

$2H\text{-}WSe_2$ with a lattice constant of 3.282A° and belonging to the P_{63}/mmc spatial symmetry group was chosen as the primary atom in the device model. A similar device was also modeled using $2H\text{-}MoS_2$ with a lattice constant of 3.282A° and belonging to the P_{63}/mmc spatial symmetry group was chosen as the primary atom. NanoDCAL device studio was used to build the devices and both the devices contained 42 atoms each. The blue, orange and gray spheres represent W, Se and Zn atoms respectively as shown in Figure 4.1 and the green, yellow and gray spheres represent M, S and Zn atoms respectively as shown in Figure 4.2. The dopant atom was introduced in the center scattering region of both the devices for a stable structure. Both the device consists of three regions: left electrode, central region, and right region. The central region of both devices was irradiated with linearly polarised light.

The required calculations and structural optimisation were performed by first principle-based quantum transport NanoDCAL software and GGA-PBE96 was chosen for the Exchange-Correlation Function. The K-point for monolayer WSe_2 and with Zn doping were set $12 \times 1 \times 1$ in the reciprocal vector space and the K-point for monolayer

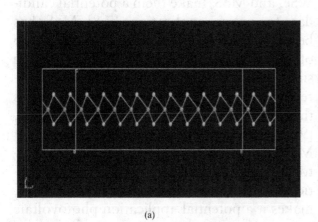
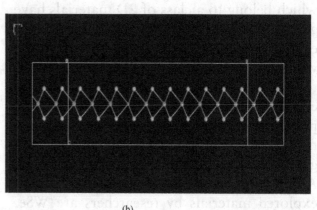

(a)　　　　　　　　　　　　(b)

Figure 4.1: (a) the side view of the WSe_2 device model (b) the side view of the WSe_2 device model with Zn dopant.

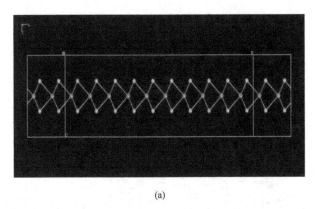
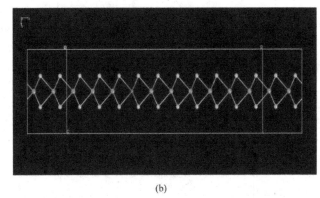

(a)   (b)

Figure 4.2: (a) the side view of the MoS$_2$ device model (b) the side view of the MoS$_2$ device model with Zn dopant.

MoS$_2$ and with Zn doping were set 1 × 1 × 1 in the reciprocal vector space. for both the devices DZP was set for the basis vector of the atoms. Non-equilibrium Green's function methodology was used to calculate the photocurrent generated by monolayer WSe$_2$ and MoS$_2$ with and without Zn dopant with the irradiation of linearly polarised light. Photon energy equation for linearly polarised light is given by, $e = \cos\theta e1 + \sin\theta e1$ where θ is the ploarisation angle and e1 and e2 represents the unit vector.

3. Simulation Results

3.1. Band Structure

Band structure monolayer 2H-WSe$_2$ and 2H-MoS$_2$ structure without and with doping of Zn atom as shown in Figure 4.3 the fermi level is denoted by green dash line. From the band structure results it can be seen that 2H-WSe$_2$ has a direct bandgap of 1.6 eV and 2H-MoS$_2$ has a direct bandgap of 1.66 eV. With Zn atom introduced as the dopant to the 2H- WSe$_2$ and 2H-MoS$_2$ the valence bands and conduction bands were shifted to a higher energy level and more impurity bands were created near the fermi level. With the addition of dopant there was an efficient regulation of electron transport properties of 2H-WSe$_2$ and 2H-MoS$_2$.

3.2. Photocurrent

The photocurrent was generated by vertical irradiated linearly polarised light with the variation in photon energy from 0.4eV to 3.6eV with an interval of 0.2eV. The polarisation angle varied from 0 to 180 degree with an interval of 30 degree.

a. under zero biasing condition:
For the results of the generated photocurrent results it can be seen that in the visible region a large photocurrent has been generated by both the device when doped with the Zn atom. The variation of normalised photocurrent with respect to photon energy with variation in the polarisation angle for both the devices with and without Zn dopant has been shown in Figure 4.4 dark blue, orange, yellow, purple, green, blue and red represents 0, 30, 60, 90, 120, 150 and 180 degrees.

It can be seen that for that the photocurrent first increases and then decreases as the photon energy varies. For monolayer WSe$_2$ the peak normailsed photocurrent was obtained at 1.2eV with polarisation angle of 30° whereas with Zn dopant atom maximum normailsed photocurrent was obtained at 0.4eV with polarisation angle of 0°. For monolayer MoS$_2$ the peak normailsed photocurrent was obtained at 3 eV

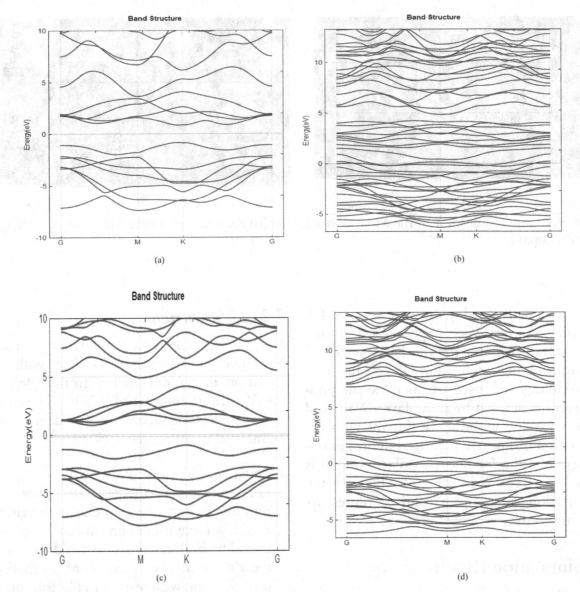

Figure 4.3: Band structure of (a) WSe$_2$ (b) WSe$_2$ with Zn (c) MoS$_2$ (d) MoS$_2$ with Zn.

with polarisation angle of 120° whereas with Zn dopant atom maximum normailsed photocurrent was obtained at 0.8 eV with polarisation angle of 60°. With the addition of Zn atom, the equilibrium of the state of 2H-WSe$_2$ get distributed and asymmetry is formed in the monolayer of 2H-WSe$_2$ which leads to a bridging of electron from valence band to conduction band which leads to an increase in the photocurrent generated. With the addition of a Zn atom for a sufficiently lower energy, photons get excited to the conduction band and result in the higher photon energy.

b. Under 0.5 V biasing condition:

For the results of the generated photocurrent results it can be seen that in the visible region a smaller photocurrent has been generated by both the device when doped with the Zn atom. The variation of normalised photocurrent with respect to photon energy with variation in the polarisation angle for both the devices with and without Zn dopant has been shown in Figure 4.5 dark blue, orange, yellow, purple, green, blue and red represents 0, 30, 60, 90, 120, 150 and 180 degrees.

It can be seen that for that the photocurrent first increases and then decreases

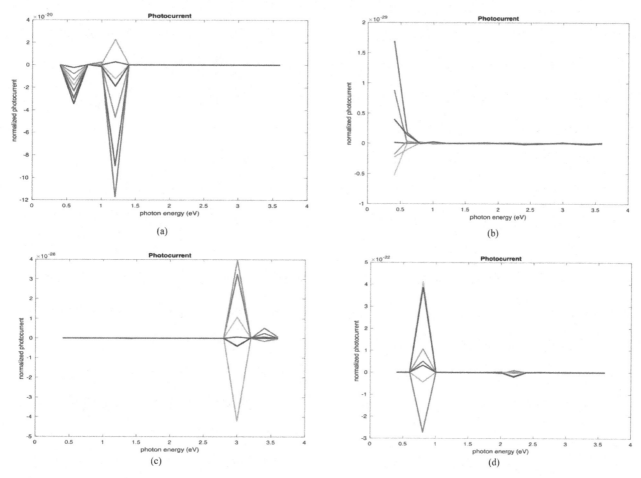

Figure 4.4: WSe₂ (b) WSe₂ with Zn (c) MoS₂ (d) MoS₂ with Zn.

as the photon energy varies. For monolayer WSe₂ the peak normailsed photocurrent was obtained at 0.6 eV with polarisation angle of 30° whereas with Zn dopant atom maximum normailsed photocurrent was obtained at 3 eV with polarisation angle of 120°. For monolayer MoS₂ the peak normailsed photocurrent was obtained at 0.6 eV with polarisation angle of 0° whereas with Zn dopant atom maximum normailsed photocurrent was obtained at 0.6 eV. When the device is supplied with 0.5eV, the electrons generated start to recombine even though addition of Zn atom creates impurity bands near the fermi level which leads to a decrease in the photocurrent generated by the devices since the electrons moving from valence band to conduction band there are insufficient to produce higher photocurrent.

4. Conclusion

The results show that the effect of Zn dopant had a positive effect on WSe2 and MoS2 under zero biasing condition over biasing condition. With addition of zinc dopant, the normalised photocurrent generated by WSe2 and MoS2 was higher than monolayer of WSe2 and MoS2, at zero biasing condition whereas the normalised photocurrent generated under biasing condition by WSe2 and MoS2 was less than monolayer of WSe2 and MoS2. In case of zero biasing condition WSe2 and MoS2 with addition of Zn atom maximum photocurrent was generated at lower photon energy when compared to without Zn dopant and the polisaration angle also decreased significantly. In contrast to zero biasing condition, at 0.5eV biasing WSe2 and MoS2 with addition of Zn atom

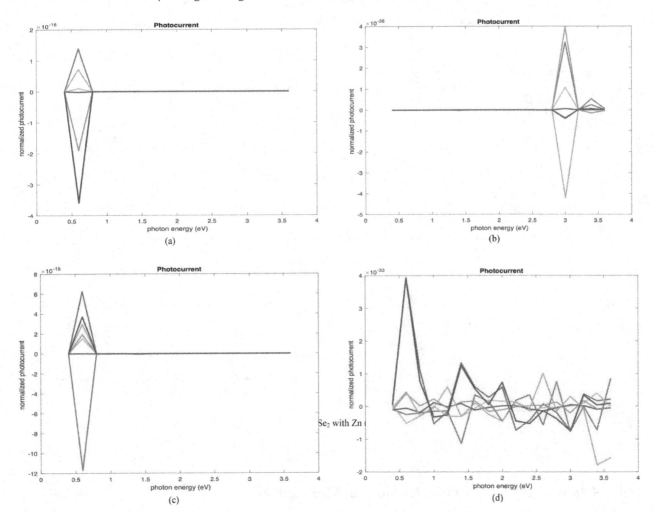

Figure 4.5: (a) WSe$_2$ (b) WSe$_2$ with Zn (c) MoS$_2$ (d) MoS$_2$ with Zn.

the maximum photocurrent was generated at higher photon energy when compared to without Zn dopant and the polisaration angle also decreased significantly. The dopant modulated the sensitivity and created an asymmetry in the structure of the device which helped to improve the performance under zero biasing condition and due to the recombining of electrons under biasing condition the maximum photocurrent generated decreased with increase in the energy required to bridge electrons from valence band to conduction band. It is expected that the results provided by this paper will provide necessary information for theoretical and practical applications of 2H-WSe2 and 2H-MoS2 in optoelectronics.

Acknowledgement

The authors gratefully acknowledge the staff, and authority of the Electrical and Electronics department of MS Ramaiah Institute of Technology for their cooperation in the research.

References

[1] Kumar, U., Mishra, K. A., Kushwaha, A. K., and Cho, S. B. (2022). Bandgap analysis of transition-metal dichalcogenide and oxide via machine learning approach. *J. Phys. Chem. Solids*, 171, 110973.

[2] Choudhary, N., Islam, M. A., Kim, J. H., Ko, T.-J., Schropp, A., Hurtado, L., Weitzman, D., Zhai, L., and Jung, Y. (2018). Two-dimensional transition metal dichalcogenide

hybrid materials for energy applications." *Nano Today* 19, 16–40.

[3] Zhan, L.-B., Yang, C.-L., Wang, M.-S., and Ma, X.-G. (2022). 2D XBiSe$_3$ (X= Ga, In, Tl) monolayers with high carrier mobility and enhanced visible-light absorption. *Spectrochim. Acta, Part A* 264, 120309.

[4] Ma, Q., Ren, G., Xu, K., and Ou, J. Z. (2021). Tunable optical properties of 2D materials and their applications. *Adv. Opt. Mater.,* 9, 2001313.

[5] Weng, Q., Li, G., Feng, X., Nielsch, K., Golberg, D., and Schmidt, O. G. (2018). Electronic and optical properties of 2D materials constructed from light atoms. *Adv. Mater.,* 30, 1801600.

[6] Qiao, H., Liu, H., Huang, Z., Hu, R., Ma, Q., Zhong, J., and Qi, X. (2021). Tunable electronic and optical properties of 2D monoelemental materials beyond graphene for promising applications. *Energy Environ. Mater.,* 4, 522–543.

[7] Cen, K., Yan, S., Yang, N., Dong, X., Xie, L., Long, M., and Chen, T. (2023). The adjustable electronic and photoelectric properties of the WS$_2$/WSe$_2$ and WSe$_2$/WTe$_2$ van der Waals heterostructures. *Vacuum* 212, 112020.

[8] Jameel, M., et al. "A comparative dft study of bandgap engineering and tuning of structural, electronic, and optical properties of 2d WS$_2$, PtS$_2$, and MoS$_2$ between WSe$_2$, PtSe$_2$, and MoSe$_2$ materials for photocatalytic and solar cell applications," in *Journal of Inorganic and Organometallic Polymers and Materials*, vol. 34, no. 1, pp. 322–335, 2024.

[9] Splendiani, A., Sun, L., Zhang, Y., Li, T., Kim, J., Chim, C.-Y., Galli, G., and Wang, F. (2010). Emerging photoluminescence in monolayer MoS$_2$. *Nano Lett.* 10, 271–1275.

[10] Si, K., Ma, J., Guo, Y., Zhou, Y., Lu, C., Xu, X., and Xu, X. (2018). Improving photoelectric performance of MoS$_2$ photoelectrodes by annealing. *Ceramics Int.,* 44, 21153–21158.

[11] Zhao, Q., Guo, Y., Zhou, Y., Yan, X., and Xu, X. (2017). Solution-processable exfoliation and photoelectric properties of two-dimensional layered MoS$_2$ photoelectrodes. *J. Colloid Interface Sci.,* 490, 287–293.

[12] Yin, Z., Chen, B., Bosman, M., Cao, X., Chen, J., Zheng, B., and Zhang, H. (2014). Au nanoparticle-modified MoS$_2$ nanosheet-based photoelectrochemical cells for water splitting. *Small* 10, 3537–3543.

[13] Yin, Z., Li, H., Li, H., Jiang, L., Shi, Y., Sun, Y., Lu, G., Zhang, Q., Chen, X., and Zhang, H. (2012). Single-layer MoS2 phototransistors. ACS Nano 6, 74–80.

[14] Lopez-Sanchez, O., Lembke, D., Kayci, M., Radenovic, A., and Kis, A. (2013). Ultrasensitive photodetectors based on monolayer MoS2. Nat. Nanotechnol., 8, 497–501.

[15] Jian, J., Chang, H., and Xu, T. (2019). Structure and properties of single-layer MoS$_2$ for nano-photoelectric devices. *Materials* 12, 198.

[16] Ma, Q., Kyureghian, H., Banninga, J. D., and Ianno, N. J. (2014). "Thin film WSe2 for use as a photovoltaic absorber material," in *MRS Online Proceedings Library* (OPL), 1670, 1614–1670.

[17] Mak, K. F. and Shan, J. (2016). Photonics and optoelectronics of 2D semiconductor transition metal dichalcogenides. *Nat. Photon.,* 10, 216–226.

[18] Bromley, R. A., Murray, R. B., and Yoffe, A. D. (1972). The band structures of some transition metal dichalcogenides. III. Group VIA: trigonal prism materials. *J. Phys. C Solid State Phys.,* 5, 759.

[19] Sahin, H., Tongay, S., Horzum, S., Fan, W., Zhou, J., Li, J., Wu, J., and Peeters, F. M. (2013). Anomalous Raman spectra and thickness-dependent electronic properties of WSe$_2$. *Phys. Rev. B* 87, 165409.

[20] Xu, Z.-H., Chen, Z., and Yuan, Q.-M. (2021). Effects of doping Ti, Nb, Ni on the photoelectric properties of monolayer 2H–WSe$_2$. *Phys. E Low-dimen. Syst. Nanostruct.,* 133, 114846.

[21] Liu, P.-P., Shao, Z.-G., Luo, W.-M., and Yang, M. (2021). Photogalvanic effect in chromium-doped monolayer MoS$_2$ from first principles. *Phys. E Low-dimen. Syst. Nanostruct.,* 128, 114577.

5 Prospective role of MoS$_2$ in Zinc Nickel flow battery system: review

A. Preethi[1,a], *S. Dawnee*[2,b], *and Victor George*[2]

[1]M S Ramaiah Institute of Technology, Bangalore, India
[2]CMR Institute of Technology, Bengaluru, India

Abstract

Integrating intermitted renewable sources with the power grid can help mitigate the power crisis and reduce dependency on fossil fuel-based energy generation. Large-scale storage systems coupled with renewable energy sources can facilitate the storage of surplus energy generated during off-peak hours and cater to the demand during peak hours. Redox flow batteries find their spot in such applications owing to their higher power density and high energy density. Low cost, low toxicity, higher cell voltage, and safe operation conditions make Zinc-Nickel single flow battery a promising choice. This system employs a single electrolyte unlike other conventional flow batteries, which enables the design to be membrane-free, less complex, and cost-effective. Commercial large-scale implementation of this battery chemistry is hindered by nonuniform Zn deposition on the cathode during the charge-discharge cycle and associated poor cycle life. This review primarily focuses on the understanding of the zinc deposition and zinc dendritic growth in zinc nickel flow batteries and proposes a prospective role for Molybdenum disulfide (MoS$_2$), a transition metal dichalcogenide. The primary factor behind the nonuniform Zn deposition is the unpaired coulombic efficiency of the electrodes. The deposited Zn on the cathode during charging is not fully consumed during discharge, and deterioration in performance is higher at high charging discharging rates, which practically reduces the cycle life and the applicability to grid-scale systems. Further instability is introduced by the gases released during the side reactions at electrodes. Electrode structural engineering with higher specific surface area (SSA) materials can promote uniform zinc deposition and reduce the effect of side reactions. The comprehensive review finds MoS$_2$ as a viable candidate as a host electrode material, interfacial layer, and passivating material for zinc nickel flow batteries owing to its tunable characteristics, higher SSA, and conductivity.

Keywords: Zinc nickel flow battery, dendrite, grid-level storage, cycling stability

1. Introduction

Net zero emission by 2050 envisions large-scale deployment of renewable energy sources all over the globe to facilitate access to clean energy on a higher scale. However, the intermitted nature of these sources introduces instability to the power grid. To overcome these instabilities and to enhance grid reliability, renewable sources must be integrated with energy storage systems. The off-peak hour higher power generated can be stored on-site in an appropriate energy storage system and fed back to the grid during high-demand scenarios [1]. Lithium-ion batteries, nickel-cadmium batteries, sodium-sulfur batteries, and vanadium redox flow batteries are the few electrochemical energy storage systems primarily explored and commercialised for grid-level renewable energy storage. The choice of a particular chemistry is largely guided by the cost, environmental concerns, capacity in terms of both energy and power and most importantly the application area. Even though the most mature storage chemistry is lithium-ion batteries, the scarcity of lithium and the associated safety factors force researchers to focus on other

[a]preethi.a@cmrit.ac.in, [b]dawnee@msrit.edu

DOI: 10.1201/9781003545941-5

storage systems. Flow batteries which decouple energy and power capability, thus much safer than lithium-ion batteries, are a promising grid-level storage solution. Compared to lithium-ion batteries, flow batteries have much lower energy density. Positively from the application point of view, since grid-level storage systems are stationary the compromise can be arrived at by increasing the electrolyte storage capacity in a cost-efficient manner [2–5]. Table 5.1 provides a comparative analysis of different battery chemistries based on their per-unit cost and performance.

Vanadium flow batteries belong to the category of redox flow batteries, where vanadium in 4 different oxidation states is employed for electrochemical energy storage. The electrolytes are stored in two external tanks, a pump is used to circulate the electrolyte to the half cells during the charging and discharging process, and a membrane is employed to separate the half cells.

The vanadium redox flow battery is characterised by its long cycling life owing to the infinite lifetime of vanadium, and the ability to decouple power and energy density. The major cost factor in vanadium flow battery is the membrane and the vanadium redox couples [6]. Hybrid flow batteries are a type of flow battery where one active mass is stored in an electrolyte tank and the other in the battery. In this battery system, a metal is plated onto the negative electrode on charging, and the same is stripped during discharging. Unlike redox flow batteries, the power and energy capacity are not decoupled in these systems which further reduces the capacity and efficiency. Single-flow batteries, a hybrid battery that does not require a membrane for operation can further reduce the cost. Other flow chemistries are explored to lower the cost per kWh storage and to enhance volumetric energy capacity. If the active material employed is abundant in nature, the per unit energy cost for the battery can be lowered to a limit. Materials like zinc, cadmium, manganese, and iron are a few active materials explored so far. Among these zinc offers better cell potential and higher volumetric capacity. Different flow chemistries with zinc (Zn) as the active materials that are explored extensively include zinc-bromine, zinc-air, zinc-iron, and zinc-nickel [7–10].

The research interest of this work is zinc-nickel single flow battery as a candidate for grid-level storage and to critically analyze the challenges hindering the commercial implementation of this battery chemistry. Further, this work tries to provide a prospective role of molybdenum disulfide (MoS_2) a transition metal dichalcogenide for performance improvement of zinc nickel single flow battery. The paper is organised as follows, Section II gives a comprehensive outlook on

Table 5.1: Comparison of Various Storage Technologies (*)

Storage Technology	Per kWh cost ($)	Life	Concerns
Lithium-ion	850–1300	7000 cycles, 5–10 years	Cost, Safety
Lead–acid	300–1500	200–1500 cycles 5–10 years	Poor energy density
Sodium–Sulfur	300–800	100–4000 cycles 5–10 years	Safety
Flow Battery	180–300	1000–15000 cycles >15 years	Moderately low energy density
Supercapacitor	60000–90000	10000–100000 cycles <1 year	Self-discharge, Lower energy density

*Adapted from publicly available online resources (accessed on 1st December 2023)

zinc-nickel flow battery and reviews the challenges hindering commercial implementation and solutions employed so far. In section III a prospective role of MoS_2 for zinc nickel battery is sought forth by critically reviewing the status of the application of the material to enhance zinc metal anode performance in alkaline electrolytes and suggesting the research gaps.

2. Zinc Nickel Flow Battery

The zinc nickel single-flow battery system was proposed in 2007 as a modification of high energy and high discharge voltage alkaline zinc nickel battery. Even though the alkaline battery demonstrated excellent voltage stability at different discharge rates and the active materials were abundant and environmentally safe, the low cyclic stability owing to the deterioration of the zinc electrode was the major hurdle in the commercial viability of this battery for reliable applications. The structure of the zinc-nickel single-flow battery is given is Figure 5.1. The battery structure consists of an inert material negative electrode and nickel hydroxide positive electrode, the electrolyte, zinc oxide in potassium hydroxide solution is stored in a separate tank. During the operation, the electrolyte is pumped into the cell using an external pumping mechanism. During charging the zinc

metal is deposited on the negative electrode and the same is removed from the electrode during discharging. Equations (5.1) and (5.2) demonstrate the cell reactions, these batteries are characterised by a cell voltage of around 1.7V higher than most flow battery chemistries.

Positive Electrode:

$$2NiOOH + 2H_2O + 2e^- \leftrightarrow 2Ni(OH)_2 + 2OH^- + 0.49V \quad (5.1)$$

Negative Electrode

$$Zn + 4OH^- \leftrightarrow Zn(OH)_4^{2-} + 2e^- - 1.24V \quad (5.2)$$

Here the active materials undergo solid-solid phase transformation and no soluble species are created during the positive electrode reaction which facilitates a membrane-less design and reduced cost. The flowing electrolyte played a substantial role in improving the cycling stability due to the mass ion transfer rate improvement [11–13]. However further studies on the battery revealed that the electrode degradation happens after few cycles and the battery eventually fails even in the presence of flowing electrolyte. To mitigate the electrode failure a clear understanding of the zinc deposition chemistry is necessary.

2.1. *Understanding Zinc Deposition and Dendrite Growth*

During the charging cycle of the zinc-nickel flow battery, the electrolyte is pumped into the cell and the zinc ions undergo reduction and get plated at the negative electrode while the nickel hydroxide undergoes oxidation at positive electrode. The reverse process occurs during the discharging process. For capacity retention and to achieve better cycle life the deposited zinc during the charging process must be removed during the discharge process. Due to the non-uniform deposition of Zn, the process is not completely reversible

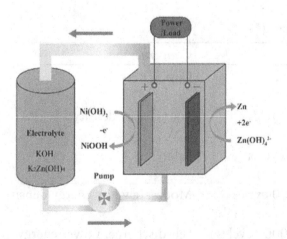

Figure 5.1: Schematic of Zinc-Nickel flow batteries [13].

which leads to capacity decay and battery failure over charge/discharge cycles [14]. To overcome this issue and to improve the performance of zinc nickel flow batteries a clear understanding of Zn deposition and growth is necessary.

The zinc deposition starts with nuclei formation on the anode, further deposition has two possibilities. One possibility is to form subsequent nuclei on the anode surface which in turn leads to uniform deposition of zinc. Another possibility is to have further depositions on the already formed nuclei which leads to uneven deposition. It is interesting to note that the overpotential for nuclei formation is much higher than the overpotential of further growth on the nuclei. Furthermore, the small nuclei with higher surface energy tend to attract the ions and facilitate deposition growth. Unfortunately, the zinc anode structure favors the growth of already formed nuclei rather than forming new nuclei, which leads to nonuniform deposition and growth of dendritic structures. These materials are not fully consumed during the discharge process and leads to capacity decay. The growing dendritic structures over the cycles eventually result in a short circuit in the battery. During the discharge process, the anode lose electron and oxidise into zinc hydroxyl ion ($Zn(OH)_4^{2-}$), as the concentration of this ion increases it finally precipitate into ZnO and contributes to further reversibility issues [15–18].

The power efficiency of a battery is directly linked to the current density, multiple theoretical and experimental studies revealed the effect of current density on zinc dendritic growth. Higher current density introduces concentration gradient between the electrode surface and electrolyte and promotes uneven nucleation and further dendritic growth, The uneven zinc deposition leads to a deviation in electrode potential which further promotes the deposition on zinc tips than on the electrode surface. The charge accumulation on the dendritic tips makes them the active sites for zinc deposition [19–23]. Figure 5.2(a) depicts the effect of current density on nucleation and growth overpotentials and on the nuclei size distribution. From the depiction it is clear, the nucleation and growth of dendritic structures are favored by higher current density operating conditions.

2.2. Solutions Employed to Improve Cycle Life

Various works to date have reported methods of inhibiting dendrite growth or slowing down the formation of deposits to increase the cycle life. The major techniques include the use of electrolyte additives, electrode modification, optimum flow rate control, and the use of an external field to control zinc ion transfer. Electrolyte additives inhibit or slow down the dendrite formation either by providing electrostatic shielding on the dendritic structures or by promoting the nucleation

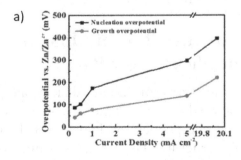 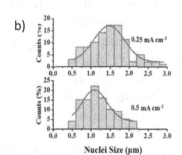 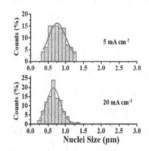

Figure 5.2: (a) Nucleation and growth overpotential dependency on current density and (b) nuclei size distribution dependency on current density, Reprinted with permission from [19]. Copyright 2021, American Chemical Society.

Table 5.2: Solutions employed in literature to inhibit dendrite growth on zinc anode

Method	Materials employed	Effect on the battery performance	Issues	Reference
Electrolyte Additive	Polymer or organic additives	Electrostatic shielding of metal deposits, dendrite inhibition, improved cyclic life, improved thermal performance	Flow rate impairment and thus capacity loss	[24–28]
	Metal additives	Promotes nucleation by uniform electrostatic field distribution, improved thermal performance	Cell chemistry may be altered, Lowering cell potential	[29–33]
Electrode Modification	Conductive Electrode- Carbon, Nickel, Tin, Chromium, Lead	Conductive hosts facilitate uniform polarisation distribution, corrosion free electrode operation, improved cycling stability, Controls side reactions	Chances of altering cell potential after few cycles.	[28,34– 42]
	Structural Modification – 3D structures, porous hosts, anode passivation with zincophilic materials	Provides easy intercalation of zinc ions thus improves volumetric energy density. Inhibition of dendritic structures by providing more nucleation sites, improved coulombic efficiency and reduced side reactions	Increase in cost, bulky structure	

of zinc. Polymer and organic additives are selectively adsorbed by protruding metal structures and act as shields against further metal deposition. On the other hand, metallic additives are found to promote nucleation and thus uniform deposition of Zn on anode [24–33].

The original battery structure utilised an inert current collector as the negative electrode and the degradation in performance was attributed to the nonuniform polarisation distribution at different current densities. The integration with intermitted renewable sources necessitates favorable charging conditions at variable current density. Anode engineering with conductive zinc hosts and modification of the anode structure is one of the plausible solutions so far to ensure uniform polarisation distribution and eventual inhibition of dendrites [34]. Conductive hosts like carbon, nickel, chromium, copper, tin, and lead were successfully employed to improve the cycle performance of zinc-nickel

flow batteries. Anode with a higher specific surface area can facilitate uniform zinc deposition at higher current densities owing to the better structural affinity to zinc ions, anode engineering with 3D and layered structures enhances cycling performance. The modified electrode designs have further reduced side reactions in battery chemistry by balancing the electrode efficiencies. [28,34–42]. Table 5.2 consolidates the different techniques employed so far to inhibit dendrite formation in batteries with zinc metal anode.

3. Prospective Role of MoS_2 in Zinc Nickel Flow Battery

MoS_2 is a transition metal dichalcogenide with a layered structure and tunable physical and electronic properties. Figure 5.3 depicts the MoS_2 structure. The material exhibits efficient electron transfer properties owing to its high electron conductivity and is a promising anode material. The device performance

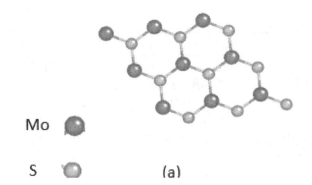

Figure 5.3: Structure of MoS$_2$.

with such anode is further enhanced due to its structural properties like high accessible surface area, porosity, and tunable properties through various synthesis techniques [43].

3.1. Application of MoS$_2$ in Zinc Ion Batteries

Due to its structural and electronic properties, the material is extensively studied as a suitable candidate for various metal anode-based batteries. The focus here is to evaluate the potential of the material for zinc storage and prevention of dendritic growth. Even though there is a significant footprint of MoS$_2$ application in zinc-based alkaline batteries the role of MoS$_2$ in zinc-based flow batteries is less explored. MoS$_2$ and its composites substantially improve the cycling stability and capacity of zinc-based metal batteries in alkaline solution when employed as a 3D host electrode material and as an interfacial coating material.

Uniform zinc deposition during charging and removal during discharging ensure the cycling stability of the battery. The transfer rate of ions on the electrode surface has direct effect on the uniform deposition and dendrite inhibition. MoS$_2$ coating on the electrode surface ensures uniform electric field strength on the electrode surface which accelerates the zinc ion migration and ensures uniform plating of zinc. Furthermore, the conductive MoS$_2$ interface disperses the nuclei and reduces the concentration polarisation, this

effectively contains the growth on nucleation sites. Delayed occurrence of dendritic growth improves the repeatability and slows down the capacity loss [44–46].

The layered structure of MoS$_2$ enhances the interfacial contact area available for zinc deposition and facilitates uniform plating on zinc ions during charging, which substantially hinders dendrite formation. The matured synthesis techniques available for MoS$_2$ allow the material to be defect-engineered or interlayer-tuned, enhancing the availability of edges and sulfur vacancies to intercalate zinc ions. These techniques provide a way of easy insertion and removal of zinc ions during charging and discharging cycles. Considering these properties, MoS$_2$-based 3D structured electrodes have demonstrated better retention and cycling performance in zinc alkaline batteries [46–51]. Furthermore, the structural defect engineering of MoS$_2$ enhances its affinity towards divalent ions, which further facilitates nucleation and uniform deposition. Nanostructured MoS$_2$ has proved to promote layer-by-layer deposition of metal ions at higher ion transfer rates, flowing electrolytes can enhance uniform metal deposition. The enhancements reported by MoS$_2$ for zinc-based alkaline batteries give a direction for the role of MoS$_2$ in zinc-based flow batteries. Table 5.3 details few significant enhancements in zinc-based batteries by employing MoS$_2$.

3.2. Challenges and Roadmap

MoS$_2$ has proven its significant contribution towards zinc alkaline batteries through various experimental verifications presented in the previous section. The major issues faced by zinc nickel flow battery is zinc dendrite formation and associated capacity loss and battery failure. Based on the literature review MoS$_2$ plays a significant role in inhibiting dendritic growth in zinc alkaline batteries. The prospective role of MoS$_2$ for zinc nickel flow battery is sought forth based on its contribution to Zinc alkaline battery enhancements.

Table 5.3: Application of MoS2 in Zinc Ion batteries

Material	Methodology	Property	Performance Enhancement	Ref
Crystalline MoS_2	Plating on the current collector	Conductive interface, uniform distribution of electric field	Specific Capacity: 200mAh/g @ 0.2A/g Cycle: 1000, no dendritic formation	[44]
MoS_2	Interfacial coating on the electrode	Conductive interface, uniform distribution of electric field	Specific capacity: 638 mAh/g @ 0.1 A/g Cycle:2000 cycles, with no dendrite formation at the Zn electrode.	[45]
Defect engineered MoS_2	Electrode/ Coating	Enhanced intercalation sites due to edges and sulfur vacancies. Enhanced affinity towards zinc ions	Specific capacity: 88.6 mAh/g @ 1 A/g Cycles: 1000 Retention: 90%	[46]
MoS_2	Intercalation Host	Engineering interlayer spacing and oxygen incorporation can tune intercalation energy	Zinc ion capacity enhancement of MoS_2:232 mAh g−1	[47]
1T phase MoS_2	Electrode	Lowering of zinc diffusion energy barrier thus enhancing capacity	Specific capacity: 120 mAh/g @ 1 A/g Cycles: 500 Retention: 98%	[48]
1T phase MoS_2	Electrode	Engineering interlayer spacing to enhance capacity retention	Specific capacity: 125 mA h/g @ 2 A/g Cycles: 500 Retention: 100%	[49]
Hydrated MoS_2 nanosheets	Intercalation Host	Crystal water molecule mediated interlayer spacing enhancement, avoids conglomeration of layers	Cycle: 2000 Retention: 98%	[51]

Changes in battery parameters like cell voltage, capacity, efficiency, temperature tolerance, active material decay are expected while introducing a new material into a well-known chemistry. The proposal must be validated with proper mathematical model-based simulation studies before approaching experimental validation. Figure 5.4 consolidates the proposed roles and challenges.

A mathematical model based on mass transfer, flow transfer, charge transfer, and reaction kinetics in the presence of MoS_2 needs to be developed to characterise the battery chemistry, along with the formulation of boundary conditions. The Zinc- Nickel flow battery with Nickel foil as the positive electrode is extensively studied by researchers using various mathematical models. Most of these studies utilised the Butler-Volmer equation for reaction kinetics, volume averaged conservation equation for electrolyte reaction kinetics modeling, the Nernst equation to model side reaction, and Navier-Stokes equations for flow transfer [52–55]. These studies can be extended to model the MoS_2 incorporated flow battery behavior. The behavior of the electrodes under different current densities and electrolyte flow rates must be carefully evaluated to arrive at an optimal operation window for the flow battery. Furthermore,

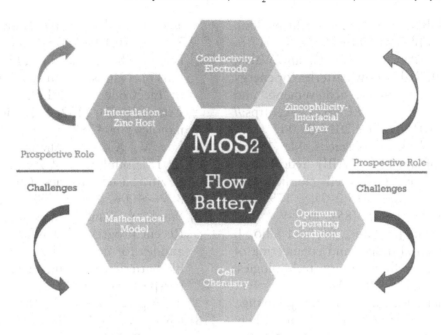

Figure 5.4: MoS₂-Prospects and challenges in Zinc-Nickel flow battery.

4. Conclusion

Zinc-based flow batteries are promising candidates for grid level energy storage. Commercial viability of grid level energy storage systems largely depends on the cost per unit energy, safe operating conditions and cycling stability than the energy density or capacity. Zinc- Nickel single flow battery has a clear advantage in terms of cost due to its membrane less structure and naturally abundant constituent elements. This battery chemistry withstands higher operating temperature and has less volumetric expansion thus offers better battery safety. Zinc dendritic growth and associated cycle life impairment and capacity loss makes this chemistry unsuitable for large scale implementation for grid level applications. Based on the understanding of zinc deposition morphology and dendrite formation, and a critical review on the performance enhancement available so far in literature this review finds MoS₂ a viable candidate to

the synthesis process favorable for the specific application also needs a thorough investigation based on literature and required material properties.

improve cycling stability of zinc nickel flow battery chemistry. Further research must focus on developing theoretical models to understand the zinc deposition morphology with modified electrodes, the variations in critical parameter due to the presence of MoS₂ and to arrive at optimal operating conditions to enhance the overall battery performance.

References

[1] https://www.iea.org/reports/net-zero-road-map-a-global-pathway-to-keep-the-15-0c-goal-in-reach/a-renewed-pathway-to-net-zero-emissions, accessed on 1st December 2023.

[2] Chen, T., Jin, Y., and Lv, H. et al. (2020). Applications of Lithium-Ion Batteries in Grid-Scale Energy Storage Systems. *Trans. Tianjin Univ.* 26, 208–217. https://doi.org/10.1007/s12209-020-00236-w

[3] Perry, M. L. and Weber, A. Z. (2015). Advanced Redox-Flow Batteries: A Perspective. *J. Electrochem. Soc.*, 163(1), A5064–A5067. https://doi.org/10.1149/2.0101601jes

[4] Tomazic, G. and Skyllas-Kazacos M. (2015). Chapter-17, Redox flow batteries, electrochemical energy storage for renewable sources and grid balancing, Elsevier,

309–336, ISBN 9780444626165, https://doi.org/10.1016/B978-0-444-62616-5.00017-6.

[5] Arenas, L. F., Ponce de León, C., and Walsh, F. C. (2019). Redox flow batteries for energy storage: their promise, achievements and challenges. *Curr. Opin. Electrochem.*, https://doi.org/10.1016/j.coelec.2019.05.007

[6] Sum, E., and Skyllas-Kazacos, M. (1985). A study of the V(II)/V(III) redox couple for redox flow cell applications. *J. Power Sour.*, 15(2–3), 179–190. doi:10.1016/0378-7753(85)80071-9

[7] Park, M, Ryu, J., Wang, W., and Cho, J. (2016). Material design and engineering of next-generation flow-battery technologies, *Nature Rev. Mater.*, 2(1), 1–18

[8] Henkensmeier, R., Ye, D,. Yoon, S. J, Huang,. Z, Kim, D. K, Chang, Z, Kim, S, and Chen, R. (2018). Redox flow batteries for energy storage: a technology review, *J. Electrochem Energy Convers. Storage*, 15, (1).

[9] Khor, A., Leung, P., Mohamed, M. R., Flox, C., Xu, Q., An, L., … Shah, A. A. (2018). Review of zinc-based hybrid flow batteries: From fundamentals to applications. *Mater. Today Energy*, 8, 80–108. https://doi.org/10.1016/j.mtener.2017.12.012

[10] Ronen, R., Gat, A. D., Bazant, M. Z., and Suss, M. E. (2021). Single-flow multiphase flow batteries: theory. *Electrochim. Acta*, 389, 138554. https://doi.org/10.1016/j.electacta.2021.1385

[11] Hu, P., Wang, T., Zhao, J., Zhang, C., Ma, J., Du, H., … Cui, G. (2015). Ultrafast Alkaline Ni/Zn Battery Based on Ni-Foam-Supported Ni3S2 Nanosheets. *ACS Appl Mater Interfaces*, 7(48), 26396–26399. https://doi.org/10.1021/acsami.5b09728

[12] Jie Cheng, Li Zhang, Yu-Sheng Yang, Yue-Hua Wen, Gao-Ping Cao and Xin-Dong (Wang 2007). Preliminary study of single flow zinc–nickel battery, *Electrochem. Commun.*, 9(11), 2639–2642. https://doi.org/10.1016/j.elecom.2007.08.016

[13] Yao, S., Liao, P., Xiao, M., Cheng, J., and Cai, W. (2017). Study on Electrode Potential of Zinc Nickel Single-Flow Battery during Charge. *Energies*, 10(8), 1101. doi:10.3390/en10081101

[14] Bass, K., Mitchell, P. J., Wilcox, G. D., and Smith, J. (1991). Methods for the reduction of shape change and dendritic growth in zinc-based

secondary cells. *J. Power Sources*, 35(3), 333–351. doi:10.1016/0378-7753(91)80117-g

[15] Pei, A., Zheng, G., Shi, F., Li, Y., and Cui, Y. (2017). Nanoscale Nucleation and Growth of Electrodeposited Lithium Metal. *Nano Letters*, 17(2), 1132–1139. doi:10.1021/acs.nanolett.6b04755

[16] Lu, W., Xie, C., Zhang, H., and Li, X. (2018). Inhibition of zinc dendrite growth in zinc-based batteries. *ChemSusChem*, 11, 3996–4006. doi: 10.1002/cssc.201801657

[17] Wang, X., Sun, C., and Wu, Z. -S. (2023). Recent progress of dendrite-free stable zinc anodes for advanced zinc-based rechargeable batteries: fundamentals, challenges, and perspectives. *SusMat.*, 3, 180–206. https://doi.org/10.1002/sus2.118

[18] Cheng, Y., Xi, X., Li, D., Li, X., Lai, Q., and Zhang, H. (2015). Performance and potential problems of high power density zinc–nickel single flow batteries. *RSC Advances*, 5(3), 1772–1776. doi:10.1039/c4ra12812e

[19] Liu, H., Zhang, Y., Wang, C., Glazer, J. N., Shan, Z., and Liu, N. (2021). Understanding and Controlling the Nucleation and Growth of Zn Electrodeposits for Aqueous Zinc-Ion Batteries. *ACS Applied Materials & Interfaces*, 13(28), 32930–32936. doi:10.1021/acsami.1c06131

[20] Yufit, V., Tariq, F., Eastwood, D. S., Biton, M., Wu, B., Lee, P. D., and Brandon, N. P. (2018). Operando visualization and multiscale tomography studies of dendrite formation and dissolution in zinc batteries. *Joule*. doi:10.1016/j.joule.2018.11.002

[21] F.R. McLarnon and E.J. Cairns(1991). The secondary alkaline zinc electrode, *J. Electrochem. Soc.* 138 645e664, https://doi.org/10.1149/1.2085653

[22] Xu, C. J., Li, B. H., Du, H. D., and Kang, F. Y. (2012). Energetic zinc ion chemistry: the rechargeable zinc ion battery, *Angew. Chem. Int. Ed.* 51 933e935, https://doi.org/10.1002/anie.201106307.

[23] Wenjia Du, Zhenyu Zhang, Francesco Iacoviello, Shangwei Zhou, Rhodri E. Owen, Rhodri Jervis, Dan J. L. Brett, and Paul R. Shearing (2023). Observation of Zn Dendrite Growth via Operando Digital Microscopy and Time-Lapse Tomography. *ACS Appl. Mater. Interfaces,* 15 (11), 14196–14205 DOI: 10.1021/acsami.2c19895

[24] Sun, K. E. K., Hoang, T. K. A., Doan, T. N. L., Zhu, Y. Y. X., Tian, Y., and Chen, P. (2017). Suppression of dendrite formation and corrosion on zinc anode of secondary aqueous batteries. *ACS Appl. Mater. Interfaces,* 9, 9681–9687. doi: 10.1021/acsami.6b16560

[25] Xu, W. N., Zhao, K. N., Huo, W. C., Wang, Y. Z., Yao, G., Gu, X., et al. (2019). Diethyl ether as self-healing electrolyte additive enabled long life rechargeable aqueous zinc ion batteries. *Nano Energy* 62, 275–281. doi: 10.1016/j.nanoen.2019.05.042

[26] Shimizu, M., Hirahara, K., and Arai, S. (2019). Morphology control of zinc electrodeposition by surfactant addition for alkaline-based rechargeable batteries. *PCCP* 21, 7045–7052. doi: 10.1039/C9CP00223E

[27] Chladil, L., Cech, O., Smejkal, J., and Vanýsek, P. (2019). Study of zinc deposited in the presence of organic additives for zinc-based secondary batteries. *J. Energy Storage,* 21, 295–300. doi: 10.1016/j.est.2018.12.001

[28] Zhang, Q., Luan, J., Fu, L., Wu, S., Tang, Y., Ji, X., et al. (2019). The three-dimensional dendrite-free zinc anode on a copper mesh with a zinc-oriented polyacrylamide electrolyte additive. *Angew. Chem. Int. Ed.* 58, 15841–15847. doi: 10.1002/anie.2019 07830

[29] Yao, S., Chen, Y., Cheng, J., Shen, Y., and Ding, D. (2019). Effect of stannum ion on the enhancement of the charge retention of single-flow zinc–nickel battery. *J. Electrochem. Soc.* 166, A1813–A1818. https://doi.org/10.1149/2.0311910jes

[30] Liu, Z.H., Ren, J.F., Wang, F.H., Liu, X.B., Zhang, Q. Liu, J., , Kaghazchi, P., Ma, D.X. and Chi, Z.Z. , L (2021). Wang Tuning surface energy of Zn anodes via Sn heteroatom doping enabled by a codeposition for ultralong life span dendrite-free aqueous Zn-ion batteries. *ACS Appl. Mater. Inter,* 13 (23), 27085–27095

[31] Dai, Y.H. , Zhang, C.Y. Zhang, W. Cui, L.M. Ye, C.M., Hong, X.F., Li, J.H., Chen, R.W., Zong, W., Gao, X., Zhu, J., Jiang, P.E., An, Q.Y., Brett, D.J.L., Parkin, I.P. and G.J. He, L.Q (2023). Mai Reversible Zn metal anodes enabled by trace amounts of underpotential

deposition initiators Angew. *Chem. Int Ed.,* 62, Article e202301192

[32] Ouyang, K.F., Ma, D.T., Zhao, N. , Wang, Y.Y., Yang, M., Mi, H.W., Sun, L.N., C.X. He, C.X. and Zhang, P.X. (2022). A new insight into ultrastable Zn metal batteries enabled by in situ built multifunctional metallic interphase *Adv. Funct. Mater.,* 32(7), 2109749

[33] Kim M.K., Shin, S.J., Lee, J.M., Park, Y.B., Kim, Y.M., Kim, H.J., and J.W (2022). Choi Cationic additive with a rigid solvation shell for high-performance zinc ion batteries Angew. *Chem. Int Ed.,* 61 (47), Article e202211589.

[34] Cheng, Y., Zhang, H., Lai, Q., and Li, X. (2014). High Power Density Zinc-Nickel Single Flow Batteries with Excellent Performance. *ECS Txn,* 59(1), 3–7. https://doi.org/10.1149/05901.0003ecst

[35] Parker, J. F., Chervin, C. N., Pala, I. R., Machler, M., Burz, M. F., Long and J. W., et al. (2017). Rechargeable nickel-3D zinc batteries: an energy-dense, safer alternative to lithium-ion. *Science,* 356, 414–417. https://doi.org/ 10.1126/science.aak9991

[36] Cheng, Y., Lai, Q., Li, X., Xi, X., Zheng, Q., Ding, C., & Zhang, H. (2014). Zinc-nickel single flow batteries with improved cycling stability by eliminating zinc accumulation on the negative electrode. *Electrochim Acta,* 145, 109–115. https://doi.org/ 10.1016/j.electacta.2014.08.0

[37] Shen, C., Li, X., Li, N., Xie, K. Y., Wang, J. G., Liu, X. R., et al. (2018). Graphene-boosted, high-performance aqueous Zn-Ion battery. *ACS Appl. Mater. Interfaces,* 10, 25446–25453. https://doi.org/ 10.1021/acsami.8b07781

[38] Zeng, Y., Zhang, X., Qin, R., Liu, X., Fang, P., Zheng, D., … Lu, X. (2019). Dendrite-free zinc deposition induced by multifunctional CNT frameworks for stable flexible Zn-Ion batteries. *Adv. Mater.,* 1903675. https://doi.org/ 10.1002/adma.201903675

[39] Yin, Y., Wang, S., Zhang, Q., Song, Y., Chang, N., Pan, Y., et al. (2020). Dendrite free zinc deposition induced by tin-modified multifunctional 3D host for stable zinc-based flow battery. *Adv Mater* 32:e1906803. https://doi.org/ 10.1002/adma.201906803

[40] Zhang, Q., Luan, J., Tang, Y., Ji, X., & Wang, H.-Y. (2020). Interfacial design of dendrite-free zinc anodes for aqueous zinc-ion batteries. *Angew. Chem. Int. Ed.* https://doi.org/10.1002/anie.202000162

[41] Zhang, L., Cheng, J., Yang, Y. S., Wen, Y. H., Wang, X. D., and Cao, G. P. (2008). Study of zinc electrodes for single flow zinc/nickel battery application. *J. Power Sources* 179, 381–387. https://doi.org/10.1016/j.jpowsour.2007.12.088

[42] Chamoun, M., Hertzberg, B. J., Gupta, T., Davies, D., Bhadra, S., Van Tassell, B., et al. (2015). Hyper-dendritic nanoporous zinc foam anodes. *NPG Asia Mater.* 7:8 https://doi.org/10.1038/am.2015.32

[43] Peng, L., Zhu, Y., Chen, D., Ruoff, R. S., and Yu, G. (2016). Two-Dimensional Materials for Beyond-Lithium-Ion Batteries. *Adv Energy Mater.*, 6(11), 1600025. https://doi.org/1002/aenm.201600025

[44] Wang, Y., Xu, X., Yin, J., Huang, G., and Guo, T. et al., (2022) MoS2 mediated epitaxial plating of Zn metal anodes. *Adv Mater.*, 2208171 https://doi.org/10.1002/adma.202208171

[45] Bhoyate, S. D., Mhin, S., Jeon, J., Park, K., Kim, J., and Choi, W. (2020). Stable and High-energy-density Zn ion Rechargeable Batteries based on MoS2 coated Zn anode. *ACS Appl. Mater. Interfaces.* doi:10.1021/acsami.0c06009

[46] Xu, W., Sun, C., Zhao, K., Cheng, X., Rawal, S., Xu, Y., and Wang, Y. (2018). Defect Engineering Activating (Boosting) Zinc Storage Capacity of MoS2. *Energy. Storage. Mater.* https://doi.org/10.1016/j.ensm.2018.09.009

[47] Liang, H., Cao, Z., Ming, F., Zhang, W., Anjum, D. H., Cui, Y., Alshareef, H. N. (2019). Aqueous Zinc Ion Storage in MoS2 by Tuning the Intercalation Energy. *Nano Lett* https://doi.org/10.10.1021/acs.nanolett.9b00697

[48] Liu, J., Xu, P., Liang, J., Liu, H., Peng, W., Li, Y., Fan, X. (2020). Boosting Aqueous Zinc-Ion Storage in MoS2 via Controllable Phase. *J. Chem. Eng.* 124405. https://doi.org/10.10.1016/j.cej.2020.124405

[49] Cai, C., Tao, Z., Zhu, Y., Tan, Y., Wang, A., Zhou, H., and Yang, Y. (2021). A nano interlayer spacing and rich defect 1T-MoS2 as cathode for superior performance aqueous zinc-ion batteries. *Nanoscale. Adv.*, 3(13), 3780–3787. https://doi.org/10.10.1039/d1na00166c

[50] Wang, L., Li, S., Li, D., Xiao, Q., & Jing, W. (2020). 3D flower-like molybdenum disulfide modified graphite felt as a positive material for vanadium redox flow batteries. *RSC Adv.*, 10(29), 17235–17246. https://doi.org/10.1039/d0ra02541k

[51] Liu, H., Wang, J.-G., Hua, W., You, Z., Hou, Z., Yang, J., Kang, F. (2020). Boosting zinc-ion intercalation in hydrated MoS2 nanosheets toward substantially improved performance. *Energy. Storage. Mater.* https://doi.org/10.10.1016/j.ensm.2020.12.010

[52] Zhang, Q., Luan, J., Tang, Y., Ji, X., and Wang, H.-Y. (2020). Interfacial design of dendrite-free zinc anodes for aqueous zinc-ion batteries. *Angew. Chemie. Int. Ed.* https://doi.org 1002/anie.202000162

[53] Xiao, M., Wang, Y., Yao, S., Song, Y., Cheng, J., and He, K. (2016). Analysis of internal reaction and mass transfer of zinc-nickel single flow battery. *J. Renew. Sustain Energy*, 8(6), 064102. https://doi.org/10.10.1063/1.4968851

[54] Liu, X., Xie, Z., Cheng, J., Zhao, P., and Gu, W. (2009). Mathematical modeling of the nickel electrode for the single flow zinc-nickel battery. *2009 WNWEC.* https://doi.org/10.10.1109/wnwec.2009.5335869

[55] Huang, X., Zhou, R., Luo, X., Yang, X., Cheng, J., and Yan, J. (2023). Experimental research and multi-physical modeling progress of Zinc-Nickel single flow battery: A critical review, *Adv. Appl. Energy*, 12, 100154, ISSN 2666–7924, https://doi.org/10.1016/j.adapen.2023.100154.

6 Removal of iron in groundwater using nanobeads of ZnO synthesized from Pseudomonas species

Gayathri Vijayakumar[1,a] and P. Dhamodhar[2]

[1]Department of Biotechnology, Rajalakshmi Engineering College, Chennai, India
[2]Department of Biotechnology, M S Ramaiah Institute of Technology, Bengaluru, India

Abstract

Iron is an essential mineral for humans, its presence in the groundwater above a certain level makes the water unusable mainly for aesthetic considerations such as discoloration, metallic taste, odor and turbidity. To overcome this problem, the concentration of iron in the groundwater must be reduced. Pseudomonas is a gram-negative bacterium with an ability to scavenge iron. These are able to reduce metal salts such as zinc to metal nanoparticles Nanomaterials can serve as an excellent support material for enzyme immobilization. To isolate Pseudomonas sp. leguminous soil samples were collected and the sample was inoculated in nutrient broth for growth. Then various biochemical tests were carried out for the confirmation of *Pseudomonas sp*. ZnO was synthesised using Pseudomonas culture and various characterization techniques are carried out. It includes UV-visible spectroscopy which confirms the synthesis of ZnO at 263.5nm and FTIR shows the ZnO presence at 545nm. EDX was performed and found the elemental composition of Zn as 35.9% and O as 40.1%. Antibacterial activity of ZnO NP showed more resistance towards gram positive bacteria than gram negative bacteria. After characterization, the ZnO NP was immobilised by the gel entrapment method. The ZnO nano beads were formed as a result. Treating the ZnO nanobeads in groundwater had resulted in lowering the iron concentration. By the AAS analysis, it was found that about 74.4% of iron removed by our study when compared to the untreated water.

Keywords: Nanomaterials, iron, groundwater, ZnO, *Pseudomonas sp*, FTIR

1. Introduction

Groundwater is essential for various purpose, including industry, agriculture, and domestic use. Human activities, especially in densely populated areas, pose significant risk to groundwater contamination [9]. Any action leading to the discharge of chemicals or waste into the environment can contaminate groundwater. Clean-up of groundwater contamination is time-consuming and expensive. Groundwater contamination arises from human activities like landfills, septic tanks and improper waste disposal as well as natural sources [10]. The presence of iron in ground water is a direct result of its natural existence in underground rock formations. Iron-rich water usually has the same appearance as pure water because the iron dissolves evenly and the water is clear [2]. However, when water enters a residential pressure vessel (well tank) or leaks into the atmosphere, the iron oxidises into an insoluble iron form. As a result, groundwater quality deteriorates and becomes visible [6]. Nanotechnology's affordable and rapid detection and treatment of water contaminants hold the potential to address the demand for clean drinking water. Microorganisms are used in the biosynthesis of nanoparticles; they take target ions from

[a]gayathriv2711@gmail.com, gayathri.vijayakumar@rajalakshmi.edu.in

DOI: 10.1201/9781003545941-6

their solutions and collect the reduced metal in its element form using enzymes produced by the activities of the microbial cell. Nanoparticles are created extracellularly by trapping metal ions on cell surfaces and reducing ions in the presence of enzymes. Cell adhesion or entrapment on nanoparticles considerably improves microbial systems for cleaning up the environment and makes regeneration easier [16].

Zinc oxide (ZnO) nanoparticles have emerged as highly promising carriers for enzyme immobilization due to their exceptional properties. These nanoparticles offer numerous advantages, including excellent biocompatibility, nanotoxicity, chemical stability, large specific surface area and cost effectiveness. Researchers have recognised their potential, leading to significant interest in utilising ZnO nanoparticles for this purpose [4]. One of the key reasons for their popularity is the simplicity of their preparation, which facilitates their widespread use in research and industrial applications. The earlier studies likely to delve into specific experimental procedures and applications, shedding further light on the advantages of ZnO nanoparticles in enzyme immobilization [7]. Hence, in this study nano-beads of ZnO synthesised from *Pseudomonas* sp., was used to remove the iron content in groundwater and its antibacterial activity was analyzed.

2. Removal of Iron from Ground Water Using Zno Nano Particles

2.1. *Collection of Soil Sample and Culturing* Pseudomonas sp

A leguminous soil sample was collected in sterile polythene bags at Sholavaram, Chennai and preserved which was then used for the isolation of *Pseudomonas sp*. 10g of soil was taken and then serially diluted for the isolation by mixing with 40ml of distilled water separately, further it was diluted by taking 10ml from each sample to make up 50ml using distilled water. *Pseudomonas* agar was

prepared for 100ml and it is sterilised and poured into 4 petri plates. 100ul of the culture was taken and spread on the plates by pour plating method. Then the plates were kept for incubation for 1 day. The growth of *Pseudomonas* was then observed in nutrient broth (Figure 6.1) after incubating it for 24–32hr. The growth of *Pseudomonas* in King's B agar and *Pseudomonas* agar plates (Figures 6.2(a), 6.2(b) were confirmed and further it was used for developing pure culture [5].

2.2. *Zno Nanoparticle Synthesis*

Nanoparticle synthesis can be performed by different methods such as Physical, Chemical, and Biological synthesis. Zinc Oxide nanoparticles were synthesised in the *Pseudomonas* broth culture. From the former study formation of white precipitate confirms the presence of ZnO nanoparticles [8]. This

Figure 6.1: Observation of color in Broth.

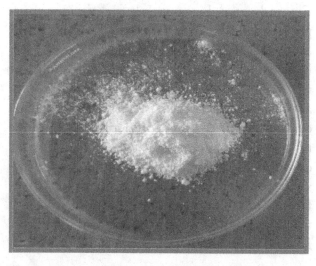

Figure 6.2: (a) King's Agar (b) Pseudomonas. Agar.

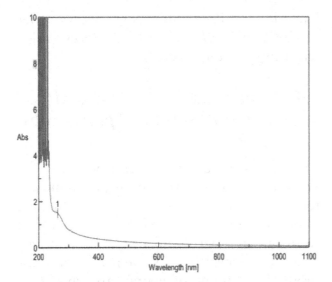

Figure 6.3: Synthesized ZnO.

precipitate is then subjected to UV-vis Spectroscopy to confirm the synthesis.

2.3. *Characterization of Nanoparticles*

UV-Visible spectrophotometer was used to analyze the formation and stability of the ZnO NPs in aqueous solution. According to earlier studies, ZnO NPs exhibit surface plasmon resonance with an absorption peak between 250 and 280 nm which was observed [12]. In the present study, the greatest plasmon surface peak for Pseudomonas culture can be seen at 263.5 nm which confirmed that at 263.5nm ZnO nanoparticles were synthesised as the result given in the graph (Figure 6.4) [11]. FTIR was performed to confirm the presence of ZnO nanoparticles and to analyze the chemical bonding between Zn and O [3]. The spectrum showed a broad peak around 1555.34 cm and a shoulder around 545 cm, which corresponds to ZnO nanoparticles (Figure 6.5).

SEM-analysis was carried out to examine the particle size. It was found that *Pseudomonas* culture produced hexagonal and crystalline ZnO nanoparticles with an average size of 50 nm. The EDX analysis confirmed the presence of zinc oxide nanoparticles grown by the biosynthesised method (Figure 6.6) [14]. Zinc was found to be 35.9% with oxide for about 40.5% and Sodium presence is due to the growth medium which contributes about 23.6% in our study.

The antibacterial assays of bio synthesised ZnO NPs were assessed by using well diffusion method against pathogenic Gram-positive (*Staphylococcus aureus, Streptococcus*

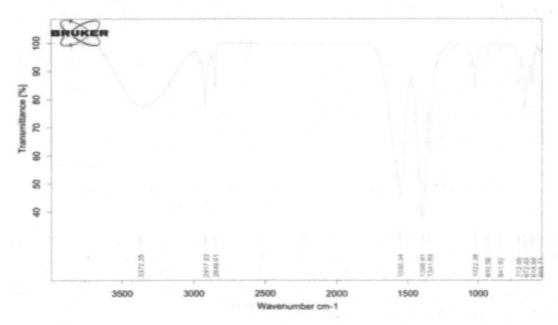

Figure 6.4: UV-Visible spectrometric analysis.

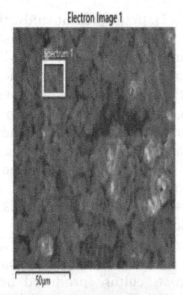

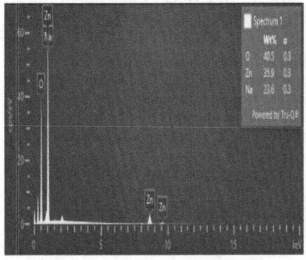

Figure 6.5: FTIR Analysis of ZnO.

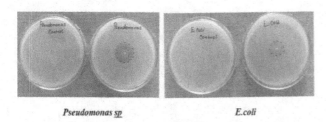

Figure 6.6: SEM and EDAX analysis.

pneumoniae) and Gram-negative (*E.coli*, *Pseudomonas aeruginosa*) bacteria grown in Nutrient broth at 37°C for 24h [1]. ZnO nanoparticles synthesised using *Pseudomonas sp.* were better against *Staphylococcus aureus* than *Streptococcus pneumoniae* (Gram +ve

bacteria) and *Pseudomonas sp.* Shows better results than *E.coli* (Gram -ve bacteria) which was shown in the Figure 6.7.

2.4. Nano Beads of ZnO Np and Atomic Absorption Spectroscopy Analysis

Atomic absorption spectroscopy was used for analysis of iron concentration in groundwater. Only gaseous mediums in the individual atoms or ions which were well separated from one another could be used for atomic absorption spectroscopy element identification. The amount of energy absorbed during this excitation was evaluated in atomic absorption spectroscopy and was proportional to the concentration of atoms present in the sample [13]. According to the AAS results, the concentration of iron is reduced from 7.72 to 1.97 mg/l i.e. 74.4% (Table 6.1 and Figure 6.8) of iron was removed from the groundwater with the help of ZnO nanobeads synthesised from *Pseudomonas sp* ZnO nanobeads were formed by the process of immobilisation technique-gel entrapment method of immobilisation [15]. These beads

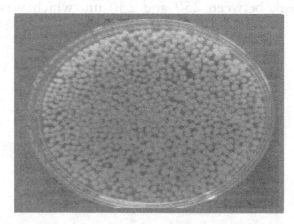

Figure 6.7: Antimicrobial activity of ZnO Np against pathogens.

Table 6.1: Iron concentration in water tested with Atomic absorption. spectroscopy

Sample	Fe Concentration (mg/L)
Control	7.72
Test	1.97

AAS Analysis

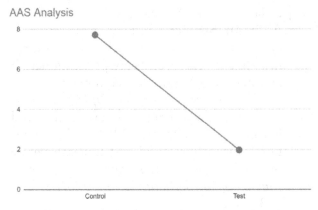

Figure 6.8: AAS Analysis of ZnO nanobead treated and untreated ground.

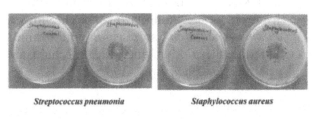

Streptococcus pneumonia   *Staphylococcus aureus*

Figure 6.9: ZnO nanobeads.

were found to be 0.8 mm in diameter. The durability test was performed for these beads and the result was, it could last for about 8 days (Figure 6.9).

The immobilisation of ZnO nanoparticles was the main reason behind the iron removal because of its stability.

3. Conclusion

The present study showed that the synthesised ZnO nanoparticle has the ability to reduce the risk of microbial contamination in groundwater. Gel Entrapment method of immobilisation which was used for produce nano beads of ZnO from *Pseudomonas* culture, effectively reduced the iron concentration of ground water. Thus the present study revealed that the biosynthesised ZnO nanobeads are efficient in removing iron from groundwater.

Acknowledgement

The authors gratefully acknowledge the students, staff, and authority of Biotechnology department for their cooperation in the research.

References

[1] Katsanou, K., and Karapanagioti, H. K. (2019). Surface water and groundwater sources for drinking water, Applications of Advanced Oxidation Processes (AOPs) in Drinking Water Treatment, 1–19.

[2] Li, P., Karunanidhi, D., Subramani, T., and Srinivasamoorthy, K. (2021). Sources and consequences of groundwater contamination. *AECT*, 80, 1–10.

[3] Abbaspour, N., Hurrell, R., and Kelishadi, R. (2014). Review on iron and its importance for human health. *JRMS*, 19, 164–174.

[4] Gad, M., Dahab, K., and Ibrahim, H. (2016). Impact of iron concentration as a result of groundwater exploitation on the Nubian sandstone aquifer in El Kharga Oasis, western desert, Egypt. *NRIAG-JAG*, 5, 216–237.

[5] Zhang, X., Yan, S., Tyagi, R. D., and Surampalli, R. Y. (2011). Synthesis of nanoparticles by microorganisms and their application in enhancing microbiological reaction rates, *Chemosphere*, 82, 489–494.

[6] Cipolatti, E. P., Valerio, A., Henriques, R. O., Moritz, D. E., Ninow, J. L., Freire, D. M., and de Oliveira, D. (2016). Nanomaterials for biocatalyst immobilization–state of the art and future trends. *RSC adv.*, 6, 104675–104692.

[7] Gudkov, S. V., Burmistrov, D. E., Serov, D. A., Rebezov, M. B., Semenova, A. A., and Lisitsyn, A. B. (2021). A mini review of antibacterial properties of ZnO nanoparticles. *Front. Phys.*, 9, 641481.

[8] Furmanczyk, E. M., Kaminski, M. A., Spolnik, G., Sojka, M., Danikiewicz, W., Dziembowski, A., and Sobczak, A. (2017). 'Isolation and characterization of Pseudomonas spp. strains that efficiently decompose sodium dodecyl sulfate'. *Frontiers in Microbiology*, 8, 1872.

[9] Jayachandran, A., Aswathy, T. R., and Nair, A. S. (2021). Green synthesis and characterization of zinc oxide nanoparticles using *Cayratia edate* leaf extract. *Biochem. Biophys. Rep*, 26, 100995.

[10] Mourdikoudis, S., Pallares, R. M., and Thanh, N. T. (2018). Characterization techniques for nanoparticles: comparison and complementarity upon studying nanoparticle properties. *Nanoscale*, 10, 12871–12934.

[11] Mohd Yusof, H., Mohamad, R., Zaidan, U. H., and Abdul Rahman, N. A. (2019). Microbial synthesis of zinc oxide nanoparticles and their potential application as an antimicrobial agent and a feed supplement in animal industry: a review. *JASB.*, 10, 1–22.

[12] Berthomieu, C. and Hienerwadel, R. (2009). Fourier transform infrared (FTIR) spectroscopy. *Photosynth. Res.*, 101, 157–170.

[13] Titus, D., Samuel, E. J. J., and Roopan, S. M. (2019). Nanoparticle characterization techniques. In Green synthesis, characterization and applications of nanoparticles. Elsevier, 303–319.

[14] Aalami, A. H., Mesgari, M., and Sahebkar, A (2020). Synthesis and characterization of green zinc oxide nanoparticles with antiproliferative effects through apoptosis induction and microRNA modulation in breast cancer cells'. *Bioinorg. Chem and Appl.*, 2020

[15] Tautkus, Stsys, Laura Steponeniene, and Rolandas Kazlauskas. (2004). Determination of iron in natural and mineral waters by flame atomic absorption spectrometry. *Journal of the Serbian Chemical Society*, 69, 393–402.

[16] Xu, P., Zeng, G. M., Huang, D. L., Lai, C., Zhao, M. H., Wei, Z., Li, N. J., Huang, C., and Xie, G. X. (2012). Adsorption of Pb(II) by iron oxide nanoparticles immobilized *Phanerochaete chrysosporium*: equilibrium, kinetic, thermodynamic and mechanisms analysis. *J. Chem. Eng.*, 203, 423–431.

7 Continuous production of monodisperse silver nanoparticles by droplet-based microreactor system for catalysis in bioethanol production

R. Halima[1,a] *and Archna Narula*[2,b]

[1]Department of Biotechnology, Sir M Visvesvaraya Institute of Technology, Bangalore, India
[2]Department of Chemical Engineering, M.S Ramaiah Institute of Technology, Bangalore, India

Abstract

Silver Nanoparticles (AgNPs), known for their exceptional catalytic properties, hold great promise in various chemical processes. However, controlling their size distribution and producing them continuously with high precision remains a significant challenge. To overcome this, we have developed a cutting-edge microreactor system, designed to synthesise monodisperse AgNPs in a continuous and efficient manner. Our approach capitalises on the advantages of droplet-based microreactors, which provide fine control over reaction conditions, resulting in nanoparticles with uniform size and enhanced catalytic activity. The system incorporates real-time monitoring and feedback mechanisms to ensure the synthesis process remains precise and reproducible. In this presentation, we will discuss the key elements of our research, including the microreactor design, reaction kinetics, and the catalytic performance of the produced AgNPs. This research work emphasis the potential applications of these monodisperse AgNPs in catalysis of lignocellulosic materials for ethanol production, offering improved efficiency and selectivity. Our work contributes to the development of advanced catalytic materials and demonstrates the potential of microreactor systems in continuous nanoparticle synthesis with greater efficiency of ethanol production of about 80%. This research work will pave the way for more sustainable and efficient catalytic processes across multiple industries.

Keywords: Green synthesis, droplet-based microreactor system, AgNPs, catalysis, ethanol production

1. Introduction

Silver Nanoparticles (AgNPs) have emerged as versatile catalysts in catalysis, for their unique properties viz. high surface area, size-dependent reactivity, and excellent conductivity. These nanoparticles exhibit catalytic activity in a wide range of reactions, from organic synthesis to environmental remediation and energy-related processes. The size of AgNPs significantly influences their catalytic behaviour, with smaller particles displaying higher activity due to increased surface area. Surface modification, green synthesis approaches, and integration with support materials further enhance their catalytic performance. The plasmonic effects of AgNPs, driven by surface plasmon resonance, contribute to improved light absorption and charge transfer reactions. Despite their promising applications, challenges like potential toxicity and stability issues require careful consideration for practical use. Overall, AgNPs stand out as promising catalysts, offering a platform for tailoring efficient and selective catalysis across diverse chemical transformations [1].

Continuous flow devices are commonly utilised in microfluidic nanoparticle production,

[a]halima_biotech@sirmvit.edu, [b]archna_71@yahoo.com

DOI: 10.1201/9781003545941-7

offering a straightforward approach with a laminar flow pattern in microchannels, ensuring uniformity and control over various process parameters. However, the inherent limitations of continuous flow, marked by lack of turbulence with low Reynolds number, pose challenges in achieving efficient mixing primarily dependent on diffusion. To surmount these limitations, the approach involves inducing turbulence through the manipulation of microchannel shapes—via bending, folding, and stretching. Passive mixing strategies, employing geometric features like spiral or zigzag channels and embedded barriers, are favoured for their simplicity, as they do not necessitate additional intricate components within the system [2–4]. Despite the advantages of passive mixing, certain reactions, particularly those involving high-viscosity fluids, may require the integration of active components into the system to enhance mixing efficiency. The microfluidic nanoparticle production process thus involves a delicate balance between the simplicity of continuous flow systems and the need for tailored mixing strategies to optimise outcomes.

Reagents and compounds are combined in small, compartmentalised quantities in different volumes created by the introduction of two immiscible phases in a droplet-based micro reactor. By containing the dispersed phase inside the continuous phase, this technique produces isolated droplets that serve as discrete micro reactors [3]. Interestingly, these produced drops—which have regulated dimensions and forms—do not depend on the device channels, although the segments that are manufactured are limited by the channel walls. Notwithstanding these differences, both technologies provide a quick and effective homogenisation procedure. Depending on the device's particular use, these attributes have differing effects on the intended result. It is beneficial to add surfactants to enhance droplet interface stabilisation. In terms of geometry, droplets or segment are able to

be generated passively in microdevices using three main microfluidic configurations: crossing flows in a T-shaped instrument (T-injector), helical flows (coflowing systems), and flow-focusing reactors. Moreover, the adjustment of fluid flow rates and channel dimensions is required for accurate control and monitoring over the creation of nanoparticles in droplet microfluidics. The study leverages droplet-based microreactor technology to achieve continuous and reproducible synthesis. The microreactor system allows for fine-tuning reaction parameters, ensuring monodispersity and enhanced catalytic activity of the resulting AgNPs [4]. This paper delves into the design and optimisation of the microreactor setup, detailing the influence of flow rates, concentrations, and reaction times on nanoparticle characteristics. Microreactors offer superior heat transfer, scalability, and safety features compared to traditional reactors, enabling precise control and efficient chemical processes in compact designs. Additionally, the research investigates the catalytic efficacy of the synthesised AgNPs in various reactions, highlighting their potential for applications in diverse catalytic processes. This work contributes to the advancement of scalable and efficient methods for producing monodisperse AgNPs tailored for catalytic applications, offering insights into the intersection of nanotechnology and continuous flow chemistry [5].

2. Materials Required

Chemicals and reagents:

Silver nitrate (1mM) is used as a contributor to the AgNPs formation. The phytochemicals present in the plant plant extract (*Piper betle*) involves in the synthesis of AgNPs by reducing the silver ions in Silver nitrate. The yeild of AgNps produced is calculated using the formula:

$$Yeild(mg) = \frac{Mass\ of\ Silver\ Nanoparticles\ Produced\ (mg)}{Mass\ of\ Silver\ Nitrate\ used\ (mg)}$$

Microreactor design

The microreactor (Figure 7.1), chosen for its droplet-based architecture, is crafted from materials such as glass or polymers, ensuring compatibility with the synthesis process and resistance to corrosion. Two syringe pumps with various flow rates were used for the synthesis of the AgNPs. Flow rates, from 0.08 to 0.2 mL min-1, were controlled by employing a syringe pump. Flow control, pressure regulation, and temperature management systems are incorporated to govern reactant input, maintaining uniform droplet formation and controlled nanoparticle synthesis. The coiled microreactor made of glass was used to enable the reaction to happen in the presence of sunlight. The Y shaped joined enable the monodisperse of the reactants enabling the better formation of the AgNps. The AgNPs formed were collected in a glass vial.

2.1. *Characterisation of AgNPs*

To study the morphological characteristics of synthesised AgNp Fourier transform infrared (FTIR), X-ray diffraction (XRD), UV-VIS, and FESEM-EDS have been employed to characterise the AgNps.

2.2.1. UV-VIS Spectroscopy

The reduction of pure silver ions was detected in the UV-VIS spectroscopy examination. By considerably diluting the sample in distilled water, this spectrum was produced. An ultraviolet-visible spectrophotometer (Shimadzu UV-2450) was used to perform the UV-VIS spectral analysis.

2.2.2. FTIR Analysis

FTIR spectroscopy was utilised to investigate the functional groups implicated in both the and stabilisation of AgNPs, which within the framework of FTIR analysis. The Bruker model Alpha ATR was used to analyze the particles that were gathered, covering the wave number range of 400 to 4000/cm.

2.2.3. XRD Analysis

Bruker D-8 equipment with Ni-filtered Cu-Kα radiation source (40 kV, 30 mA) was used for X-ray photoelectron diffraction (XRD) investigation of AgNPs produced by Piper betle leaf extract. A Cu-Kα source of radiation was used in XRD spectroscopy, having a scattered range of 20–80. The device ran at 45 kV of voltage and 40 mA of current. The purpose of the XRD spectroscopy was to verify the synthesised AgNPs' existence, crystal nature, phase variety, and grain size. Scherrer's equation, which is described as follows, was utilised to determine the particle size of the produced samples.

$$D = \frac{0.94 * \lambda}{\beta * cos\theta}$$

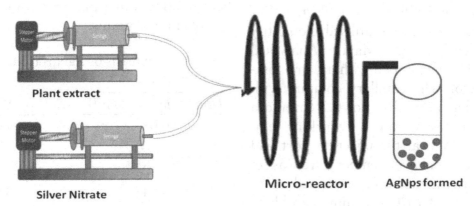

Figure 7.1: Microreactor design for the synthesis of AgNPs.

Here D stands for the mean crystallite size, β for the line broadening (full width on half maximum at the peak) in radians, λ for the X-ray wavelength, θ for the Bragg's angle, & the geometric factor, which is the constant 0.94.

2.2.4. FESEM-EDS

A little amount of the sample was deposited on a grid of carbon-coated copper (for the membrane) in order to form a thin sample film for the FESEM-EDS analysis. In order to ensure conductivity, a smear containing the synthesised AgNPs were created, dried, and gold-coated using the Oxford EDS detector and the Carl Zeiss ultra 55 SEM.

2.2.5. AgNPs in Catalysis

The AgNPs produced through the mono-disperse continous reactor process was used in the catalysis reaction to produce ethanol through the hydrolysis of the substrate (saw dust) using Nano-enzymatic method. The AgNps were tagged with the enzyme α-Amylase and used in catalysis process [6]. The fermentation was carried out for 48 hrs using *Sacchromyces cerevisea* to the hydrolysed substrate in Erhlemeyer flask.

2.2.6. Estimation of Ethanol

The fermented flasks were subjected to distillation and the distillate was collected. Gas chromatography (GC) was utilised to quantify bioethanol levels using a Mayura Analytical GC Model 1100 equipped with a flame ionisation detector. The capillary column, with an Internal Diameter (I.D.) of 0.2 cm and a length of 2.0 m, maintained a working temperature of 225°C at the injection port and detector. The initial oven temperature was set to 100°C for 2 minutes, then ramped up to 225°C at a rate of 10°C per minute and held for 9.5 minutes. Each GC analysis involved a sample injection of 200 µL. Helium served as the carrier gas, and 2-pentanone was used as the internal standard at a concentration of 0.5 percent (v/v).

2.2.7. Estimation of Glucose

1 mL of the sample was periodically collected from the flasks undergoing hydrolysis under various conditions such as different time intervals, pH levels, temperatures, substrate concentrations, and biomass concentrations. The glucose content in the hydrolyzed substrate was determined using the Ortho Toluidine method. Standard glucose solution (100 mg/mL, AR grade, Sigma Aldrich) was diluted and dispensed into different test tubes to create a concentration range from 0 to 100 mg/mL. Ortho Toluidine reagent was added to each tube and heated in a water bath for 20 minutes. A standard graph was constructed using the known concentrations. The samples obtained from the hydrolyzed flasks were then subjected to the Ortho Toluidine test, and the concentration of glucose in the hydrolyzed flasks was determined by referencing the standard graph.

The non-hydolysed flasks were also maitained for comparison of ethanol produced and the glucose produced.

2.2.8. Results and discussion

AgNp synthesis from Microreactor
The Table 7.1 below presents the relationship between flow rate (mL min^{-1}) and the yield of AgNps in milligrams.

Table 7.1: Synthesis of AgNps from Microreactor

Flow Rate (mL min^{-1})	Yield of AgNps (mg)
0.08	18
0.10	23
0.12	28
0.14	32
0.18	37
0.20	42

The data indicates a distinct correlation between the flow rate and the yield of AgNps. With the increase in flow rate from 0.08mL/min to 0.20 mL/min, a consistent and significant rise in the AgNps yield is observed. This suggests that a higher flow rate positively influences the synthesis process, potentially enhancing the efficiency of nanoparticle production. The observed trend aligns with common principles in nanoparticle synthesis, where the reaction kinetics is influenced by the rate of reactant introduction [6]. In this case, a higher flow rate could result in more efficient mixing of reactants, promoting a faster and more effective nucleation and growth process, leading to an increased yield of AgNps. However, it's crucial to note that while higher flow rates appear beneficial in terms of yield [7].

2.2.9. UV-VIS spectroscopy

The conversion of Ag⁺ ions into AgNps, initiated by exposure to *Piper betle* extract, is manifested through a noticeable change in color. This color change is explained as the result of a surface plasmon resonance (SPR) effect. The free electrons displayed by the functional nanoparticles contribute to the SPR absorption band. This absorption band is the outcome of the metal nanoparticles'

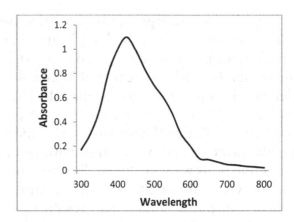

Figure 7.2: UV-Vis Spectrum of AgNPs.

electrons vibrating in unison with the incoming light wave.

Notably, sharp absorption bands of AgNps were observed around 430 nm for *Piper betle*. These unique spectral characteristics highlight the distinct SPR behavior of AgNps synthesised using *Piper betle* extracts, underscoring the influence of plant-specific constituents on the optical properties of the resulting nanoparticles.

2.2.10. FTIR Analysis

FTIR spectroscopy has identified certain biomolecules in *Piper betle* leaf extract serving as reducing agents for silver ions, facilitating their conversion into AgNps (Figure 7.3).

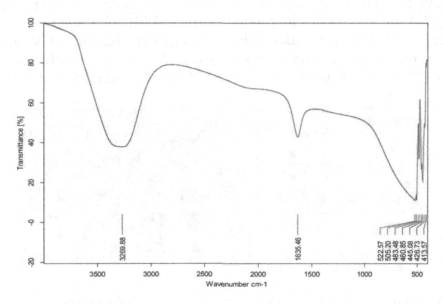

Figure 7.3: FTIR spectra of AgNps.

The synthesised AgNps exhibit a distinctive FTIR pattern, as illustrated in Figure 7.3. Notably, the broad peak observed at 3270.23 cm −1 signifies O-H stretching of intermolecular bonds, while the band around 1635.68 cm −1 (C=N) indicates the presence of primary amide. Within the proximity of AgNps, specific proteins and metabolites with designated functions were identified. Furthermore, the study revealed that carbonyl groups in proteins, including amino acid residues, are inclined to bind metallic ions. This suggests that proteins play a crucial role in capping AgNps, ensuring medium stability, and preventing agglomeration [8]. The process of generating and stabilising AgNps in an aqueous medium may involve two distinct biological molecular processes.

2.3. XRD Analysis

The structure and crystal size of the synthesised AgNps were examined by XRD analysis (Figure 7.4). Characteristic diffraction peaks in AgNps derived from Piper betel leaf extract were identified at angles I_s = 28A°, 32A°, 38A°, 46A°, 54A°, and 68A°. The XRD spectra clearly confirmed the nanocrystalline and crystalline properties of the synthesised AgNps compared with the sample. These peaks are associated with the (122), (111), (200), (220) and (311) crystallographic planes of silver [9–11]. The pure peaks in the XRD analysis show a strong silver component supporting the index (Figure 7.4). This

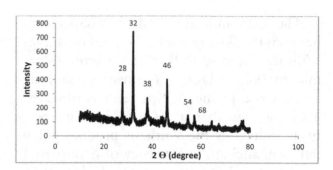

Figure 7.4: XRD of AgNps synthesised.

observation is consistent with the findings reported by Jose et al. and show that AgNps exhibit face-centered, spherical, and crystalline structures identified by JCPDS card number 04-0783. Determination of nanoparticle size is based on the width at half maximum (FWHM) value, using the Scherrer equation and choosing a constant value of 0.94 due to the spherical shape and crystal material of the nanoparticles. The average size of AgNps synthesised from bean leaf extract was calculated as 42 nm. Broadening of the Bragg peak near its base indicates the formation of small AgNps [13]. The invisible peaks can be attributed to bioorganic compounds or proteins present in the leaf extract and crystallised on silver.

The surface morphology, size and shape of AgNps synthesised from leaf extract were examined using scanning electron microscopy (SEM) as shown in Figure 7.5a) at 75000x magnification SEM images show that single AgNps are mostly spherical in shape and some clusters do not have well-defined

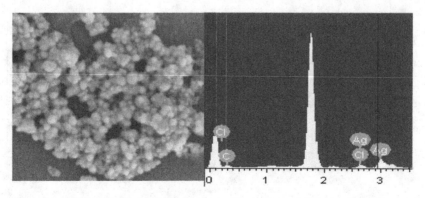

Figure 7.5: a) SEM Image of AgNp at 7 5KX magnification b) EDX spectrum.

morphology. The presence of biomolecules in piper bean leaf extract affects global AgNps synthesis, and the aggregation can be attributed to the presence of other metabolites in the leaf extract. The sizes of AgNps observed in SEM images range from 40 to 50 nm. X-ray energy dispersive spectroscopy (EDS) was performed to determine AgNps formation. The EDS peaks corresponding to silver confirmed the presence of silver as the major element, confirming the formation and purity of AgNps synthesised from bean leaf extract, as shown in Figure 7.5b. The presence of AgNps is evidenced by the difference in the peak at 3 keV in the silver region, indicating surface plasmon resonance [13]. In general, silver nanocrystals show an optical peak around 3 keV due to surface plasmon resonance [14]. EDS elemental analysis of the synthesised AgNps showed the majority of silver followed by C and O. The weak oxygen signal may come from X-ray emission from carbohydrates/protein/enzymes in the extract [12] or may indicate strong reactivity. Silver oxide nanoparticles are produced with chemicals due to the high surface area/volume ratio. Water resistance. In addition, X-ray diffraction analysis (Figure 7.4) shows that the peaks at 27.97 and 32.29 correspond to the (110) and (111) planes of silver oxide Ag2O. These peaks follow the silver oxide standard (JCPDS 76-1393), confirming the formation of Ag2O nanoparticles. The carbon peak can be attributed to biomolecules attached to the AgNps surface.

2.4. AgNps in Catalysis

The Figure 7.6 illustrates the variations in percentatge of ethanol produced over time (hours) for both non-hydrolyzed and hydrolyzed conditions by nano-enzymatic method.

The data depicts the impact of hydrolysis on glucose concentration over a period of time. Notably, under both non-hydrolyzed and hydrolyzed conditions, the initial glucose concentration remains constant at 13.3 mg/L.

In the non-hydrolyzed scenario, the glucose concentration remains relatively stable throughout the entire time span, indicating minimal changes in the absence of hydrolysis. However, in the hydrolyzed condition, there is a substantial increase in ethanol production ranging from 63% at 2 hours to 80% at 6 hours.

The observed rise in glucose concentration in the hydrolyzed condition can be attributed to the enzymatic breakdown of complex carbohydrates into simpler sugars, particularly glucose. The hydrolysis process, catalyzed by enzymes, enhances the release of glucose from larger polysaccharides, resulting in an augmented concentration of glucose in the solution [15].

These findings underscore the pivotal role of hydrolysis in the conversion of complex carbohydrates into more readily available sugars. This has implications in various applications, such as biofuel production or bioprocessing, where the efficient hydrolysis of carbohydrates is crucial for obtaining higher concentrations of fermentable sugars. Further investigations into the specific enzymes and conditions optimising hydrolysis can contribute to the development of more efficient and sustainable processes in biotechnological applications.

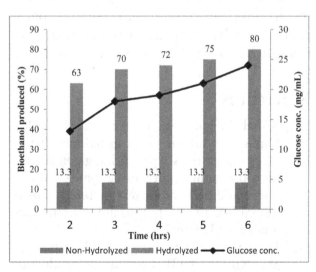

Figure 7.6: Ethanol production by catalysis of AgNPs.

3. Conclusion

In conclusion, the continuous microreactor system designed for the synthesis of monodisperse AgNps has demonstrated significant promise in achieving precise control over nanoparticle production. The integration of a droplet-based microreactor, carefully selected materials, and optimised fluid dynamics has allowed for the continuous and efficient synthesis of AgNps. The system's ability to regulate key synthesis parameters, including reactant concentrations, flow rates, and residence time, has resulted in the consistent production of monodisperse nanoparticles. In-line monitoring techniques, such as UV-Vis spectroscopy, have facilitated real-time analysis, ensuring the quality and uniformity of the synthesised nanoparticles. The safety features, automation, and control mechanisms have enhanced the reliability and user-friendliness of the system. Overall, this innovative microreactor system not only provides a robust platform for academic research but also holds potential for scalable industrial applications, offering a versatile and efficient means of producing monodisperse AgNps for catalytic processes. The current research signifies the enhancement catalytic ability of the monodispersed AgNP in ethanol production as 80%. Further research and development in this direction could yield broader applications and advancements in the field of continuous nanoparticle synthesis and catalysis.

References

[1] Robab J., Munir, I., and Yesiloz, G. (2023). One-step synthesis of ultrasmall nanoparticles in glycerol as a promising green solvent at room temperature using omega-shaped microfluidic micromixers. *Anal. Chem.* 28;95(47), 17177–17186. doi: 10.1021/acs.analchem.3c01697.

[2] Huang, X., Xia, S., Lee, S., Jia, Y., Chen, Z., and Xu, J. (2023). Continuous production of monodisperse AgNps suitable for catalysis in a droplet-based microreactor system. *ACS Appl. Nano Mater.*, 6(10), 8574–8583. https://doi.org/10.1021/acsanm.3c00943

[3] Kale, A. R., Barai, D. P., Bhanvase, B. A., and Sonawane, S. H. (2021). An ultrasound-assisted minireactor system for continuous production of TiO2 nanoparticles in a water-in-oil emulsion. *Indus. Eng. Chem. Res.*, 60(41), 14747–14757. https://doi.org/10.1021/acs.iecr.1c02413

[4] Shi, Y., Lyu, Z., Zhao, M., Chen, R., and Nguyen, Q. N., Xia, Y. (2021). Noble-metal nanocrystals with controlled shapes for catalytic and electrocatalytic applications. *Chem. Rev.*, 121 (2), 649–735. https://doi.org/10.1021/acs.chemrev.0c00454

[5] Adhikari, L., Larm, N. E., and Baker, G. A. (2020). Batch and flow nanomanufacturing of large quantities of colloidal silver and gold nanocrystals using deep eutectic solvents. *ACS Sustainable Chem. Eng.*, 8(39), 14679–14689. https://doi.org/10.1021/acssuschemeng.0c04244

[6] Halima, N. A. (2023). Nano-enzymatic hydrolysis and fermentation of waste starch sources for bioethanol production: an optimization study. *J. Mines Metals Fuels.* 71(3), 439–445. https://doi.org/10.18311/jmmf/2023/33756

[7] Zhang, H., Lu, K., Li, B., Liu, Y., Su, Y., and Wang, R., Cheng, Y. (2020). Microfluidic, one-batch synthesis of Pd nanocrystals on N-doped carbon in surfactant-free deep eutectic solvents for formic acid electrochemical oxidation. *ACS Appl. Mater. Interfaces*, 12(38), 42704–42710. https://doi.org/10.1021/acsami.0c10136

[8] Ahmad, A., Mushtaq, Z., Saeed, F., Afzaal, M., and Al Jbawi, E. (2023). Ultrasonic-assisted green synthesis of AgNps through cinnamon extract: biochemical, structural, and antimicrobial properties. *Int. J. Food Prop.*, 26(1), 1984–1994.

[9] Michael, A. Bruckman, VanMeter, A., and Steinmetz, N. F. (2015). Nanomanufacturing of tobacco mosaic virus-based spherical biomaterials using a continuous flow method. *ACS Biomater. Sci. Eng.*, 1 (1), 13–18. https://doi.org/10.1021/ab500059s

[10] Quinsaat, J. E. Q., Testino, A., Pin, S., Huthwelker, T., Nüesch, F. A., Bowen, P.,

Hofmann, H., Ludwig, C., and Opris, D. M. (2014). Continuous production of tailored silver nanoparticles by polyol synthesis and reaction yield measured by X-ray absorption spectroscopy: toward a growth mechanism. *J. Phys. Chem. C* 118(20) , 11093–11103. https://doi.org/10.1021/jp500949v

[11] Zhang, T., Lu, Y., Liu, J., Wang, K., and Luo, G. (2013). Continuous ammonium silicofluoride ammonification for SiO_2 nanoparticles preparation in a microchemical system. *Indus. Eng. Chem. Res.* 52(16), 5757–5764. https://doi.org/10.1021/ie400547z

[12] Lazarus, L. L., Riche, C. T., Marin, B. C., Gupta, M., Malmstadt, N., and Brutchey, R. L. (2012). Two-phase microfluidic droplet flows of ionic liquids for the synthesis of gold and silver nanoparticles. *ACS Appl. Mater. Interf.,* 4(6), 3077–3083. https://doi.org/10.1021/am3004413

[13] Ali, M. H., Azad, M. A., Khan, K. A., Rahman, M. O., Chakma, U., and Kumer, A. (2023) Analysis of crystallographic structures and properties of silver nanoparticles synthesized using PKL extract and nanoscale characterization techniques. *ACS Omega.* 8(31), 28133–28142.

[14] Naveed, M., Batool, H., Javed, A., Makhdoom, S. I., Aziz, T., Mohamed, A. A., Sameeh, M. Y., Alruways, M. W., Dablool, A. S., Almalki, A. A., and Alamri, A. S. (2023). Characterization and evaluation of the antioxidant, antidiabetic, anti-inflammatory, and cytotoxic activities of silver nanoparticles synthesized using Brachychiton populneus leaf extract. *Processes.* 10(8), 1521.

[15] Singh, R., Langyan, S., Sangwan, S., Gaur, P., Khan, F. N., Yadava, P., Rohatgi, B., Shrivastava, M., Khandelwal, A., Darjee, S., Sahu, P. K. (2022). Optimization and production of alpha-amylase using Bacillus subtilis from apple peel: Comparison with alternate feedstock. *Food Biosci.,* 49, 101978.

8 Analyzing the performance of dynamic factors of abrasive water jet cutting and topography of ZA 27 alloy using L27 orthogonal array

S. Hamritha[a], G. S. Prakash, C. A. Niranjan, S. Akshaya, and P. Harshitha

Department of Industrial Engineering and Management, Ramaiah Institute of Technology, Bengaluru, India

Abstract

The results of an experimental investigation on the abrasive water jet machining (AWJM) process parameters in ZA 27 alloy are presented in this paper. The process parameters, including mass flow rate, traverse speed, and water pressure, are analyzed and optimised using an L27 orthogonal array. The topography of cut surfaces were examined through SEM images and Ra was assigned as a performance indicator. The Ra value of the cut surface was determined using an optical profile which is of non-contact type. Based on experimental investigations it was observed that the surface quality of ZA-27 was affected by the impact of sharp edge particles in the SCZ and spherical abrasive particles in the RCZ. It was observed that, traverse speed has significant influence on the Ra. Minimum Ra was observed in SCZ of 1.85 µm and 1.778 µm in RCZ for lower traverse speed conditions. From the SEM analysis, SCZ was influenced by impact of sharp abrasive particles. From the topography studies, micro cutting was found dominant in SCZ and ploughing was more pronounced in RCZ region of the cut surface. It can be concluded that, AWJC is one of the best and safest cutting techniques for ZA-27 alloy to enhance the performance in industrial applications.

Keywords: AWJC, ZA-27, traverse speed, topography, surface roughness

1. Introduction

ZA-27 alloy is one of the popular and most utilised zinc-aluminum alloys. Owing to their good ductility, wear resistance and high damping properties ZA-27 alloy are predominantly used in journal bearing applications [1–3]. Due to their good machinability, low cost availability and light weight property ZA-27 alloy provides competitive advantage to other lightweight materials such as aluminum and magnesium alloys. In most cases, ZA-27 alloy is now a superior substitute for cast iron, copper, brass, and aluminum [4–7]. Apart from bearing applications ZA-27 alloys are also used in bushing, thrust washers, wear resistant parts, pulleys, valves farm equipment and automobile parts [2,7]. Although ZA alloy is easily machined, its poor performance at high temperatures (over 100°C) limits its usage since it quickly deteriorates qualities that are critical for bearing applications, such as strength, thermal stability, and creep resistance [8,9]. Therefore, while machining ZA 27 alloy using traditional methods, great care should be given because it may change the microstructure of the cut surface and result in structural deformation due to changes in hardness. This

[a]hamritha.shankar@gmail.com

DOI: 10.1201/9781003545941-8

is due to the possibility that significant qualities may potentially deteriorate as a result of traditional machining [10–12] ZA-27 finds its application in bearing components such as bushes and thrust washers, it is essential to focus on cutting quality to enhance the efficiency of bearing. Conventional machining or any other heat generating processes will induce residual stresses that can affect the performance the component [13–15]. Along with the heat generation, geometric imperfections such as shoulders and grooves can also, introduce stress concentration and also results in reduced fatigue life of the component [16].

Numerous studies on the surface integrity of variety of ductile materials cut by an abrasive water jet have been conducted till date [17–23]. Understanding the influence of dynamic control factors such as water pressure (WP), traverse speed (TS), mass flow rate (MFR) and constant factors such as standoff distance, nozzle diameter (focusing nozzle), impact angle, number of passes, orifice diameter etc., [24,25] on the behavior of target material is considered as most important criteria for enhancement of cutting performance of AWJC process. Even though researchers have used a variety of optimisation techniques to extensively examine the effects of both individual and combined parameters on surface quality, there is still a dearth of knowledge regarding the intricacies of process parameters and how these parameters interact to affect the cutting quality. As a result, the current study focuses on examining the abrasive water jet's machining capability in ZA-27 alloy. The primary objective of this study is to study the impact of process parameters such as water pressure, traverse speed and mass flow rate (as far as surface quality is concerned these parameters influences most on surface quality) [26] on surface roughness of ZA-27 alloy. A regression model was developed to correlate the relationship between input parameters and surface quality of ZA27 alloy. Erosion nature of cut surfaces was analyzed using SEM images [27].

Therefore, in the present study machining performance of ZA-27 is evaluated through Abrasive water jet cutting process. Surface roughness is assigned as performance indicator, as surface roughness remains one of the essential surface attributes to evaluate the cutting process's quality [28].

2. Experimental Setup

The manufacturing issues are increasing daily as a result the outputs of manufacturing materials are composites, ceramics, and so forth. The manufacturing industries use AWJM to evaluate variety of machining materials. The combination of abrasive jet machining (AJM) and water jet machining (WJM) is known as the AWJM method. In this machining process, abrasive particles such as garnet, silicon carbide, aluminum oxide, sand, etc. are used from the material surface to cause the materials to erode quickly. In this study, OMAX 1515 AWJ cutting system equipped with 30 Hp direct drive pump, gravity feed type Hooper, pneumatically controlled (X-Y-Z directions) head and with a pressure range of 100MPa to 345 MPa was used to conduct linear cutting experiments. The nozzle is made of Tungsten carbide focusing with 0.76 mm diameter and 0.35 mm sapphire nozzle was used to speed up the water jet containing abrasive and air mixture. ZA-27 alloy with cross section 200 × 50 × 20mm was used to

Figure 8.1: AWJC setup.

Table 8.1: Elemental composition of ZA alloy

Element	Composition %
Zn	46
Mg	0.5
Al	33.2
Cd	0.8
Sn	0.6
Fe	1.5
Cu	12
Pb	6

Table 8.2: Process parameters for linear cutting experiments

Level	Water Pressure	Transverse Speed	Mass Flow Rate
	(Psi)	*(mm/min)*	*(g/min)*
1	30000	30	200
2	35000	40	300
3	40000	50	400

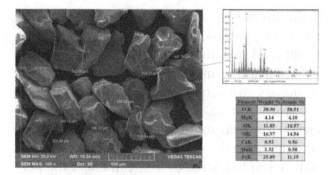

Figure 8.2: Standard garnet used for cutting experiments.

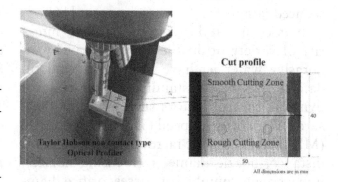

Figure 8.3: Non-contact type optical profiler.

conduct 27 cutting experiments as shown in Figure 8.1. Elemental composition of ZA-27 alloy obtained through Optical Emission Spectroscopy (OES) is shown in Table 8.1. Table 8.2 shows the AWJC process variables and their level used to conduct experiments.

All cutting experiments were conducted using L27 orthogonal array and using standard garnet, since 80% of the industries uses garnet as abrasive for cutting process [27]. Garnet containing multiple cutting edges (80 mesh) obtained through SEM is shown in Figure 8.2. Using a non-contact optical profiler of the Taylor Hobson type, the surface roughness of cut surfaces was measured. The cut surface in AWJC is separated into three zones: the rough cutting zone (RCZ), the smooth cutting zone (SCZ), and the initial damage zone (IDZ). In the present study IDZ was neglected and SCZ & RCZ are considered. Surface roughness Ra was measured at two spots of both SCZ and RCZ and average of 2 readings were considered. Figure 8.3 shows equipment used to measure Ra and cut surface spots where Ra is measured.

3. Result and Discussions

The influence of the AWJM process parameters and Ra values while machining ZA 27 was invesigated in this study. Table 8.3 shows Ra values obtained for L27 conditions in both SCZ and RCZ. Minimum *Ra* 1.71 μm & 1.854 μm and maximum *Ra* 2.827 μm & 3.099 μm were observed in SCZ and RCZ respectively. It can be observed that lower Ra values were observed for lower traverse speeds and higher Ra values for higher traverse speeds. This indicates that, the traverse speed has significant role in deciding the surface quality. Analysis of variance and main effects plots have been used to confirm the specific impact of process parameters on surface roughness in SCZ and RCZ.

Table 8.3: L27 orthogonal array with Ra values in SCZ and RCZ

Sl No	Water Pressure (Psi)	Transverse Speed (mm/min)	Mass Flow Rate (g/min)	SCZ Ra (μm)	RCZ Ra (μm)
1	30000	30	200	1.778	1.854
2	30000	30	300	2.475	2.664
3	30000	30	400	2.676	2.855
4	30000	40	200	1.717	1.953
5	30000	40	300	2.529	3.038
6	30000	40	400	2.424	2.824
7	30000	50	200	2.525	2.816
8	30000	50	300	2.474	3.099
9	30000	50	400	2.504	2.879
10	35000	30	200	2.387	2.724
11	35000	30	300	2.399	2.577
12	35000	30	400	2.701	2.744
13	35000	40	200	2.437	2.827
14	35000	40	300	2.699	2.91
15	35000	40	400	2.539	2.876
16	35000	50	200	2.827	3.013
17	35000	50	300	2.641	2.707
18	35000	50	400	2.494	2.765
19	40000	30	200	2.446	2.513
20	40000	30	300	2.44	2.446
21	40000	30	400	2.5	2.59
22	40000	40	200	2.286	2.486
23	40000	40	300	2.595	2.603
24	40000	40	400	2.414	2.613
25	40000	50	200	2.349	2.481
26	40000	50	300	2.353	3.031
27	40000	50	400	2.291	2.68

3.1. Main Effects Plot

3.1.1. Effect of WP on Surface Roughness

Figure 8.4 shows main effects plot for SCZ and RCZ *Ra* mean values in both zones Figure 8.4. (a) and Figure 8.4(b). It can be observed that, initially surface quality diminishes with increase in surface roughness and further increase in water pressure leads to improves the surface quality. This can be attributed to increase in abrasive kinetic energy at high pressures, which allows portion of excess energy to smoothen the surface hence minimum Ra was observed at high water pressure. However, influence of water pressure on surface quality is also depends

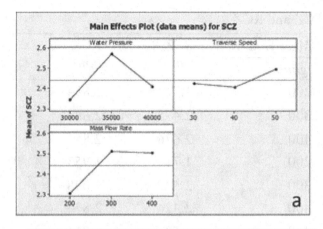

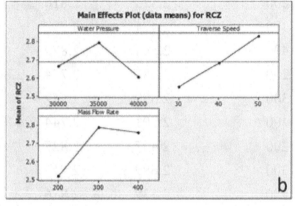

Figure 8.4: Main effects plot for (a) SCZ and (b) RCZ Ra values.

Traverse speed plays significant role in deciding surface quality.

3.1.3. Effect of MFR on Surface Roughness

It was discovered that the effects of mass flow rate in SCZ and RCZ were comparable. Surface roughness rises with increased mass flow. The number of abrasive particles that impact the surface increases with an increase in mass flow rate. This increase in particle number also results in an increase in kinetic energy, which enhances surface quality. However, after a certain point, an additional increase in the mass flow rate of abrasives causes abrasive particles to collide, which lowers surface quality because the kinetic energy of the abrasives decreases. Similar findings were noted in relation to aluminum.

3.2. Surface Topography

SEM images were used to analyze the surface morphology produced under various cutting conditions. Four modes of material removal mechanisms, such as melting, brittle fracture, cutting and fatigue are typically used in traditional machining or cutting processes to identify the cut surfaces. When cutting with an abrasive water jet, the material removal mechanism is identified by two pre-dominant modes of micro cutting and ploughing deformation. The surface quality of any material is depend on the type of deformation mode that occurs on the surface. Figure 8.5 shows the impact of abrasive particles with sharp edges on the upper surface of ZA alloy cut by Abrasive water jet. From the Figure 8.5 it can be observed that, the top portion of SCZ is severely affected by the impact of sharp edged abrasive particles at low angles. Although, the randomly distributed wear tracks does not affect the surface profile, in SCZ the quality of cut surface is governed by influence of traverse speed and water pressure. From the Figure 8.5 formation of micro continuous chips supports the assumption of ductile erosion documented in earlier studies.

on traverse speeds and it is significant only for lower traverse speeds. Even though when water pressure is set to higher values the quality of surface become poor at higher traverse speeds.

3.1.2. Effect of TS on Surface Roughness

From the Figures 8.4(a) and 8.4(b), effect of traverse speed on surface roughness almost shows linear relationship. Increase in traverse speed increases the surface roughness. This observation agrees with the results obtained for aluminum. As the traverse speed increases the interaction time between water jet and contact materials reduces and thus reduces the surface quality. At lower traverse speeds due to higher interaction time, more number of abrasive particles involve in improving the surface. Therefore, minimum surface roughness was observed at lower traverse speed.

Spherical abrasive particles had an impact on the surface quality in RCZ. Ploughing traces can be observed on the surface of RCZ as shown in Figure 8.5, this is because of the abrasive particles' sharp edges and loss of kinetic energy. At this point, the sharp-edged abrasive particle becomes spherical and embeds itself in the surface because it lacks the energy to make a significant impact. Using energy-dispersive X-ray spectroscopy (EDX), the embedding of the abrasive particle is confirmed as shown in Figure 8.6. Because of the high impact angle of the spherical abrasive, the surface roughness at RCZ is typically higher than SCZ, resulting in a more regular wavy zone. Here jet kinetic energy governs the surface characteristics.

4. Conclusion

The current study examined the effects of abrasive water jet cutting process constraints on the topography and cutting performance of ZA-27 alloy, and the findings are as follows:

- AWJC offers better machining possibilities for ZA-27 alloy with acceptable surface quality.
- Traverse speed was found to be the most important factor for determining the surface quality. SCZ was found to be affected

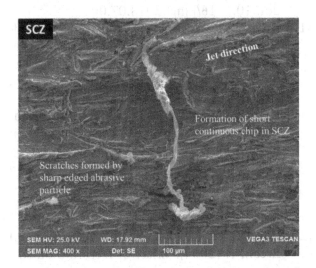

Figure 8.5: Topography of SCZ (Water pressure 40,000 psi; 30 mm/min Traverse speed; 400 mm/min mass flow rate).

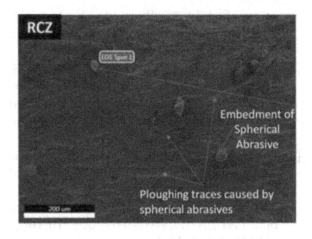

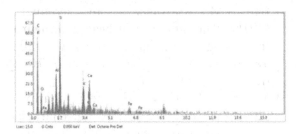

Figure 8.6: Embedment of Spherical abrasive particle in RCZ (Water pressure 40,000 psi; 30 mm/min Traverse speed; 400 mm/min mass flow rate).

by the impact of sharp edged particles and RCZ was found affected by spherical abrasive particles.

- This study provides the basis for machinability of ZA-27 and better machining possibilities can be exploited using different optimisation and modeling techniques.
- AWJC can be effectively adopted for thermally sensitive materials such as ZA alloys without sacrificing the physical properties.

References

[1] Purcek, G., Altan, B. S., Miskioglu, I., and Patil, A. (2005). Mechanical properties of severely deformed ZA- 27 alloy using equal channel angular extrusion, Mater. *Sci. Technol.*, 21(9), 1044–1048, https://doi.org/10.1179/174328405X51

[2] Algur, V., Hulipalled, P., Lokesha, V., Nagaral, M., and Auradi, V. (2022). Machine learning algorithms to predict wear behavior

of modified ZA-27 alloy under varying operating parameters. *J. Bio- Tribo- Corrosion*, 8(1), 7. doi: 10.1007/s40735-021-00610-8

[3] Ares, A. E. and Schvezov, C. E. (2011). The effect of structure on tensile properties of directionally solidified Zn-based alloys, *J. Cryst. Growth* 318 (1), 59–65. doi: 10.1016/j.jcrysgro.2010.11.112

[4] Babic, M., Slobodan, M., Džunic, D., Jeremic, B. and Ilija, B. (2010). Tribological behavior of composites based on ZA-27 alloy reinforced with graphite particles. *Tribol. Lett.*, 37(2), 401–410. doi: 10.1007/s11249-009-9535-2

[5] Babic, M., Mitrovic, S., and Jeremic, B. (2010). The influence of heat treatment on the sliding wear behavior of a ZA-27 alloy, *Tribol. Int.*, 43(1–2), 16–21. doi: 10.1016/j.triboint.2009.04.016

[6] Mishra, S. K., Biswas, S., and Satapathy, A. (2014). A study on processing, characterization and erosion wear behavior of silicon carbide particle filled ZA-27 metal matrix composites, *Mater. Des.*, 55, 958–965. doi: 10.1016/j.matdes.2013.10.069

[7] Girish, B. M., Prakash, K. R., Satish, B. M., Jain, P. K., and Devi, K. (2011). Need for optimization of graphite particle reinforcement in ZA-27 alloy composites for tribological applications. *Mater. Sci. Eng. A*, 530, 382–388. doi: 10.1016/j.msea.2011.09.100

[8] Gangwar, S., Patnaik, A., Yadav, P. C., Sahu, S., and Bhat, K. (2019). Development and properties evaluation of marble dust reinforced ZA-27 alloy composites for ball bearing application, *Mater. Res. Express*, 6, 7. doi: 10.1088/2053-1591/ab1362

[9] Sharma, S. C., Somashekar, D. R., and Satish, B. M. (2001). A note on the corrosion characterization of ZA- 27/zircon particulate composites in acidic medium. *J. Mater. Process. Technol.*, 118(1–3), 62–64. doi: 10.1016/S0924-0136(01)00864-0

[10] Akkurt, A. (2015). The effect of cutting process on surface microstructure and hardness of pure and Al 6061 aluminium alloy. *Eng. Sci. Technol. Int. J.*, 18(3), 303–308. https://doi.org/10.1016/j.jestch.2014.07.004

[11] Akkurt, A. (2013). The cutting front side geometry in the applications of D3 cold work tool steel material via abrasive water jet. *Gazi Univ. J. Sci.*, 26(2), 225–239.

[12] Boud, F., Loo, L. F., and Kinnell, P. K. (2014). The impact of plain waterjet machining on the surface integrity of aluminium 7475. *Procedia CIRP* 13, 382–386. https://doi.org/10.1016/j.procir.2014.04.065

[13] Duan, Z. et al. (2021) Milling force model for aviation aluminum alloy: academic insight and perspective analysis. *Chinese J. Mech. Eng.*, 34(1), 18. https://doi.org/10.1186/s10033-021-00536-9

[14] Santos, M. C., Machado, A. R., Sales, W. F., Barrozo, M. A. S., and Ezugwu, E. O. (2016) Machining of aluminum alloys: a review. *Int. J. Adv. Manuf. Technol.*, 86(9–12), 3067–3080. doi: 10.1007/s00170-016-8431-9

[15] Chithirai Pon Selvan, M., Mohana Sundara Raju, N., and Sachidananda, H. K. (2012). Effects of process parameters on surface roughness in abrasive waterjet cutting of aluminium. *Front. Mech. Eng.*, 7(4), 439–444. doi: 10.1007/s11465-012-0337-0

[16] Anandatirthachar, N. C., Srinivas, S., and Ramachandra, M. (2018) Abrasive water jet cutting: a risk-free technology for machining mg-based materials *Intech Open* 61–82. doi: 10.5772/intechopen.85209

[17] Niranjan, C. A., Srinivas, S., and Ramachandra, M. (2018). Effect of process parameters on depth of penetration and topography of AZ91 magnesium alloy in abrasive water jet cutting. *J. Magnes. Alloy.*, 6(4), 366–374. doi: 10.1016/j.jma.2018.07.001

[18] Kovacevic, R. and Fang, M. (1994). Modeling of the influence of the abrasive waterjet cutting parameters on the depth of cut based on fuzzy rules. *Int. J. Mach. Tools Manuf.*, 34(1), 55–72.

[19] Kalpana, K., Mythreyi, O. V., and Kanthababu, M. (2015). Review on condition monitoring of Abrasive Water Jet Machining system In *2015 International Conference on Robotics, Automation, Control and Embedded Systems (RACE)*, 1–7. doi: 10.1109/RACE.2015.7097254

[20] Akkurt, A., Kulekci, M. K., Seker, U., and Ercan, F. (2004). Effect of feed rate on surface roughness in abrasive waterjet cutting applications. *J. Mater. Process. Technol.*,

147(3), 389–396. https://doi.org/10.1016/j.jmatprotec.2004.01.013

[21] Natarajan, Y., Murugesan, P. K., Mohan, M., and Liyakath Ali Khan, S. A. (2020) Abrasive water jet machining process: a state of art of review. *J. Manuf. Process.*, 49, 271–322.

[22] Begic-Hajdarevic, D., Cekic, A., Mehme-dovic, M., and Djelmic, A. (2015). Experimental study on surface roughness in abrasive water jet cutting. *Procedia Eng.*, 100, 394–399. doi:10.1016/j.proeng.2015.01.383

[23] Meng, H. C. and Ludema, K. C. (1995). Wear models and predictive equations: their form and content. *Wear*, 181–183, 443–457. doi: https://doi.org/10.1016/0043-1648(95)90158-2

[24] Yuvaraj, N. and Kumar, M. P. (2016). Cutting of aluminium alloy with abrasive water jet and cryogenic assisted abrasive water jet: A comparative study of the surface integrity approach. *Wear*, 362–363, 18–32. doi:10.1016/j.wear.2016.05.008

[25] Zhao, W. and Guo, C. (2014). Topography and microstructure of the cutting surface machined with abrasive waterjet. *Int. J. Adv. Manuf. Technol.*, 73(5–8), 941–947. doi:10.1007/s00170-014-5869-5

[26] Kunaporn, S., Chillman, A., Ramulu, M., and Hashish, M. (2008) Effect of waterjet formation on surface preparation and profiling of aluminum alloy. *Wear*, 265(1–2), 176–185. doi:10.1016/j.wear.2007.09.008

[27] Hlaváčová, I. M., Sadílek, M., Váňová, P., Szumilo, Š., and Tyč, M. (2020). Influence of steel structure on machinability by abrasive water jet. *Materials (Basel).*, 13(19), 4424. https://doi.org/10.3390/ma13194424

[28] Mu, M., Feng, L., Zhang, Q., Zang, W., and Wang, H. (2022). Study on abrasive particle impact modeling and cutting mechanism. *Energy Sci. Eng.*, 10(1), 96–119. doi:10.1002/ese3.1012

9 Facile click approach for the synthesis of innovative 1,2,3-triazole derivatives for structural determination and its characterisation

D. H. Harshitha[1,a], M. K. Kokila[1,b], M. R. Nithin Kumar[2], and B. Paul[2]

[1]Department of Physics, Bangalore University, Bangalore, India
[2]Department of Studies in Chemistry, Bangalore University, Bangalore, India

Abstract

Triazoles are biologically active heterocyclic scaffolds with the molecular formula $C_2H_3N_3$ and are considered a mimic of the peptide bond. Two possible isomeric forms are 1,2,3-triazole and 1,2,4-triazole, which contain two carbon and three nitrogen atom in their five-membered ring.

1,2,3-Triazole   **1,2,4-Triazole**

The synthesis of the above derivatives will use site-specific click chemistry, which results in high yields at mild aerobic conditions and achieves selectivity, orthogonality, simplicity, and biocompatibility by carbon-hetero bond formation reactions. This study reports the synthetic approach of novel 1,2,3-triazole derivatives. Regiospecific copper-catalyzed azide and terminal alkyne reactions provide a synthetic pathway for triazole derivatives. The bromo-substituted starting material was stirred with sodium azide (NaN_3) at a low temperature to obtain the azide compound. A copper-catalyzed click reaction between the obtained azide compound and the terminal alkyne resulted in the desired product. It mainly projects the catalytic activity, resulting in good yields. The reaction mechanism was studied, and the molecular structures of the compounds were confirmed using analytical methods for structural integrity. Spectroscopic investigations are confirmed through characterisation. These compounds are interested in diverse scientific domains like medicinal and material chemistry. These derivatives at positions [1–3] have potential scientific prospects in agriculture, medical chemistry, material science, biochemistry, and pharmaceuticals. As a future direction, one can study the ability of the triazole compounds' biological aspects and optimisation conditions of the reaction by enhancing the computational stability and structure-activity relationship.

Keywords: Azide; alkyne, click chemistry, CuAAC, regiospecific, triazole

1. Introduction

K.B. Sharpless introduced click chemistry in 2001 to outline the reactions that result in high yields at mild aerobic conditions and to accomplish selectivity by carbon-hetero bond formation reactions [1,2]. The production of a 5-membered heteroatom ring using copper catalysis reaction between an azide (Az) product and an alkyne product is popularly

[a]harshetha.dh@gmail.com, [b]drmkkokila@gmail.com

DOI: 10.1201/9781003545941-9

known as the click reaction. Cu-catalyzed-(3+2)-cycloaddition between an organic azide and terminal alkynes, well known as the famous click reaction [10] for producing heterocyclic compounds [3].

Click chemistry is an important mechanism in the synthesis of simple, convenient, and potent triazole derivatives due to their robust reliability and specificity. Two aspects of click chemistry involve SPAAC and CuAAC. Triazole formation can be catalyzed by Cu catalyst or by using strained alkyne [4,5]. In our research, we focus on CuAAC and its prominence in its quantitative functionalisation [6]. Cu(I)-Catalysed azide-alkyne cycloaddition that affords 1,4-disubstituted triazole is a prototypical example of a bio-orthogonal click reaction. This reaction helps to model and implement the synthesis of cyclic peptides and peptidomimetic scaffolds of triazole moieties [7,8].

Figure 9.1(a): Characteristics of azide (Az)-alkyne cycloaddition reaction.

Azide-alkyne catalyzed by copper is the 1,3-dipolar cycloaddition reaction mentioned by Huisgen [4] which is utilized for the synthesising 1,4- and 1,5-disubstituted 1,2,3-triazole [9,10]. The Sharpless and Meldal groups found this Nobel-winning concept, which completely changed the research on biologically significant organic molecules [11]. Metal catalysts like copper sulfate accelerate the click reaction, mediating the cyclisation of terminal akynes and azide product to give only 1,4-disubstituted-[1H]-1,2,3-triazoles product [12, 13]. The combination of Az (azide) and Alk (alkyne) serves as the driving force behind my research endeavors, aiming to implement and advance these methodologies and techniques for future progress [14,15]. Moving forward, an intriguing avenue for exploration involves delving into the biological aspects of triazole compounds, and assessing their capabilities. Additionally, optimising reaction conditions is crucial, enhancing computational stability, and establishing the intricate relationship between the compound's structure and its activity [16–20]. This multifaceted approach promises to contribute significantly to the field, unlocking new dimensions in the synthesis and understanding of innovative 1,2,3-triazole

Figure 9.1(b): Site-specific reactivity in click approach between an Az(azide) and (Alk) alkyne product for the outcome of an isomer triazole at 1,2,3-position.

Source: Jane Totobenazara, Anthony J. Burke, "New click-chemistry methods for 1,2,3-triazoles synthesis: recent advances and applications", Tetrahedron Letters, Volume 56, Issue 22, 2015, Pages 2853–2859.

derivatives for structural determination and characterisation.

2. Materials and Methods

The experimental procedure employed commercially available reagents were used for the reactions. Necessary solvents procured from commercial sources (Spectrochem, Rankem). The reaction monitoring was observed by thin-layer chromatography (TLC) using (F254, Merck Silica gel 60) aluminium sheets and chromatogram accompanied by UV-visualisation. ^{1}H-NMR and ^{13}C-NMR spectroscopy in CDCl$_3$, performed on a Bruker AMX-400 (400 MHz) spectrometer. HRMS (ESI) measurements carried out using an Impact HD (Bruker) ESI QTOF high resolution mass

spectrometer from Indian Institute of Science (IISc), Bengaluru.

3. Experimental Procedure

3.1. Synthetic Scheme: General Procedure for 1,2,3-triazole Derivative

(a) Displacement reaction: An efficient two-step reaction in which the first step consists of 2-(2-bromo-acetylamino)-benzoic acid methyl ester (I) (1.0 eq) dissolved in 10mL DMF [17] and stirred at low temperature (10–20°C). Sodium azide (1.1eq) was weighed, added and stirred for more than 2hr for the completion [21,22]. Thin layer chromatography (TLC) (Stahl. *et al.*, 1969) is

Table 9.1: Experimental data conditions

Sl.no	X (Terminal-alkyne)	Obtained Product	Time (Hrs)	Yielda (%)
1	≡⟨-OH	H1	02	48.56
2	≡C-CH₂-Ph	H2	02	46.74
3	≡—OH	H3	02	49.47
4	≡—Ph-CF₃	H4	02	74.47
5	≡—OH	H5	02	19.68
6	H₃C,O-CH₃,CH₃	H6	02	75.83
7	≡—CH₃	H8	02	37.20

Yield^a: Isolated yield after purification process

used to monitor the reaction process for total consumption. The resultant reaction mixture underwent quenching, leading to the formation of white precipitates, which were subsequently rinsed, filtered, and desiccated under vacuum conditions. This process resulted in the synthesis of the anticipated azide compound (II) [23].

(b) **Click approach reaction:** A 1,2,3-triazole derivative was produced through click approach during the second stage. The azide compound(II) was treated with 10 mL N, N-dimethylformamide and started stirring. Terminal alkyne(III) in the presence of $CuSO_4$ (5 eq) along with sodium ascorbate (10 eq) was added slowly and kept stirring at RT for two hours [3]. TLC monitored the reaction for the total consumption. Extracted from dichloromethane (40mL × 3) and EtOAc & brine wash (50 mL) and dried using Na_2SO_4 (Sodium sulfate) and evaporated. The purification process was carried out using column chromatography (SiO_2; 100% hexane to 60/40 EtOAc/hexane), depending upon the polarity of the compounds to facilitate 1,4-disubstituted-1,2,3-triazole(IV). Diethyl ether was used for non-polar wash, filtered, and dried to yield the desired compound. The synthesized triazole compound has been characterized by nuclear magnetic spectroscopy (NMR) for structural integrity high resolution mass spectroscopy for mass confirmation [24].

Synthesis of H1: Preparation of the title compound from 2-methyl-3-butyn-2-ol (0.469 mmol, 1.1 eq) acquiring the general procedure described above has been followed. The reaction mixture was quenched, filtered, and kept for stirring for more than an hour. TLC monitored the reaction for the total consumption. Extraction with dichoromethane (40mL × 3) and washed with EtOAc & brine (50mL × 50 mL), dried and distilled. Purification was conducted via column chromatography (SiO_2; 100% hexane to 60/40 EtOAc/hexane) depending on the polarity of the compound. Diethyl ether was used for nonpolar wash, filtered, & dried. White solid of

the triazole compound was observed. (Isolated yield: 48.56%).

The nuclear magnetic resonance (HNMR) was recorded at 400 MHz in $CDCl_3$, revealing distinct chemical shifts: a singlet at 1.75 (6H; S; CH3(2)), a singlet at 2.69 (1H; S;-OH), a singlet at 3.89 (3H; S; -CH3), a singlet at 5.30 (2H; S; -CH2), a triplet at 7.17 (1H; T; -Aromatic-H; J:8Hz), a triplet at 7.59 (1H; T; -Aromatic-H; J:8Hz), a singlet at 7.73 (1H; S; -Tri-H), a doublet at 8.04 (1H; D; -Aromatic-H; J:8Hz), a doublet at 8.70 (1H; D; -Aromatic-H; J:8Hz), and a peak shift at δ 11.10 (1H; S; -NH)

13C-NMR(100MHz, $CDCl_3$): 168.47, 164.13, 156.52, 140.14, 134.82, 130.99, 123.68, 120.96, 120.46, 115.60, 68.51, 53.82, 52.60, 30.26

HRMS: m/z calculated for $[M+H]^+$ $[C_{15}H_{18}N_4O_4 + H]^+$ = 319.1406

Synthesis of H2: The general technique mentioned above has been followed to prepare the title chemical from 4-phenyl-1-butyne (0.345 mmol, 1.2 eq). To extract, DCM (Dichloromethane) (40 mL × 3), rinsed with EtOAc 5 mL), brine (50 mL), dried and distilled. Diethyl ether was utilized for purification, while n-pentane was employed for a non-polar wash, filtering, and drying. A polar impurity was detected and tracked using TLC (10%EtOAc+Hexane). As a result, the solid crude obtained was recrystallized, it was heated to between 50–60°C, and then dissolved in 40 milliliters of ethanol while being stirred to room temperature. Filtered over Buckner funnel and dried to yield the triazole compound as an off-white solid (Isolated yield-46.74%).

The nuclear magnetic resonance (HNMR) was recorded at 400 MHz in $CDCl_3$, revealing distinct chemical shifts: a triplet at 3.12 (4H; T; -CH2; J:8Hz), a singlet at 3.92 (3H; S; -CH3), a singlet at 5.25 (2H; S; -CH2), a triplet at 7.19 (2H; T; J:8Hz), a triplet at 7.32 (4H; T; J:8Hz), a singlet at 7.42 (1H; S; -Tri-H), a triplet at 7.59 (1H; T; J:8Hz), a doublet at 8.06 (1H; D; -Aromatic-H; J:8Hz), a doublet at 8.67 (1H; D; -Aromatic-H; J:8Hz), and a peak shift at δ 11.25 (1H; S; -NH)

13C-NMR (100 MHz, CDCl₃): 164.15, 148.24, 141.19, 134.70, 130.96, 128.45, 126.17, 123.61, 122.45, 120.53, 53.84, 52.55, 35.57, 27.62

HRMS: m/z calculated for [M+Na]⁺ [C₂₀H₂₀N₄O₄ + Na]⁺ = 387.1433

Synthesis of H3: Preparation of the title compound from Propargyl alcohol (0.4696 mmol, 1.1 eq) acquiring the general procedure described above has been followed. Purification was conducted via column chromatography (SiO2; 100% hexane to 10/90 up to 90/10 EtOAc/hexane), but the compound was highly polar. [At 40/60 EtOAc/hexane combination, TLC indicated the solvents washed off impurities]. Since the mixture was exceptionally polar, flash chromatography using methanol and dichloromethane (SiO2; 100% DCM to 10/90 MeOH/DCM) was used for column purification. Filtered and dried the final product. (Isolated yield: 49.47%)

The nuclear magnetic resonance (HNMR) was recorded at 400 MHz in CDCl₃, revealing distinct chemical shifts: a triplet at 2.65 (1H; T; -OH; J:8Hz), a singlet at 3.91 (3H; S; -CH3), a doublet at 4.90 (2H; D; -CH2; J:4Hz), a singlet at 7.16 (1H; T; -Aromatic-H; J:8Hz), a triplet at 7.58 (1H; T; J-Aromatic-H; J:8Hz), a singlet at 7.82 (1H; S; -Tri-H), a doublet at 8.04 (1H; D; -Aromatic-H; J:8Hz), a doublet at 8.66 (1H; D; -Aromatic-H; J:8Hz), and a peak shift at δ 11.14 (1H; S; -NH)

13C-NMR(100MHz, CDCl₃): 168.64, 163.93, 148.58, 140.12, 134.82, 131.01, 123.69, 123.49, 120.48, 115.61, 56.68, 53.83, 52.62

HRMS: m/z calculated for [M+H]⁺ [C₁₃H₁₄N₄O₄ + H]⁺ = 291.1093

Synthesis of H4: Preparation of the title compound from 4-(Trifluoromethyl) phenylacetylene (0.4696mmol, 1.1 eq) acquiring the above mentioned procedure in the presence of saturated CuSO₄ solution (1.5 eq) along with sodium ascorbate (2.5 eq) added and kept for stirring at RT for two hours. The

recrystallisation process was conducted using ethanol by heating up to 80°–90°C for complete solubility. The solution is then cooled for about 1 hr and left to settle down. It was then filtered and dried to obtain 1.28 mg of the final product. (Isolated yield: 74.47%).

The nuclear magnetic resonance (HNMR) was recorded at 400 MHz in CDCl₃, revealing distinct chemical shifts: a singlet at 3.90 (3H; S; -CH3), a singlet at 5.34 (2H; S; -CH2), a triplet at 7.20 (1H; T; J:8Hz), a triplet at 7.73 (1H; T; -Aromatic-H; J:8Hz), a singlet at 7.74 (2H; S; -Tri-H), a triplet at 8.04 (3H; T; J:8Hz), a singlet at 8.14 (1H; S; -Aromatic-H), a doublet at 8.68 (1H; D; -Aromatic-H; J:8Hz), and a peak shift at δ 11.47 (1H; S; -NH)

13C-NMR(100MHz, CDCl₃): 163.54, 147.19, 140.16, 134.81, 133.95, 130.99, 126.09, 125.99, 125.80, 123.75, 121.97, 120.51, 115.62, 53.99, 52.57

HRMS: m/z calculated for [M+H]⁺ [C₁₉H₁₅F₃N₄O₃ + H]⁺ = 405.1175

Synthesis of H5: Following the basic technique mentioned above, 3-butyn-1-ol (0.3586 mmol, 1.2 eq) was used to prepare the triazole chemical. Column chromatograph (SiO₂; 100% dichloromethane to 05/95 MeOH/DCM) was used for purification, and 100% ethyl acetate was used to elute the product. Ethanol was used to process recrystallisation, it was heated to 80°C to achieve total solubility. The final white product is then obtained by cooling, filtering and drying the solution. (19.68% : Isolated yield)

The nuclear magnetic resonance (HNMR) was recorded at 400 MHz in CDCl₃, revealing distinct chemical shifts: a singlet at 2.97 (1H; S; -OH), a multiple at 3.05 (2H; M; -CH2; J:8Hz), a singlet at 3.88 (3H; S; -CH3), a singlet at 4.03 (2H; S; -CH2), a singlet at 5.28 (2H; S; -CH2), a triplet at 7.15 (1H; T; -Aromatic-H; J:8Hz), a triplet at 7.58 (1H; T; -Aromatic-H; J:8Hz), a singlet at 7.68 (1H; S; -Tri-H), a doublet at 8.01 (1H; D; -Aromatic-H; J:8Hz), a doublet at 8.70 (1H; D; -Aromatic-H; J:8Hz) and a peak shift at δ 10.95 (1H; S; -NH)

13C-NMR(100MHz, CDCl₃): 168.97, 164.25, 146.33, 140.14, 134.92, 131.06, 123.68, 120.41, 115.45, 61.50, 53.73, 52.80, 29.18

HRMS: m/z calculated for [M+Na]⁺ [C₁₄H₁₆N₄O₄ + Na]⁺ = 327.1069

Synthesis of H6: Preparation of the title compound from 1-butyne-3-methoxy-3-methyl (322.6 mg, 3.287 mmol) acquiring the general procedure described above has been followed. The reaction was quenched using ice and water, filtered, and kept stirring for more than an hour. Further purification was carried out in column chromatography (SiO₂; 100% hexane to 45/50 ethyl acetate/hexane) to obtain the desired product. Diethyl ether and n-pentane non-polar wash were carried out for about 1 hour, filtered, and dried to yield the pure white solid. The reaction mixture underwent extraction with ethyl acetate (30 mL × 3) followed by washing with 20 mL of brine, desiccation using sodium sulfate, and subsequent filtration. (Isolated yield: 75.83%)

The nuclear magnetic resonance (HNMR) was recorded at 400 MHz in CDCl₃, revealing distinct chemical shifts: a triplet at 1.70 (6H; S; -(CH3)2), a singlet at 3.24 (3H; S;-OCH3), a singlet at 3.88 (3H; S; -CH3), a singlet at 5.30 (2H; S; -CH2), a quartet at 7.19-7.14 (1H; M; -Aromatic-H), a triplet at 7.58 (1H; T; -Aromatic-H; J:8Hz), a singlet at 7.71 (1H; S; -Tri-H), a doublet at 8.04 (1H; D; -Aromatic-H; J:8Hz), a doublet at 8.69 (1H; D; -Aromatic-H; J:8Hz), and a peak shift at δ 11.27 (1H; S; -NH)

13C-NMR (100 MHz, CDCl₃): 168.37, 164.03, 153.54, 140.20, 134.72, 130.92, 123.61, 122.21, 120.48, 115.60, 72.97, 53.84, 52.48, 50.72, 26.65

HRMS: m/z calculated for [M+Na]⁺ [C₁₆H₂₀N₄O₄ + Na]⁺ = 355.1382.

Synthesis of H8: Preparation of the title compound from 1-Pentyne (318.9 mg, 4.687 mmol), the general procedure described above has been followed. The reaction was quenched using ice and water, & it was then filtered over a Buckner funnel. 980 mg of the crude was recrystallised using ethanol under stirring, washed, and filtered to obtain an off-white, solid final product. (Isolated yield: 37.20%)

The nuclear magnetic resonance (HNMR) was recorded at 400 MHz in CDCl₃, revealing distinct chemical shifts: a triplet at 1.05 (3H; T; J:8Hz), a quartet at 1.8-1.71 (2H; M; -Aromatic-H), a singlet at 3.91 (3H; S; -CH3), a singlet at 5.27 (2H; S; -CH2), a triplet at 7.18 (1H; T; J:8Hz), a singlet at 7.53 (1H; S; -Tri-H), a triplet at 7.60 (1H; T; J:8Hz), a doublet at 8.05 (1H; D; -Aromatic-H; J:8Hz), a doublet at 8.68 (1H; D; -Aromatic-H; J:8Hz), and a peak shift at δ 11.27 (1H; S; -NH)

13C-NMR (100 MHz, CDCl₃): 168.22, 164.15, 149.10, 140.19, 134.67, 130.93, 123.58, 122.20, 120.52, 115.72, 53.84, 52.51, 27.72, 22.65, 13.81

HRMS: m/z calculated for [M+H]⁺ [C₁₅H₁₈N₄O₃ + H]⁺ = 303.1457

3.2. Results and Discussion

One of the key advancements in chemical synthesis and an effective click chemistry technique is the CuAAC [25, 26]. 1,4-disubsituted-1,2,3-triazole compounds are more resistant to biochemical degradation, making them stable and potentially effective therapeutic agents [27]. We have synthesised a sequence of novel 1,2,3-triazole derivatives [28,29]. NMR spectroscopy to analyze the solution-phase confirmation of the compounds by evaluating ¹H (proton) & ¹³C (carbon) NMR in solution-phase CDCl₃ solvent. In the H1 compound spectrum, the chemical shift δ 11.10 ppm indicates the secondary amide proton indicating potential hydrogen bonding with side chain carbonyl of the ester group. The signal appearing at 1.75 ppm indicates two methyl groups present near the 5-membered triazole ring. Signals at 8.70, 8.04, 7.59, & 7.17 ppm are the aromatic protons on the benzene ring. One proton singlet at 7.73 ppm clearly signifies triazole ring formation. 2.69 ppm singlet proton indicates OH group present in the structure. The formation of the compound is further confirmed using mass spectrometry. Calculated $[M+H]^+$ for the H1 compound is at 319.33 and the observed mass is found to be at 319.14. Furthermore, all other compounds are characterised by the conformation of the molecular structure [30].

4. Conclusion

Effective click reaction techniques led to synthesising a library of new compounds with different functional groups. Click chemistry is a significant reaction to obtain the desired substituted triazole compound. The click refers to facile, efficient, and selective chemical transformation under mild aerobic conditions. Considerable research has been conducted on regiospecific azide-alkyne cycloaddition with copper catalytic system. The copper needed for the catalyzed Huisgen cycloaddition is added as cuprous salts with stabilising ligands, or it can be produced from copper (II) salts using sodium ascorbate as the reducing agent, as represented in our study. This paper details a facile production of derivatives containing the 1,2,3-triazole moiety from primary amine synthesis. The aforementioned site-specific reactions provide a synthetic pathway for triazole derivatives. It mainly projects that the catalytic activity results in good yields. The reaction mechanism was studied, and for structural integrity, the confirmation of the molecular structures of the compounds was validated through analytical methodologies [31,32].

Acknowledgements

The author(s), acknowledge that the work reported here was supported in part by the I-STEM (Indian Science, Technology and Engineering Facilities Map) program, funded by the Office of the Principle Scientific Advisor to the Govt. of India. We thank DST-SAIF at the Institute NMR facility (INF), IISC, for providing NMR data & IISC organic chemistry dept for HRMS data. We sincerely thank Bangalore University, JB Campus, the Department of Physics, and the Department of Studies in Chemistry for supporting the scientific research & for providing the laboratory.

References

[1] Hartmuth C. K., Finn, M. G., and Sharpless, K. B. (2001). Click chemistry: diverse chemical function from a few good reactions. *Angew. Chem. Int. Ed.*, 40(11), 2004–2021.

[2] Ji Ram, V., Sethi, A., Nath, M., and Pratap, R. (2019). Five-membered heterocycles. *The Chemistry of Heterocycles*, Elsevier, 149–478.

[3] Rostovtsev, V. V., Green, L. G., Fokin, V. V., and Sharpless, K. B. (2002). A stepwise huisgen cycloaddition process: copper(I)-catalyzed regioselective "ligation" of Azides and terminal alkynes. *Angew. Chem. Int. Ed.* 41(14), 2596–2599.

[4] Huisgen, R. (1961). In: *Proceedings of the Chemical Society*. Royal Society of Chemistry (RSC), 357.

[5] Albert, A. and Taylor, P. J. (1989). The Tautomerism of 1,2,3-triazole in aqueous solution. *J. Chem. Soc. Perkin Trans.*, 2(2), 1903–1905.

[6] Meldal, M. and Tornoe, C. W. (2008) Cu-catalyzed azide-alkyne cycloaddition. *Chem. Rev.*, 108(8), 2952–3015.

[7] Liang, L and Didier, A. (2011). The copper (I)-catalyzed alkyne-azide cycloaddition (CuAAC) "click" reaction and its applications. *Coord. Chem. Rev.*, 255(23–24), 2933–2945.

[8] Demko, Z. P., Sharpless, K. B. (2001). Preparation of 5-substituted 1H-tetrazoles from nitriles in water. *J. Org. Chem.*, 66(24), 7945–7950.

[9] Horst, H. J. and Scharff, H.-D. (2005). Hydrazoic acid and azide. Ullmann's Encyclopedia of Insdustrial Chekmistry, Wiley-VCH, Weinheim.

[10] Kolb, H. C. and Sharpless, K. B. (2003). The growing impact of click chemistry on drug discovery. *Drug Discovery Today*, 8(24), 1128–1137.

[11] Tornøe, C. W., Christensen, C., and Meldal, M. (2002). Peptidotriazoles on solid phase: [1,2,3]-triazoles by regiospecific

copper(I)-catalyzed 1,3-dipolar cycloadditions of terminal alkynes to azides. *J. Org. Chem.*, 67(9), 3057–3064.

[12] Worrell, B., Malik, J., and Fokin, V. (2013). Direct evidence of a dinuclear copper intermediate in Cu (I)-catalyzed azide-alkyne cycloadditions. *Science*, 340(6131), 457–460.

[13] Kumar, V., Lal, K., Naveen, and Tittal, R. K. (2022). The fate of heterogeneous catalysis & click chemistry for 1,2,3-triazoles: Nobel prize in chemistry. *Catal. Commun.* 176(2023), 1066–1029.

[14] Wang, L., Peng, S., Danence, L. J. T., Gao, Y., and Wang, J. (2012). Amine-catalyzed [3 + 2] huisgen cycloaddition strategy for the efficient assembly of highly substituted 1,2,3-triazoles. *Chem. - Eur. J.*, 18(19), 6088–6093.

[15] Hong, L., Lin, W., Zhang, F., Liu, R., and Zhou, X. (2013). Ln[N(SiMe3)2]3-catalyzed cycloaddition of terminal alkynes to azides leading to 1,5-disubstituted 1,2,3-triazoles: new mechanistic features. *Chem. Commun.*, 49(49), 5589–5591.

[16] Okuda, Y., Imafuku, K., Tsuchida, Y., Seo, T., Akashi, H., and Orita, A. (2020). Process-controlled regiodivergent copper-catalyzed azide–alkyne cycloadditions: tailor-made syntheses of 4- and 5-bromotriazoles from bromo(phosphoryl)ethyne. *Org. Lett.*, 22(13), 5099–5103.

[17] Phan, T. B. and Mayr, H. (2006). Nucleophilic reactivity of the azide ion in various solvents. *J. Phys. Org. Chem.* 19, 706–713.

[18] Dunbrack, R. L. Jr. and Karplus, M. (May 1994). Conformational analysis of the backbone-dependent rotamer preferences of protein sidechains. *Nat. Struct. Biol.*, 1(5), 334–340.

[19] Woolley, D. W. (1944). Some biological effects produced by benzimidazole and their reversal by purines. *J. Biol. Chem.*, 152(2), 225–232.

[20] Jie, L., Wei, L., Zhang, L., Liu, Z., and Han, B.-H. (2017). Triazole derivatives: versatile building blocks for enhanced materials properties. *Chem–. Asian J.,* 12(24), 3133–3150.

[21] Paul, B., Butterfoss, G. L., Boswell, M. G., Renfrew, P. D., Yeung, F. G., Shah, N. H., Wolf, C., Bonneau, R., Kirshenbaum, K. (2011). A rotamer library to enable modeling and design of peptoid foldamers *J. Am. Chem. Soc.* 133, 10910–10919.

[22] Purushotham, M. et al. (2022). Ortho-halogen effects: n→π* interactions, halogen bonding, and deciphering chiral attributes in N-aryl glycine peptoid foldamers, *J. Mol. Struct.*, 1264(133276), 0022–2860.

[23] Lu, R., Chang, X., Zhang, J., and Xiang, J. (2017). Coordination chemistry of triazole derivatives: recent advances and perspectives. *Coord. Chem. Rev.*, 337, 1–26.

[24] Lebeau, A., Abrioux, C., Bénimèlis, D., Benfodda, Z., and Meffre, P. (2016). Synthesis of 1,4-disubstituted 1,2,3-triazole derivatives using click chemistry and their src kinase activities. *Mc* 13, 40–48.

[25] Chen, K., Qian, X., Wang, Q., and Lai, Z. (2015). Coordination chemistry of triazole and tetrazole derivatives: metal-organic frameworks, coordination polymers and luminescence. *Coord. Chem. Rev.*, 285, 50–74.

[26] Farooq, T. (2021). *Advances in Triazole Chemistry*. Elsevier, 21–27.

[27] Potts, K. T. (1961). The chemistry of 1,2,4-triazoles. *Chem. Rev.*, 61(2), 87–127.

[28] Hitchcock, C. A., Dickinson, K., Brown, S. B., Evans, E. G. V., and Adams, D. J. (1990). Interaction of azole antifungal antibiotics with cytochrome P-450-dependent 14α-sterol demethylase purified from candida albicans. *Biochem. J.*, 266, 475–480.

[29] Kumar, S., Khokra, S. L., and Yadav, A. (2021). Triazole analogues as potential pharmacological agents: a brief review. *Futur. J. Pharm. Sci.*, 7, 106.

[30] Pedersen, S. L. et al. (2019). Triazoles as bioisosteres in drug design. *Chem. Rev.*, 119(11), 6597–6641.

[31] Noor, A., Lewis, J. E., Cameron, S. A., Moratti, S. C., and Crowley, J. D. (2012). A multi-component CuAAC 'click' approach to an exo functionalised pyridyl-1, 2, 3-triazole macrocycle: synthesis, characterisation, Cu (I) and Ag (I) complexes. *Supramol. Chem.*, 24(7), 492–498.

[32] Brittain, W. D. G., Buckley, B. R., and Fossey, J. S. (2016). Asymmetric copper-catalyzed azide–alkyne cycloadditions. *ACS Catal.*, 6(6), 3629–3636.

10 Synthesis and characterisation of brass-Molybdenum composite

S. Jasper[1,a], N. Mohanrajhu[2,b], M. Durairaj[3,c], D. Vijayakumar[4,d], N. Muthuselvakumar[5,e], and P. Chandramohan[6,f]

[1]Department of Mechanical Engineering, Jeppiaar Institute of Technology, Chennai, India
[2]Department of Mechanical Engineering, R.M.K. Engineering College, Chennai, India
[3]Department of Mechanical Engineering, Tagore Engineering College, Chennai, India
[4]Department of Mechanical Engineering, Vel Tech Multi Tech Dr. Rangarajan Dr. Sakunthala Engineering College, Chennai, India
[5]Department of Ocean Engineering, Indian Institute of Technology, Madras, Chennai, India
[6]Department of Mechatronics Engineering, Rajalakshmi Engineering College, Chennai, India

Abstract

To obtain the microstructural research of the brass-Molybdenum composite, powder particles were repeatedly cold-welded and broken in a high energy ball mill to form a homogeneous material. This study synthesised mechanical alloys with varying molybdenum weight percentages (Mo) with brass matrix composites. SEM investigations were performed on milled composite powders to determine their characterisation of the brass-molybdenum composite. The SEM study demonstrated that Molybdenum was spread equally throughout the Brass matrix. It ensured that Molybdenum was distributed uniformly across the matrix, apart from its pores. When Molybdenum is added to the brass matrix, it increases hardness and density of composite material when sintering temperature goes up to 800° degrees Celsius while decreasing porosity until three hours later.

Keywords: Brass, molybdenum (MO), mechanical alloying, powder metallurgy

1. Introduction

Mechanical milling and densification of brass/molybdenum powder blends resulted in brass/molybdenum composites with 5, 10, and 15 weight percent Mo. Separate applications of the pressure less sintering and spark plasma techniques were used to solidify the ground particles. The pressure less sintered pellets were compressed at 350, 450, and 650 MPa before sintering. According to the findings, the relative densities of sintered materials containing 70 and 85 weight percent Mo rose up to 700 MPa of cold pressure. Despite this, samples cold crushed below 450 MPa. When compared to other samples the Molybdenum particles are mixed with the base brass –30% matrix composite were not as much as fine to the other samples. according to microstructural studies employing energy dispersive spectroscopy and scanning electron microscopy. Spark plasma sintered samples were harder and had a greater relative density than pressure less sintered materials, according to the results. Furthermore, as the molybdenum volume percent grew, the relative density of the SPS samples decreased. 397,376 and 114Mpa are the flexural strengths of SPS composite with weight percent of 5, 10, 15% respectively. The vacuum switches are

[a]joeljasper7@gmail.com, [b]nmj.mech@rmkec.ac.in, [c]durairajtagore@gmail.com, [d]dvijayakumar@veltechmultitech.org, [e]muthuselvakumar22@gmail.com, [f]Chandramohan.p@rajalakshmi.edu.in

DOI: 10.1201/9781003545941-10

the most commonly used applications of reinforced copper metal refractory composites such as WCu & Mo Cu Composites. Metal matrix composites (MMCs) of this type are frequently created through infiltration, which entails injecting copper into a refractory metal framework [1–3]. low relative densities of composites are generated particularly for tungsten and molybdenum due to molten copper's poor wettability of refractory metals. Press-sinter-repressed (PSR) powder combinations have also been used to create W/Cu and Mo/Cu composites [3,7]. Because of the repression of the sintered samples by this procedure, the relative density of the sintered specimen increases [5,6]. It is also used in the production of composite powders. There have been numerous investigations undertaken on the characterisation, density, and behaviour of Mo/Cu composites. [11–13]. a special consolidation method to increase the densification of bronze/W powder mixes that simultaneously involves milling, compacting, and sintering. [14,15]

The density was increased by increasing the temperature of the pre-centrifugation. [16]. The majority of research has been on the differentiation and synthesis of copper alloys/W composites, with minimal focus on the fabrication and characteristics of molybdenum- containing alloys. Furthermore, no study has been undertaken on the densification of these composites using existing technologies like as spark plasma sintering (SPS). In powder metallurgy, Spark plasma sintering method is well-known for its low-temperature production capabilities. [8,18]. The influence of pressure-less sintering technique and compaction pressure on the manufacturing of composite powders such as brass and molybdenum was examined in this study. The composite powders were created mechanically through high energy ball milling of brass and molybdenum combinations. By SPS method the density of brass/W composite powders were determined. In this study, the microstructural behavior of the finished

pellets, porosity and the energy dispersive x-ray analysis, were also investigated.

2. Materials and Methods

Copper (Cu) and zinc (Zn) in varying proportions are combined to create brass, an alloy with unique mechanical, electrical, and chemical characteristics. Since the alloy is substitutional, atoms from the two constituents can interchange within the same crystal structure. The Brass - Mo powders are meticulously weighed using electronic weighing equipment in accordance with the material composition, as shown in Table 1, and then hand blended. Planetary Ball Milling Machine with High Energy. It's used in a number of composites, including 100 grammes of brass with no Molybdenum, 95 grams of brass with 5 grams of Molybdenum, 90 grams of brass with 10 grams of Molybdenum, and 85 grams of brass with 15 grams of Molybdenum. To reduce wear, run the machine dry with 1:3 powder to ball ratio and the speed of the drum is 350 rpm.

The material took five hours to mill. Following ball milling, intermetallic bonding and homogeneity of a brass-molybdenum composite were noted. After mixing, the die is subjected to high pressure, crushing the metal powder. The powder combination is formed into the necessary shape before being ejected from the die chamber [9,10,19]. Sintered green compacts are presently on the market. In the regulated argon environment, the green compacts were sintered for 3 hours using muffle furnace.

Table 10.1: Material composition

Sample	Brass wt % in grams	Molybdenum wt% in grams
1	100	0
2	95	5
3	90	10
4	85	15

3. Result Analysis

3.1. *Microstructural Analysis*

Figure 10.1 shows brass with a weight percentage of 100grams with nil amount of Molybdenum, Figure 10.2 shows brass with a weight percentage of 95 grams and 5grams of Molybdenum, Figure 10.3 shows brass with a weight percentage of 90grams and Molybdenum of 10 grams, and Figure 10.4 shows brass with a weight percentage of 85grams and increase in molybdenum of 15grams as reinforcement. Without any agglomeration, the reinforcement particles are strongly bonded within the matrix – brass material and the size of the particles gets reduced from micro to nano level is shown in Figure 10.2(a)

revealed that due to 5hours of ball milling. Figure 10.2(b) depicts a brass matrix with evenly distributed Molybdenum-reinforced particles. The strength coefficient of brass + 8 wt% Molybdenum improves when Molybdenum is used as a reinforcement. Even with the extra grammes of reinforcing Molybdenum (Molybdenum) and the 12 weight percent Molybdenum, there will surely be a difficulty. In most circumstances, it eliminates a negative impact on mechanical properties. In the absence of agglomeration, Figure 10.2(d) shows the evident link between the reinforcement and the matrix. Using the most advanced PM route, the structural properties of the Brass-Molybdenum were improved [13,14].

Figure 10.1: Brass 100 grams + 0 grams of molybdenum.

Figure 10.3: Brass 90 gram + 10 grams of molybdenum.

Figure 10.2: Brass 95 grams + 0 grams of molybdenum.

Figure 10.4: Brass 85 grams + 15 grams of molybdenum.

3.2. Porosity Analysis

It is difficult to obtain zero porosity. But, in this work, the pellets were made of less porosity by mixing reinforcement molybdenum with the matrix brass composite, The parameter called porosity in the composite materials can be formed by various reasons. When the specimen is filled with molybdenum (Mo) particles and mixed with the composite brass matrix, the porosity and number of cavities decrease [21,22]. While in the first attempt, very low amount of porosity level is attained. The porosity of the brass-molybdenum composite is depicted in Figure 10.5. Adding molybdenum to the brass matrix increases the hardness of the composites, as shown in Figure 10.6. The porosity increases as the percentage of reinforcement increases. Furthermore, heat energy absorption decreases significantly as porosity increases due to friction.

4. Conclusion

In a high velocity ball mill, in order to create a homogenous material for the microstructural analysis of the brass-Molybdenum composite, powder particles were repeatedly cold-welded and broken. Mechanical alloys with different molybdenum weight percentages (Mo) and brass matrix composites were created as part of this project. To evaluate the characteristics of the brass-molybdenum composite, SEM examination of milled composite powders was performed. The SEM showed that molybdenum is distributed uniformly throughout the matrix of Brass. In addition, it ensured that Molybdenum was evenly distributed throughout the matrix, with the exception of the pores. Molybdenum increases the density and hardness of the composites while lowering their porosity when added to the brass matrix.

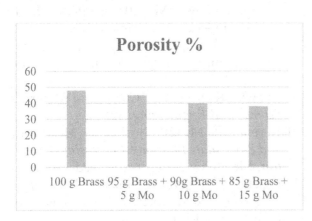

Figure 10.5: Porosity % of different samples.

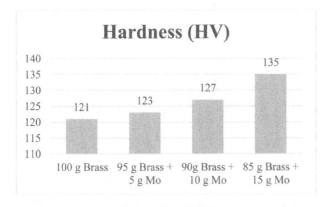

Figure 10.6: Hardness % of different samples.

References

[1] Chawla, N. and Chawla, K. K. (2013). *Metal Matrix Composites*, second ed., Springer.

[2] Balasubramanian, M. (2014) *Composite Materials and Processing*, CRC Press

[3] *Powder Metal Technologies and Applications* (ASM Handbook Vol. 7), ASM International, USA, 1998.

[4] Ardestani, M., Rezaie, H. R., Arabi, H., and Razavizadeh, H. (2009). The effect of sintering temperature on densification of nanoscale dispersed W–20–40wt% Cu composite powders. *Int. J. Refract. Metals Hard Mater.*, 27, 862–867.

[5] Costa, F. A., Silva, A. G. P, and Gomes, U. U. (2003). The influence of the dispersion technique on the characteristics of the W–Cu powders and on the sintering behavior. *Powder Technol.*, 134, 123–132.

[6] Maneshian, M. H. and Simchi, A. (2008). Solid state and liquid phase sintering of mechanically activated W–20 wt. % Cu

powder mixture. *J. Alloys Compd.*, 463, 153–159.

[7] Elsayeda, A., Omayma, W. L., Kady, A.E., Daoush, W. M., Olevsky, E. A., and German, R. M. (2015). Experimental investigations on the synthesis of W–Cu nanocomposite through spark plasma sintering, *J. Alloys Compd.*, 639, 373–380.

[8] Dolatmoradi, A., Raygan, S., Abdizadeh, H. (2013). Mechanochemical synthesis of W–Cu nanocomposites via in-situ co-reduction of the oxides. *Powder Technol.*, 233, 208–214.

[9] Cheng, J., Song, P., Gong, Y., Cai, Y., Xia, Y. (2008). Fabrication and characterization of W–15Cu composite powders by a novel mechano-chemical process. *Mater. Sci. Eng.*, 488A, 453–457.

[10] Ardestani, M., Arabi, H., Rezaie, H. R., and Razavizadeh, H. (2009). Synthesis and densification of W– 30wt%Cu composite powders using ammonium meta tungstate and copper nitrate as precursors, Int. J. Refract. Metals Hard Mater. 27 (2009) 796–800.

[11] Benavides, B. S. and Palma, R. H. (2017). Liquid phase sintering of mechanically alloyed Mo-Cu powders, *Mater. Sci. Eng.*, 701A, 237–244.

[12] Aydinyan, S. V., Kirakosyan, H. V., and Kharatyan, S. L. (2016). Cu– Mo composite powders obtained by combustion–coreduction process. *Int. J. Refract. Metals Hard Mater.*, 54, 455–46.

[13] Ardestani, M., Rafiei, M., Salehian, S., Raoufi, M. R., and Zakeri, M. (2015). Compressibility and solid-state sintering behavior of W-Cu composite powders. *Sci. Eng. Compos. Mater.*, 22, 257–261.

[14] Mohammed, K. S., Rahmat, A., andAziz, A. (2013). Self-compacting high density tungsten–bronze composites. *J. Mater. Process. Technol.*, 213, 1088–1094.

[15] Mohammed, K. S., Rahmat, A., Ahmad, and K. R. (2013). Sintering behavior and microstructure evolution of mechanically alloyed W–bronze composite powders by two-step ball milling process. *J. Mater. Sci. Technol.*, 29, 59–69.

[16] Gowon, B., Mohammed, K. S., Baharin, S., Jamaluddin, B., Hussain, Z., and Evarastics, P. (2015). The effects of sintering temperature on the densification of mechanically alloyed W-brass composites. *Open J. Metals*, 5, 19–26.

[17] Mohammed, K. S., Gowon, B., Baharin, S., Jamaluddin, B., Hussain, Z., and Evarastics, P. (2015). Synthesis and densification of tungsten-brass composite by mechanical alloying. *Open J Metals*, 5, 27–36.

[18] Asl, M. S., Namini, A. S., Motallebzadeh, A., and Azadbeh, M. (2018). Effects of sintering temperature on microstructure and mechanical properties of spark plasma sintered titanium. *Mater. Chem. Phys.*, 203, 266–273.

[19] Chawla, K. K.(2012). *Composite Materials Science and Engineering*, Springer.

[20] German, R. M. (2005). *Powder Metallurgy and Particulate Materials Processing, Princeton*, Metal Powder Industries Federation.

[21] Lenel, F. V. (1980). *Powder Metallurgy: Principles and Applications*, Metal Powder Industries Federation.

[22] Campbell, F. C. (2008). *Elements of Metallurgy and Engineering Alloys*, ASM International.

11 Green synthesis, characterisation and antimicrobial activity of calcium nanoparticles and their efficacy against drug resistant Streptococcus pyogenes

N. Ahalya[1], Gayathri Vijayakumar[2], P. Dhamodhar[1,a], and Idhaya Kumar[1]

[1]Department of Biotechnology, M S Ramaiah Institute of Technology, Bangalore, India
[2]Department of Biotechnology, Rajalakshmi Engineering College, Chennai, India

Abstract

Streptococcus pyogenes is a major upper respiratory tract pathogen, that causes a wide variety of diseases from bacterial pharyngitis in children, to life-threatening diseases such as rheumatic fever, rheumatic heart disease and scarlet fever. The post-pandemic of 2022–2023 shows 89 cases of *S. Pyogenes* than before the pandemic period. There are multifactorial reasons for the infection hence this study provides a sustainable nanomaterial-based approach to the treatment. In the present study, clinical isolates of *Streptococcus pyogenes* were collected and identified by routine biochemical tests such as Gram staining, catalase test, blood agar hemolysis test etc. Antibiotic susceptibility towards the class of macrolides, penicillin derivatives and fluoroquinolones antibiotics in clinical isolates showed sensitivity to penicillin G and streptomycin. However, resistance to antibiotics like Tetracycline, Erythromycin, Lincomycin and Ampicillin was observed among them, indicating an increasing pattern of resistance. The functional peaks like 2.83A, 2.78A, and 2.22A in XRD confront the values of nanoparticles whereas the FTIR peak of 1035 cm-1 and spherical shape in SEM results confirm the calcium nanoparticle. The comparative study of different concentrations of the Punica granatum extract and calcium nanoparticles against the *S. pyogenes* shows the zone of inhibition of 10–15 in plant extract and 15–20 mm Ca NPs. These zones formed by the calcium nanoparticle show relevant inhibition to the commercial drug. Both pomegranate epicarp extract and calcium nanoparticles showed higher antibacterial activity against the antibiotic-resistant isolates of *S. pyogenes*. Thus, the calcium nanoparticles synthesized using pomegranate epicarp can be explored in the management of increasing antibiotic resistance in *Streptococcus pyogenes* causing pharyngotonsillitis.

Keywords: *S. pyogenes,* calcium nanoparticles, antibiotic resistance, pomegranate epicarp

1. Introduction

The never-ending concept of antimicrobial resistance in the microbiome has been a critical side of antibiotic treatment. The incidence of death caused by antimicrobial resistance is high as it is estimated that it will reach two billion in 2050 [1]. Notable and common epidemiological infections like *Streptococcus pyogenes* have been the focus as they cause severe diseases like endocarditis, rheumatic heart disease, and scarlet fever. The intrusive infection may cause flesh-eating disease, a necrotising fasciitis infection, and fatal Streptococcal toxic shock syndrome. The patients are usually treated with penicillin derivatives

[a]dhamu_bio@msrit.edu

DOI: 10.1201/9781003545941-11

but β-lactamase and penicillin hypersensitivity can be treated with first-gen cephalosporins [2]. Improper use of broad-spectrum antibiotics has led to a change in resistance antibiotic resistance patterns.

Antibiotics have been used in treating microbial disease and infection for many decades. Microbes are more flexible life forms that exist in a world with a long-term evolutionary background due to their high genome plasticity and capacity for genetic information. Indiscriminate use of antibiotics has led to antibiotic resistance. Oxidative damage and nucleotide pool unbalancing caused by certain drugs have increased the mutation in microorganisms [3]. Anti-microbial resistance caused by this mutation has led to demand to produce a new generation of antibiotics.

The *Punica granatum* is known for its biological activity and numerous pharmacological applications. Pomegranate has various phenolic compounds and flavonoids that are high in antioxidant and anti-atherogenic property properties [4]. The contents of Pomegranate are 16.4–69% of Phenolic acid and ellagitannins [5], 1.6–24.4% of flavonoids, and 17–82.0% of anthocyanins [6]. fruit peel, which contains a high amount of antimicrobial activity is used as a preservative [7]. study studies to study studies on pomegranate extract have shown high inhibitory action on certain Enterobacter and mucus bacteria like *Klebsiella pneumonia, Escherichia coli*, and *Pseudomonas aeruginosa* [8]. Due to its high phenolic content, it has the potential to rupture the bacterial cell wall and inhibit its proliferation.

Nanotechnology has gained attraction in recent years due to its high surface area and targeted dynamics [9]. The recent applications of nanoparticles include drug delivery and targeted cancer therapy [10]. Calcium carbonate and calcium phosphate nanoparticles are best known for their biological activity as they are used as drug carriers for encapsulating the biomolecule and in bone replacement graft material for osseous deformities [11]. Calcium nanoparticles are used extensively due to their biocompatibility and degradability.

In the present study, calcium nanoparticles were synthesised using pomegranate epicarp and tested for their efficacy against resistant antibiotic-resistant *Streptococcus pyogenes* [12]. The activity was compared with the efficacy of various commonly used antibiotics [13].

2. Materials and Methods

2.1. Identification of Streptococcus pyogenes

The isolates of *Streptococcus pyogenes* were obtained from Ramaiah Metropolis, Bengaluru. Biochemical tests such as Gram staining, Blood agar hemolysis test, and Catalase test were carried out to identify *Streptococcus pyogenes* [14].

2.2. Antibiotic Susceptibility Testing

Muller Hinton Agar of (AR grade)(MHA) plates were used to detect the antibacterial activity of the overnight culture of S. pyogenes and inoculated on MHA plates using sterile swap buds. The antibiotic ring was placed using sterile forceps onto the inoculated agar plate and incubated for 24 hours at 37°C under aseptically. The zone of inhibition was measured in mm to compare results. The susceptibility of the isolates towards each antibiotic was analysed and interpreted as per CLSI guidelines [15].

2.3. Phytochemical Analysis

Antibacterial activity of pomegranate extract Muller Hinton Agar (MHA) plates were used to detect the antibacterial activity of the overnight culture of S. *pyogenes* and inoculated on MHA plates using sterile swap buds. The antibiotic ring was placed using sterile forceps onto the inoculated agar plate and incubated for 24 hours at 37°C under aseptically.

The zone of inhibition was measured in mm to compare results. The susceptibility of the isolates towards each antibiotic was analysed and interpreted as per CLSI guidelines. The epicarp of pomegranate was washed and dried in the shade at room temperature and blended finely. The 50 g powder was mixed with 40 absolute ethanol, diluted with 160 ml of sterile Water (20% of extract), and incubated in the dark for 36 hrs at room temperature. later they are stored under 4°C for later use.

Using a Soxhlet apparatus, the powdered pomegranate epicarp was put in a thimble and extracted for ethanol (1:50, w/v). After filtration, the filtrate was heated for half an hour at 60°C. The filtrate was stored at 4°C for further use [16].

The pomegranate extract antimicrobial proficiency was measured by testing it against *S. pyogenes* in Muller Hinton Agar (MHA) plates.

2.4. *Phytochemical Analysis*

The pomegranate epicarp Extract was analysed for bioactive compounds like saponins, steroids, tannins, flavonoids, carbohydrates, terpenoids, alkaloids, and anthocyanin. Tested [17].

2.5. *Calcium Nanoparticles Synthesis Characterization and Antibacterial Activity*

The pomegranate epicarp extract sample was ground into fine powder and stored for further use. 5gm of powdered samples were centrifuged at 1000 rpm for 10 mins after adding it to the 10 ml distilled water and the filtrate was removed by Wattman filter paper. The green synthesised CPNP was prepared by adding 7.5 ml of 12.5 mM Cacl2 and 7.5 ml of 12.5 mM of Na2HPO4 to the Epicarp extract supernatant and stirred for 10–15mins. The dispersion of the mixtures was filtered after settling for 30 minutes and stored under 4°C.

This was subjected to XRD, SEM, and FTIR characterization studies. The antibacterial activity of synthesised calcium nanoparticles against the isolates of *S. pyogenes* was tested and the zone of inhibition was recorded.

3. Result

3.1. *Identification of Streptococcus pyogenes*

The clinical isolates of *Streptococcus pyogenes* were identified using routine biochemical tests. The isolates were Gram-positive, Gram-positive, Gram-positive, and Gram-positive, exhibited Beta Hemolysis patterns in blood agar plates, and were negative catalase-negative. Thus, the routine biochemical tests confirmed that the isolates obtained were *S. pyogenes*. Figure 11.1 shows the different antimicrobial resistance shows in different patient samples.

3.2. *Antibiotic Susceptibility Testing*

The clinical isolates of *Streptococcus pyogenes* were tested against commonly used antibiotics, and the zone of inhibition was measured. The diameters of the zone of inhibition were recorded after 24, 48, and 72 hours. The clinical isolate culture of *S. pyogenes* has shown diverse effects on different antibiotics that are commercially available and exhibited different zone sizes. The zone of inhibition for each antibiotic was recorded and the susceptibility of *S. pyogenes* towards each antibiotic was analysed as per CLSI guidelines [18]. The isolates are susceptible to **ampicillin** if they exhibit a zone size ≥24 mm. All the clinical isolates I1 to I9 exhibited intermediate sensitivity. For the antibiotic **tetracycline**, the zone size of almost all the isolates was observed within the range of 12–17mm, and hence they were categorised as resistant. For the antibiotic Ciprofloxacin, I1, I2, I5, I7, I8, and I9 had the zone of inhibition values less than 15mm and hence were categorised

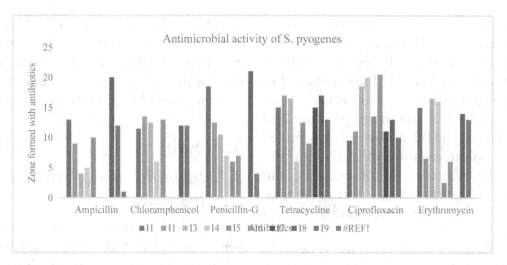

Figure 11.1: The graphical representation of the zone of inhibition formed by each sample (I1–I9 and ATCC) when they are treated with 13 different antibiotics that are commercially used.

as resistant. I3, I4, I6 exhibited intermediate sensitivity to **ciprofloxacin**. All the isolates of *S. pyogenes* were susceptible to **Penicillin**. The susceptibility testing with the antibiotic **erythromycin** revealed that I1, I2, I5, I6, I8, and I9 were resistant, whereas the isolates I3 and I4 exhibited intermediate sensitivity.

3.3. Phytochemical Analysis of Pomegranate Epicarp Extract and Its Antibacterial Activity

The qualitative analysis of the pomegranate epicarp extract revealed the presence of Tannins, Steroids, Alkaloids, carbohydrates, Flavonoids, and Emodin. Further to confirm the antibacterial activity of the extract, varying concentrations of the pomegranate epicarp extract were used and the zone of inhibition was observed around 8–16mm.

Phytochemicals present in the extract have to be separated and its antibacterial activity has to be assessed. This would reveal if a single phytochemical or is it the synergistic effect of two or more Phytochemicals that is responsible for the antibacterial activity.

From Figure 11.2, it is evident that the zone of inhibition obtained for 30µl and 40µl is almost the same. It is also that clinical isolates labelled I4, I5, I6, I9, and ATCC strain did not show any zone of inhibition when the

concentration of pomegranate extract loaded on the disc was 4µl. The maximum diameter of the zone of inhibition recorded was for sample I8 when the concentration of the plant extract loaded in the disc was 30µl and 40µl, for sample I1 and I4 when the concentration was about 40µl. The extract obtained by maceration showed a greater zone of inhibition and there was no zone of inhibition observed when the concentration of the extract was less than 10 µL for exhalation exhalation. The diameter of the zone of inhibition obtained by using the maceration procedure was greater than that of exhalation

3.4. Characterisation of Calcium Phosphate Nanoparticles

The nanoparticles synthesised from the liquid extraction method were analysed for morphological and molecular construction by X-ray diffraction

3.4.1. X-Ray Diffraction

The surface analysis with XRD (Figure 11.3) showed varying peaks in nanoparticles compared with the standard JCPDS (896438) pattern. The d-spacing values of the peak in 2.83A, 2.78A, and 2.22A in the hydroxyapatite were observed. These patterns of d

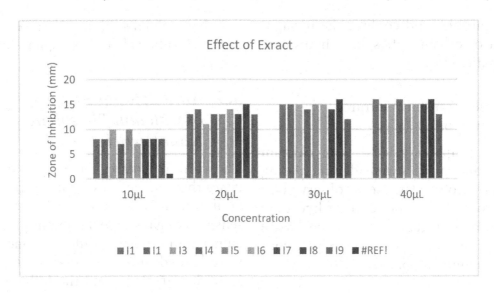

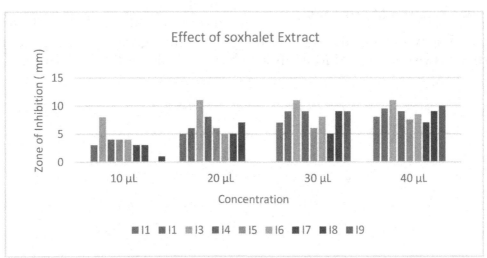

Figure 11.2: Graphical comparison of antimicrobial activity in Extract and Soxhlet extract.

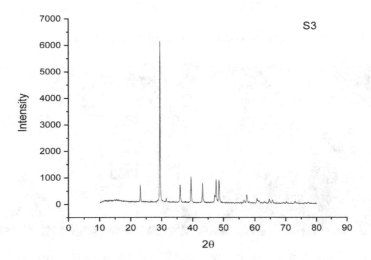

Figure 11.3: The XRD analysis of calcium nanoparticles.

values synchronise and confirm the hexagonal system of calcium phosphate hydroxide hydroxyapatite [19].

3.4.2. Fourier Transform Infrared Spectroscopy

Functional groups associated with calcium phosphate nanoparticles were identified using FTIR spectroscopy. The peak observed at $1035 cm^{-1}$ confirms the presence of hydroxyl group are seen in Figure 11.4. The peaks are compared with the early studies which confirms the calcium nanoparticle [19].

3.4.3. Scanning Electron Microscopy

The Figure 11.5 shows clumped distribution of nanoparticles is visible from SEM analysis

and their partial spherical shape confirms the material to be calcium phosphate nanoparticles [20].

3.5. *Antibacterial Activity of Calcium Phosphate Nanoparticles*

The clinical isolates of *Streptococcus pyogenes* that showed resistivity to the standard antibiotics used were selected for testing their sensitivity against calcium nanoparticles. calcium nanoparticles of different concentrations were loaded into wells that were punctured in MH agar plates and the following results were observed.

From the above Figure 11.6 data, the plant extract gives the synergistic effect of the high zone above 10mm whereas the Soxhalet

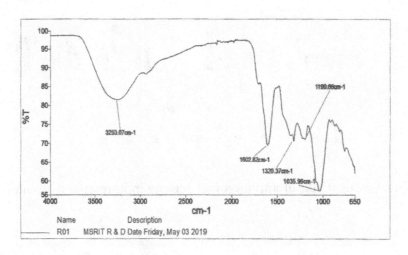

Figure 11.4: The FTIR representation of calcium phosphate nanoparticles.

Figure 11.5: S4canning electron microscopy was used to determine the morphological features of calcium phosphate nanoparticles. Figures A and B show the images of calcium nanoparticles at different magnifications.

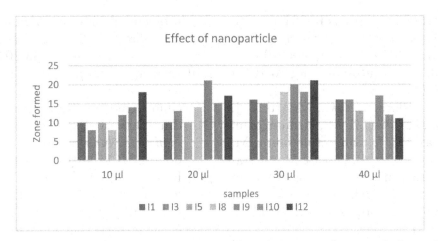

Figure 11.6: S4canning electron microscopy was used to determine the morphological features of calcium phosphate nanoparticles. Figures A and B show the images of calcium nanoparticles at different magnifications.

shows mostly above 10mm but the green synthesised calcium nanoparticle shows most of the above 20mm which intersects with the inhibition zone formed by the commerical drugs used in treating *S. pyogenes* infection in humans. This comparison study gives clear data that the synthesised nano-drug could be considered a better alternative to the treatment of antimicrobial *S. pyogenes*. The Plant-based nanoparticles have less toxicity and biodegradability is high so these calcium phosphate nanoparticles show a prominent effort towards AMR.

4. Conclusion

The AMR in the clinical diagnostic sector causes the development of new antibiotics. The literature and early research data show a promising solution through a plant-based drug that induces necrosis without stimulating genetic abnormalities. This study clearly states the efficacy of green synthesised Calcium phosphate nanoparticles from pomegranate epicarp extract towards the AMR *S. pyogenes*. The trilogy ideology to compare the crude extract, Soxhlet extract, and nanoparticle extract clearly defines the drug proficiency towards AMR. The effects of these extracts are visualised with the microbial plating technique and the zone formed is compared with CLSI standards. The Calcium

nanoparticle synthesised from the pomegranate extract was characterised by XRD, FTIR, and SEM which give the static surface of nanoparticle morphology. The comparison of extract, Soxhlet and nanoparticle establishes a holistic outcome that the Green synthesised calcium nanoparticle has higher antimicrobial activity. the zone of inhibition formed by the nanoparticle is in the range of 15–20 mm which coincides with the commercial drug used in treating *S. pyogenes* infection. Above all these data we could suggest the calcium nanoparticle could be a better alternative to an antibiotic drug that is currently used.

Acknowledgement

The authors thank the Centre for Advanced Materials Technology (CAMT), Ramaiah Institute of Technology, for providing characterization facilities.

References

[1] Ikuta, K. S., Swetschinski, L. R., Robles Aguilar, G., Sharara, F., Mestrovic, T., Gray, A. P., Davis Weaver, N., Wool, E. E., Han, C., Gershberg Hayoon, A., Aali, A., Abate, S. M., Abbasi-Kangevari, M., Abbasi-Kangevari, Z., Abd-Elsalam, S., Abebe, G., Abedi, A., Abhari, A. P., Abidi, H., ... Naghavi, M. (2022). Global mortality associated

with 33 bacterial pathogens in 2019: a systematic analysis for the Global Burden of Disease Study 2019. *Lancet*, 400(10369), 2221–2248. https://doi.org/10.1016/S0140-6736(22)02185-7

[2] Walker, M. J., Barnett, T. C., McArthur, J. D., Cole, J. N., Gillen, C. M., Henningham, A., Sriprakash, K. S., Sanderson-Smith, M. L., and Nizet, V. (2014). Disease Manifestations and Pathogenic Mechanisms of Group A Streptococcus. *Clin. Microbiol. Rev.*, 27(2), 264–301. https://doi.org/10.1128/CMR.00101-13

[3] Rodríguez-Beltrán, J., Couce, A., and Blázquez, J. (2013). Antibiotics and antibiotic resistance: a bitter fight against evolution. *Int. J. Med. Microbiol.*, 303(6–7), 293–297. https://doi.org/10.1016/j.ijmm.2013.02.004

[4] Li, Y., Guo, C., Yang, J., Wei, J., Xu, J., and Cheng, S. (2006). Evaluation of antioxidant properties of pomegranate peel extract in comparison with pomegranate pulp extract. *Food Chem.*, 96(2), 254–260. https://doi.org/10.1016/j.foodchem.2005.02.033

[5] Yasoubi, P., Barzegarl, M., Sahari, M. A., and Azizi, M. H. (2007). Total phenolic contents and antioxidant activity of pomegranate (Punica granatum L.) Peel Extracts. *J. Agric. Sei. Technol.*, 9. J. Agric. Sci. Technol. (2007) Vol. 9: 35–42

[6] Gómez-Caravaca, A. M., Verardo, V., Toselli, M., Segura-Carretero, A., Fernández-Gutiérrez, A., and Caboni, M. F. (2013). Determination of the major phenolic compounds in pomegranate juices by HPLC-DAD-ESI-MS. *J. Agric. Food Chem.*, 61(22), 5328–5337. https://doi.org/10.1021/jf400684n

[7] Al-Zoreky, N. S. (2009). Antimicrobial activity of pomegranate (Punica granatum L.) fruit peels. *Int. J. Food Microbiol.*, 134(3), 244–248. https://doi.org/10.1016/j.ijfoodmicro.2009.07.002

[8] Al-Hassnawi, A. A.-R. A. (2017). Evaluation of antibacterial activity of aqueous and methanolic extracts of pomegranate peels (punica granatum lin.) Against some bacteria. *World J. Pharm. Res.*, 2426–2436. https://doi.org/10.20959/wjpr20178-9193

[9] Lindquist, E., Mosher-Howe, K. N., and Liu, X. (2010). Nanotechnology … What is it good for? (Absolutely everything): a problem definition approach. *Rev. Policy Res.*, 27(3), 255–271. https://doi.org/10.1111/j.1541-1338.2010.00441.x

[10] Render, D., Rangari, V. K., Jeelani, S., Fadlalla, K., and Samuel, T. (2014). Bio-based calcium carbonate ($CaCO_3$) nanoparticles for drug delivery applications. *Int. J. Biomed. Nanosci. Nanotechnol.*, 3(3), 221. https://doi.org/10.1504/IJBNN.2014.065464

[11] Biradar, S., Ravichandran, P., Gopikrishnan, R., Goornavar, V., Hall, J. C., Ramesh, V., Baluchamy, S., Jeffers, R. B., and Ramesh, G. T. (2011). Calcium carbonate nanoparticles: synthesis, characterization and biocompatibility. *J. Nanosci. Nanotechnol.*, 11(8), 6868–6874. https://doi.org/10.1166/jnn.2011.4251

[12] Wu, V. M., Tang, S., and Uskoković, V. (2018). Calcium Phosphate Nanoparticles as Intrinsic Inorganic Antimicrobials: The Antibacterial Effect. *ACS Appl. Mater. Interfaces*, 10(40), 34013–34028. https://doi.org/10.1021/acsami.8b12784

[13] Dianat, O., Saedi, S., Kazem, M., and Alam, M. (2015). Antimicrobial Activity of nanoparticle calcium hydroxide against enterococcus faecalis: an in vitro study. *IEJ Iran. Endod. J.*, 10(1). PMCID: PMC4293579; Pg.No: 5 : x

[14] Wat, L. L., Fleming, C. A., Hodge, D. S., and Krishnan, C. (1991). Selective medium for isolation of Arcanobacterium haemolyticum and Streptococcus pyogenes. *Eur. J. Clin. Microbiol. Infect. Dis.*, 10(5), 443–446. https://doi.org/10.1007/BF01968026

[15] Camara, M., Dieng, A., and Boye, C. S. B. (2013). Antibiotic susceptibility of *Streptococcus Pyogenes* isolated from respiratory tract infections in Dakar, Senegal. *Microbiol. Insights*, 6, MBI.S12996. https://doi.org/10.4137/mbi.s12996

[16] Talaat, R., El-Toumy, S. A., & Samaka, R. (2014). Anti-angiogenic and Anti-Inflammatory Activity of Punica granatum Peel on Experimentally-Induced Gastric Ulcer in Rats. *Article in International Journal of Biological and Chemical Sciences*. https://doi.org/10.13140/2.1.4458.7522

[17] Sharma, S., V Rai, D., and Rastogi, M. (2021). Compositional characteristics of rudraksha (*elaeocarpus ganitrus* roxb.).

Plant Arch., 21(1). https://doi.org/10.51470/plantarchives.2021.v21.no1.087

[18] Hombach, M., Bloemberg, G. V., and Bottger, E. C. (2012). Effects of clinical breakpoint changes in CLSI guidelines 2010/2011 and EUCAST guidelines 2011 on antibiotic susceptibility test reporting of Gram-negative bacilli. *J. Antimicrob. Chemother.*, 67(3), 622–632. https://doi.org/10.1093/jac/dkr524

[19] Ahmad, W., Kamboj, A., Banerjee, I., and Jaiswal, K. K. (2022). Pomegranate peels mediated synthesis of calcium oxide (CaO) nanoparticles, characterization, and antimicrobial applications. *Inorg. Nano-Metal Chem.*, 1–8. https://doi.org/10.1080/24701556.2021.2025080

[20] Pakravanan, K., Rezaee Roknabadi, M., Farzanegan, F., Hashemzadeh, A., and Darroudi, M. (2019). Amorphous calcium phosphate nanoparticles-based mouthwash: preparation, characterization, and antibacterial effects. *Green Chem. Lett. Rev.*, 12(3), 278–285. https://doi.org/10.1080/17518253.2019.1643412

12 Preparation of papercrete bricks using agro wastes with a superficial layer of microbially synthesised silver nanoparticles

K. Kannan[1,a], K. Sakthimurugan[2], and G. Pugazhmani[3]

[1]Department of Agricultural Engineering, Mohamed Sathak Engineering College, Ramanathapuram, India
[2]Department of Civil Engineering, C.K College of Engineering and Technology, Cuddalore, India
[3]Department of Civil Engineering, Mohamed Sathak Engineering College, Ramanathapuram, India

Abstract

The quest to determine the alternative sustainable material or technology that will improve our environment and simultaneously decrease the ecological footprints resulting in the development of innumerable new materials actually helps in attaining the current generation's sustainability goals. Concrete is more prevalent as construction material across the globe. Increased utilisation of concrete as construction material leads to increased demand. Carbon dioxide emission from cement is a huge global concern. Meanwhile, the abundant quantities of paper waste produced by different countries at the global level caused serious environmental problems. Hence, the current work focuses on the preparation of papercrete bricks using waste paper, agro wastes, food wastes and lime. Research was performed in order to find out the mechanical properties, compressive strength, capacity to absorb water, dry weight of the prepared papercrete bricks. Standard quality comparisons of the prepared papercrete bricks with the normal bricks were performed using different tests such as fire resistance, soundness and hardness. The cost effective papercrete bricks were exposed to 35 days sun drying and 14 days air respectively prior to the conduction of tests. Papercrete bricks are more economic, possess very low weight and are best suited for non-load bearing walls and partition walls. Papercrete bricks are more readily exposed to contaminating microorganisms. Hence, the application of silver nano particles synthesised from microorganisms was carried out on the surface of the papercrete bricks in order to enhance their quality and purity. Silver nano particles produced from microorganisms were studied using Fourier Transform Infrared Spectroscopy, X-ray Diffraction Analysis, UV-visible spectrophotometry, and Transmission Electron Microscopy. These papercrete bricks are replenishable, eco-friendly, cost effective and strong thereby resulting in the development of a much more efficient formulation of brick that could serve as an appropriate alternative construction material.

Keywords: Papercrete, compressive strength, water absorption, agro wastes, paper waste, food waste, lime

1. Introduction

Harmful greenhouse gases like carbon dioxide and methane are produced during disposal of agricultural wastes and food wastes in landfills, which contributes to climate change and global warming. Wastage of human resources and natural resources like fossil fuel, freshwater and land occurs at a larger extent [1]. The impact of paper on the environment is massive, leading to drastic changes in several industries. Harvesting of food wastes, disposable paper and wood has become inexpensive thereby leading to an increased

[a]kannankbiotech@gmail.com

DOI: 10.1201/9781003545941-12

level of consumption and waste generation. Paper production and its intensive usage pose multiple deleterious effects on the environment thereby leading to paper pollution. One important part of it is discarded paper [2]. Papercrete, a building material, is created with this problem in mind. Papercrete is a novel aggregate material composed of paper, water, and cement that was created to make an environmentally friendly home. It has been said to be a low-cost substitute material for building construction, to be lightweight, fire resistant, and to have superior absorption and thermal insulation. The paper that is utilised in the process is recyclable. Due to the significant amount of recycled components, it is thought to be an environmentally friendly material. To make it, slurry is formed by combining the drying components with water. Then, the slurry is poured into blocks and let to dry [3].

The original development of papercrete material was done 80 years ago but it has been rediscovered only recently. Waste paper and Portland cement together constitute the fibrous cementious compound, known as Papercrete [4]. Both these components are mixed in water resulting in the formation of a paper cement pulp. The prepared pulp was then transferred into a mould and subjected to drying, followed by its utilisation as a construction material of enhanced durability. Papercrete is an absolutely novel idea with very little scope. Papercrete possesses different variants such as fidobe, padobe and fibrous concrete. A mixture of water, portland cement and paper constitutes the fibrous concrete [5].

Production of Papercrete bricks is devoid of any harmful by-products or excessive usage of energy. Padobe derivative of papercrete is characterised by the absence of Portland cement. It usually consists of earth with clay, water and paper, wherein clay is used as the binding material. Earth is used in papercrete brick instead of cement. The earth should possess more than 30% clay content. Extremely high clay content in the normal brick leads to formation of cracks during drying, while incorporation of paper fibre in the earth mix leads to strengthening of the drying block. This provides flexibility thereby leading to prevention of cracking. Even though fidobe is quite similar to padobe, it contain so the fibrous materials as well [6].

Papercrete bricks are more readily exposed to contaminating micro organisms. Therefore, mixing of nanoparticles and paper, water, agro wastes, food wastes and earth with clay can be used as an advanced option to synthesise high purity as well as high quality papercrete bricks [7]. Silver nano particles have been more prevalent because of the most exquisite physicochemical properties like increased antagonistic activity against most fungal and bacterial cells caused by high reactivity. This is attributed to enhanced surface to volume ratio [8]. Silver nanoparticles possess unique characteristics which enables them to possess numerous myriad applications like wound healing, catalytic, larvicidal, anti-cancerous and antimicrobial activities [9]. Therefore, a type of efficient, novel and eco-friendly bricks called papercrete bricks can be prepared by the use of paper, water, food wastes, earth with clay & lime and by applying a thin layer of silver nanoparticles on their surface. The papercrete bricks are innovative in the sense that they are synthesised using agro wastes and food wastes, they do not undergo burning as in the case of conventional bricks and they have a superficial layer of silver nanoparticles which enhances their durability and endurance.

2. Materials and Methods

The chemicals employed in the investigation were of reagent grade quality, procured from Sigma Aldrich.

2.1. Constituents of Papercrete Bricks

2.1.1. Paper

Made of wood cellulose, paper is a naturally occurring polymer and the most prevalent organic substance on Earth. Glucose mono- meric units, a polysaccharide, make up cel- lulose. The sugar ß-D-glucose is present in several links within the cellulose chain. Despite having several hydroxyl groups, cellulose remains insoluble in water due to the stiffness of the chains and the hydrogen bonding between two OH groups on nearby chains. The creation of strong, stable crystal- line areas as a result of regular chain packing gives the bundled chains increased strength and stability. Hydrogen bonding is the basis for the strength of papercrete bricks. Apply- ing force breaks the hydrogen bond that binds the cellulose molecule to the water. By exert- ing pressure on the paper, the bond between the cellulose molecule and water is broken. When Portland cement is applied on top of cellulose fibers, a cement matrix is created, strengthening the fibers and adding more strength to the mixture.

2.1.2. Earth or Brick Earth

The brick earth used for making good quality bricks should contain the constituents such as 50% silica, 20 to 30% alumina, 5 to 10% lime, 5 to 9% iron oxide and 1% magnesium.

2.1.3. Sand

The sand particle is made up of tiny silica (SiO_2) particles. The several weather-related factors cause sand stones to break down and produce sand particles. Based on the natural resources from which the sand is obtained, sand particles are referred to as sea, river, or pit sand. Based on the size of the grains, the sand is categorised as gravel, coarse, or fine. The qualities were exam- ined in accordance with the guidelines pro- vided by the Bureau of Indian Standards (BIS). When compared to other types of concrete, the main drawbacks of fine-aggregate concrete are increased binder usage, usual creep, and increased shrinkage. M sand that was readily accessible locally and that had gone through a 4.75 mm IS screen was employed.

2.1.4. Water

Water, the primary element of papercrete, is discovered to be actively involved in the chemical interaction with cement. Water should be absolutely free of organic debris, and its pH should always be between 6 and 7.

2.1.5. Agricultural Wastes

Agricultural wastes or agro wastes are com- prised of animal wastes such as animal car- casses and manures, crop wastes such as prunings, sugarcane bagasse, culls and drops from vegetables and fruits, corn stalks, food processing wastes and toxic and hazardous agro wastes such as pesticides, insecticides and herbicides. The agricultural wastes used for the study include paddy straw, wheat straw, sugarcane stalks and corn stalks.

2.1.6. Food Wastes

Food wastes are made up of traces of inor- ganic compounds, lipids, proteins and carbo- hydrates. The composition of food wastes is dependent on its constituents and the type. Abundant carbohydrates are found in food wastes consisting of rice and vegetables whereas large quantities of lipids and pro- teins are found in food wastes containing meat and eggs. The food wastes used for the study include discarded food, food residues, vegetable wastes and market wastes.

2.1.7. Lime

The primary components of lime, an inor- ganic calcium-containing mineral, are oxides and hydroxides, most commonly calcium oxide and/or calcium hydroxide. These mate- rials are widely employed as chemical feed stocks, in engineering and construction, and in the refinement of sugar. Lime used in con- struction materials can be artificial or natural, and is often categorised as "pure," "hydrau- lic," and "poor." Depending on how much magnesium is present, lime can be classified as dolomitic or magnesium lime. The properties of the various processed lime varieties have an impact on how lime is used.

2.1.8. Preparation of Papercrete Bricks

Papercrete bricks were made by immersing large quantities of wastepaper in water overnight thereby allowing the fibers to undergo softening. Then a homogeneous pulp was obtained by agitating the mixture thoroughly. The agricultural wastes, food wastes, brick earth, water and sand are then incorporated in to the pulp and thoroughly mixed. The bricks are then prepared by transfer of the mixture in to the moulds. The bricks were ready for further testing after subjecting them to 14 days of air curing.

2.1.9. Curing of Papercrete Bricks

Curing is the act of preserving enough heat and moisture in the brick to guarantee that its hydration happens at a regular pace. Additionally, curing guarantees that the brick will remain at the proper temperature and moisture content after being transferred into the mould. This makes it possible to keep hydrating brick until the necessary qualities are sufficiently established. Brick quality will be lost if curing is skipped during the first hydration interval. Papercrete bricks can be cured using one of two techniques. They are wet curing and dry curing. Papercrete bricks made with food and agricultural wastes underwent air dry curing, a dry curing technique.

2.2. Tests Performed on Papercrete Bricks

2.2.1. Compressive Strength Test

The unevenness found in papercrete bricks is eliminated to provide two parallel, smooth faces. After that, the papercrete bricks were submerged in room-temperature water for 48 hours. This is followed by the bricks being removed and any excess moisture

being drained at room temperature. After being kept in damp jute bags for 48 hours, the bricks were submerged in clean water for four days. The specimen is carefully positioned between the testing machine's plates and sandwiched between three to four sheets of plywood, each having 4 mm thickness [7]. Application of stress occurred axially at a consistent rate of 15 N/mm2 per minute until the event failed. The maximum load which produced no enhancement in the reading of the machine was recorded as the load during failure.

The compressive strength was determined using the formula [12]:

$$\text{Compressive Strength (N/mm}^2\text{)} = \frac{\text{Maximum load failure in N}}{\text{Average area of bed faces in mm}^2}$$

2.2.2. Test to Determine Absorption of Water by the Bricks

Papercrete bricks were placed in a vented oven between 115 and 125 degrees celsius in order to dry them. The bricks were first allowed to cool to ambient temperature, and their mass was noted down. After that, submerging of the bricks in clean water was carried out for 48 hours at ambient temperature (28 ± 200 degrees C). After wiping the bricks with a damp cloth, the weight was recorded once more [7].

2.3. Microbial Synthesis of Silver Nanoparticles from Actinomycetes

After making 200 ml of Starch Casein Nitrate Broth, it was autoclaved for 15 minutes at 15 pounds per square inch to ensure sterility. The culture of actinomycetes was added to the Starch Casein Nitrate broth that had been prepared. This culture was kept alive for 120 hours at 140 rpm on a rotary shaker set at 300 degrees Celsius. Whatmann's filter paper was used to separate the cells after the incubation time. The mycelia was recovered from the filtrate by running it through distilled water. 200 ml of distilled water were used to resuspend 10 grams of the acquired wet mycelia, which were then kept on a rotary shaker for five days. The filtrate from the flask above was poured into 50 ml of aqueous 1 mM AgNO3. This was put on a shaker that rotated at 150 rpm in the dark for 30 degrees celsius. The nano particles obtained were then characterised.

2.4. *Characterisation of Synthesised Silver Nanoparticles*

Sophisticated techniques such as Energy Dispersive X-ray Detector (EDX), Transmission Electron Microscopy (TEM), Fourier Transformation Infrared Spectroscopy (FTIR) and UV Visible Spectro photometry were used to characterise the silver nano particles produced from microorganisms by exploring their constituents and accurate size [11].

2.5. *Antibacterial Activity of Synthesised Silver Nanoparticles*

2.5.1. Minimum Inhibitory Concentration (MIC)

2.5.1.1. *Tube Diffusion Method*
After being prepared, Mueller Hinton broth was autoclaved to achieve sterility. The nano

particles were made in various dilutions and transferred to the tubes holding the microorganisms. After that, the tubes were placed in an incubator at 320°C for 72 hours. Following observation, the tubes were contrasted with the control group (Christopher Woolverton et al., 2017).

2.6. *Antifungal Activity of the Silver Nano Particles*

Various strains of fungi were put onto Mueller-Hinton agar plates after they had been prepared and autoclaved to sanitise them. The produced nanoparticles were put onto a paper disc. Following three days of incubation at 320C, the plates were compared to the control (Ana Alastruey et al., 2015).

2.7. *Application of Silver Nano Particles Superficially on the Papercrete Bricks*

To improve the effectiveness of created papercrete bricks, the acquired silver nanoparticles were added to the bricks' surface.

3. Results and Discussion

3.1. *Tests Conducted on Papercrete Bricks*

3.1.1. Test to Determine Absorption of Water by the Bricks

Table 12.1: Test for absorption of water by the papercrete bricks

S. No	Type of Brick	Mass before absorption of water (W1)	Mass after absorption of water (W2)	Absorption of water in % (W2/W1)
1	Papercrete brick	1.8 kg	3.9 kg	21.7
2	Normal brick	2.86 kg	3.7 kg	12.9

3.1.2. Analysis of Compressive Strength

The compressive strength of traditional and papercrete bricks was examined. In terms of compressive strength, the papercrete

brick was inferior to the conventional brick. Numerous scholars have tested papercrete bricks in their studies. Compressive strength of papercrete bricks was between 7.2 and 7.5 N/mm², as shown by Fuller et al. [1].

Table 12.2: Analysis of compressive strength of papercrete bricks

S. No	Type of Brick	Compressive strength (N/mm²)
1	Papercrete brick	7.68
2	Conventional brick	11.58

3.1.3. Dry Weight Test

Table 12.3: Dry Weight of Papercrete bricks

S. No	Type of Brick	Dry weight (kg)
1	Papercrete brick	3.54
2	Conventional brick	3.96

3.2. *Silver Nanoparticles' Synthesis from Microorganisms–Actinomycetes*

Several actinomycetes strains grown in potato dextrose broth (PDB) were used to produce silver nanoparticles. The medium's colour changed from yellow to brown, signifying the production of nanoparticles.

3.3. *Silver Nanoparticles' Characterisation*

3.3.1. Analysis by Transmission Electron Microscopy (TEM)

Reduced form of silver nitrate solution produced by the bio-reduction process could be distinguished easily thanks to size differences,

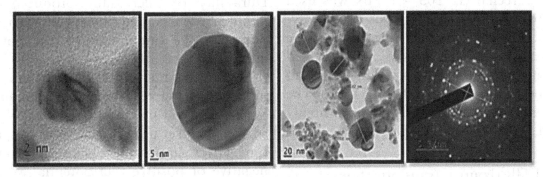

Figure 12.1: Graphical comparison of antimicrobial activity in Extract and Soxhlet extract.

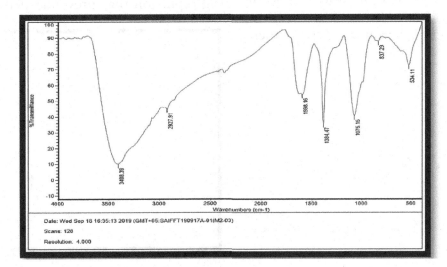

Figure 12.2: Silver nano particles characterisation using FTIR to reveal the functional groups.

Table 12.4: Illustration of FTIR results

Absorption cm⁻¹	Functional group	Compound class
3550–3200	O–H stretching	Alcohol
3000–2840	C–H stretching	Alkane
1650–1580	N–H bending	Amine
1380–1385	C–H bending	Alkane
1085–1050	C–O stretching	Primary alcohol
840–790	C=C bending	Alkene (trisubstituted)
600–500	C–I stretching	Halo compound

according to TEM analysis of the silver nanoparticles. Tecnai G2 20S – TWIN was used for transmission electron microscopic analysis. The TEM image makes it clear that the bio-reduced colloidal suspensions possessed silver nanoparticlesin the size rangeof 2 nm to 51 nm (2 nm, 10 nm, 20 nm, 51 nm). The particles had a spherical form, were highly defined, and were spaced apart.

3.3.2. FTIR (Fourier Transformation Infrared Spectroscopy) Analysis

FTIR analysis was used to characterise the produced silver nano particles and identify the various functional groups present in them.

The Perkin Elmer spectrophotometer (Model Perkin Elmer 1600 series) was utilised to investigate the FTIR spectra of the microbially synthesised silver nanoparticles. Various functional groups, such as halo compounds, alkene, alkane, amine, and alcohol, were shown by the existence of distinct peaks in the wave number vs. percentage transmittance graph.

3.3.3. Analysis of Silver Nanoparticles by EDX

The silver nanoparticles contained 70.56% silver, according to the EDX spectrum. NOVA 450 apparatus was used to carry out the EDX analysis of the produced silver nanoparticles.

3.3.4. Determination of Zeta Potential of Silver Nanoparticles

Durability of microbially produced silver nanoparticles is determined by their zeta potential. We can investigate the potential and usual stability of silver nanoparticles thanks to the intensity of zeta potential. The influence of the zeta potential technique has been linked to variations in the charge on the surface of silver nanoparticles with respect to time. Silver nanoparticles possessing immense negative or positive zeta potential are characterised by their repulsion and inability to assemble. However, the lack of the force causing repulsion that prevents such aggregation

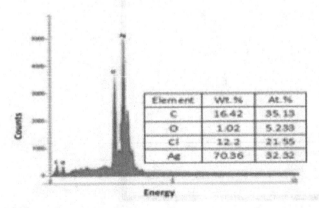

Figure 12.3: EDX Spectrum of the synthesised silver nanoparticles.

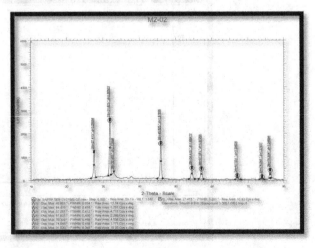

Figure 12.4: XRD analysis of microbially synthesised silver nanoparticles.

makes low zeta potential nanoparticles incapable of flocculating and aggregating. Values of zeta potential more positive than +30 mV or more negative than –30 mV are indicative of stable nanoparticles. The higher stability of actinomycetes-derived silver nanoparticles was shown by their zeta potential of –26 mV. Using a Zetasizer nano ZS particle size analyser from Malvern Instruments Limited in the UK, the zeta potential of the silver nanoparticles was examined at 250C and a 90-degree detection angle.

3.3.5. Silver Nanoparticles' Analysis by X-ray Diffraction (XRD)

A Philips PW 11/90 diffractometer was used to examine silver nanoparticles. The key distinctive peaks that correlate to the crystalline structure are clearly visible in the XRD pattern of silver nanoparticles derived from actinomycetes. Along with two unassigned peaks at 28.840 and 32.260, the XRD pattern shows large peaks at (2θ) 45.14, 53.86, 54.48, 67.86, 74.24, and 76.4 that belong to different planes. Based on the breadth of Bragg's reflection, the nanoparticle size is 25 nm. Anonymous peaks were brittle in comparison to silver. This was explained by the bioorganic chemicals that were found on the silver nanoparticles' surface.

3.3.6. Silver Nanoparticles' Analysis by UV Visible Spectroscopy

Silver nanoparticles made from microorganisms–actinomycetes were characterised by the use of a UV Visible Perkin Elmer 25 double beam spectrophotometer. The peak found in the range of 350–400 nm proved the occurrence of silver nanoparticles.

3.4. Determination of Silver Nanoparticles' Antibacterial Activity

Comparing the actinomycetes-derived silver nanoparticles to the positive control, streptomycin and the negative control, double distilled water, they demonstrated stronger antibacterial properties. The silver nanoparticles possessed higher activity towards *Staphylococcus aureus*, moderate activity towards *Bacillus subtilis*, and much lower activity

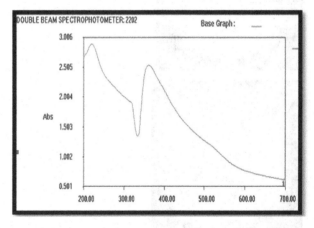

Figure 12.5: Silver nanoparticles' analysis by UV visible spectroscopy.

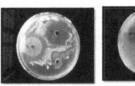

Against *S. aureus*   Against *E. coli*   Against *B. subtilis*

Figure 12.6: Silver nanoparticles' antibacterial activity.

Table 12.5: Determination of silver nanoparticles' antibacterial activity

	ZONE OF INHIBITION (mm)		
Particulars	Staphylococcus aureus	Escherichia coli	Bacillus subtilis
Silver Nanoparticles	32	19	30
Streptomycin (positive control)	30	29	26

towards *Escherichia coli*, according to the zone of inhibition.

3.5. *Determination of Silver Nanoparticles' Antifungal Activity*

Comparing the actinomycetes-derived silver nanoparticles to the positive control, fluconazole and the negative control, double distilled water, the latter demonstrated stronger

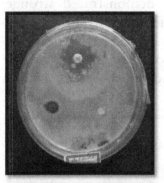

Against Candida sp.

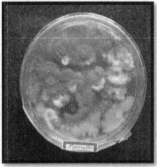

Against Penicillium sp.

Against Aspergillusniger

Figure 12.7: Silver nanoparticles' antifungal activity.

antifungal properties. The silver nanoparticles' zone of inhibition showed that they were most active against *Candida* sp., somewhat active against *Penicillium* sp., and significantly less active against *Aspergillus niger*.

3.6. *Application of Silver Nanoparticles on the Papercrete Bricks*

Using a brush, the actinomycetes-derived silver nanoparticles were applied to the papercrete bricks' surface.

3.7. *Cost of Papercrete Bricks*

Papercrete brick is more cost-effective than traditional brick, costing just Rs. 4, as opposed to Rs. 8 for regular brick. The market still does not have papercrete bricks.

3.8. *Discussion*

In direct sunshine, the exceptional insulating qualities of papercrete bricks were shown. Over the past few years, there has been a noticeable growth in demand for bio-based papercrete bricks with improved self-insulating qualities [12,13]. Lime is used as a binding element along with food and agricultural wastes in the construction of bricks [14]. Good binding and insulating qualities can be found in papercrete bricks made from food and agricultural wastes [15].

4. Conclusion

A novel kind of recyclable biomaterials created from food and agricultural wastes are called papercrete bricks. In order to solve the shortcomings offered by cemented bricks, papercrete bricks can be utilised as an alternate building material. These bricks are

Table 12.6: Determination of silver nanoparticles' antifungal activity

| Particulars | ZONE OF INHIBITION (mm) | | |
	Candida sp.	Penicillium sp	Aspergillus niger
AgNO$_3$ Nanoparticle	15	13	10
Streptomycin (positive control)	22	-	-

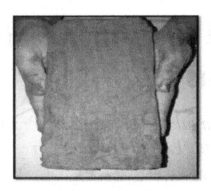

Figure 12.8: Papercrete bricks with a surface layer of silver nanoparticles.

affordable and undoubtedly helpful for building projects. Tables, chairs, and insulating boards are also made using them. Papercrete bricks have the potential to revolutionise the construction business.

Compared to regular bricks, they are rather lightweight. They can be used to build interior partition walls and non-load bearing walls. However, they are not appropriate for exterior or wet walls. Due to the papercrete bricks' reduced weight, the building's dead load may be reduced [16]. Applying silver nanoparticles made from microorganisms topically may improve the papercrete bricks' purity and fend off bacterial and fungal contamination. Therefore, a low-cost, extremely pure, environmentally friendly, and highly effective papercrete brick might be utilised as a substitute for concrete.

References

[1] Fuller, B J, Fatifis, A., and Santamaria, J. L. (2006). The paper alternative. *Civ. Eng. Mag. Arch.*, 76 (5), 72–78.

[2] Rohit Kumar, A. and Rajeev, K. (2016). Utilization of waste papers to produce eco-friendly bricks. *Int. J. Sci. Res.*, 5 (8), 92–96.

[3] Shewit, B., Mikyas, M., and Werku, K. (2017). Experimental study on some mechanical properties of papercrete concrete. *Adv. Mater.*, 6 (1), 1–6.

[4] Gallardo, R. S. and Adajar, M. A. (2006). Structural performance of concrete with paper sludge as fine aggregates partial replacement enhanced with admixtures. *Symposium on Infrastructure Development and the Environment.* 2006.

[5] Gunarto, A., Satyarno, I., and Tjokrodimuljo, K. (2008). Newsprint Paper Waste Exploiting for Papercrete Panel. *Institute of Research Center, Gadjah Mada University.*

[6] Malthy, R. and Jegatheeswaran, D. (2011). Comparative study on papercrete bricks with conventional bricks. *ICI J.* January–March 2011.

[7] Yun, H., Jung, H. and Choi, C. (2007). Mechanical properties of papercrete containing waste paper. Architectural Institute of Korea. *18th International Conference on Composite Materials.* 2007.

[8] Slawson, R., Van Dyke, M. I., Lee, H., and Trevors, J. T. (1992). Germanium and silver resistance, accumulation and toxicity in microorganisms. *Plasmid.* Elsevier. 27 (1)72–79.

[9] Firdhouse, J. M. and Lalitha, P. (2015). Biosynthesis of silver nanoparticles and its applications. *J. Nanotech.*, 2015 (1), 1 – 18. DOI: 10.1155/2015/829526.

[10] IS3495 – 1 to 4. (1992). Method of test of burnt clay building bricks,1. 1992.

[11] Ravichandran, V., Tiah Zi, X., Subashini, G., Tarrence Foo Wei, Z., Eddy F., Chou Y., Nelson J., and Sokkalingam, A. D. (2010). Biosynthesis of silver nanoparticles using mangosteen leaf extract and evaluation of their antimicrobial activities. *J. Saudi Chem. Soc.*, 15, 113–120.

[12] Moser, F., Trautz, M., Beger, A. L., Lower, M., Jacobs, G., Hillringhaus, and Reimer, J. (2017) Fungal mycelium as a building material. In *Proceedings of IASS Annual Symposia* (vol. 2017, no. 1, pp. 1–7).

[13] Dahmen. J. (2017). Soft futures: mushrooms and regenerative design. *J. Arch. Edu.* 71 (1), 57–64.

[14] Shewit, B., Mikyas, M., and Werku, K. (2017). Experimental study on some mechanical properties of papercrete concrete. *Adv. Mater.*, 6 (1), 1–6.

[15] Fuller, B J, Fatifis, A., and Santamaria, J. L. (2006). The paper alternative. *Civ. Eng. Mag. Arch.*, 76 (5), 72–78.

[16] Joo-Hong, C., Byoung-Hoon, K., Hyun-Ki, C., and Chang-Sik, C. (2015). Development of papercrete due to paper mixing ratio. *Int. Conf. Sustain. Build. Asia.*, 317–320.

13 Bioabsorbable scaffold for bone tissue engineering applications

Roshni Ramchandran[a], Ananya N. Nayak, Keerthana D. Praveen, Anusha S. Shetty, Divyashree T. K., and Lavanya Bhushan

Department of Biotechnology, Ramaiah Institute of Technology, Bengaluru, India

Abstract

The development of biological replacements to restore tissue function is the goal of tissue engineering. Bone tissue engineering is a branch of tissue engineering used to restore bone defects caused by malignancies, congenital diseases and accidents. Implantable scaffolds can facilitate the cellular growth, thereby enhancing endogenous tissue regeneration. Currently various metallic and steel substitutes are used as scaffolds for bone replacements. These implants connect well with the bones hence extensively used in bone replacement and tooth implants Over time this becomes corrosive and requires removal from body to avoid unnecessary toxicity also the Young's modulus of titanium and steel rods are higher than that of natural bone which directly reduces the degree of bone resorption around the prosthesis during daily activities, prompting the need for a better suited material for scaffold fabrication.

The ability of gel formation and its biocompatibility, alginate, an anionic polymer, is becoming an ideal candidate in the field of bone tissue engineering. Alginate scaffolds have shown to have better properties like porosity, cell adhesion capability, biocompatibility and also enables a better differentiation of stem cells into osteogenic lineage. Hydroxyapatite is an important bio ceramic that finds application in medical and dental field as it is a calcium-phosphate compound that naturally occurs in the human body.

In this study, nanocomposite scaffold was synthesized using nano-hydroxyapatite loaded alginate scaffold using lyophilization technique. The prepared nanocomposite scaffolds will be characterized using XRD. Furthermore, an assessment was conducted on the composite scaffolds' swelling, degradation, in order to determine whether or not they were a good candidate for tissue engineering applications.

Keywords: Alginate, hydroxyhapatite, nanocomposite, scaffold, Tissue engineering, Bone tissue engineering

1. Introduction

The multidisciplinary discipline of tissue engineering uses concepts from biology and engineering to create biologic substitutes that can protect, boost, or restoration tissue function. The primary role of tissue engineering is to address the limitations of traditional medical treatments by creating functional tissues and organs. [1] Bone Tissue Engineering uses techniques and strategies for the regeneration of damaged or diseased bone using a combination of cells, growth factors, and biomaterial and has various applications including for treatment of bone defects due to trauma, disease, or congenital abnormalities, to create biological substitutes for bone grafts. Scaffolds are three-dimensional (3D) porous biomaterials that provide adequate mechanical support, and biochemical cues for cell growth and tissue remodelling. Alongside features like mechanical strength, porosity, and biocompatibility, scaffolds carrying biomaterials for bone tissue engineering have the capacity to boost bone regeneration and may disintegrate after a specific amount of time following implantation. [2–5,15] Alginate is one of the naturally occurring polymers, consisting

[a]roshniramu@msrit.edu

DOI: 10.1201/9781003545941-13

of anionic polysaccharides mainly found in brown seaweeds. Alginate is made from two copolymers, glucuronic acid, and mannuronic acid which provide strength and flexibility. The use of alginate hydrogels as biodegradable and biocompatible materials has gained widespread acceptance for their role in wound healing, drug delivery, and tissue engineering applications. Use of Alginate scaffolds however, is still far from ideal due to its poor mechanical properties especially as a bone tissue engineering scaffold. With the formation of new bone tissue, the scaffold is expected to withstand the weight-bearing capacity of the bone which polymeric scaffold like alginate has not been able to achieve. [6,7] One of the most important bioceramic used for Bone Tissue Engineering is Hydroxyapatite (HA), as it is the main inorganic constituent of hard tissues like bone and teeth. With its exceptional mechanical, chemical, biological, and physical qualities, nano-hydroxyapatite (nHAp) is a relatively new material that provides a reliable framework for the engineering of artificial bone substitutes. Due to its capacity to directly connect with bone mineral and to stimulate the production of new bone through osteo-conduction, this material is highly sought after in the areas of odontology and traumatology. Alginate can be reinforced with hydroxyapatite in alginate/Hap composite scaffolds as an osteoconductive element and inorganic reinforcement.

Hydroxyapatite (HAp), $Ca_{10}(PO4)6(OH)_2$, has a chemical composition similar to that to human bone tissue making it a favourable implant material. The most favourable features of this material is the excellent biocompatibity [8,9]. With its robust osteoconductivity, high biocompatibility, and biological activity, it is one of the biomaterials that is anticipated to be used in bone tissue engineering and regeneration [10–12]. NanoHAp (1–100nm in size) enables better close contact with surrounding tissues [13,14]. However the n-HAP's a brittle structure makes it a lesser promising candidate to

be used alone for Bone reconstruction, however using it in combination with various biocompatible biomaterials can open the avenue for fabrication of nano composite biomaterials for bone tissue engineering [15,16,25]. In this study we will be fabricating alginate nanohydroxyapatite composite scaffolds to evaluate its candidature as a tissue engineering scaffold

2. Materials and Methods

2.1. Materials

Sodium Alginate salt ($NaC_6H_7O_6$) was purchased from Nice Chemicals Pvt. Ltd. Calcium Chloride ($CaCl_2$) and anhydrous Monosodium phosphate (NaH_2PO_4) was purchased from Central Drug House Pvt. Ltd. Anhydrous Disodium phosphate (Na_2HPO_4), Sodium Chloride (NaCl), Di-Ammonium Phosphate($(NH_4)_2HPO_4$), Calcium hydroxide ($Ca(OH)_2$) and Sodium Bicarbonate (Na2HCO3) was purchased from Priya Research Labs. Potassium Chloride (KCl), Potassium Hydrogen Phosphate Trihydrate ($K_2HPO_4·3H_2O$), Magnesium Chloride ($MgCl_2·6H_2O$) and Hydrochloric acid (HCl) was purchased from Avra Synthesis Pvt.Ltd. Sodium Sulphate (Na_2SO_4), tris(hydroxymethyl)aminomethane $[(CH_2OH)_3CNH_2]]$ was purchased from S.D.Fine- Chem Pvt.Ltd.

2.2. Preparation of Nanohydroxyapatite Particles

1M calcium chloride solution was prepared by adding 7.35g of Calcium chloride in 50ml distilled water and stirred using magnetic stirrer. This mixture was then stored in freezer until it forms a chilled solution. For preparation of nanohydroxyapatite particles, 0.37g of Di-Ammonium phosphate and 0.52g of Calcium chloride was weighed and added in 10ml of distilled water each in separate beakers. The solutions were stirred separately using Magnetic Stirrer until homogenous solution was obtained and both the solutions were mixed together. The beaker was put in a microwave until all the moisture content was removed and the beaker was placed in a Hot Air Oven at 40°Celsius till fine powder was obtained. The powder was scraped out and collected [18–20].

2.3. *Preparation of Alginate/nHAp Composite Scaffolds*

Sodium Alginate solution of 4%(w/v) and 6%(w/v), was prepared. Then 0.1g of previously prepared nHAp was added to each of the above solutions until uniform dispersion was obtained. The solutions were then transferred into separate petri plates and approximately 2ml CaCl2 solution was added to each under continuous stirring conditions until a solid mass of gel was obtained. The 4% and 6% gels were pre-freezed at –20° Celsius overnight. These frozen composite gels were lyophilized in freeze dryer for 24hrs to obtain the required scaffolds [21].

2.4. *Characterization*

The structural morphology of the composite scaffold was examined using X-ray Diffraction studies (XRD). XRD patterns of the composite scaffolds were analyzed at room temperature using a Panalytical diffractometer (XPERT PRO powder). (Cu K_radiations) operating at a voltage of 40 kV. XRD was taken 2_angle range of 5–60° and the process parameters were: scan step size 0.02 and scan step time 0.05 s.

2.5. *Swelling Studies*

Swelling capacity and limits of these scaffolds were carried out in PBS at pH 7.4. The prepared gels (4% and 6%) were weighed (Wo) and divided approximately into three equal sections and was labelled as A, B and C. Each of the sections was submerged in PBS solution separately. The wet weight of each section (Ww) was noted on Day 0, 2 and 4 and the swelling ratio was calculated as per the equation given below (13.1) (21)

$$\text{Swelling Ratio} = (Ww - Wo)/Wo \qquad (13.1)$$

2.6. *In vitro Degradation Studies*

The degradation trend of the composite scaffolds was performed using PBS at pH 7.4 along with Trypsin. A section of pre-weighed gels was weighed (Wo) and submerged in PBS solution. 10ul of Trypsin was added to the above solution. The wet weight of the section was noted on Day 0, 4, and 8(Wt) and the percentage degradation was calculated using the equation given below (13.2) [21]

$$\text{Degradation \%} = (Wo - Wt)/Wo * 100 \qquad (13.2)$$

2.7. *Cytocompatibility of the Scaffolds*

The colorimetric MTT test was applied to assess the cytocompatibility of the cells following interaction with the scaffolds.. The MTT assay determines the extent to which viable cells can cause the reduction of the tetrazolium of MTT. The assay was carried out on, MG 63 (Osteosarcoma cell lines), L929 (Mouse fibroblast cell lines), and POB (Primary Osteoblast cells) were cultured under conventional culturing conditions Following a seeding density of 104 cells per well into 96-well plates. Under aseptic circumstances, the scaffolds were left immersed in full medium for 48 hours at 37°C with agitation. The medium containing the leachable was collected in a falcon tube following the incubation period. After 24 hours, the extract (media containing the leachables) was added to the culture media of the seeded cells. For a whole day, the cells were cultured on the extract. Following the incubation period, the extract was replaced with new medium containing 10% MTT solution. After that, the plates were incubated for four hours at 37°C in a humidified atmosphere. After removing the medium, each well received 100μl of the solubilization buffer, which included Triton-X 100, 0.1N HCl, and isopropanol, to dissolve the formazan crystals. At 570 nm, the absorbance was determined using a Biotek microplate reader. Utilized were leachables from alginate scaffolds cultured in a medium [21–23].

3. Results and Discussions

3.1. *Characterization*

3.1.1. XRD Characterization

The alginate and nanohydroxyapatite were characterized by XRD studies. The XRD spectrum of alginate is shown in Figure 13.1 nano Hap is shown in (Figure 13.1) and the composite scaffold of alginate with

nano-hydroxyapatite is shown in Figure 13.1. Alginate being a polymer is showing amorphous peaks in XRD and Nanohydroxyapatite shows the characteristic peak at 32.6Θ. (JCPDS No 09-0432). This peak is shown in XRD patterns of both NanoHap samples as well as the composite scaffold that indicates complete and homogenous mixing of nano Hap with alginate scaffold.

3.2. *Swelling Studies*

According to research on in vitro cell culture, where swelling and pore size increases facilitate three-dimensional cell attachment and growth, a scaffold's capacity to swell facilitates the three-dimensional migration of cells into scaffolds. (2.3) The weight of the scaffolds (both 4% as well as 6%) taken on specified days. (Figure 13.2) shows swelling capacity of the scaffolds, increase in a linear pattern till day 4 and then stabilizes after day 4 upon reaching its maximum swelling capability. The swelling ratio calculated as per equation-1 gives 19.7 for 4% Alginate nano-HAp scaffold and 20 for 6% Alginate nano-HAp scaffold.

3.3. *Degradation Studies*

Degradation trend of the scaffolds is a predominant characteristic of a scaffold that is used for tissue engineering applications. For the newly produced tissue to properly function, the scaffolds should ideally break down as new tissue forms [17,22]. The in vitro

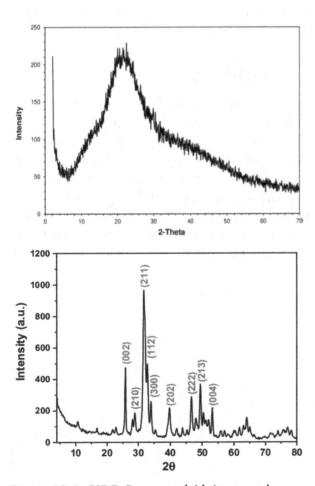

Figure 13.1: XRD Pattern of Alginate and (b) XRD Pattern of nHAp particle.

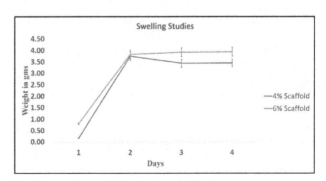

Figure 13.2: Swelling graph of Alginate nano Hap Scaffold.

degradation of the alginate/nHAp composite scaffold when it is submerged in PBS containing trypsin is demonstrated in this study. The initial weight was calculated on Day 0 and subsequent weight of sample places in PBS with trypsin was noted on Day 4 and 8. Figure 13.4 shows the degradation pattern of both 4% as well as 6% alginate nano-HAp scaffolds. As seen from the graph there is a uniform dip in the weight of the scaffolds till day 5 with a downward trend. This steady degradation of the composite scaffold ensures that the newly formed tissue will have the support of the scaffold for a period of 4 weeks which will be sufficient for the new cells to form a tissue and start functionalisation of the tissue. Post this time frame the scaffold will degrade

completely for the newly formed tissue to function effectively filling the defect [21].

3.4. Cytocompatibility Studies

It has already been demonstrated that alginate is a biocompatible polymer [22–23]. Using an MTT assay akin to the previously described methodology, we evaluated the cytocompatibility of the alginate/nano-hydroxyapatite scaffolds in this study. According to Figure 13.5's results, 93% of the various cell lines were compatible with the nanocomposite scaffolds following a 48-hour incubation period. Comparing the alginate scaffolds used as reference to the alginate/nanohydroxyapatite composite scaffolds, this result implies that there are no appreciable hazardous leachables. These findings indicate that all varieties of osteoblastic lineage cells are biocompatible with the nanohydroxyapatite-Alginate composite scaffold.

4. Conclusions

Nanocomposite scaffold was prepared using nanohydroxyapatite and alginate gel (varying concentrations) and characterized for tissue engineering applications, these scaffolds demonstrated the optimum levels of, swelling capabilities, and degradation rate, needed for effective cell adhesion and proliferation. Cytotoxicity tests demonstrated that the nanocomposites are safe for use with a

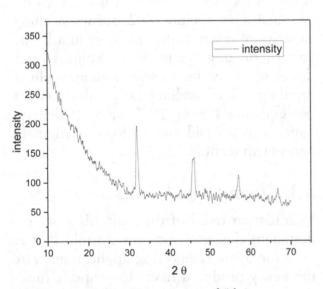

Figure 13.3: The XRD pattern of Alginate nanohydroxyapatite scaffold.

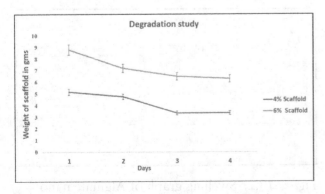

Figure 13.4: Degradation graph of Alginate nano Hap Scaffolds.

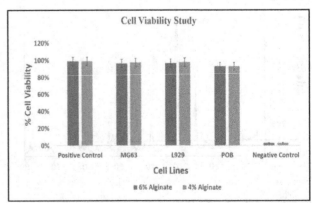

Figure 13.5: Cell Viability Studies using different cell lines.

variety of cell lines, including POB, L-929, and MG-63 cells. These studies suggested that the fabricated nanocomposite scaffold would be a scaffold candidate for bone tissue regeneration.

References

[1] Zmora, S., Glicklis, R., and Cohen, S. (2002). Tailoring the pore architecture in 3-Dscaffolds by controlling the freezing regime during fabrication. *Biomaterials*, 23, 4087–4094.

[2] Shanmugasundaram, N., Ravichandran, P., Reddy, P. N., Ramamurty, N., Pal, S., and Rao, K. P. (2001). Collagen–chitosan polymeric scaffolds for the in vitro cultureof human epidermoid carcinoma cells. *Biomaterials*, 22(14), 1943–1951.

[3] Boccaccini, A. R., Blaker, J. J., Maquet, M., Chung, W., Jerome, R., and Nazhat, S. N. (2006). Poly (dl-lactide) (PDLLA) foams with TiO2 nanoparticles and PDLLA/TiO2—Bioglass foam composites for tissue engineering scaffolds. *J. Mater. Sci.*, 41, 3999–4008.

[4] Dalby, M. J, McCloy, D., Robertson, M., Wilkinson, C. D. W., and Oreffo, R. C. (2006). Osteoprogenitor response to defined topographies with nanoscale depths. *Biomaterials*, 27, 1306–1315.

[5] Torres, F. G., Nazhat, S. N., Sheikh, M. D., Fadzullah, S. H., Maquet, M., and Boccaccini, A. R. (2007). Mechanical properties and bioactivity of porous PLGA/TiO2nanoparticle-filled composite for tissue engineering scaffolds. *Compos. Sci. Technol.*, 67, 1139–1147.

[6] Majeti, N. V. and Ravi Kumar. (2000). A Review of Chitin and Chitosan Applications.

[7] Sutherland, I. W. (1991). Novel materials from biological sources. In David Byrom (ed.), *Reactive and Functional Polymer*, Volume 46, Biomaterials Stockton Press, New York, 1–27.

[8] Monroe, E., Bass, D., Votava, W., Mc Mullen, J. (1971). New Calcium Phosphate Ceramic Material for Bone and Tooth Implants. *J. Dent. Res.*, 50, 860–861.

[9] Jarcho, M., Bolen, C. H., Thomas, M. B., Bobick, J., Kay, J. F., and Doremus, R. H. (1976). Hydroxylapatite synthesis and characterization indense polycrystalline form. *J. Mater. Sci.*, 11, 2027–2035.

[10] Rodrigues, C., Serricella, P., Linhares, A., Guerdes, R., Borojevic, R., Rossi, M., Duarte, M., and Farina, M. (2003). Characterization of a bovinecollagen–hydroxyapatite composite scaffold for bone tissue engineering. *Biomaterials* 24, 4987–4997.

[11] Zhu, Y., Cao, N., Zhang, Y., Cao, G., Hao, C., Liu, K., Li, X., and Wang, W. (2022). The Ability and Mechanism of nHAC/CGF in PromotingOsteogenesis and Repairing Mandibular Defects. *Nanomaterials*, 12, 212.

[12] Du, Z., Feng, X., Cao, G., She, Z., Tan, R., Aifantis, K. E., Zhang, R., and Li, X. (2020). The effect of carbon nanotubes on osteogenic functionsof adipose-derived mesenchymal stem cells in vitro and bone formation in vivo compared with that of nano-hydroxyapatite and

[13] the possible mechanism. *Bioact. Mater.*, 6, 333–345.

[14] Shahrezaie, M., Moshiri, A., Shekarchi, B., Oryan, A., Maffulli, N., Parvizi, J. (2017). Effectiveness of tissue engineered three-dimensional bioactive graft on bone healing and regeneration: An *in vivo* study with significant clinical value. *J. Tissue Eng. Regen. Med.*, 2018 Apr;12(4):936–960.

[15] Kubasiewicz-Ross, P., Hadzik, J., Seeliger, J., Kozak, K., Jurczyszyn, K., Gerber, H., Dominiak, M., and Kunert-Keil, C. (2017). New nanohydroxyapatite in bone defect regeneration: A histological study in rats. *Ann. Anat. Anat. Anz.*, 213, 83–90.

[16] Chen, X., Wu, D., Xu, J., Yan, T., and Chen, Q. (2021). Gelatin/Gelatin-modified nano hydroxyapatite composite scaffolds with hollow channel

[17] arrays prepared by extrusion molding for bone tissue engineering. *Mater. Res. Express*, 8, 015027.

[18] Bal, Z., Korkusuz, F., Ishiguro, H., Okada, R., Kushioka, J., Chijimatsu, R., Kodama, J., Tateiwa, D., Ukon, Y., Nakagawa, S., et al. (2021). A novel nano-hydroxyapatite/synthetic polymer/bone morphogenetic protein-2 composite for efficient bone regeneration. *Spine J.*, 21, 865–873

[19] Mikos, A. G., Sarakinos, G., Lyman, M. D., Ingber, D. E., Vacanti, J. P., and Langer, R. (1993). Pre-vascularisation of porous biodegradable polymers. *Biotechnol. Bioeng.*, 42, 716–723.

[20] Rameshbabu, N., Prasad Rao, K., Sampath Kumar, T. S. (2005), Acclerated microwave processing of nanocrystalline hydroxyapatite. *J. Mater. Sci.*, 40, 6319–6323

[21] Basha, R. Y., Sampath Kumar, T. S., and Doble, M. (2015). Design of biocomposite materials for bone tissue regeneration *Mater. Sci. Eng.*, 57, 452–463.

[22] Sampath Kumar, T. S., Siddharatan, A., Seshadri, and S. K. (2004). Microwave accelerated synthesis of nanosized calcium deficient hydroxyapatite *J. Mater. Sci.: Mater. Med.*, 15, 1279–1284

[23] Divya Rania, V. V., Roshni Ramachandrana, Chennazhia, K. P., Tamurab, H., Naira, S. V., and Jayakumar, R. (2011). Fabrication of alginate/nanoTiO$_2$ needle composite scaffolds for tissue. *Carbohydr. Polym.*, 83 (2011) 858–864.

[24] Vacanti, J. P. and Langer, R. (1999). Tissue engineering: the design and fabrication of living replacement devices for surgical reconstruction and transplantation. *Lancet*, 354, 32–34.

[25] Zmora, S., Glicklis, R., and Cohen, S. (2002). Tailoring the pore architecture in 3-D scaffolds by controlling the freezing regime during fabrication. *Biomaterials*, 23, 4087–4094.

14 Efficacy of dye removal using mesoporous silica nanoparticles synthesised from coconut husk

R. Jayasree[a], *M. Monica, and J. Irine Briny Hepzibha*

Department of Biotechnology, Rajalakshmi Engineering College, Chennai, India

Abstract

The increasing focus on utilising waste materials and byproducts has spurred research across various disciplines, including ecology, science, economics, technology, and society. Coconut husk, which is abundant in silica and readily accessible through coir industries worldwide, has emerged as a sustainable resource. The controlled combustion of coconut husk produces coconut husk ash, a crucial material for deriving silica gels via alkaline treatment and acid precipitation methods. This study specifically targets the synthesis of mesoporous silica nanoparticles using cetyltrimethylammonium bromide as a surfactant, with a primary focus on its efficacy in removing dyes from aqueous solutions. Methylene blue, chosen as a model dye due to its prevalence in industrial wastewater, serves as a benchmark for evaluating nanoparticle adsorption capabilities. The developed mesoporous silica nanoparticles were analyzed structurally and chemically using Fourier transform infrared spectroscopy (FT-IR), energy dispersive X-ray spectroscopy (EDX), and scanning electron microscopy (SEM). Additionally, systematic experiments manipulating contact time, pH, and dye concentration were carried out using a UV-visible spectrophotometer to optimise dye adsorption conditions. The results unequivocally showcase the remarkable efficiency of mesoporous silica nanoparticles in adsorbing methylene blue from aqueous solutions, highlighting their potential as highly effective adsorbents for dye removal processes. This study not only introduces a novel application of waste-derived materials but also underscores the crucial role of sustainable practices in mitigating environmental pollutants. This research provides a foundation for developing sustainable methods for wastewater treatment and pollution control through the utilisation of innovative green materials, aligning with global sustainability goals and promoting environmental stewardship and resource conservation.

Keywords: Waste utilisation, coconut husk ash, mesoporous silica nanoparticles, dye removal, sustainable practices

1. Introduction

The sustainability of life on Earth depends on crucial resources such as water, air, and soil, with water playing a central role due to its significance for human survival and ecosystem health [1]. However, increasing global water pollution necessitates continuous assessments of water policies to mitigate severe consequences, including disease outbreaks resulting in approximately 14,000 daily fatalities worldwide [2]. Various factors, including climate, soil composition, vegetation, and human activities such as agriculture, urbanisation, and industrial processes, contribute to shaping water quality [3]. Notably, industrial and municipal pollution, coupled with nonpoint sources such as mining, agriculture, and urban expansion, poses significant threats by introducing pollutants into the marine environment. The coconut palm (Cocos nucifera) holds a

[a]jayasreerohith@gmail.com

DOI: 10.1201/9781003545941-14

special status as the "tree of life," owing to its multifaceted contributions to human life and industry across tropical Southeast Asia and the Indian-Pacific islands [4]. Every part of the coconut palm, from its trunk to its fruit, serves valuable purposes, making it an indispensable resource for various human needs. For instance, coconut husk, previously considered a byproduct, has applications in agriculture and industry, including in carpets, geotextiles, ropes, and plant growth support media [5]. Coconut water, which is renowned for its isotonic properties, is also a widely consumed beverage that contributes to economic activities in countries such as Brazil and the Philippines through the craft industry. Due to its unique structural features, it possesses vast surface dimensions adjustable pore size, and superior adsorption capacity, mesoporous silica nanoparticles (MSNs) present promising opportunities for dye removal applications [6]. Mesoporous silica nanoparticles (MSN) synthesis using coconut husk not only addresses waste management challenges but also presents an eco-friendly solution for water treatment in alignment with sustainable development goals. This introduction highlights the importance of adsorption as an eco-friendly approach for dye removal from industrial wastewater, focusing on utilising agricultural waste as an effective adsorbent. High durability against abrasion, thermal endurance, and small pores to boost surface area and adsorption capacity are requirements for effective adsorbents [7]. These adsorbents are chosen according to their particular dye types, wastewater conditions, and intended results. They are categorised into groups such as carbon-based substances, oxygen-containing compounds, and polymer-based compounds [8,9]. Efforts to develop efficient dye removal methods have explored physical, chemical, and combined approaches [10]. Interestingly, the application of agricultural waste materials as environmentally friendly and sustainable adsorbents has gained traction, addressed problems with dye removal, and enhanced the value of waste products

[11]. This approach not only addresses environmental issues but also aligns with global sustainability goals. The novelty of this research lies in utilising waste-derived materials for advanced adsorbent development, contributing to green materials and pollution control technologies and offering promising avenues for sustainable wastewater treatment practices [12]. Thus, this study explored the potential of coconut husk waste in nanoparticle synthesis and dye removal, contributing to the evolving field of sustainable materials and pollution mitigation technologies.

2. Materials and Methods

2.1. Synthesis of Mesoporous Silica for Environmental Remediation

Coconut husks from a local Tamil Nadu marketplace were processed rigorously to create mesoporous silicon nanoparticles (MSNs). After giving the husks a good wash with water that was distilled to remove any impurities, they were dried out by baking them for six hours. After the husks were cleaned, they were chopped finely and burned for three hours at 400°C in a furnace. The ash was then allowed to air dry and weighed exactly. Boiling 30 g of ash in 1 L of 1 M NaOH solution until the ash was totally dissolved yielded a sodium silicate solution. Silica was created by this procedure. After the residue was filtered out, 1 M HCl and 1 M H_2SO_4 were added to the filtrate to lower its pH to 7. After inducing gelation, a xerogel was created. It was then cleaned to get rid of contaminants, aged for eighteen hours, centrifuged, and dried for 11 hours at 80°C. The resulting 6 g of coconut husk silica powder was treated with 100 ml of 1 N NaOH to produce a sodium silicate solution. After that, a 2% CTAB solution was gradually added to this solution to provide an alkaline pH of 10–11. pH neutralisation with 5 M CH3COOH resulted in the formation of a white precipitate comprising MSNs and maintained the pH at 7–8 [13–15].

2.2. *Characterisation Studies*

2.2.1. Energy Dispersive X-ray Spectroscopy

Energy-dispersive X-ray spectroscopy (EDX), a chemical analysis technique associated with electron microscopy, provides information about the composition of samples using characteristic X-rays. Based on the energy and peak position of every component in the spectrum—the peak region representing the number of molecules—each element is precisely identified. This technique was applied to thoroughly analyze the elemental composition of granulated mesoporous silicon dioxide nanoparticles [16].

2.2.2. FTIR Analysis

Molecular data on the material was obtained through the application of an infrared Fourier-transformed (FT-IR) spectrometer. It makes it possible to analyze substances qualitatively and quantitatively using the appropriate standards. Using a JASCO FTIR-4700 instrument, samples combined with KBr were examined and spectra were acquired between 4000 and 400 cm-1. The device is capable of measuring samples in the upper layer or bulk up to approximately 11 mm in diameter [13].

2.2.3. Scanning Electron Microscopy

Using electron scanning microscopy (SEM), the precise morphology and structure of the nanoparticles were investigated. The Hitachi S4500 scanning electron microscope was used to evaluate the images from the SEM of the silica samples, which were obtained using Phenom Pro at a predetermined magnification. A proper amount of material was dropped onto carbon-coated copper grids, and any surplus was removed using paper to make thin sample films. To guarantee optimal imaging conditions, the grids containing sample films were dried out under the light of a mercury-filled light bulb for five minutes [17].

2.2.4. Batch Adsorption

In batch studies, variables like adsorbent dosage (0.5–2.5 g/L), pH (3–11), contact time (30–150 minutes) and dye concentration (10–50 mg/L) were varied to remove dyes from water. Optimising the environment for optimal dye adsorption was the aim. 50 ml of the dye solution was placed in Erlenmeyer flasks, and the resulting mixture underwent shaking for varying amounts of time. The formula used to calculate the removal efficiency was $E = (Co - Cf)/Co$, where UV-Vis spectroscopy was utilised to estimate the dye concentration. The results are expressed as the proportion of dye removed from the solution [18].

2.2.5. UV-Visible Spectrophotometer

A device used for determining light intensity after it passes through a sample is a UV-Vis spectroscopy. The absorbance of the methylene blue and crystal violet aqueous solutions was measured using this apparatus both before and after adsorption. At a wavelength of 655 nm for methylene blue and 550 nm for crystal violet, the absorbing capacity of the aqueous solution was measured. These numbers can be used to estimate the dye elimination proportion [19].

3. Results

3.1. *Synthesis of Mesoporous Silicon Dioxide Nanoparticles*

Different agricultural residues had different concentrations of silica. To synthesise silica from husks, coconut husk ash (CHA) was utilised as the initial material. The process yielded approximately 1.1 grams of silica from approximately 50 grams of coconut husks, with the CHA generating approximately 3 grams of ash. This synthesis pathway resulted in a silica content of approximately 1.1 weight % and an oxygen content of 18.5 weight %, both of which are significant fractions of the total coconut husk ash (CHA) weight [20]. About 1.1 grams of silica from the synthesis from the ash of coconut husk significantly aids in the production of mesoporous silicon dioxide nanoparticles (MSNs) at an amount of 3.6 weight percent silica and 38.1 weight percent oxygen.

3.2. EDAX Analysis of Mesoporous Silica Nanoparticles

The composition of the chemicals of the silica-coated ferrous magnetic nanoparticles (Figure 14.1) was investigated using energy-dispersive X-ray spectroscopy (EDX). The presence of silica-coated ferrous magnetic nanoparticles was confirmed by the EDX results, which showed peaks linked to silica, oxide, and ferrous elements. Specifically, the study revealed that the oxide content was

40.1% and the silica level was 8.2%. These values collectively account for 100% of the total elemental composition, according to a previous study [21], suggesting a high degree of sample purity devoid of identifiable elemental contaminants.

3.3. FTIR Characterisation

The FTIR spectrum examination of the produced mesoporous silica nanoparticles, as illustrated in Figure 14.2, revealed unique peaks at various wavenumbers. The peaks located at 618.08 cm^{-1}, 796.14 cm^{-1}, and 960.96 cm^{-1} reflect the bending vibrations of Si-O-Si (siloxane) bonds, whereas the peak at 1054.72 cm^{-1} represents the stretching vibrations of Si-OH (silanol). Additionally, the peak at 1479.12 cm^{-1} indicates CH bending, while

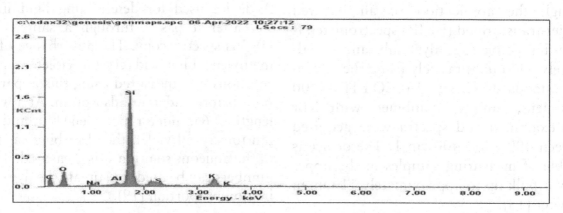

Figure 14.1: Mesoporous silica nanoparticles: an analysis using energy-dispersive X-ray spectroscopy.

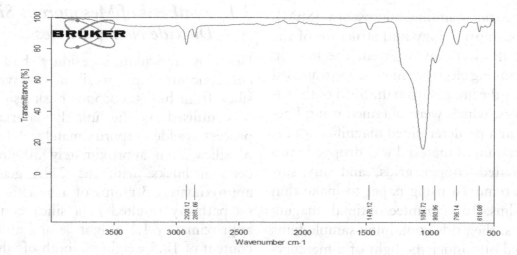

Figure 14.2: FTIR spectra of mesoporous silica nanoparticles (MSNs).

the peaks at 2851.06 cm⁻¹ and 2920.12 cm⁻¹ reflect CH asymmetric bending vibrations. Notably, a large peak observed at 3500 cm⁻¹ is associated with Si-OH stretching vibrations. These FTIR results are consistent with the findings of a previous investigation [22].

3.4. Scanning Electron Microscopy

The particles produced by this process were thoroughly analyzed using scanning electron microscopy (SEM); the resulting pictures are shown in Figure 14.3. Particle morphology with structural homogeneity is surprisingly consistent, exhibiting a distinct pattern in the micrographs. These particles range in diameter from 50 to 80 nm, with an average size of roughly 80 nm (Figure 14.3). The smooth and uniform surface seen in the images—which suggests that the particles are well-crystallised—is especially noteworthy. This illustration emphasises the existence of large particles with a rough surface that shows a remarkable degree of consistency in the distribution of both size and shape. These important particles play a critical role by increasing the mass transfer rate and making it easier for metal ions to adhere to the surface of the produced mesoporous silica nanoparticles. This feature plays a major role in improving the ability to remove dye.

3.5. Optimisation Studies of Methylene Blue

This research investigated various factors influencing the absorption of methylene blue by mesoporous silica nanoparticles. The initial dye concentration notably affected the uptake, with optimal removal observed at lower concentrations (10 mg/L), achieving an impressive 87% removal rate. The greatest removal rate of 90.9% was achieved by using an adsorbent dosage of 2.0 g, which was found to be optimum. pH exerted a significant impact, with a neutral pH of 7 yielding the highest removal rate at 90.6%. Furthermore, the duration of contact time proved to be crucial, with 120 minutes demonstrating the highest removal percentage of 90.9% (Figure 14.4). These findings underscore the intricate interplay of these factors in optimising the adsorption process without altering the core content [23].

3.6. Optimisation Studies of Crystal Violet

This study investigated the factors influencing the optimisation of crystal violet adsorption onto mesoporous silica nanoparticles. The initial dye concentration emerged as a critical factor, with the most effective concentration determined to be 10 mg/L, resulting in the highest removal rate of 89.1%. Fine-tuning the adsorbent dose revealed that 2.0 g of adsorbent yielded optimal results, achieving an 88.74% removal rate. pH significantly impacted the process, with a neutral pH of 7 producing the highest removal rate of 88.9% (Figure 14.5). Moreover, the duration of contact time played a crucial role, with 120 minutes proving to be the most effective period, resulting in a maximum removal rate of 90.9% [24].

Figure 14.3: Mesoporous silica nanoparticles captured in a SEM (scanning electron microscopy).

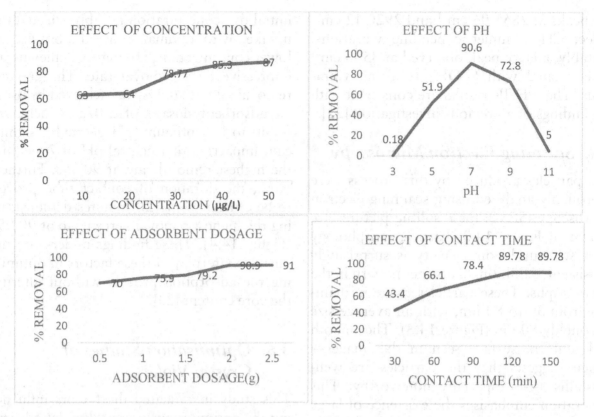

Figure 14.4: Adsorption of methylene blue is influenced by the concentration of dye, the dosage of adsorbent, pH levels, and the duration of contact.

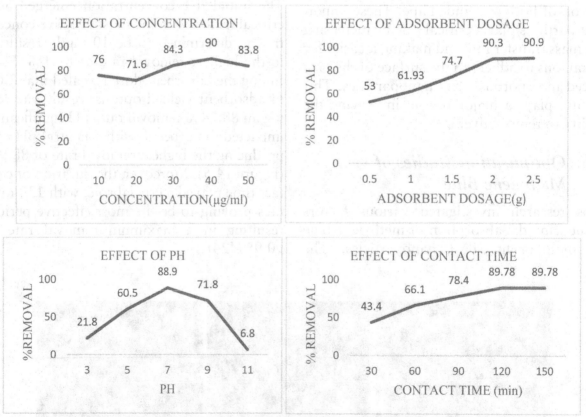

Figure 14.5: Adsorption of crystal violet is impacted by the concentration of dye, the amount of adsorbent used, pH conditions, and the duration of contact.

4. Conclusion

Using methylene blue and crystal violet as model dyes, we successfully created mesoporous silica nanoparticles (MSNs) from coconut husk ash (CHA) in this study and showed their effectiveness as effective adsorbents for dye removal. With methodical testing, we were able to maximise important variables like contact time, pH, adsorbent dosage, and dye concentration, leading to high removal rates—above 88% for crystal violet and over 87% for methylene blue—when optimised. Characterisation investigations using SEM, EDX, and FTIR confirmed the synthesised MSNs' high purity, uniform shape, and well-crystallised organisation, indicating their suitability for usage in dye adsorption applications. Our results highlight the significance of optimising parameters to optimise MSN adsorption capacities and highlight their potential applications in environmentally friendly water treatment methods. By highlighting the importance of using waste-derived substances in addressing environmental challenges, this research adds to the expanding field of environmentally friendly substances and pollution prevention technologies. To confirm MSNs' effectiveness in real-world wastewater treatment scenarios and to further the goals of sustainable development and the promotion of environmentally friendly solutions for mitigating water pollution, future research can concentrate on increasing the production of MSNs, investigating different dye types, and carrying out practical evaluations.

Acknowledgements

The authors would like to express their gratitude to the biotechnology department's authorities, staff, and students for their assistance with the research.

References

[1] Postel, S. (2007). Facing water scarcity. New York: Norton.

[2] Letchinger, M. (2000). Pollution and water quality, neighborhood water quality assessment Project oceanography. *J. Indian Chem. Soc.*, 80, 267–270.

[3] Tripathi, B.D., Misra, K., Pandey, V.S., and Srivastva, J. (2000). Effect of tissue-N content on decomposition of water hyacinth (Eichhornia Crassipes) (Mart.) Solms. *Geobios*, 17 (2–3), 67–69.

[4] Rencoret, J., Ralph, J., Marques, G., Gutierrez, A., Martínez, A.T., and Del Rio, J.C. (2013). Structural characterization of lignin isolated from coconut (Cocos nucifera) coir fibers. *J. Agric. Food Chem.*, 61, 2434–2445.

[5] Magtoto, K. B. V., Salcedo, T. M. C., Amongo, R. M. C., and Capareda, S. C. (2019). Gasification: a study on fouling and slagging tendencies. *Philipp. J. Agric. Biosyst. Eng.*, 15, 32.

[6] Anuar, M. F., Fen, Y. W., Zaid, M. H. M., Matori, K. A. M., and Khaidir, R. E. M. (2018). Synthesis and structural properties of coconut husk as a potential silica source. *Results Phys.*, 11, 1–4.

[7] Songolzadeh, M., Ravanchi, M. T., and Soleimani, M. (2012). Carbon dioxide capture and storage: a general review on adsorbents. *World Acad. Sci. Eng. Technol.*, 70, 225–232.

[8] Salleh, M. A. M., Mahmoud, D. K., Karim, W. A. W. A., and Idris, A. (2011). Cationic and anionic dye adsorption by agricultural solid wastes: a comprehensive review. *Desalination*, 280, 1–13.

[9] Onal, Y., Akmil-Başar, C., Eren, D., Sarıcı-Özdemir, C., and Depci, T. (2006). Adsorption kinetics of Malachite green onto activated carbon prepared from Tuncbilek lignite. *J. Hazard. Mater.*, 128, 150–157.

[10] Bhatnagar, A., Vilar, V. J. P., Boaventura, R. A. R., and Botelho, C. M. S. (2011). A review of the use of red mud as an adsorbent for the removal of toxic pollutants from water and wastewater. *Environ. Technol.*, 32(3), 231–249.

[11] Kapdan, I. K., and Kargi, F. (2002). Simultaneous biodegradation and adsorption of textile dye stuff in an activated sludge unit. *Process Biochem.*, 37, 973–981.

[12] Namasivayam, C., and Kavitha, D. (2002). Removal of Congo red from water by

adsorption on activated carbon prepared from coir pith, an agricultural solid waste. *Dyes Pigm.*, 54, 47–58.

[13] Aldona, B., Valdas, S., Maryte, K., Remigijus, J., and Aivaras, K. (2004). FTIR, TEM and NMR investigations of stober silica nanoparticles. *Mater. Sci.*, 10, 4. https://doi.org/ 10.5755/j01.ms.10.4.26643

[14] Norzahir, S., Nor, S. O., Mohd, Z. Z., Syed Amirul, S. S., and Muhammad Amir, M. A. (2018). Synthesis of green silica from agricultural waste by sol-gel method. *Mater. Today.*, 5(10), 21861–21866.

[15] Purwaningsih, H., Ervianto, Y., Pratiwi, V., Susanti, D., and Purniawan, A. (2019). Effect of cetyl trimethyl ammonium bromide as template of mesoporous silica MCM-41 from rice husk by sol-gel method. *IOP Conf. Ser. Mater. Sci. Eng.*, 515, 1–9.

[16] Dang Thi, T. N., Tran, H., Nguyen Thuy, A. T., Dang Van, P., Phan Dinh, T., and Nguyen, Q. N. (2017). Synthesis of silica nanoparticles from rice husk ash. *Sci. Technol. Dev. J.*, 20, 50–54.

[17] Vladar, A. E., and Hodoroaba, V.D. (2020). Characterization of nanoparticles by scanning electron microscopy. In *Characterization of Nanoparticles*, 7–27. https://doi.org/ 10.1016/b978-0-12-814182-3.00002-x

[18] Salami, S., Bahram, M., and Farhadi, K. (2024). Improving the coremoval efficiency of dyes from water by a novel four-component chitosan flexible film containing graphene. *J. Polym. Environ.* Advance online publication. https://doi.org/10.1007/s10924-024-03207-7

[19] Patel, K., Parangi, T., Solanki, G. K., et al. (2021). Photocatalytic degradation of methylene blue and crystal violet dyes under UV light irradiation by sonochemically synthesized CuSnSe nanocrystals. *Eur. Phys.* *J. Plus*, 136, 743. https://doi.org/10.1140/ epjp/s13360-021-01725-0

[20] Mohamad, D. F., Osman, N. S., Nazri Mohd, K. H. M., Mazlan, A. A., Hanafi, M. F., Esa, Y. A., Rafi, M. I. I. M., Zailani, M. N., Rahman, N. N., Rahman, A. H. A., and Sapawe, N. (2019). Synthesis of mesoporous silica nanoparticles from banana peel ash for removal of phenol and methyl orange in aqueous solution. *Mater. Today*, 19, 1119–1125. https://doi.org/10.1016/j.matpr.2019.11.004

[21] Ranjbakhsh, E., Bordbar, A. K., Abbasi, M., Khosropour, A. R., and Shams, E. (2012). Enhancement of stability and catalytic activity of immobilized lipase on silica-coated modified magnetite nanoparticles. *Chem. Eng. J.*, 179, 272–276. https://doi.org/10.1155/2014/705068

[22] Teresa, O., and Chi Kyu, C. (2010). Comparison between SiOC thin films fabricated by using plasma-enhanced chemical vapor deposition and SiO2 thin films by using Fourier transform infrared spectroscopy. *J. Korean Phys. Soc.*, 56(4), 1150–1155. https://doi.org/10.3938/jkps.56.1150

[23] Upendar, G., Biswas, G., Adhikari, K., and Dutta, S. (2017). Adsorptive removal of methylene blue dye from simulated wastewater using shale: Experiment and modeling. *J. Indian Chem. Soc.*, 94, 1–12.

[24] Mehmood, Y., Khan, I. U., Shahzad, Y., Khan, R. U., Iqbal, M. S., Khan, H. A., Khalid, I., Yousaf, A. M., Khalid, S. H., Asghar, S., Asif, M., Hussain, T., and Shah, S. U. (2020). In-vitro and in-vivo evaluation of velpatasvir-loaded mesoporous silica scaffolds: a prospective carrier for drug bioavailability enhancement. *Pharmaceutics*, 12(4), 307.

15 Rapid solidification of aluminium alloys – a review

K. T. Kashyap[1,a], Lakshya Nahar[2,b], S. N. Nagesh[2,c], and C. Siddaraju[2,d]

[1]Department of Mechanical Engineering, Siddaganga Institute of Technology, Tumkur, India
[2]Department of Mechanical Engineering, Ramaiah Institute of Technology, Bangalore, India

Abstract

Rapid solidification refers to solidification at very high cooling rates of the melt. The cooling rates during this process are of the order of 10^4 to 10^6 K s^{-1}. In this paper, aluminium and its alloys are considered to be solidified. The Solid–Liquid (S–L) interfaces move with very high velocities because of the high cooling rates. In accordance with the perturbation theory of Mullins and Sekerka, perturbations form on S – L interfaces, which break down the flat interfaces, thereby leading to the formation of cells and dendrites. The high velocities of the S – L interfaces cause a very fine secondary dendrite arm spacing. Due to the local solidification time being small, there is no coarsening of the secondary dendrite arms. This leads to a very fine dendrite arm spacing and the mechanical properties of the cast structure are very high. Amorphous solids like metallic glasses, are formed because of the extremely high cooling rates. Also, other metastable structures such as -quasi-crystals, are produced. The techniques of rapid solidification, which will be addressed in this study are melt spinning (splat cooling), gas atomisation, laser surface melting and cooling. This work discusses the fundamental aspects of rapid solidification phenomenon.

Keywords: High cooling rates, metallic glasses, metastability, rapid solidification, quasi-crystals

1. Introduction

Rapid solidification is one of the processes which produces metastable microstructures far away from equilibrium. The cooling rates are of the order of 10^4 to 10^6 K s^{-1}, wherein metastability i.e. processes that occur far away from equilibrium, are prevalent. The metastable phases include metallic glasses, quasi–crystals and extended solid solutions. Alloys produced by employing this technique exhibit an extension of solid solubility in alloys (essentially in aluminium alloys). Modelling and simulation of the competitive growth kinetics inherent in such metallic systems pose a challenge for solidification theory.

The processes of rapid solidification are melt spinning, splat cooling, atomisation and laser surface treatment. The laser could be CO_2 gas laser or Nd: YAG (Neodymium-doped Yttrium Aluminium Garnet) laser. The major objective of the present study is to review the processes of rapid solidification and discuss the metastable phases developed at high cooling rates.

2. Literature Survey

The classic book that presents an unorthodox view of phase transformations and provides a multitude of insights into solid-state physics is 'Stability of Microstructures in Metallic Systems' by J. W. Martin et al. [1]. In the chapter on highly metastable alloys, the phase diagram of complete solid solutions (solid solubility) and associated free energy

[a]ktkashyap@yahoo.com, [b]lakshyanahar71@gmail.com, [c]nagesh@msrit.edu, [d]siddaraju80@gmail.com

DOI: 10.1201/9781003545941-15

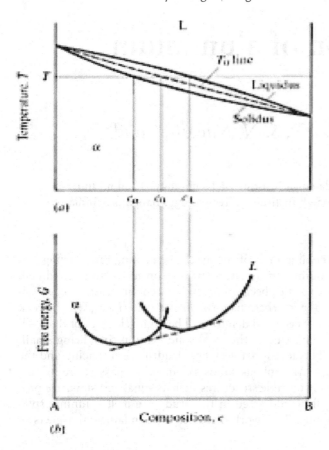

(a)

(b)

Figure 15.1: (a) Phase diagram and (b) free energy–composition curves showing T_0 lines for segregation–free solidification [1].

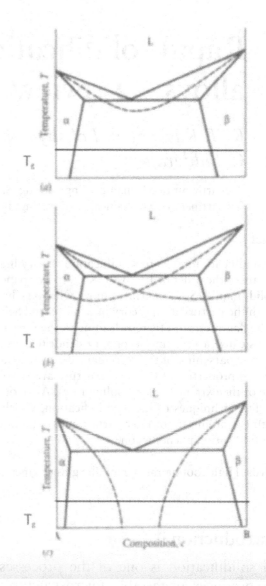

Figure 15.2: Phase diagrams showing T_0 lines (dashed) and T_g in eutectic alloys [1].

composition diagram have been illustrated in Figure 15.1(a) and (b).

The T_0 line is a locus of the points where the free energy of the solid and free energy of the liquid intersect. It always lies between the liquidus and solidus in the phase diagram. Undercooling is a phenomenon by virtue of which the thermal energy possessed by the melt is utilised in the creation of new surfaces during homogeneous and heterogeneous nucleation in the liquid-solid transformation. If the undercooling is below the liquidus and below the T_0 line, then segregation–free solidification takes place. Segregation is not formed between dendrite arms since solute rejection occurs ahead of the flat S–L interface.

Figure 15.2 shows the T_0 lines in the eutectic alloy phase equilibria. The T_0 lines with deep eutectics and glass transition temperature (T_g) have also been indicated.

If the alloy is undercooled below the T_0 lines, then segregation–free nucleation and growth of dendrites form on cells. In case the alloy undercools below Tg, metallic glass is formed. Figure 15.3 is an image of the cells captured by a Transmission Electron Microscope (TEM). The figure shows wonderful cells in bright-field imaging, without any segregation [1]. The cells possess low- angle grain boundaries with a misorientation of 2–3 degrees, because of the high cooling rates. Undercooling occurs to a great extent and further, the local solidification time is small, which is the reason for the formation

Figure 15.3: Segregation–free solidification microstructure in melt–spun Fe–35 wt.% Ni [1].

of very fine dendrites either by splat cooling or atomisation. During this short period of local solidification time, dendrite arm coarsening does not occur due to the curvature effects.

The local solidification time is given by [2]

$$T_f = \Delta T_s / GR \qquad (1)$$

where ΔT_s is the non-equilibrium temperature range of solidification, GR is the product of the thermal gradient and growth rate.

If the undercooling is below the glass transition temperature, then a metallic glass forms. There will be a short-range order of atoms. However, long-range order and periodicity will not be observed in the structure. This happens when the T_0 lines plunge and is

accompanied by deep eutectics. Apart from that, intermetallic compound formation will be suppressed.

A large undercooling below the liquidus presupposes extensive homogeneous nucleation. This happens when heterogeneous catalysts are not present or activated in an aluminium alloy melt. When this large undercooling occurs, the melt can cross the glass transition temperature and metallic glasses are the result. It is also possible to form quasi–crystals with five-fold symmetry axis.

They are also referred to as incommensurate structures or structures possessing icosahedral symmetry. These structures were also modelled as Penrose tilings. They are also taken to be Fibonacci numbers. Electron diffraction patterns revealed by TEM are evidence for five-fold or ten-fold symmetry. They have been presented in Figures 15.4 and 15.5.

Now, the techniques employed for rapid solidification include gas atomisation, splat

Figure 15.4: Five-fold diffraction pattern from melt spun Al6CuLi3 [1].

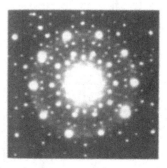

Figure 15.5: Ten-fold diffraction pattern from melt spun Al4Fe [1].

cooling, melt spinning and laser surface treatment. Figures 15.6 and 15.7 are the schematic diagrams for melt spinning and gas atomisation. In the former (splat cooling), strips or ribbons are formed. In the latter, powders are formed and the principle has been illustrated in Figure 15.8.

In melt spinning or splat cooling, liquid aluminium melt is poured onto a rotating copper wheel, where rapid cooling takes place. The solidified structure will be in the form of strips or ribbons. In gas atomisation, liquid aluminium alloy is passed through a convergent-divergent nozzle in an inert atmosphere such as argon, wherein powder particles are formed. These powders can be processed by powder metallurgy route into shaped components.

In laser treatment, the surface melts and resolidifies with a very high rate of cooling. The surface layer will consist of many metastable phases. The growth velocity during rapid solidification can be synchronised with the velocity of a source of thermal energy like a laser or electron beam covering the surface of the material shown in Figure 15.9. In this scenario, a stable melt-pool is formed. It serves to melt the new material at its leading edge at a rate that is equal to that of refreezing of the previously melted material at its trailing edge. In fact, the influence on the resulting microstructure of systematic variations in front velocity up to approximately 1 m s^{-1} can be found experimentally due to the coupling [3].

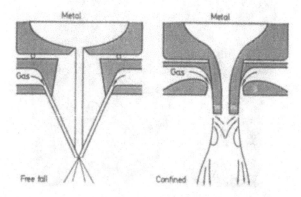

Figure 15.8: Principle of spray droplet (atomisation) by impingement of high velocity gas jets on to a free falling or emergent melt–stream [3].

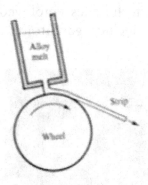

Figure 15.6: Melt spinning process for manufacturing strip [3].

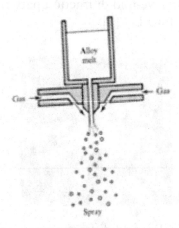

Figure 15.7: Gas atomisation process for manufacturing powder [3].

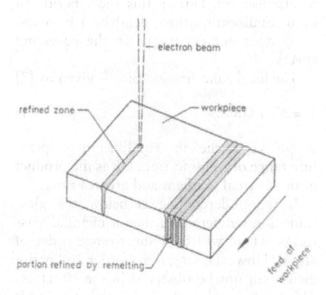

Figure 15.9: Electron beam melting [3].

According to Noga et al. [4], Scanning Electron Microscopy has confirmed the homogeneity of particle distribution and it's proven, rapid solidification is an effective method of increasing the strength of Al-Si alloys. Zhou et al. [5] has provided evidence supporting the fact that after carrying out precipitate coarsening on A1-Fe-V-Si (AA8009), it still possesses high mechanical strength in the range 150–450 MPa, a region where traditional wrought Al-Cu (AA2219) and Al-Zn-Mg-Cu (AA7075) are unable to retain a large percentage of their high strength at room temperatures shown in Figure 15.10. The notable resistance offered to the coarsening of the high-volume fraction of the nanoscale silicides raises the maximum operating temperature by around 200 K as dispersoid is refractory in nature. Apart from that, creep resistance and corrosion resistance are the other mechanical parameters, whose values exhibit an appreciable increase when compared to the established wrought, ingot counterparts of comparable strength. Enhanced thermal stability is another consequence of the new chemical compositions of the alloy that have become a possibility because of rapid solidification, such as in the AA8009 alloy.

In this paper, all the issues pertaining to rapid solidification, have been explained.

3. Conclusions

This study presents a succinct overview of rapid solidification processes. It discusses the microstructure with the theoretical background of solidification at high cooling rates, where the S–L interface moves with extremely high velocities. The rapid solidification processes such as melt spinning, atomisation and laser surface melting have been described.

Acknowledgement

The authors wish to thank the Principal, CEO and Director of Siddaganga Institute of Technology, for their support. The authors express their gratitude to the Head of the Department of Mechanical Engineering, Principal and management of Ramaiah Institute of Technology, Bangalore, for their cooperation.

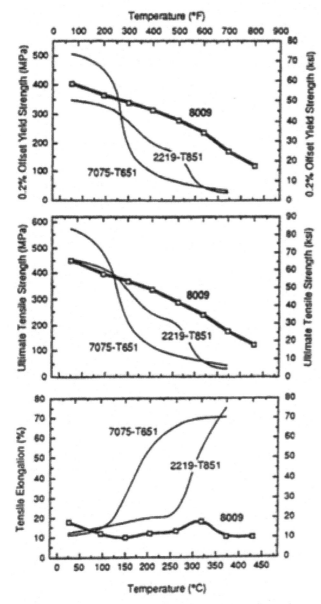

Figure 15.10: Tensile properties of rapidly solidified AA8009 (Al-Fe-V-Si) alloy v/s test temperature compared with high–strength wrought alloys AA2219–T851 and AA7075–T651 [3].

References

[1] Martin, J. W., Doherty, R. D., and Cantor, B. (1997). Stability of Microstructure in

Metallic Systems, 2nd ed. Cambridge University Press.

[2] Flemings, M. C. (1974). Solidification Processing. McGraw Hill.

[3] Jones, H. (1999). Non-equilibrium processing of materials. In C. Suryanarayana and R. W. Cahn (eds.). Pergamon, Amsterdam, 1999, *Pergamon Material Series*, Volume 2, 1st ed., 23–48.

[4] Noga, P., Skrzekut, T., and Wedrychowicz, M. (2023). Microstructure and Mechanical Properties of Al-Si Alloys Produced by Rapid Solidification and Hot Extrusion Materials. *Materials*, 16(15), 5223: MDPI. https://doi.org/10.3390/ma16155223.

[5] Zhou, J., Duszczyk, J., and Korevaar, B. M. (1992). Mechanical response and structural development during the hot extrusion of a rapidly solidified AI-20Si-7.5Ni-3Cu-1 Mg alloy powder. *J. Mater. Sci.*, 27, 3856–3868. Chapman and Hall. https://doi.org/10.1007/BF00545468.

16 Enhancing wear resistance of aluminum alloy wheel hubs through aluminum metal matrix composite (MMC) materials

S. Suthagar[1,a], *Gooty Rohan*[2,b], *G. Boopathy*[1,c], *and N. Ramanan*[3,d]

[1]Aeronautical Engineering, Vel Tech Rangarajan Dr. Sagunthala R& D Institute of Science and Technology, Chennai, India
[2]Aeronautical Engineering, Ajeenkya D Y Patil University, Pune, India
[3]Mechanical Engineering, Sri Jayaram Institute of Engineering and Technology, Chennai, India

Abstract

The importance of materials and their inherent strength cannot be emphasised in the current era of lightweight applications in the industrial sphere. The behavior of materials under different stress circumstances, as well as their applicability for specific applications, is critical. Unfortunately, a significant number of the materials available do not meet the requirements of these applications. We used aluminium Metal Matrix Composite (MMC) materials as our main point in this research project. Our primary focus is on improving the wear resistance of aluminium alloys used in wheel hub applications. Wear and tear is a common concern in this setting, with average alloy wheel hubs lasting roughly 50,000 km in motorcycle use before showing indications of wear. To solve this issue, we investigated the enhancement of material hardness with the goal of reducing wear and, as a result, improving wear resistance. Our research comprises putting aluminium MMC to various levels of modification and analyzing their performance in the particular application. Finally, our research explores into a thorough examination of the fracture surfaces of these materials in order to gain insights and draw beneficial inferences.

Keywords: Lightweight applications, aluminium metal matrix composite (MMC), wheel hub, wear resistance, fracture surface analysis

1. Introduction

Metal matrix composites (MMCs) with an aluminium basis are frequently utilised to improve the wheel hubs' resistance to wear. These composites have enhanced mechanical and chemical characteristics, such as excellent electrical conductivity, fatigue resistance, and strength-to- weight ratio. [1] The efficiency of several reinforcements, including as SiC, CNT, Al2O3, and globular carbon, in boosting the strength of the aluminium matrix has been investigated. [2,3] Aluminium alloy-based Metal Matrix Composites (MMCs) offer significant benefits in terms of low weight and high specific strength.[4] Consequently, these composites find extensive use in many applications such as pistons, gears, and bearings. [5] The stir casting technique is used in the manufacture of composites. The Box-Behnken design is used for experiment implementation. Because of the higher density, the use of 3% Mica results in less wear. The load and Mica percentage are the most important metrics. [6]

The effective application of a revolutionary process to the manufacture of Aluminium Metal Matrix Composite (AMCs). An

[a]suthagars@veltech.edu.in, [b]rohan.gooty@adypu.edu.in, [c]boopathyg81@gmail.com, [d]ramananinjs@gmail.com

DOI: 10.1201/9781003545941-16

analysis to the application of Spent Alumina Catalyst (SAC) and Scrap Aluminium Alloy Wheels (SAAWs) as matrix and reinforcing elements, respectively. [7] If it is feasible to use waste resources and abandoned aluminium alloy. [8] Microstructure and mechanical qualities affected by reinforcement. [9] The study concentrated on composites made of aluminum and tungsten carbide in order to investigate their potential for better qualities. More ceramic phase content was discovered to increase composite density, hardness, and strength. Powder metallurgy techniques were used in the preparation of the composites. [10] The testing was conducted in compliance with ASTM guidelines. [11–13]

For automotive applications, metal matrix composites (MMCs) have the potential to replace non-reinforced aluminium alloys and ceramic reinforced composites. [14] Hybrid composites based on aluminium offer benefits like reduced costs, a good strength-to-weight ratio, and enhanced wear resistance. [15,16] Aluminum matrix composites (Al- MMCs) with hybrid reinforcement and based on DPS (Discontinuously Reinforced Precipitation Hardened) exhibit outstanding wear characteristics. The results obtained using the specified parametric setups show a considerable 6% increase in wear resistance. [17] When compared to the base alloy, Al7075 matrix composites supplemented with tungsten carbide and cobalt exhibit improved wear resistance. The surface morphology of the worn composites was examined using scanning electron microscopy (SEM). [18] The Al-WC-Co composite has better wear resistance. These composites could find application in many different kinds of fields, such as protection materials and automotive components. [19,20]

Composites outperformed metals in terms of wear resistance under high loads. Adhesion was the principal source of wear under these conditions. The wear characteristics of Al6061 hybrid metal matrix composites were created and analyzed. Coated or uncoated MWCNTs with varying graphene weight

percentages were used. [21] The integration of Beryl particles into aluminum-based composites has been shown to successfully improve wear resistance and reduce friction levels. The incorporation of Beryl particles into the aluminum 2024 alloy reduced wear loss and coefficient of friction. Aluminum-based composites have a lower density and a higher rigidity. Various reinforcements or fillers have been reported to improve wear resistance and reduce friction. [22] The addition of ilmenite to the AA6061 alloy increased its hardness while decreasing its wear rate and coefficient of friction. Furthermore, composites containing 10% ilmenite have increased pitting corrosion resistance. Stir casting is the process used to make these composites. Next, a detailed evaluation is conducted to determine the composites' hardness, wear properties, coefficient of friction, and pitting corrosion behaviour. [23,24]

AlMMCs, or aluminium metal matrix composites, are employed in many different industries, such as telecommunications, automotive, marine, and aviation. The incorporation of silicon carbide (SiC) as a reinforcing element in AlMMCs has the potential to improve their intrinsic properties. [25,26] A technique known as the Representative Volume Element (RVE) homogenisation approach is used for simulating the properties of a material. The automobile, aerospace, and defense industries all make extensive use of lightweight structures. These structures have better qualities including rigidity, strength, and resistance to wear. [27] Coating aluminum alloys with ceramic powders improves wear resistance. B4C is highly recommended as the best reinforcement material. [28] Hybrid metal matrix composites (HMMCs) were investigated for their mechanical, wear, microstructural, and corrosion characteristics. The mechanical properties were enhanced by the use of Ti, Si3N4, and TaC reinforcements. [29]

The study was done on a dynamic subsurface crack by the researchers. They employed Finite Element analyses (FEA) and

Elastic-Plastic Fracture Mechanics (EPFM) analyses. Comparing MnS inclusion to Al2O3, they discovered that the J-Integral values consistently increased. Additionally, MnS incorporation caused wear fractures in the wheels. For an angled crack, all inclusion types exhibited high J-Integral values, which fell as the crack's depth grew. [30] Aluminium wheel forging was the subject of research. Various scenarios were simulated. A little model was used for the tests. Results and the simulation were compared. A full-scale replica of an aluminium wheel underwent simulations. Various parameters were recorded. These results are crucial for streamlining the forging procedure. [31]

The investigation outlines ways to reduce stress concentration and strengthen the wheel rim's durability. The study examines techniques, highlighting the significance of improving vibration control and lowering noise production by the wheel rim. It describes how to increase the wheel rim's thermal resistance, which is essential for achieving the best possible performance in a variety of temperature conditions. With an emphasis on improving corrosion resistance—a critical component of long-term robustness and dependability—the study's main goal to increase the lifespan to be wheel rim. In order to accomplish as goal, deliberate changes in material composition and design are required. [32]

2. Materials and Methods

The tyre rims have the role of securely fastening the rubber tyre in its assigned location, which makes the vehicle easier to operate. These rims, also called wheels, are available to be broad range of sizes and styles to fit a different of car types, from luxury sedans to commercial vehicles. Apart from their visual appeal, car rims play a crucial function in improving vehicle performance and guaranteeing a more comfortable ride. Important factors including alignment, manoeuvrability, braking effectiveness, acceleration, and speedometer calibration are directly affected

by changes made to the rim's size. Wheel rims consist of many components, such as the bolt pattern, valve stem, concave structure, plate, central aperture, external board face, and support spokes. Moreover, they are made from a range of materials, including steel, aluminium alloy, and carbon fibre.

Everything in the world is made of a variety of components, either mixed together or wholly composed of one type of substance. Similar to this, contemporary wheel rims are made of composite or a single material. Nowadays, materials like carbon steel 1008, forged steel, and magnesium alloy are used to create wheel rims. Each material has unique qualities; the two most widely used alloys for wheels are aluminium and magnesium. Aluminium alloy is stronger than magnesium alloy and lighter than both of them, making it the material of choice for automotive purposes. Because of its lightweight and sturdy properties, we use aluminium alloy in our project. Modifications are also made to the reinforcing materials, which are essential for improving impact strength and thermal resistance while lowering rim stress. In this sense, silicon carbide and zircon sand have been chosen as reinforcing materials due to their strength and high melting point, while aluminium has been chosen as the basis material. [33]

2.1. Aluminium 6061

Composite materials find extensive application in numerous industries, including aerospace, automotive, and structural components. This is mostly because of the noteworthy benefits these materials provide in terms of material and energy savings. The primary focus of this work is the advantages of particulate-reinforced aluminium metal matrix composites over unreinforced alloys. The materials' high specific stiffness, specific strength, fatigue resistance, coefficient of thermal expansion, and dimensional stability are all highlighted. In order to comply with ASTM standards, the research project entails the production of composites using Al 6061 as the matrix material and Zircon particles reinforced by stir casting. Through the use of an optical microscope for microstructural research, it was shown that the addition of Zircon particles at a concentration of 9% produced a more refined grain structure and a uniform dispersion within the aluminium matrix. In addition, the assessment of the composites' tensile strength and hardness characteristics showed that they exceeded the unreinforced Al 6061 matrix. Additionally, by increasing the concentration of zircon—9% produced the greatest results—tensile strength and hardness were further enhanced.

The heat-treatable aluminium alloy 6061 stands out due to its strength, which is higher than that of 6005A and is moderate to high. It's remarkable how well it resists corrosion and is weldable, even if the strength slightly decreases in the weld area. The strength against fatigue of this alloy is moderate. The following provides a complete list of the Al 6061 alloy's key attributes.

Al 6061 alloy	
Physical Properties	
Density in ρ	2.70 g/cm³
Mechanical Characteristics	
Ratio of Young's modulus E	68.9 GPa (9,990 ksi)
Strength in Tensile σ_t	124–290 MPa (18.0–42.1 ksi)
At break, elongation in ε	12–25%
The Poisson's ratio v	0.33
Thermal Properties	
Melting temperature in T_m	585 °C (1,085 °F)
Conductivity of heat in k	151–202 W/(m-k)
Linear thermal expansion coefficient in α	2.32×10^{-5} K⁻¹
Heat capacity specific in c	897 J/(kg-k)
Electrical Characteristics	
Volume resistivity in ρ	32.5–39.2 nOhm·m

2.2. Silicon Carbide (SiC)

Carbon and silicon combine to form the unusual chemical substance known as silicon carbide. Sand and carbon performed a high-temperature electrochemical process to create it initially. Silicon carbide has shown to be a superior abrasive material throughout the course of more than a century of use in the production of grinding wheels and other abrasive goods. A semiconductor composed of silicon and carbon is called silicon carbide, another name for it is carborundum. While moissanite is an uncommon naturally occurring stone, since 1893, SiC powder has been commonly produced artificially for abrasive uses. Sintering is a technique that brings silicon carbide grains together to form extraordinarily resistant ceramics. These ceramics find extensive use in high-end applications requiring extreme endurance, such as ceramic plates used in bulletproof vests and automobile brakes and clutches. Additionally, silicon carbide has been utilised in electronics, most famously in the form of detectors and light-emitting diodes (LEDs) in early radio equipment dating back to about 1907. SiC is utilised in semiconductor electronics for high-temperature, high-voltage, or both operating devices. The Lely process makes it easier to produce large single crystals of

Characteristics of Silicon Carbide	
Chemical formula	SiC
Molar mass	40.096 g.mol⁻¹
Density in ρ	3.16 g.cm⁻³ (hex.)
Melting point	2,830°C
Solubility	Insoluble in water, Soluble in alkalis and molten iron
Electron mobility	~900 cm2/ V-s
Magnetic Susceptibility in χ	-12.8×10^{-6} cm³/mol
Refractive index in nD	2.55

silicon carbide, which can then be made into synthetic moissanite gemstones.

2.3. *Zircon Sand*

Zirconium, sometimes referred to as zircon sand (Zr4), is extracted from old mineral sand deposits, mostly as crystalline sands. Zircon exhibits a spectrum of colours, mostly brown but also yellow-golden, pink, red, blue, and green. Zircon sand is one of the primary products made from mineral sands; most of it is mined in Australia and Africa, and more than a million tonnes are produced there each year. India, Europe, North America, and the Asia- Pacific area all consume a sizable amount of zircon, but China produces almost half of the mineral globally. The primary uses of zircon sand include the production of flour, fused zircon, chemical zirconia, zirconium compounds, and zirconium metal. The direct uses of zircon sand include refractories, foundries, and other small-scale uses. Zirconia can be produced by treating the sand to melt it at a high temperature and produce molten zirconia, also known as zirconium oxide (ZrO2). The chemical element zircon (Zr) is represented in the periodic table as a silvery-grey metal. Zirconium is generated from zircon. Zircon was the most common element in the Earth's crust in the 20th century, usually found in the silicate form (though it was also occasionally found in the oxide form as baddeleyite).

In ceramics, zircon sand is essential, especially when it comes to specialist casting and a range of uses including heat- and wear-resistant materials. This is because to its exceptional resistance to abrasion and high temperatures. The significance of this material is evident in the fields of manufacturing and ceramics, including the use of glazes for tiles and sanitary items. Zirconium sand is chosen based on its distinct qualities and requirements, which are listed in the following table format.

3. Fabrication of Specimen and Testing

The manufacturing process of the composite materials resulted in the production of three different samples with different compositions. The base material is aluminium 6061, while the reinforcing is zirconium sand and silicon carbide. These elements are mixed using the stir casting method in three distinct ratios to produce a molten metal matrix composite (MMC). The elements' exact compositions are displayed in the Table 16.1.

Stir casting has been shown to be a cost-effective way to create aluminium matrix composites Figure 16.1. Numerous factors ultimately impact the mechanical characteristics and eventual microstructure of these composites. [34] In this study, Al-3 weight percent SiC composites were made using

Specification of Zirconium Sand	
Molecular weight in g/mol	183.1
Bulk density inlbs./ft.3)	115–130 & 170–180
Specific gravity	4.5 to 4.8
Solid specific heat	0.132
Melting point in °C	2200 to 2550
Dissociation temperature in °F	3200
Thermal expansion coefficient in cm/cm/°C	4.2
Mohs hardness @20°C	7.5–8.0
Angle of repose (°)	30
L.O.I	0.15 to 0.25
Colour	Off-white
Thermal conductivity in BTU/sq.ft/hr/°F/in.	14.5

Table 16.1: Fabrication compositions

Elements	Sample 1	Sample 2	Sample 3
Al 6061	97.5%	97.5%	95%
SiC	2.5%	0%	2.5%
ZrSO4	0%	2.5%	2.5%

micron-sized SiC particles as reinforcement. Two casting temperatures of 680 and 850°C were considered, as well as two stirring times of two and six minutes. This study looked at a number of characteristics, including porosity, ceramic integration, particle agglomeration, and the response at the matrix-ceramic interface. High-resolution transmission electron microscopy (HRTEM) and scanning electron microscopy (SEM) were used to study these elements. A shorter stirring time is necessary for the ceramic to integrate well and form a solid connection with the metal at the interface, according to the microstructure study. [35] Moreover, enhanced ceramic inclusion is facilitated by stirring temperatures up to 850°C. Additionally, shrinkage porosity and widespread Al4C3 production at the metal/ceramic contact were observed in certain situations. The study concludes with an evaluation of the mechanical properties of the composites and an exploration of their relationship to the related microstructure and manufacturing parameters. [35]

3.1. Tensile Test

Tensile testing is a crucial procedure in materials science and engineering where a specimen is put under controlled strain until failure. Uniaxial tensile testing is a widely used technique to evaluate isotropic materials' mechanical properties. Important information on tensile strength, yield strength, and ductility is provided by this technique. A basic test measures the ultimate tensile strength of a material by

Figure 16.1: Sample for tensile test & impact test.

pulling a specimen until it breaks. This characteristic shows the highest tensile stress that a material can bear before failing. A universal testing apparatus that measures compressive and tensile strength is usually used for the assessment. Tensile force is measured by the device using a load cell.

3.2. Impact Test

The impact test assesses toughness, temperature-dependent brittle-ductile transitions, and the amount of energy absorbed when a material breaks. Through the calculation of energy absorbed during fracture, the test assesses the toughness or impact resistance of materials at varying temperatures. The maximal kinetic energy of the drop weight at the time of impact or the energy delivered into the specimen determines the impact energy required to break material. A pivoting arm is lifted to a certain height and then released to strike and fracture a sample with a notch in it. This is the Izod test, a common ASTM test. Based on the height arm achieves after striking the sample, absorbed energy is computed.

3.3. Flexural Test

For this test, a three-point load or a center-point load may be utilised. Flexure tests are marginally less expensive and have slightly different findings than tensile testing. The sample's flexural strength is shown by the maximum force measured during the test. For stress-bearing restorations, particularly under high pressure or stress, significant flexural strength is essential. It establishes a material's range of uses since stronger materials can be used in more repair units. When an object, such as a steel rod or hardwood beam, is bent during a flexural test, it receives various loads along its depth. The highest tensile stress at the extreme fibres is found outside the bend, while the highest compressive stress is found inside the bend. Since tensile stress usually leads to material breakdown before compressive stress, flexural strength is the highest tensile stress endured before failure.

3.4. *Heart Treatment Methods*

This technique, which uses heat treatment to increase a material's strength, is also utilised to achieve manufacturing requirements. After a cold working procedure, heat treatment techniques alter the properties of the metal by heating it to a specified temperature and maintaining it there for a predetermined amount of time. They can also be used many times to regulate the metal's cooling. The method of heating and cooling metals to alter its microstructure and highlight the mechanical and physical qualities that increase metal's appeal is known as heat treatment. The temperature at which metals are heated and the rate at which they cool adhering to heat treatment can alter a metal's characteristics significantly. Heat treatment process has been classified annealing, hardening and tempering.

3.5. *Particle Hardening*

Precipitation hardening, also known as age hardening or particle strengthening, is a heat treatment method that increases the durability and toughness of the original material by uniformly dispersing minuscule particles of copper or aluminum known as precipitates throughout the material. The process of particle hardening depends on temperature-dependent variations in the solubility of solids, which result in the production of small impurity phase particles that obstruct the sliding of dislocations, or lattice faults, in crystals. This helps to harden the material since dislocations

Table 16.2: Parameters of the hardening process

	Parameters		
Process	*Tempering*	*Quenching*	*Ageing*
Temperature	565°C	79°C	180°C
Durations	3 Hours	25 Minutes	6 Hours

are frequently the primary carriers of plasticity. Depending on the thermal history of a particular area of the atmosphere, the development of ice in the air might result in clouds, snow, or hail. Precipitation in solids can result in a wide variety of particle sizes with drastically different characteristics. Also, the precipitate hardening process can be separated & examined into three phases: quenching, ageing, and tempering. The Table 16.2 also shows the various temperatures at which precipitate hardening occurs.

3.6. *Measurement of Wear*

Wear is defined as the material gradually being removed or deformed at solid surfaces. Wear can be caused by chemical (like corrosion) or mechanical (like erosion). Wear on machine parts and other factors like fatigue and creep lead to a degradation of functional surfaces, which in turn results in material failure or loss of functionality. Wear on metals results from the separation of particles that contribute to wear debris and the plastic displacement of material near and on the surface. There is a range in particle size from millimetres to nanometers. Wear rate is dependent on several factors, such as temperature, type of motion (rolling or sliding), impact, static and dynamic loading, and lubrication (especially the deposition and wearing out of the border lubrication layer).

3.7. *Wear Technique Pin on Disc*

The pin-on-disc sliding wear test is an established tribological characterisation technique used to ascertain the wear mechanism and coefficient of friction of diamond films. Wear in machine elements causes functional surfaces to deteriorate in addition to processes like fatigue and creep. This, in turn, results in material failure or loss of functionality. [36] Consequently, the wear rate is affected by the following factors: temperature, lubrication (especially the deposition process and the wearing out of the boundary lubrication layer), motion type (such as sliding, rolling), impact, static, and dynamic loading. Wear measurement techniques include surface profiling, quantifying the weight (mass) loss with a precision scale, and measuring the deformation thickness or cross-sectional region of a wearing track under a microscope to determine the wear volume loss or linear dimensional change. [37]

4. Wheel Rim Model

The wheel rim that has to be represented in order to simulate stress distribution and deformation. Thus, we have built the Bajaj Pulsar 150 cc wheel rim here using a variation of diameters. Below is a list of the wheel rim modelling measurements a) 475 mm for the external diameter; b) 130 mm for the internal diameter; c) 107 mm for the length of the rim leg; and d) 72 o for the rim leg angle. Modelling was done with CATIA V5 and the ANSYS workbench. It is possible to determine the stress deformation from this. After the wheel is modelled, it is stated to be simulated in ANSYS Workbench, whereupon the load is tested and the stress distribution is also simulated. The finite element analysis [FEM] approach was used, which can quickly determine the stress distribution on the wheel rim.

4.1. Testing Procedure

This paper is being carried out in order to demonstrate that the alternative metal matrix composite material exhibits greater strength than the current wheel rim material. We have chosen the three composite materials in order to do this. [38] The foundation material for the wheel rim is aluminium 6061, while reinforcement elements like silicon carbide and zricona sand are employed. Following the three separate samples that have been cast at the three various proportions that are already mentioned in the above sections, these materials are next cast using stir casting at a certain temperature. Next, each of the three samples is machined in preparation for different testing processes, including tensile, impact, and flexural tests. All three mechanical tests use the first two samples as direct carriers. Subsequently, the initial two samples' result values are acquired and set aside for a comparative analysis with sample 3. In order to begin the heat treatment procedure, which involves precipitate hardening, sample number three has been taken. This precipitate hardening process is divided into three stages: quenching, ageing, and tempering. This heading also included a list of the temperatures for these three sessions. Subsequently, sample three is further machined to create three more specimens, which are used only for the thread test.

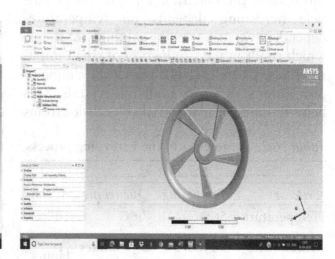

Figure 16.2: Parameters of the hardening process.

Because other mechanical testing properties increase with a higher tensile test. Lastly, the sample 3 results are also acquired. Afterwards, the values from the various samples are compared to one another. Based on this, sample 3 exhibits a greater tensile strength and has undergone a wear test. Additionally, when they undergo pin-on-disc wear testing, it demonstrates the reduced wear quantity. Additionally, sample 3 is said to have been provided for SEM testing. The particles are closer together than typical MMCs, according to the SEM. It unequivocally demonstrates that sample 3 has more strength when it comes to replacing the wheel rim.

5. Results and Discussion

5.1. Analysis of Tensile Testing

A tensile test has been performed on Sample 3 of the specimen. The only test employed for this is the tensile test since it yields greater impact, flexural, and wear strength. Sample 3 has been machined so that the three specimens can be examined once the precipitate hardens. We then found that sample 3 had a higher tensile strength than samples 1 and 2, which were previously reported. This also applies to sample 3, where specimen 3's tensile strength is higher than that of sample 3's

other two specimens. The sample 3 value for each of the three specimens is thus displayed in the table. Additionally, it provides an explanation of the values for the specimens' yield strength, UTS, and elongation Figure 16.3. These values are also compared in the Table 16.3.

5.2. Analysis of Flexural Value

Each specimen samples 1 and 2, which are cast according to the proportion specified in the casting methods, have their impact values

Table 16.3: After the effects of heat treatment

Test Parameter	Observed Values		
Sample Id	T 1	T 2	T 3
Yield strength N/mm^2	102	108	117
Ultimate tensile strength MPa	124	119	144
% of elongation in 25mm GL	3.50	4.50	4.50

Table 16.4: Before to the heat treatment's effects

Testing	Sample 1	Sample 2
Tensile strength in MPa	58.54	143.82
Yield strength in MPa	55.50	99.59
Elongation in %	2.60	6.16
Flexural load in KN	4.43	7.17

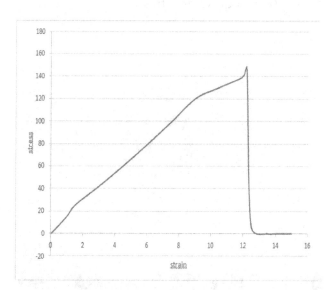

Figure 16.3: Tensile test (Sample 3).

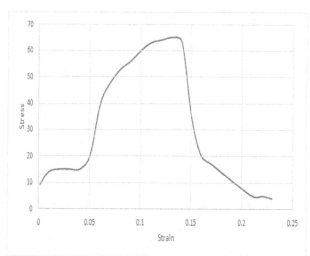

Figure 16.4: Flexural values (Sample 1).

tested. When the impact values are evaluated and analysed in this case, it is evident that, when compared to sample 2, sample 1 has a higher impact value. Additionally, the two samples' graphs are presented so that we may quickly analyse the impact value. The table below shows the results of Specimens 1 and 2 for the various tests described below the Table 16.4. Below are the graphs for the two examples Figures 16.4 and 16.5.

5.3. *Analysis of Impact Value*

The impact values are examined for specimens 1 and 2, which are cast according to the proportion specified in the casting methods. When the impact values are compared

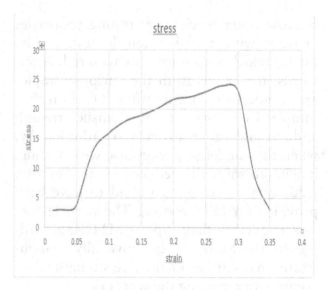

Figure 16.7: Impact values (Sample 2).

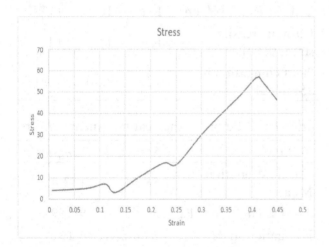

Figure 16.5: Flexural values (Sample 2).

Figure 16.8: Equivalent stress.

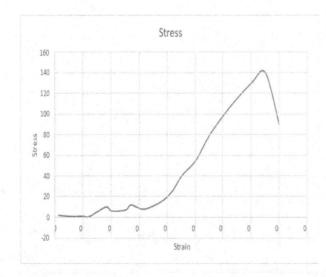

Figure 16.6: Impact values (Sample 1).

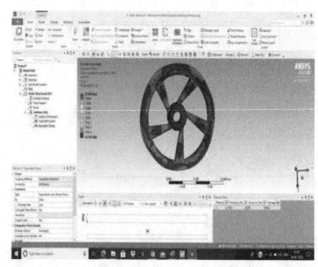

Figure 16.9: Total deformation.

between samples 1 and 2, it is evident from the analysis and testing that sample 2 has a higher impact value. We can readily analyse the impact value by looking at the graphs of the two samples that are also listed below Figures 16.6 and 16.7.

5.4. Results of Simulation

As a result, the ANSYS simulation result illustrates the variation in both equivalent stress and total deformation, and the images of these results are displayed below Figures 16.8 and 16.9.

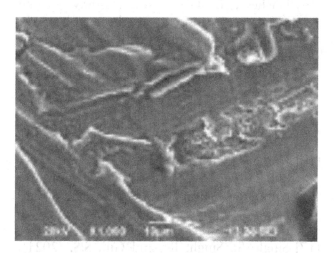

Figure 16.10: SEM image of the flawlessly cast.

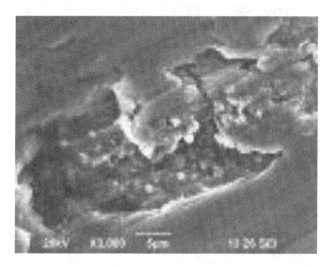

Figure 16.11: Part of the sample that was improperly cast.

5.5. Analysis of SEM

SEM analysis is an effective research method that creates intricate, high magnification images of a sample's surface topography using a concentrated electron beam Figures 16.10 and 16.11. A SEM is a kind of electron microscope that uses a concentrated electron beam to scan a surface in order to create an image of the sample. [37] When electrons interact with atoms in a sample, they produce a variety of signals that provide details about the material's composition and surface topography. An image is created by combining the position of the electron beam and the strength of the detected signal as it scans in a raster scan pattern. Our metallurgical specialists thoroughly examine the material properties using SEM analysis, offering manufacturers insightful information. Using a concentrated electron beam, SEM analysis is a strong research tool that creates intricate, exquisite photographs of a sample's surface topography.

6. Conclusions

Following extensive study and simulation using ANSYS 2020 R2, the wheel rim's load impact, stress, and strain were examined. This leads us to the conclusion that our design, the material, and the combination we selected are accurate. Furthermore, we have discovered that the recently created composite material has higher yield strength, impact strength, and wear value. And as a result, the material's elongation is decreased. Additionally, the wheel rim's lifespan has been extended in comparison to its current state. Therefore, the following is our conclusion.

- To reduce the stress placed on the wheel rim and increase impact strength.
- To reduce the noise generated in the rim and enhance vibration control.
- To increase the tyre rim's resistance to heat.
- To increase the rim's resistance to corrosion.
- To lengthen the wheel rim's lifespan.

References

[1] Munnur, H., Nagesh, S. N., Siddaraju, C., Rajesh, M. N., and Rajanna, S. (2021). Characterization & tribological behaviour of aluminium metal matrix composites–A review. *Mater. Today: Proc.* 47, 2570–2574.

[2] Jayavelu, S., Mariappan, R., and Rajkumar, C. (2021). Wear characteristics of sintered AA2014 with alumina and titanium di-Boride metal matrix composites. *Int. J. Ambient Energy.*, 42(2), 173–178.

[3] Popov, V., Borunova, A., Shelekhov, E., Khodos, I., Senatilin, B., Matveev, D., and Versinina, E. (2022). Peculiarities of chemical interaction of some carbon nanoreinforcements with aluminum matrix in metal matrix composite (MMC). *Materialwiss. Werkstofftech.*, 53(5), 602–607.

[4] Anand, A., Tiwari, S. K. (2022). Recent advancements in the production of hybrid metal matrix composites (HMMC): A Review, IOP Conference Series: Materials Science and Engineering, 2022

[5] Prakash, M., Badhotiya, G. K., and Chauhan, A. S. (2019). A review on mechanical and wear characteristics of particulate reinforced Al-alloy based MMC. In AIP Conference Proceedings, vol. 2148, no. 1. AIP Publishing.

[6] Velavan, K. and Palanikumar, K. (2020). 'Analysis on sliding wear behavior of Al+ B4C+ mica hybrid metal matrix composites. Mater. Exp. 10(7), 986–997.

[7] Arunachalam, R. and Krishnan, P. K. (2021). Compressive response of aluminum metal matrix composites, Elsevier BV.

[8] Tian, S., Li, J., Zhang, J., Wulabieke, Z., Lv, D. (2019). Effect of Zr and Sc on microstructure and properties of 7136 aluminum alloy, *J. Mater. Res. Technol.*

[9] Krishnan, P. K., Christy, J. V., Arunachalam, R., Mourad, A.-H. I., Muraliraja, R., Al-Maharbi, M., Murali, V., and Chandra, M. M. (2019). Production of aluminum alloy-based metal matrix composites using scrap aluminum alloy and waste materials: Influence on microstructure and mechanical properties. *J. Alloys Comp.*, 784, 1047–1061.

[10] AbuShanab, W.S., Moustafa, E. B. (2020). Effects of friction stir processing parameters on the wear resistance and mechanical properties of fabricated metal matrix nanocomposites (MMNCs) surface, *J. Mater. Res. Technol.*

[11] Chinnapand, M., Boopathy, G., and Vijayakumar, K. R. (2017). Fabrication and fatigue analysis of laminated composite plates. *Int. J. Mech. Eng. Technol.* 8(7), 388–396.

[12] Kumar, G. B. V., Pramod, R., Shivakumar Gouda, P.S., and Rao, C. S. P. (2019). Effect of tungsten carbide reinforcement on the aluminum 6061 alloy. *J. Testing Eval.*, 47(4), 2613–2629.

[13] Boopathy, G., Vijayakumar, K. R., Chinnapandian, M., and Gurusami, K. (2019). Development and experimental characterization of fibre metal laminates to predict the fatigue life. *Int. J. Innov. Technol. Explor. Eng.*, 8(10), 2815–2819.

[14] Hashem, E. G.S. (2023). Study of the mechanical properties of the metal matrix composites developed with alumina particles, Elsevier BV.

[15] Khelge, S., Kumar, V., Shetty, V., and Kumaraswamy, J. (2022). Effect of reinforcement particles on the mechanical and wear properties of aluminium alloy composites. *Mater. Today Proc.*, 52, 571–576.

[16] Kumaar, R. K., Kannan, G., Boopathy, G., and Surendar, G. (2017). Fabrication and computational analysis of cenosphere reinforced aluminum metal matrix composite disc brakes. *Technology*, 8(6), 553–563.

[17] Singh, S., Singh, R., and Gill, S.S. (2021). Investigations for wear characteristics of aluminium-based metal matrix composite prepared by hybrid reinforcement. *Proc. National Acad. Sci. India Sect. A Phys. Sci.* 91, 569–576.

[18] Abouzeid, R., Dardeer, H., Mahgoub, M., Abdelkader, A. (2021). Adsorption of cationic methylene blue dye on polystyrene sulfonic acid composites from waste: Kinetics and equilibrium., *Egypt. J. Chem.*

[19] Krishna, U.B. G., Vasudeva, B., Auradi, V., and Nagaral, M. (2021). Effect of percentage variation on wear behaviour of tungsten carbide and cobalt reinforced Al7075 matrix composites synthesized by melt stirring method. *J. Bio-and Tribo-Corrosion*, 7(3), 89.

[20] Boopathy, G., Gurusami, K., Chinnapandian, M., and Vijayakumar, K. R. (2022).

Optimization of process parameters for injection moulding of nylon 6/SiC and nylon 6/B4C polymer matrix composites. *Fluid Dyn. Mater. Process.*, 18(2), 223–232.

[21] Kumar, V., Nagegowda, K. U., Boppana, S. B., Sengottuvelu, R., and Kayaroganam, P. (2021). Wear behavior of Aluminium 6061 alloy reinforced with coated/uncoated multiwalled carbon nanotube and graphene. *J. Metals Mater. Minerals*, 31(1).

[22] Sagar, K. G., Suresh, P. M., and Sampathkumaran, P. (2021). Addition of beryl content to aluminum 2024 alloy influencing the slide wear and friction characteristics. *J. Inst. Eng. (India) Ser. C*, 102(1), 27–39.

[23] Morampudi, P., Venkata Ramana, V. S. N., Bhavani, K., Reddy, C. K., and Sri Ram Vikas, K. (2022). Wear and corrosion behavior of AA6061 metal matrix composites with ilmenite as reinforcement. *Mater. Today Proc.*, 52, 1515–1520.

[24] Boopathy, G., Vanitha, V., Karthiga, K., Gugulothu, B., Pradeep, A., Pydi, H. P., & Vijayakumar, S. (2022). Optimization of tensile and impact strength for injection moulded Nylon 66/Sic/B 4 c Composites. *Journal of Nanomaterials*, 2022.2022.

[25] Sharma, A. K., Bhandari, R., and Pinca-Bretotean, C. (2021). Impact of silicon carbide reinforcement on characteristics of aluminium metal matrix composite. *J. Phys. Conf. Ser.*, 1781(1), 012031. IOP Publishing.

[26] Boopathy, G., Udaya Prakash, J., Gurusami, K., and Sai Prasanna Kumar, J. V. (2022). Investigation on process parameters for injection moulding of nylon 6/SiC and nylon 6/B4C composites. *Mater. Today Proc.*, 52, 1676–1681.

[27] Deshmukh, S., Joshi, G., Ingle, A., and Thakur, D. (2021). An overview of aluminium matrix composites: Particulate reinforcements, manufacturing, modelling and machining. *Mater. Today Proc.*, 46, 8410–8416.

[28] Shaik, D., Sudhakar, I., Bharat, G. C. S. G., Varshini, V., and Vikas, S. (2021). Tribological behavior of friction stir processed AA6061 aluminium alloy. *Mater. Today Proc.*, 44, 860–864.

[29] Smart, D. S. R., Pradeep Kumar, J., and Periasamy, C. (2021). Microstructural,

mechanical and wear characteristics of AA7075/TaC/Si3N4/Ti based hybrid metal matrix composite material. *Mater. Today Proc.*, 43, 784–794.

[30] Nagavendra K., Sharma, S. C., Harsha, S. P. (2014). EPFM analysis of subsurface crack beneath a wheel flat using dynamic condition. *Sci. Direct Proc. Mater. Sci.*, 6, 43–60.

[31] Kim, Y. H., Ryou, T. K., Choi, H. J., and Hwang, B. B. (2002). An analysis of the forging processes for 6061 aluminum-alloy wheels. *J. Mater. Process. Technol.*, 123(2), 270–276.

[32] Huang, M., Chen, J. (2023). A novel proactive soft load balancing framework for ultra dense network, *Dig. Commun. Netw.*

[33] Khalid, M. Y., Umer, R., Khan, K. A. (2023). Review of recent trends and developments in aluminium 7075 alloy and its metal matrix composites (MMCs) for aircraft applications, *Results Eng.*

[34] Kumar, N., Vasanth Kumar, H. S., Hemanth Raju T., Nagaral, M., Auradi, V., Veeresha, R. K. (2022). Microstructural characterization, mechanical and taguchi wear behavior of micro-titanium carbide particle-reinforced Al2014 alloy composites synthesized by advanced two-stage casting method, *J. Bio- Tribo-Corrosion.*

[35] Soltani, S., Azari Khosroshahi, R., Taherzadeh Mousavian, R., Jiang, Z. Y., Fadavi Boostani, A., & Brabazon, D. (2017). Stir casting process for manufacture of Al–SiC composites. *Rare Metals*, 36, 581–590.

[36] Dionysopoulos, D. and Gerasimidou, O. (2021). Wear of contemporary dental composite resin restorations: a literature review, *Restorat. Dentis. Endodon.*

[37] Gokhan Zengin, N., Yildiztugay, E., Bouyahya, A., Cavusoglu, H., Gevrenova, R., Dimitrova, D. Z. (2023). A comparative study on UHPLC HRMS profiles and biological activities of Inula sarana different extracts and its beta cyclodextrin complex: effective insights for novel applications, *Antioxidants.*

[38] Kumaraswamy, J., Anil, K. C., Shetty, V. (2023). Development of Ni-Cu based alloy hybrid composites through induction furnace casting, *Mater. Today Proc.*

17 A survey on hydrogen energy

V. S. Sukshith[a], K. L. Rajeev, S. H. Prashant, and S. B. Halesh

Department of Mechanical Engineering, Sir M Visvesvaraya Institute of Technology, Bengaluru, India

Abstract

Hydrogen energy is a clean and sustainable form of energy that uses hydrogen as fuel. It involves using hydrogen to generate electricity through the fuel cells or combustion, with water and the only by product. It's a promising alternative to fossil fuels because it doesn't produce greenhouse gas emissions, and it is produced the renewable energy sources like water and sunlight. Hydrogen energy is its own potential to revolutionize our energy systems and help combat climate change. The energy changes not something that expect us for the next 10 years. For opposing, the process where we are already intensely enlisted. The leading step is to neutralize a carbon society for the carrying out the process of renewable energy sources (RES) in a replacement of the fossil fuels. It gives an alternate form of RES is a form of energy storage which produce an essential role for this transition. The technology of hydrogen energy to have more advancements, which is recognized the energy to be most suitable choice. It is more as a hydrogen energy applications are researched in the recent time, in this current situation to obtain the hydrogen energy technology was not yet a large-scale implementation situation level. When there are increasing numbers of works based on studies and particular projects, in that usage of hydrogen energy is the sustainable potential was expected in the next few decades. They rapidly changing era of this technology and this technological advancement brings the fourth public discussions, that decide the factor of society should be able to alter the hydrogen energy and should accept the new technology or reject it. In this present situation the hydrogen energy is the best related with fuel cell electric vehicles they emits water vapors and dry air is released and not producing any harmful emissions. There are many scientists who are stressed on the concept of gravity, they produces the effects that is stretching out the physical environment and ecosystems are the general humanity, which is concerned with the future, for becoming a global topic. As a result the government implementing the new suitable products that promote renewable energy sources as a replacement for the fossil fuels. In the increased development of hydrogen energy technology in the country and worldwide, with the incorporate the hydrogen energy regulations and the national development plans are results in the numbers of national hydrogen energy strategies.

Keywords: Renewable energy sources (RES), fuel cell electric vehicles

1. Introduction

Hydrogen is simple and the most common element in the universe, which holds the exciting potential. The world is in need of more energy available in the past years. The growing demand, which means we need a more-sustainable supply of energy and the generation system, one that will meet those demands, whereas addressing the carbon dioxide emissions and the overall impact on the hydrogen energy generation in the environment [1].

Hydrogen energy can be produced using various methods, including the natural gas forming, electrolysis, solar-driven, and biological process. When hydrogen is consumed in fuel cells, the only emission that use is the water vapor, making it a clean and green energy source. However, the production of hydrogen from fossil fuels can lead to release the carbon dioxide into the atmosphere.

Hydrogen energy is rapidly growing area as the alternate for the both petrol and natural gases, this makes a prospect of low or

[a]sukvimath@gmail.com

DOI: 10.1201/9781003545941-17

even zero carbon energy grid which is genuine reality. The use of hydrogen energy as a fuel source as the potential energy to reduce the greenhouse gas emission and to mitigate climate changes.

Hydrogen energy systems which offers a great potential energy for the solution to ever increasing in the demand for a sustainable energy system. For the long term and ultimate technological challenge that is large-scale hydrogen production energy from renewable sources, a critical and the practical issue which stores the hydrogen energy efficiently and safely, for the particular purpose for obtaining hydrogen fuel cell vehicles. The tremendous efforts which have been devoted to the research and development of new systems that can hold sufficient hydrogen energy to make a satisfactory driving range [2].

2. Mathematical Model

2.1. Mathematical Model of a Survey on Hydrogen Energy

2.1.1. Production

Hydrogen energy is produced the various methods, with most common being electrolysis and steam methane reforming. For example, in electrolysis, the water is splits into the hydrogen and oxygen using the electric current. In which the steam methane reforming, hydrogen is extracted from methane in natural gas.

2.1.2. Storage

The storing hydrogen efficiently is a crucial aspect. It's often a stored as gas in which high-pressure tanks or as the liquid at very low temperature. Recently, there's been exploration into materials like metal hydrides and carbon-based materials for solid-state storage.

2.1.3. Utilization

Hydrogen energy is used in the fuel cells which produces electricity through an electrochemical reaction with the oxygen. This electricity can power various applications, from vehicles to industrial processes. The byproduct of this reaction is water.

The mathematical models involved in these processes which makes a quite complex and especially when considering factors like efficiency, thermodynamics, and reaction kinetics. Engineers and scientists use various equations to optimize these processes and improve the overall efficiency of a hydrogen energy systems.

3. Objectives

The objectives of hydrogen energy revolve around harnessing hydrogen as a clean and sustainable energy carrier. Here are some key objectives:

3.1. Green Energy Production

One primary objective is to produce a hydrogen by using a renewable energy sources, for example solar, wind, or hydropower. This ensures that the hydrogen production process itself is environmentally friendly and doesn't contribute to carbon emissions.

3.2. Energy Storage

Hydrogen can serve as a means of storing surplus energy generated from irregular renewable sources. By converting surplus energy into hydrogen during periods of abundance, it can also be stored and later utilized when the demand is more or when renewable sources are not actively generating power.

3.3. Clean Transportation

Hydrogen fuel cells which are used in power vehicles, offering an alternative to traditional internal combustion engine which rely on fossil fuels. The objective is to reduce carbon emission from the transportation sector and decrease dependence on non-renewable energy sources.

3.4. Industrial Level

Hydrogen energy is the useful fuel that can be used in various industrial processes, such as manufacturing and refining. The objective is to replace the traditional fossil fuels in these applications, reducing carbon emissions and promoting sustainability in industries.

3.5. Power Generation

Hydrogen is used in fuel cells to generate electricity for both stationary and portable applications. The objective is to make easy to provide a clean and efficient energy source for powering houses, businesses, and electronic devices.

3.6. Decentralized Energy Systems

Hydrogen energy can also helpful to the development of decentralized energy systems, allowing communities to generate and use their own energy locally. This can enhance energy security and resilience.

3.7. Research and Innovation

Ongoing research and innovation are essential objectives is to upgrade the efficiency and cost-effectiveness for a producing the hydrogen energy and utilization of technologies. Advancements in materials science, engineering, and chemistry which are successful and crucial for carrying through the hydrogen energy solutions [2].

3.8. Economic Viability

To make a hydrogen energy economically viable, ensuring that it can also compete with the traditional energy sources in-terms of cost. This involves reducing production costs, improving efficiency, and creating incentives for the acquisition of hydrogen technologies.

By achieving these objectives, the goals is to establish hydrogen as plays critical key in global transition for the suitable and low-carbon energy systems.

4. Advantages

Hydrogen energy which offers the several advantages, making a promising approach for the sustainable and clean energy. When energy produced using the renewable energy sources such as solar, wind or hydropower, hydrogen becomes the clean and renewable energy carrier. Its combustion of its use in fuel cells they produce only water vapor as a byproduct, contributing to significant reduction in greenhouse gas emission systems.

Hydrogen is a sufficient energy that can be used in the various sectors, they are including the transportation, industry, and power generation. Its adaptability makes it a valuable component in addressing energy needs across different applications. It provides an alternative to traditional fossil fuels, helping to reduce dependence on non-renewable resources. This is particularly important for mitigating climate change and ensuring long-term energy security [3].

The pursuit of hydrogen energy has driven technological innovation in areas such as electrolysis, fuel cells, and storage technologies. Continued research and development in these fields can lead to further improvements in efficiency and cost-effectiveness.

Hydrogen has garnered international attention, leading to collaborative efforts between countries and industries to advance hydrogen technologies. This global collaboration can accelerate the development and adoption of hydrogen energy solutions.

While hydrogen energy holds significant promise, it's important to address challenges such as production efficiency, storage solutions, and infrastructure development to fully realize its potential (Figure 17.1).

5. Overview

The overview of hydrogen energy reveals a modern approach to sustainable and clean energy solutions. At its core, hydrogen serves as a versatile of energy carrier with potential to

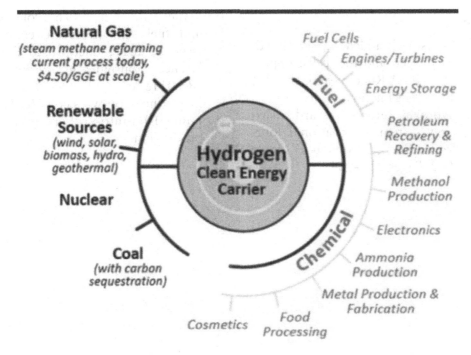

Figure 17.1: The hydrogen clean energy carrier [4].

a revolution of various sectors. Through methods like electrolysis and steam methane reforming, hydrogen production has become a focal point, especially when combined with renewable energy sources for environmental benefits.

Storage mechanisms, ranging from high-pressure tanks to advanced materials, aim to make energy accessible when and where it's needed. The utilization of hydrogen, notably in fuel cells, presents an opportunity for zero-emission power generation, transforming transportation, industry, and residential energy use.

While challenges such as production efficiency and infrastructure development persist, international collaboration and ongoing research signify a shared commitment to realizing hydrogen's potential. The comprehensive overview of hydrogen energy encapsulates its role of key player in the environmental pursuit of a sustainable and low-carbon energy future.

6. Conclusion

In conclusion, hydrogen energy stands at the forefront of transformative advancements in the search for sustainable and cleaner energy solutions. Its unique qualities, including versatility, zero-emission potential, and efficiency in energy storage, position hydrogen as a key player across various industries. The global focus and collaborative efforts in research and development emphasize the shared commitment to harnessing hydrogen's potential. However, challenges such as production efficiency, infrastructure development, and cost-effectiveness remain critical hurdles. Overcoming these challenges requires continued innovation and concerted efforts from governments, industries, and the scientific community. As we navigate the complexities of the energy transition, hydrogen emerges not only as a clean fuel source but as a catalyst for a paradigm shift toward a resilient, low-carbon, and sustainable energy future.

Acknowledgement

The authors are gratefully of acknowledge the Professor, Head of the Department of Mechanical Engineering for their cooperation in the research of new technologies.

References

[1] Hart, D., Howes, J., Madden, B., and. Boyd, E. (2016). Hydrogen and fuel cells: opportunities for growth. A Roadmap for the UK, E4Tech and Element Energy.

[2] Dunn, S. Hydrogen futures: toward a sustainable energy system, International Journal of Hydrogen Energy, Volume 27, Issue 3, March 2002, Pages 235–264.

[3] Staffell, I., et al., "The role of hydrogen and fuel cells in future energy systems:, Issue 2, 2019, Energy and Environmental Science Journal, https://doi.org/10.1039/c8ee01157e

[4] https://neutronbytes.com/2020/11/12/feds-pump-10m-into-xcel-energy-for-hydrogen-production.

18 Machinability performance investigation and increasing tool life of TiAlN

K. Karthik[1,a], K. Bravilin Jiju[2], B. Madhu[3], N. Sathiyapriya[4], P. Duraimurugan[5], and B. Murali[1]

[1]Department of Mechanical Engineering, Vel Tech Rangarajan Dr. Sagunthala R&D Institute of Science and Technology, Chennai, India

[2]Department of Mechanical Engineering, ICCS College of Engineering and Management, Thrissur, India

[3]Department of Mechanical Engineering, Sri Sairam College of Engineering, Bengaluru, India

[4]Department of Electronics and Communication Engineering, Karpagam Institute of Technology, Coimbatore, India

[5]Department of Mechanical Engineering, Sri Sairam Engineering College, Chennai, Tamil Nadu, India

Abstract

This paper examines the Interaction Streamlining to increment device life by accomplishing Six Sigma. The target of the cycle is to acquire an expanded apparatus life utilising ideal setting of interaction boundaries—cutting rate, feed, profundity of cut and rpm which might bring about augmenting device life of TiAlN covered embeds for reaming. Information is gathered from CNC machines which were controlled by 12 examples of analyses. The contributions to the cycle comprises of feed, cutting velocity, profundity of cut and rpm while the result from the machine for apparatus life and instrument wear was estimated utilising Picture Analyzer Magnifying lens. The result is approved through an examination of the exploratory qualities with their anticipated partners. The direct and connection impact of the machining boundary with apparatus wear were examined and plotted utilising variable hunt (six sigma device), which assisted with choosing process boundary to diminish instrument wear which guarantees nature of reaming and counter exhausting.

Keywords: Tool life, TiAlN tool, six sigma, CNC program, experimental technique

1. Introduction

Device life is a significant perspective ordinarily viewed as in assessing the exhibition of a machining cycle [1–3]. It is notable that during the machining, the cutting boundaries, for example, cutting velocity, feed rate and profundity of cut frequently present a deviation from the determined qualities. Besides, it has been seen that the improvement in the result factors might bring about a huge efficient execution of machining tasks [4–6]. The angle should be considered for the instrument life is cutting power on the work piece and the temperature during machining that is oppressed by the shaper devices and apparatus wear can be estimated by utilising magnifying lens [7]. Ideal device life is an extraordinary worry in assembling conditions where, economy of machining activity assumes a vital part in seriousness on the lookout. The reaming system contrasted with the other metal machining process is slow, consequently having a low creation rate. Despite

[a]karthikmeed@gmail.com

DOI: 10.1201/9781003545941-18

the fact that NC machines' capability is to lessen lead times significantly, the machining time is practically equivalent to regular machining where machining boundaries are chosen from machining information bases or handbooks [8–14]. Mistake of slicing apparatus add to unfortunate surface completion, instrument harm, gab, layered incorrectness and numerous different issues that add to low efficiency and much opportunity to be squandered. One of the well-known cutting apparatuses that are utilised is covered supplements. This study assists with working on the exhibition of reaming process by utilising covered TiAlN device as a shaper for the ideal presentation. This undertaking can likewise assist with massing creation machining in the business [15–19].

2. Methodology

Problem Definition: The following factors initiated the selection of this research. To achieve reduction of overall tool cost/ component, improving profitability. Objective of the project: Finish bore reamer tool life improvement by optimising bore operation process parameters [20–21].

Tool Cost per Component, Existing tool life: A No's (Average)

- Tool cost : Rs.B
- Existing tool cost / comp : Rs C
- Expected tool cost / comp: Rs 1.76 considering 10,000 No's as target.
- To achieve above target life we have to run 60 shifts considering 172 No's / shift is designed output / shift (as per cycle time).

Hence we are establishing the quality requirements and tool life is being monitored in the mean time due to time constraints. Parameters Identification, In this stage measured values has been analyzed and listed below Table. 18.1.

Product Specification
RESPONSES (Y)
Y1 - bore diameter: 54.000/54.050 mm
Y2 - bore finish: 5 microns.
Y3 - tool life: 10,000 No's (Target)

Table 18.1: Validation of causes

Possible causes	Current condition	Recommendation	Remarks
Speed	707 rpm	Optimum to be arrived	To be experimented
Feed	1060 rpm	Optimum to be arrived	To be experimented
Amount of material removal by reamer tool	0.9 mm	0.4 mm	Manufacturing standard and recommended by tool manufacturer also. Implemented in shop floor and tool drawing regularised.
Tool run out at spindle	60 micron	10 micron	Implemented in shop floor. To attain the lowest run out , SOP updated for spindle orientation position with respect to shop floor operator to ensure the same during tool change
Coolant flow	Adequate	Nil	Nil
Coolant concentration	3 ~ 6%	6.5 ~ 8.5%	Tool manufacturer recommendation. Implemented in shop floor.
Coating on TiAlN tool	Nil		Nil

The Figures 18.1 and 18.2 shoes the Tool holder orientation position A and B.

The normally used N95 respirators are of negative pressure variant i.e., they require the wearer's lungs to inhale air through the resistive membranes of the filter layers. This is strenuous and uncomfortable to wear for a long duration. This is non-existent in positive air pressure respirators as they use external filters and has a motorised air supply system. The pandemic in recent scenario also necessitates respiration apparatus as a part of its treatment. Respirators that are in commonly used are negative pressure system which require the power of lungs to draw-in purified air which is not suitable and sometimes not possible if the person lacks sufficient lungs strength, or if they suffer from respiratory illness. This work proposes a forced air (positive air pressure) solution to the problem.

The present scenario has shown the increased threats represented by respiratory illness like chronic obstructive pulmonary disease (COPD), asthma etc. That risk has increased due to increase in air pollutants like PM2.5, PM10 etc. a respirator can be utilised as an immediate countermeasure on an individual level safety measure as bringing down pollution levels require much longer time than the severity of the problem is allowing. The normally used N95 respirators are of negative pressure variant i.e., they require the wearer's lungs to inhale air through the resistive membranes of the filter layers. This is strenuous and uncomfortable to wear for a long duration. This is non-existent in positive air pressure respirators as they use external filters and has a motorised air supply system. The pandemic in recent scenario also necessitates respiration apparatus as a part of its treatment. Respirators that are in commonly used are negative pressure system which require the power of lungs to draw-in purified air which is not suitable and sometimes not possible if the person lacks sufficient lungs strength, or if they suffer from respiratory illness. This work proposes a forced air (positive air pressure) solution to the problem.

Above the Figures 18.3 and 18. 4 shows the input parameter of cutting tools and different

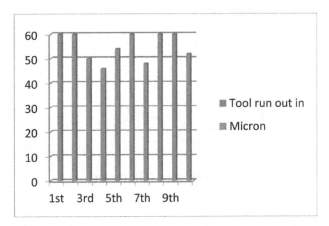

Figure 18.1: Tool holder orientation position A.

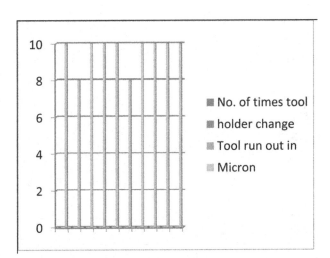

Figure 18.2: Tool holder orientation position B.

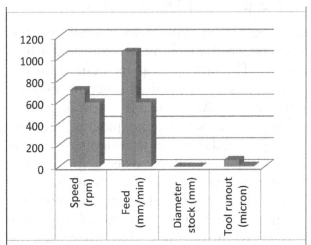

Figure 18.3: Initial parameters.

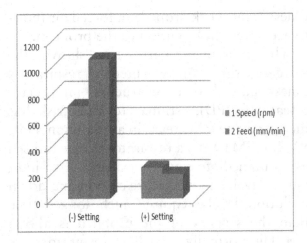

Figure 18.4: Optimal settings.

Feed pace of 590 RPM and 590 mm/min individually, an ostensible worth of bore measurement of the chamber of circle brakes are gotten and recorded.

Figure 18.6 shows the Main effects diameter of cutting tool. The different qualities (breadth) acquired at different various settings are recorded and plotted in the above given plot chart. The mean worth of each setting is recorded and plotted on a diagram. It tends to be seen that as the speed builds the pace of material evacuation likewise increments and as the feed increments not a very remarkable enormous change is viewed as contrasted and

definitions and hypothetical information the ideal settings have been characterised. The excellent boundaries which were influencing the apparatus life of the reamer instrument were distinguished as Cutting Velocity and FEED rate. The above given table outlines two different settings, one being the underlying setting ((-) setting) and the other being the ongoing setting ((+) setting). The underlying setting was characterised by the device creator for the proposed number of hardware life. The ongoing setting is the most un-conceivable setting under which machining can occur. The proposed number of hardware life as far as parts produced per instrument was 10,000.

3. Results and Discussions

A shape plot is utilised to characterise the reach between which an ostensible worth is wanted. The different variety designs on the chart portrays the scope of different distance across values. An ostensible worth is wanted in light of the fact that the qualities between starting boundaries and the ongoing boundaries end up being above ostensible and thus the worth between this reach is liked to expand the device life of reamer instrument.

The above Figure 18.5 shows the chart portrays of the qualities which are gotten for the chose boundaries. At improved Speed and

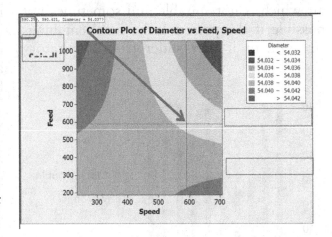

Figure 18.5: Contour plot (Diameter).

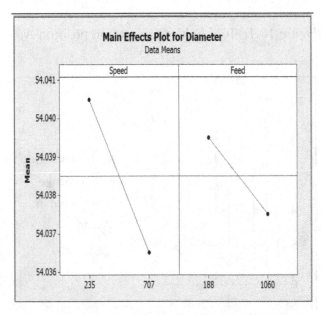

Figure 18.6: Main effects plot (DIAMETER).

the qualities got for cutting rate. An ideal setting can be gotten from this diagram by choosing a reach between which the breadth of the completed item ought to be. This should be possible with the assistance of the form plot graph (bore distance across).

The Figure 18.7 shows the Main effects diameter of cutting tool. Cooperation plot is utilised to give out the between connection between the Cutting Velocity and the FEED rate. The mean upsides of all breadth readings are taken and plotted on the diagram for both starting setting and the other different setting preliminaries. In the above chart the speed in beginning setting (707 RPM) and the speed in current setting (235 RPM) are plotted concerning bore distance across and FEED rate. The subsequent lines collaborate at one point to portray the ideal drag distance across concerning the ideal speed settings. The red specked line addresses the underlying pace settings and the dark line addresses the ongoing rate settings.

The Pareto Figure 18.8 shows the fondness of the reason with the impact. To find what factor for the most part influences the drag width of the item, a Pareto diagram is drafted. In the above diagram terms An and B address SPEED and FEED separately. On plotting the outcomes acquired from past diagrams, SPEED has greater partiality towards influencing bore measurement on examination with FEED rate.

In any case, both these variables set up have a more noteworthy fondness towards influencing the drag measurement on examination with SPEED and FEED independently. Subsequently it is clarified from the above chart that both SPEED and FEED assume a significant part in influencing the drag breadth.

The above given Figure 18.9 contour diagram portrays the qualities which are acquired for the chose boundaries. At enhanced Speed and Feed pace of 590 RPM and 590 mm/min individually, an ostensible worth of bore finish of the chamber of plate brakes are gotten and recorded. A form plot is utilised to characterise the reach between which an ostensible worth is wanted. The different variety design on the diagram portrays the scope of

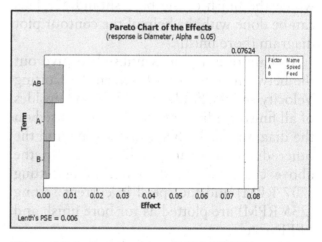

Figure 18.8: Pareto chart (DIAMETER).

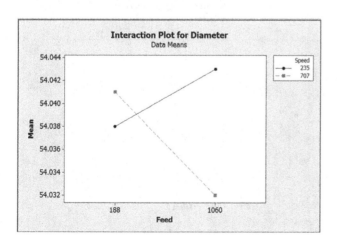

Figure 18.7: Interaction plot (DIAMETER).

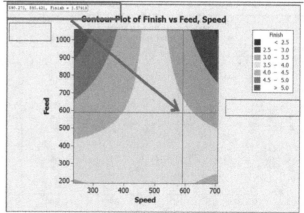

Figure 18.9: Contour plot (FINISH).

different completion values (in microns). An ostensible worth is wanted in light of the fact that the qualities between starting boundaries and the ongoing boundaries end up being above ostensible and subsequently the worth between this reach is liked to expand the device life of reamer apparatus.

The Figure 18.10 Main effects the various values (diameter) obtained at various different settings are recorded and plotted in the above given plot graph. The mean value of each setting is recorded and plotted on a graph. It can be seen that as the speed increases the rate of finish decreases and as the feed increases not much of a large change is seen as compared with the values obtained for cutting speed. An optimum setting can be obtained from this graph by selecting a range between which the finish of the bore should be. This can be done with the help of the contour plot diagram (bore finish).

Collaboration plot is utilised to give out the between connection between the Cutting Velocity and the FEED rate. The mean upsides of all finish readings are taken and plotted on the diagram for both starting setting and the other different setting preliminaries. In the above chart the speed in beginning setting (707 RPM) and the speed in current setting (235 RPM) are plotted as for bore finish and FEED rate.

The subsequent lines collaborate at one point to portray the ideal drag get done regarding the ideal speed settings. The red specked line addresses the underlying pace settings and the dark line addresses the ongoing rate settings.

The Figures 18.11 and 18.12 shows the Interaction plot and Pareto outline shows the liking of the reason with the impact. To find what factor generally influences the drag finish of the item, a Pareto diagram is drafted. In the above diagram terms An and B address SPEED and FEED separately. On plotting the outcomes acquired from past charts, SPEED has greater fondness towards influencing bore finish on correlation with FEED rate, which is practically unimportant.

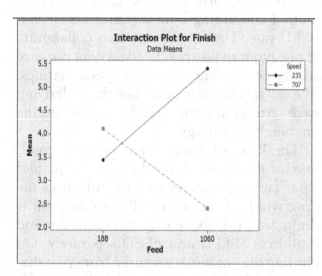

Figure 18.11: Interaction plot (FINISH).

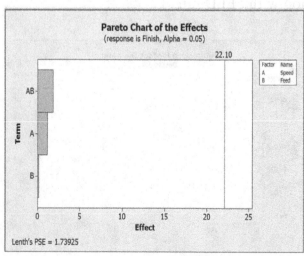

Figure 18.12: Pareto chart (FINISH).

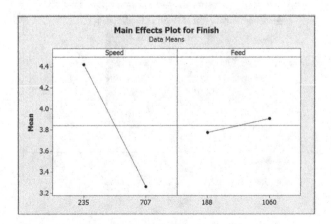

Figure 18.10: Main effects plot (Finish).

However, both these variables set up have a more noteworthy liking towards influencing the drag width on examination with SPEED and FEED independently.

Consequently it is clarified from the above diagram that both SPEED and FEED assume a significant part in influencing the drag finish. The further develop stage managed concluding the ideal and ideal settings to figure out the main boundaries for example Cutting Pace and FEED rate. As given in the above table, the finished ideal settings were 590 rpm (Speed) and 590 mm/min (Feed).

The cycles and information gathered from past advances are recorded and the best strategy is utilised or placed into use in this stage. The outcomes got out of executing the ideal settings for both drag breadth and bore finish are as per the following.

By quantifying the improvements for both bore diameter and bore finish, the tool life of reamer tool was found to be increased. The above graph clearly shows the increase in tool life as a result of achieving optimum settings. The reason for this stage is to support the additions. Screen the enhancements to guarantee proceeded and practical achievement. Make a control plan. Update reports, business cycle and preparing records as required. A Control outline can be useful during the Control stage to assess the constancy of the redesigns for a really long time by filling in as (1) a manual for continue to actually look at the cycle and (2) give a response plan to all of the activities being seen if the collaboration becomes unstable. The CONTROL stage is the completion of the endeavor. The Figures 18.13 and 18.14 tool life improvement. The last not permanently set up and the end execution and all associated changes are accounted for. This stage isn't by and large so truly thought as the Activity, Explore and Additionally foster stages. Updation of Related Archives

- CNC program for reamer instrument activity changed for improved boundaries.

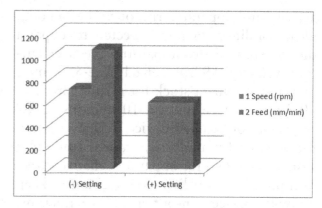

Figure 18.13: Significant parameters.

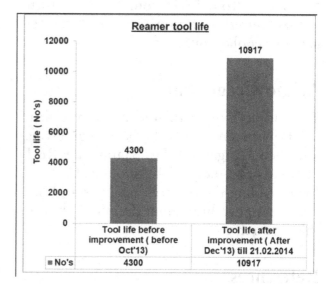

Figure 18.14: Tool life improvement.

- Standard working strategy refreshed for instrument change direction position concerning axle position to keep up with device run-out inside 10 microns.
- Harsh drag instrument drawing refreshed for 53.60 mm from 53.10 mm.
- Coolant focus graph refreshed for expansion in convergence of coolant oil. or if they suffer from respiratory illness. Figures 18.15 and 18.16 shows CNC program before improvement.

4. Conclusion

It is shown that the device wear in reaming diminishes with the abatement in feed rate, decline in measurement stock for reaming activity and apparatus run out. The ongoing ideal condition that is expected to expand the covered TiAlN apparatus life are: cutting velocity 590rpm, feed rate-590 mm/min, measurement stock for reaming activity –0.5 mm, device run out-10 microns. CNC program for reamer device activity altered for enhanced boundaries. Standard working system refreshed for apparatus change direction position regarding axle position to keep up with device run-out inside 10 microns. Harsh drag apparatus drawing refreshed for 53.60 mm from 53.10 mm. Coolant focus outline refreshed for expansion in convergence of coolant oil.

Acknowledgement

The authors express their sincere gratitude to the Department of Mechanical Engineering, Vel Tech Rangarajan Dr. Sagunthala R&D Institute of Science and Technology, Chennai, for providing the research facilities and necessary support throughout the entire phase of this research work.

References

[1] Hasçalık, A. et al. (2008). Optimization of turning parameters for surface roughness and tool life based on the Taguchi method. *Int. J. Adv. Manuf. Technol.*, 38, 896–903

[2] Ravikumar, M. M. et al. (2009). Implementation of lean manufacturing in automotive manufacturing plant. *Int. J. Appl. Eng. Res.*, 4 (10), 2041–2051.

[3] Vijayakumar, M. D. et al. (2020) Experimental investigation on single point incremental forming of IS513Cr3 using response surface method. *Mater. Today*, 21 (1), 902–907.

[4] Adithiyaa, T., et al. (2020). Flower pollination algorithm for the optimization of stair casting parameter for the preparation of AMC. *Mater. Today*, 21 (1), 882–886.

[5] Gnanavel, B. K. et al. (2020). Effect of interfacial contact forces and lay ratio in cardiac lead outer insulation due to internal cable motion. *ASME Int. Mech. Eng. Congress Expo. (IMECE)*, 12, V012T12A038.

[6] Banu, M., Madhavan, V. R. B., Manickam, D., and Devarajan, C. (2021). Experimental investigation on stacking sequence of Kevlar and natural fibres/epoxy polymer composites. *Polimeros: Ciencia e Tecnologia*, 31 (1), 1–9.

[7] Murali, B., Vijaya Ramnath, B. M., Rajamani, D., Nasr, E. A., Astarita, A., and Mohamed, H. (2022). Experimental investigations on dry sliding wear behaviour of Kevlar and natural fiber-reinforced hybrid composites through an RSM–GRA hybrid approach. *Materials*, 15 (3), 1–16.

[8] Murali, B., Yogesh, P., Karthickeyan, N. K., and Chandramohan, D. (2022). Multipotency of bast fibers (flax, hemp and jute) as composite materials and their mechanical properties: a review. *Mater. Today.* 62, 1839–1843.

[9] Karthik, K., and Manimaran, A. (2020). Wear behaviour of ceramic particle reinforced hybrid polymer matrix composites. *Int. J. Ambient Energy*, 41 (14), 1608–1612.

[10] Ganesh, R., Karthik, K., Manimaran, A., and Saleem, M. (2017). Vibration damping characteristics of cantilever beam using piezoelectric actuator. *Int. J. Mech. Eng. Technol.*, 8 (6), 212–221.

[11] Karthik, K., Ganesh, R., and Ramesh, T. (2017). Experimental investigation of hybrid polymer matrix composite for free vibration test. *Int. J. Mech. Eng. Technol.*, 8 (8), 910–918.

[12] Ahmed, I., Renish, R. R., Karthik, K., and Karthik, M. (2017). Experimental investigation of polymer matrix composite for heat distortion temperature test. *Int. J. Mech. Eng. Technol.*, 8 (8), 520–528.

[13] Murali, B., and Nagarani, J. (2013, April). Design and fabrication of construction helmet by using hybrid composite material. In 2013 International Conference on Energy Efficient Technologies for Sustainability (pp. 145–147). IEEE.

[14] Karthik, K., Prakash, J. U., Binoj, J. S., & Mansingh, B. B. (2022). Effect of stacking sequence and silicon carbide nanoparticles on properties of carbon/glass/Kevlar fiber reinforced hybrid polymer composites. *Polym. Compos.*, 43 (9), 6096–6105. https://doi.org/10.1002/pc.26912.

[15] Fayaz, H., Karthik, K., Christiyan, K. G., Kumar, M. A., Sivakumar, A., Kaliappan, S., Mohamed, M., Subbiah, R., and Yishak, S. (2022). An investigation on the activation energy and thermal degradation of biocomposites of Jute/Bagasse/Coir/Nano TiO 2/epoxy-reinforced polyaramid fibers. *J. Nanomater.*, 2022, 5. ArticleID 3758212. https://doi.org/10.1155/2022/3758212.

[16] Karthik, K., Rajamani, D., Venkatesan, E. P., Shajahan, M.I., Rajhi, A. A., Aabid, A., Baig, M., and Saleh, B. (2023). Experimental investigation of the mechanical properties of carbon/basalt/sic nanoparticle/polyester hybrid composite materials. *Crystals*, 13 (3), 415.

[17] Ramesh, V., Karthik, K., Cep, R., and Elangovan, M., 2023. Influence of stacking sequence on mechanical properties of basalt/ ramie biodegradable hybrid polymer composites. *Polymers*, 15 (4), 985.

[18] Sathish, T. et al. (2021). A facile synthesis of Ag/ZnO nanocomposites prepared via novel green mediated route for catalytic activity. *Appl. Phys. A*, 127 (9), 1–9.

[19] Logendran, D. et al. (2020). Microstructural analysis of Friction stir welded AA 7010–titanium diboride (TiB2) reinforced composites. *Mater. Today*, 33, 4663–4665.

[20] Rajesh, S., et al. (2020). Machining parameters optimization of surface roughness analysis for AA5083 - Boron carbide (B4C) composites. Mater. Today, 33, 4642–4645.

[21] Gurusami, K. et al. (2021). A comparative study on surface strengthening characterisation and residual stresses of dental alloys using laser shock peening. *Int. J. Ambient Energy*, 42(15), 1740–1745.

19 A comparative study on performance of RC beam retrofitted with CFRP sheets and Nitowrap matrix

M. Nayana[1], B. Nagashree[1], R. Mourougane[1], and Manoj Vasudev[2]

[1]Department of Civil Engineering, MSRIT, Bengaluru, India
[2]Business Development Manager, Repairs and Protective Coatings, Fosroc, Bengaluru, India

Abstract

The use of composite materials for strengthening and retrofitting concrete structures has advanced recently, allowing concrete structures to be strengthened with retrofitting materials. Due to the economic benefits, Fiber Reinforced Plastic (FRP) applications have increasingly been applied to concrete constructions. This paper presents the experimental results of RC beam retrofitted to enhance its flexural strength using Nitowrap CFRP and Nitowrap CF(CS) matrix is used as retrofitting material.

The objective of this experimental work is to investigate the effect of variation in retrofitting material and its configuration on ultimate load carrying capacity, deflection and failure pattern and to compare results between both the retrofitting material. Fourteen rectangular RC beams is casted with M30 grade self-compacting concrete (SCC) with a span of 2.1m and is categorised into seven beams each for flexure beams in which cross section is uniform but the reinforcement is varied. In flexural strengthening of RC beam, the retrofitting material is varied as 0%, 70%, 100% and 200%.

Keywords: Retrofitting, wrapping, Nitowrap CFRP, Nitowrap CF (CS) matrix

1. Introduction

Concrete is a material widely used in construction of various structures. Damage and failure of these structures results in injuries, death and monetary loss. Concrete structures should withstand the designed load without deterioration or damage throughout its intended lifetime. Concrete structures should be durable in terms of resistance to variety of physical or chemical attacks due to internal or external causes [10]. Internal causes may be due to alkali aggregate reaction, corrosion of reinforcement, volume changes die to variation in thermal properties of cement paste and aggregate and mainly due to permeability of concrete [1].

Strengthening or retrofitting of structure is done when the structure needs to accommodate increased loads for various reasons like change in use of structural utility and in cases where structure is deteriorating from its normal usage and due to environmental factors [3,8].

The research work focuses on the techniques of retrofitting of RC beam with respect to flexure strengthening in which the most suitable and sustainable method of retrofitting and its behavior post retrofitting is analyzed. Further the comparison between the retrofitting materials based on ultimate load carrying capacity, deflection and energy absorption is intended.

Hamdy M. Afefy et al. [1] in his research focuses on both analytical and experimental

msritrm@gmail.com

DOI: 10.1201/9781003545941-19

investigations of the full-scale precast pre-tensioned DT Girder behaviour to evaluate strength and ductility enhancements with CFRP strengthening. At around 30% of the load capacity, Girder A revealed hair flexural cracks, which spread in the direction of the support as the load increased. At about 77% of the ultimate load, the end of the CFRP sheet developed significant shear cracks. With the use of 0°–90° approach, or in two orthogonal directions, Girder B was solely strengthened for shear.

Renata Kotynia et al. [2] aimed strengthen RC beam with externally bonded carbon fiber-reinforced polymer (CFRP) strips. To delay CFRP debonding and to increase its efficiency additional U-jacket strips are provided. Different configuration of U-strips was tried to check its efficiency in strengthening. In the pure bending zone, Series I Type S specimens abruptly debonded, and cracks spread rapidly in the directions of one of the supports. Without transverse laminates, the debonding of the beam shifted to the support from the center. The cracked area was relocated toward support using U-shaped sheets. CFRP with lower axial stiffness (Type S strips) had only flexural cracks before debonding, while CFRP with higher stiffness caused more cracks with smaller spacing.

Needa Lingga et al. [3] focuses on the analysis of concrete beams with defects that simulate totally corroded steel beams that have undergone CFRP laminates. The main objective is to investigate the behavior of a totally corroded reinforced concrete beam that has been reinforced with CFRP laminates that are externally joined.

Er. Gurpreet Singh et al. [4] aims on experimental study on RC beams retrofitted using unidirectional and bi-directional wrapping of CFRP sheet to investigate the behavior. Similar concept is explained another research paper by Murad [7]. Six beams were retrofitted with unidirectional wrapping using 0.2mm and 0.3mm sheets and other three beams were retrofitted with bi-directional wrapping using 0.4mm thickness sheets. When compared with control beams, beams retrofitted with unidirectional and bi-directional CFRP sheets had an increased ultimate load carrying capacity of 93% and 120% respectively.

P. Vijaya Kumar et al. [5] objective is to strengthen beam in terms of shear capacity using CFRP. Six beams were casted using M20 grade concrete and Fe 500 grade steel with a cross section of 150 × 300mm and a span of 2200mm. beams were designed to be weak in flexure by reducing reinforcement from 100% to 70% and 50%.

P. Polu Raju et al. [6] aims on strengthening RC beams using Carbon fiber reinforced polymer (CFRP) wrapping at bottom and sides of the concrete beam to enhance the flexural strength and its stiffness. Similar concept is explained another research paper by Yalburgimath [9]. Beams retrofitted with CFRP was bonded using Notiwrap25 and Nitowrap410 as epoxy. The beam designed to be weak in flexure started cracking at 39kN and flexure cracks started appearing at 82kN. After retrofitting its strength was improved and failure load was 125kN where spalling of concrete was observed with delamination of FRP.

2. Experimental Program

2.1. Materials Used

Cement: Birla Super 53 Grade OPC is used. The cement was tested as per IS 12269-1987 [11] to determine its properties.

Fly ash: Fly ash was found to be Class F type. The Specific Gravity was found to be 2.15 and the Specific Surface was found to be 427.18 m^2/ kg. Tested as per IS 1727 (1967) [13].

Aggregates: The coarse aggregate for the concrete mixtures, which had a nominal size of 20 mm in crushed granite stone, procured from Bidadi along with fine aggregate, manufactured sand. According to Indian Standard [12], the sand was verified to be in Zone-II. In order to estimate Specific gravity and

Fineness modulus for the coarse aggregate, tests in accordance with IS 2386–1963 [15] were performed.

Admixtures: Super plasticisers are used to enhance the workability property and to reduce water cement ratio. Admixtures used in this study is sulphonated naphthalene formaldehyde based Conplast SP430 DIS as supplied from FOSROC chemicals, Bangalore.

Nitowrap CFRP [21]: Nitowrap CFRP (Figure 19.1) is a high-quality carbon Fiber fabric is used with epoxy laminating resins. It can be used to strengthen structural member in both flexure and shear. Master Seal P 2525 primer [23] is used to seal the pinholes and defects and to provide flat surface for application of saturant. Nitowrap 410 saturant [24] consist of base and hardener and it is used to develop bond between concrete surface and CFRP fabric.

Nitowrap CF(CS) Matrix [22]: Nitowrap CF(CS) (Figure 19.2) is a high-strength bi-directional carbon fiber grids made with Renderoc SP40 polymer [25] as modified cementitious matrix to create Fiber Reinforced Cementitious Matrix (FRCM) Composite used for repairs and strengthening for concrete structures. Renderoc SP40 is applied on cleaned concrete surface onto to which Nitowrap CF(CS) is spread and the second layer is applied on it.

2.2. Methodology

Material characterisation includes testing of cement, fly ash, aggregates to determine various properties like specific gravity, density and fineness modulus etc. Trial mix to obtain M30 Grade self-compacting concrete is done as per IS 10262:2019 [14] with cement partially replaced fly ash. Designing RC beam as per IS456(2000) [12] and SP 16:1980 [16] in two categories as flexure beam (weak in flexure) and casting 7 beams as flexure beams.

Figure 19.1: Nitowrap CFRP with Master Seal P 2525 primer and Nitowrap 410 saturant.

Figure 19.2: Nitowrap CF(CS) and renderoc SP40 polymer.

RC beam is subjected to two-point loading to evaluate flexural strength. After initial crack RC beam is retrofitted with Nitowrap CFRP fabric and Nitowrap CF(CS) to evaluate the enhanced strength.

2.3. Experimental Investigation

Material characterization is carried out to collect the data required for trial mix design. IS 10262:2019 [14] Annex E is followed while preparing the mix design for M30 Grade Self Compacting Concrete (SCC) [17–19].

Fresh properties of SSC are also measured [20]. Cement is replaced with fly ash by 20%, and a water to cement ratio of 0.43 was chosen to obtain the mix proportioning shown in Table 19.1.

2.4. Hardened Properties of Concrete

Compression test is conducted using compression testing machine (UTM) in which load is applied at constant rate of 14 N/mm^2/

Table 19.1: Mix proportioning

Cement	353.49 kg/m^3
Fly ash	88.37 kg/m^3
Water	190 kg/m^3
Fine aggregate	976.74 kg/m^3
Coarse aggregate	776.85 kg/m^3
Chemical Admixture	2.65 kg/m^3

min on the specimen of size 150 × 150 × 150mm confirming to IS 516 Part 1. Compressive strength of 39.64 MPa was achieved at 28days of testing.

2.5. Retrofitting and Wrapping Methods

RC beams with respect to flexure are designed with a dimension of 150 × 250 × 2100mm considering two-point loading condition.

Flexure beam FB (Weak in flexure) is provided with two nos. of 12mm diameter bar as tension reinforcement and two nos. of 8mm diameter bar as compression reinforcement by proving 25% less flexural reinforcement. Stirrups of 2 legged 8mm diameter bar is provided at 280mm center to center. Different wrapping methods enforced for flexure strengthening of RC beam along with beam configuration are shown in Table 19.2.

In case of flexural strengthening of RC beam, wrapping is done at the soffit of the beam for its full clear span whereas the other beams are provided with variation of retrofitting material as 70%, 100% and 200% as shown Figure 19.3.

3. Results and Discussion

This section entails the analysis and results for the test used in the current study. This includes results for the experiments conducted

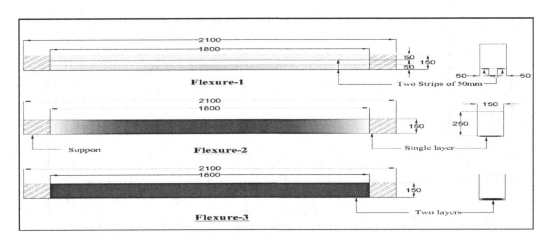

Figure 19.3: Wrapping configuration for flexure beams.

Table 19.2: Beam configuration

SL No.	Wrapping Method		Percentage of material used	Beam ID	
				Nitowrap CFRP	Nitowrap CF(CS) (matrix)
Flexure Beams	Control Beam		0%	CFB	
	At soffit of the beam	Two stripes (single layer)	70%	CFB1	NFB1
		Single layer	100%	CFB2	NFB2
		Two layers	200%	CFB3	NFB3

in various phases of project as discussed in previous chapter. The study also includes comparison between the retrofitting methods. In testing of control flexure beam (CFB), it was observed that up to 17.4kN the beam was very stiff with no deflection. Initial flexure cracks appeared on the side faces of the beam at around 135kN and the response is almost linear up to 181.3kN. The specimen continued with formation of other flexure cracks at the peak load of 203.4kN (considered as ultimate capacity of beam) and the maximum deflection is 11mm as shown in Figure 19.4.

In flexural strengthening of RC beam Flexure-1, Flexure-2, Flexure-3 type configuration is used as represented in Figure 19.3 i.e., with 70%, 100% and 200% variation of retrofitting material by provide Two strips of 50mm and single layer and double layer wrapping respectively as shown in Table 19.3.

In Flexure-1 type retrofitting, two strips of 50mm wide along the length at soffit of the beam. In NF1 specimen after a load of 243.2kN plastic deformation starts in which at a constant load large deformation was observed. When compared with control beam (CFB), ultimate load carrying capacity of CF1 is 33.18% and NF1 is 26.84%. CF1 specimen failed due to combination of flexure and shear cracks and debonding of FRP due to flexure cacks (Figure 19.5a) while NF1 also failed due to flexure cracks (Figure 19.5b). The retrofitting materials provided at the soffit of beam in two strips were insufficient to seize flexure failure even after flexural strengthening.

In Flexure-2 type retrofitting, a single layer of FRP is wrapped at the soffit for full length of the beam and compared with control beam. In NF2 same plastic deformation is observed while in CF2 (Figure 19.6a)

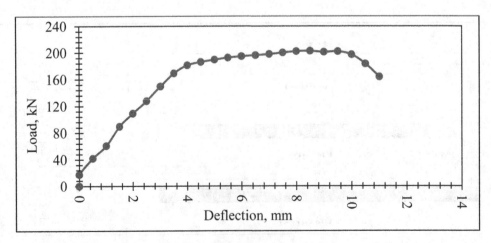

Figure 19.4: Load vs deflection for Control Flexure Beam (CFB).

Table 19.3: Ultimate load carrying capacity and maximum deflection of flexure strengthened beam

Beam Configuration	Percentage of material used	Ultimate Load Carrying Capacity, kN	Maximum Central Deflection, mm
CFB	0%	203.4	11
CF1	70%	270.9	10
CF2	100%	335.7	8.2
CF3	200%	403.4	12.8
NF1	70%	258	26.32
NF2	100%	297.1	33.2
NF3	200%	371.1	23.2

Figure 19.5a: Crack pattern of CF1.

Figure 19.5b: Crack pattern of NF1.

sudden failure was observed. In comparison with control beam, CF2 has about 65.04% and NF2 (Figure 19.6b) has about 46.07% increase in load carrying capacity. CF2 specimen had a shear failure crack with debonding of FRP while NF2 had a combination of shear and flexure cracks with crushing of concrete at load cell point.

In Flexure-3 type retrofitting, double layer of FRP is wrapped at the soffit for full length

Figure 19.6a: Crack pattern of CF2.

Figure 19.6b: Crack pattern of NF2.

of the beam and compared with control beam. In NF3 same plastic deformation is observed while in CF3 sudden failure was observed. In comparison with control beam, CF3 has about 98.33% and NF3 has about 82.45% increase in load carrying capacity.CF3 specimen failed due to shear cracks and debonding of FRP (Figure 19.7a) while NF3 failed due to flexure cracks along with debonding (Figure 19.7b).

Figure 19.8 shows comparison of ultimate load carrying capacity for flexure beam configuration i.e., CFB, CF1, CF2, CF3, NF1, NF2 and NF3 in which CF3 beam specimen recorded maximum peak load of 403.4kN with a maximum deflection of 8.9mm. In the below that as percentage of retrofitting

material is increased load carrying capacity is also increased.

FOSROC has introduced the FRCM system under its brand name Nitowrap CF (CS) - a high strength carbon fiber grid embedded with specially designed cementitious matrix and formulated to prevent the formation of plastic hinges and to improve the overall capacity. From below results it can be perceived that NF series specimens exhibit plastic deformation after yield point until failure hence direct comparison with respect to ultimate deflection cannot be made between both the material.

The influence of variation of material used in retrofitting on the load carrying capacity of RC beam for flexure configuration is shown in Figure 19.9.

Figure 19.7a: Crack pattern of CF3.

Figure 19.7b: Crack pattern of NF3.

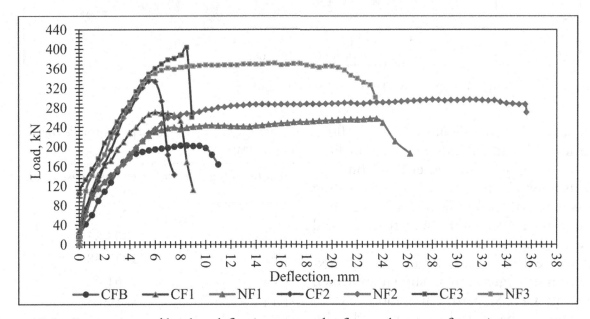

Figure 19.8: Comparison of load vs deflection curves for flexure beam configuration.

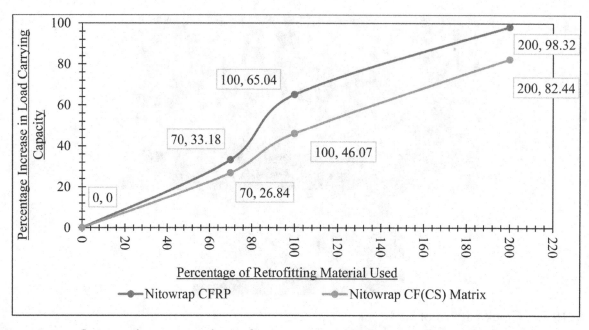

Figure 19.9: Influence of variation of retrofitting material for flexure beam configuration.

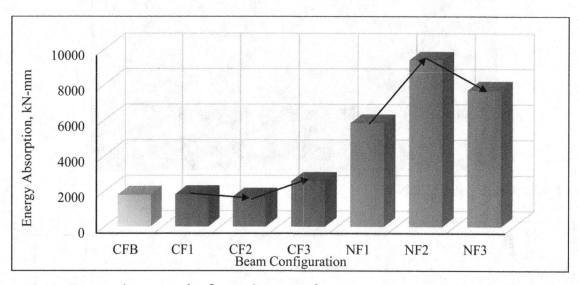

Figure 19.10: Energy absorption for flexure beam configuration.

In flexure beam configuration, up to 100% variation in material almost steep inclination can be observed but after that point the slope is approaching a slant line representing excess material used in flexure strengthening i.e., beyond 100% will have increase in load carrying capacity but on a lower rate.

In Table 19.4, energy absorption of flexure beam configuration is shown in which Nitowrap CF(CS) matrix has more energy absorption capacity when compared with Nitowrap CFRP as it is formulated to

Table 19.4: Energy absorption for flexure beam configuration

Beam Configuration	Energy Absorption, kN-mm
CFB	1782.85
CF1	1804.87
CF2	1678.38
CF3	2561.07
NF1	5819.09
NF2	9361.98
NF3	7673.04

prevent the formation of plastic hinges and to improve the overall capacity and exhibit plastic deformation after yield point until failure. Graphical representation of energy absorption of flexure beam configuration is shown in Figure 19.10.

4. Conclusion

- For flexural strengthening Nitowrap CFRP is found to have more strengthening capacity when compared to Nitowrap CF(CS) matrix
- The ultimate load carrying capacity of CF1, CF2 and CF3 is 270.9kN, 335.7kN and 403.4kN which is 33.18%, 65.04% and 98.32% more than the control flexure beam respectively.
- The ultimate load carrying capacity of NF1, NF2 and NF3 is 258kN, 297.1kN and 371.1kN which is 26.84%, 46.07% and 82.44% more than the control flexure beam respectively.
- RC beam specimen retrofitted with Nitowrap CFRP exhibited sudden failure after reaching peak load value while NF series specimens (Nitowrap CF(CS)) exhibit plastic deformation after yield point until failure hence direct comparison with respect to ultimate deflection cannot be made between both the materials.
- In flexure strengthening as excess material is used, strength varying is observed but the variation is very slant that is for 200% variation in material used about 80–90% variation in load carrying capacity is obtained.
- Nitowrap CF(CS) matrix has more energy absorption capacity when compared with Nitowrap CFRP as it is formulated to prevent the formation of plastic hinges and to improve the overall capacity and exhibit plastic deformation after yield point until failure.

References

[1] Afefy, H. M. (2015). Retrofitting actual-size precracked precast prestresses concrete double-tee girders using externally bonded CFRP sheets. *J. Perform. Constr. Facil.*, ASCE, https://doi.org/10.1061/(ASCE)CF.1943-5509.0000763, ISSN 0887-3828/04015020(18)

[2] Kotynia, R. et al. (March/April 2008). Flexural strengthening of RC beam with externally bonded CFRP systems: tear results and 3D nonlinear FE analysis. *J. Compos. Constr.*, ASCE. Vol. 12, No. 2, April 1, 2008. ©ASCE, ISSN 1090-0268/2008/2-190–201 DOI-10.1061/ASCE1090-0268200812:2190

[3] Lingga, N. (January 2019). Concrete beams with fully corroded steel repaired with CFRP laminates. *ACI Symp. Pap.*, 333, 136–158

[4] Bhat, M. A., and Singh, E. G. (2018). Retrofitting of reinforced concrete beams by using carbon fiber reinforced polymer sheets. *Int. J. Civ. Eng. Technol.*, 9 (9), 1782–1790.

[5] Vijaya Kumar, P. et al. (April 2019). Retrofitting of flexural deficient beams using CFRP. *Int. J. Recent Technol. Eng.* (IJRTE), 7 (6C2). ISSN: 2277-3878. Volume-7, Issue-6C2, April 2019

[6] Polu Raju, P. (October 2011). Experimental evaluation of retrofitted concrete beams using CFRP. *Int. J. Earth Sci. Eng.*, 04 (06 SPL), 812–818. ISSN 0974-5904.

[7] Murad, Y. (September 2018). The influence of CFRP orientation angle on the shear strength of RC beam. *Open Constr. Build. Technol. J.* https://doi.org/10.2174/1874836801812010269

[8] Sudhakar, R. et al. (November 2017). Strengthening of RCC column Using Glass Fiber Reinforced polymer (GFRP). *Int. J. Appl. Eng. Res.*, 12, 4478–4483. ISSN 0973-4562.

[9] Yalburgimath, C. et al. (October 2018). Retrofitting of reinforced concrete beam using Carbon Fiber Reinforced Polymer (CFRP) fabric. *IRJET*, 05 (10) -ISSN: 2395-0056.

[10] DeRose, D. (July–August 2002). Retrofitting of concrete structures for shear and flexure with fiber-reinforced polymers. *ACI Struct. J.*, V. 99, No. 4, July-August 2002.

[11] IS 12269 (1987): 53 Grade Ordinary Portland Cement) CED 2: Cement and Concrete), Bureau of Indian Standards, New Delhi.

[12] IS 456(2000), Plain and Reinforced Concrete – Code of Practice (CED 2: Cement

and Concrete), reaffirmed 2021, Bureau of Indian Standards, New Delhi.

[13] IS 1727 (1967): Methods of Test for Pozzolanic Materials [CED 2: Civil Engineering], Bureau of Indian Standards, New Delhi.

[14] IS 10262:2019, Concrete Mix Proportioning-guidelines (Second Revision), Bureau of Indian Standards, New Delhi.

[15] IS 2386-1:1963, Methods of Test for Aggregates for Concrete, Part I: Practice Size and Shape (CED 2: Cement and Concrete), Bureau of Indian Standards, New Delhi.

[16] SP 16:1980, Design Aid for Reinforced Concrete to IS:456-1978, IS 456(2000), Bureau of Indian Standards, New Delhi.

[17] Domone, P. L., and Chai, H. W. Design and testing of self-compacting concrete. Thesis (Doctoral) https://discovery.ucl.ac.uk/id/eprint/1317644

[18] Okamura, H., Ozawa, K., & Ouchi, M. (2000). Self-compacting concrete. *Struct. Concr.*, 1 (1), 317.

[19] Vengala, J. et al. (July 2003). Experimental study for obtaining self compacting concrete. *Indian Concr. J.*, 77 (8), 1261–1266.

[20] Bartos, P. J. M. (5–6 June 2000). Measurement of key properties of self compacting concrete. In *CEN/STAR PNR workshop, measurement, testing and standardization.*

[21] Nitowrap CW (Formerly known as Nitowrap EPCF), FOSROC constructive solutions, www.fosroc.com

[22] Nitowrap CF (CS), FOSROC Constructive Solutions, www.fosroc.com

[23] MasterSeal P 2525 (formerly known as Concresive 2525), Master Builders Solution Construction Products,

[24] https://www.master-builders-solutions.com

[25] Nitowrap 410 Saturant, FOSROC constructive solutions, www.fosroc.com

[26] Renderoc SP40, FOSROC constructive solutions, www.fosroc.com

20 An experimental study on the influence of construction demolition waste and sustainable materials on the properties of concrete

E. Gopichand[a], B. Narendra Kumar[b], and N. Karthik

Department of Civil Engineering, VNR Vignana Jyothi Institute of Engineering & Technology, Hyderabad, Telangana, India

Abstract

Concrete is the primary source for construction works. Vertical expansion is becoming unavoidable due to a paucity of suitable land area, leading to tall buildings and skyscrapers. Vertical growth is a problem for engineers in two ways: one from the angle of design, and the other from the prospective of developing the appropriate building material and technology. As such, the building industry today faces a huge difficulty in manufacturing concrete that is both strong and durable and ductile. The vertical growth in the structures results in congested reinforcement, demanding concretes that can readily pass through the area between the reinforcement. As a result, customized concretes that meet specific site requirements are becoming more prevalent. Some specific concretes must be able to flow, pass through narrow gaps, and detour through congested places in structural elements. This study is a major advance in understanding the potential and efficacy of using recycled resources in many aspects of concrete production. The utilization of recycled coarse and fine aggregates from construction demolition waste processing in concrete provides a sustainable approach to construction. We lessen the environmental impact of typical concrete production by integrating recycled resources. Recycled fine aggregate and C&D waste contribute to the sustainable economy by encouraging resource efficiency and waste reduction. Conventional Concrete (CC) using sustainable material such as fly ash mixes are developed by replacing natural river fine aggregate and natural coarse aggregate with recycled fine and coarse aggregates of construction demolition waste respectively in dosages of 20, 40, 50% individually and with combination of 50% of FRA and 50% C&D waste.

Keywords: Sustainable material, recycled aggregate material, construction demolition waste

1. Introduction

The quest for durable and ecologically friendly building techniques has resulted in the creation of novel materials for the production of concrete by using the Recycled aggregate of coarse and fine from construction demolition waste has gotten a lot of instance. Concrete [8], a vital component of construction, generally depends on limited natural resources, causing environmental issues [5]. The use of recycled materials tackles this issue, providing an opportunity to lessen the environmental impact of concrete manufacturing while also promoting an economy that is circular in the construction industry [7].

Natural aggregates offer better workability due to smoother surfaces and are often more rounded whereas C&D waste aggregates

[a]gopichandeedarala@gmail.com, [b]narendrakumar_b@vnrvjiet.in

DOI: 10.1201/9781003545941-20

affects the workability due to irregular shapes and rough surfaces as [11] Construction demolition waste aggregates are affected by continues loading and whether conditions due that its properties are changed that results in requirement of increased porosity resulting in a high water to cement ratio of some recycled materials and it contributes high level of bleeding and segregation due to variations in particle sizes and shapes [5].

Bond strength plays an important role in strength of concrete, CD waste aggregates vary widely in quality and some recycled materials may have lower strength [13] potentially affecting the overall performance of the concrete whereas natural aggregates offer good bond strength due to inherent properties [7].

With a focus on how strength attributes are affected, this study explicitly examines the usage of RFA and C&D waste in concrete mixtures

Table 20.1: Blend proportions of standard concrete all measurements in kg/m³

Designation of mix	Cement	Fly ash	Fly-ash	CA	RFA	C&D	SP	Water/cement ratio
SC1	400	80	810	1020	0	0	1.5	0.42
SC2	400	80	648	1020	162	0	1.5	0.42
SC3	400	80	486	1020	324	0	1.5	0.42
SC4	400	80	405	1020	405	0	1.5	0.42
SC5	400	80	810	816	0	204	1.5	0.42
SC6	400	80	810	612	0	408	1.5	0.42
SC7	400	80	810	510	0	510	1.5	0.42
SC8	400	80	405	510	405	510	1.5	0.42

Table 20.2: Primary test results

Material	Specific gravity	Fineness modulus
Natural FA	2.27	2.86
RFA	2.32	2.69
Natural CA	2.7	7.2
RCA	2.5	6.85

Table 20.3: Fresh properties of standard concrete

Grade	Mix ID	Slump(mm)
	SC1	125
	SC2	168
	SC3	159
	SC4	150
M 40	SC5	146
	SC6	148
	SC7	138
	SC8	130

Figure 20.1: Slump test.

[4]. Understanding the structural consequences and benefits of using recycled components is becoming increasingly important as the need for both structural reliability and ecological accountability in building develops [3].

2. Materials Used

Fly ash, natural FA and natural CA [9], Ordinary Portland Cement (OPC), and C&D demolition debris are the constituent materials employed in the research project to develop standard concrete [12].

3. Formulation of Blend Rations of Sc

3.1. Mixture Proportions

Eight concrete mixes were prepared using OPC 0f 53 grade, fly ash, natural aggregates and construction and demolition waste SC1 indicates 050 & of C&D aggregates, the mixes represented by SC2, SC3 and SC4 indicates 20, 40 and 50% replacement of C&D

fine aggregate mixes respectively, SC5, SC6 and SC7 indicates 20, 40, 50% replacement of recycled coarse aggregate mixes respectively, S8 indicates that both find and coarse of CD waste get replaced in 50 & 50%.

3.2.1. Crushing Strength

Table 20.4 displays the strength for compression loading results for a standard size of cube specimen after 7, 28, and 90 days. 1 shows the mix proportions of all eight mixes.

3.2. Rheological Properties

According to EFNARC criteria, the different mix proportions obtained, as indicated in Table 20.3, are checked or the fresh properties. Table 20.3 displays the fresh characteristics' specifics [9].

Table 20.4: Crushing strength of SC

Designation of mix	Crushing strength of cubes specimen (MPa) (Cube 150 mm × 150mm × 150 mm) After no. of days of curing		
	7	28	90
SC1	27.64	43.95	48.46
SC2	22.03	36.90	39.77
SC3	22.98	39.20	42.23
SC4	24.62	38.03	40.92
SC5	22.42	37.10	39.80
SC6	24.01	38.24	41.03
SC7	25.94	39.23	42.05

Figure 20.2: Pan mixture.

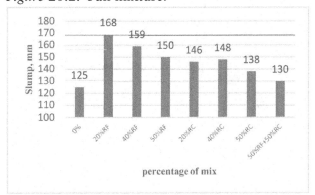

Figure 20.3: Slump flow variation for standard concrete.

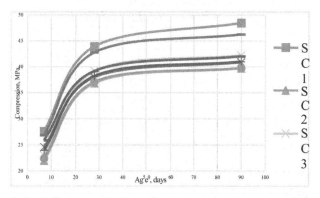

Figure 20.4: Crushing strength of standard concrete.

The matured properties are crushing strength, split tensile strength, and bending strength are ascertained, as previously discussed in the articles [2].

The details of age versus crushing strength for the different mixes SC1–SC8 are shown in Figure 20.4. About 7, 28, and 90 days were determined to be the curing ages [10].

The details of the ratio of crushing strength at various ages in relation to the crushing strength after 28 days for various natural aggregate replacements with CD waste are displayed in Figure 20.4.

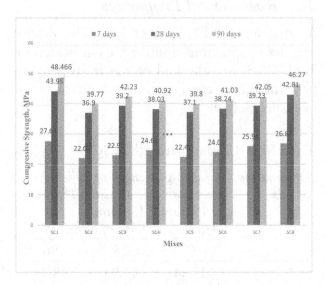

Figure 20.5: Plot of crushing strength of SC.

Table 20.5: Cylinder splitting tensile strength of SC

Designation of mix	Split tensile strength (MPa) (Cylinder 150 mm diameter × 300 mm height) After no. of days of curing		
	7	28	90
SC1	1.93	3.23	3.55
SC2	1.67	2.92	3.15
SC3	1.54	2.77	2.98
SC4	1.37	2.6	2.89
SC5	1.43	2.65	2.84
SC6	1.51	2.71	2.9
SC7	1.59	2.76	2.92
SC8	1.75	3.03	3.27

Details of the cylinder splitting tensile strength are shown in Table 20.5 for a cylinder which is 150 mm diameter & 300 mm in height.

The details of age vs. cylinder splitting tensile strength for various mixes SC1–SC8 are displayed in Figure 20.6. 7, 28, and 90 days were determined to be the curing ages. The ratio of cylinder splitting tensile strength at various ages in relation to the cylinder splitting tensile strength after seven days for various replacements of RFA [6] for river sand is detailed in Figure 20.7.

Figure 20.8 shown the cylinder splitting tensile strength against crushing strength for several SC1–SC8 mix combinations after 28 curing days.

3.2.2. Bending Strength

Table 20.6 displays the bending strength details based on specimens measuring 100 × 100 × 500 mm.

The details of age vs. bending strength for various mixes SC1–SC8 are displayed in Figure 20.9. It was determined that a cure would take 7, 28, or 90 days [1].

The details of the ratio of bending strength at various ages in relation to the flexural strength after seven days for various quartz sand replacements for river sand are displayed in Figure 20.10 [2].

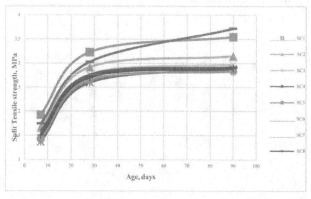

Figure 20.6: Cylinder splitting tensile strength of SC.

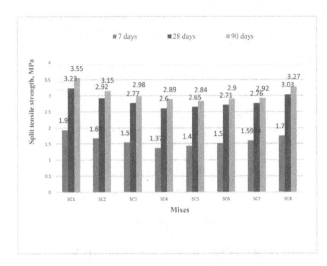

Figure 20.7: Plot of cylinder splitting tensile strength of SC.

Table 20.6: Bending Strength of Standard Concrete

Designation of mix	Bending strength (MPa) (Prism 500 mm × 100 mm × 100 mm) After no. of days of curing		
	7	28	90
SC1	4.08	6.83	7.53
SC2	3.92	6.60	7.14
SC3	3.57	6.30	6.80
SC4	3.04	5.90	6.30
SC5	3.16	5.92	6.36
SC6	3.24	5.94	6.37
SC7	3.33	5.96	6.38
SC8	3.85	6.67	7.18

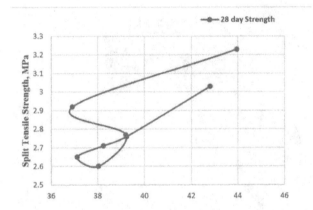

Figure 20.8: Graph plots between cylinder splitting tensile strength vs the SC crushing strength.

The plot of bending strength against crushing strength for various mixes of SC1–SC8 after 28 days of curing is displayed in Figure 20.11.

It is evident from the results above that the mechanical properties of the mixes SC5 through SC8 are persistently increasing.

4. Discussion of Test Results

The experimental outcomes of the study conducted on M40 Standard Concrete (SC) mixes

Studies on optimization of SC by replacing natural FA and CA with C&D waste.

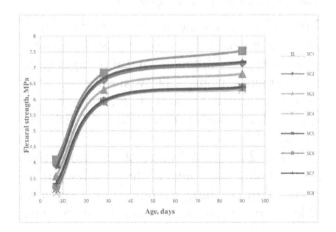

Figure 20.9: Bending strength of SC.

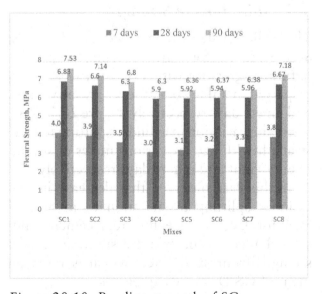

Figure 20.10: Bending strength of SC.

4.1. Fresh Properties of SC

As previously explained, SC2–SC4 refer to the substitution of C&D waste for natural fine aggregates, whereas SC5–SC7 deal with the substitution of construction and demolition waste for natural coarse aggregates. SC8 is a 50 and 50% combination of construction demolition fine and coarse aggregates.

Table 20.1 displays the specifics of the mixture proportions, and Table 20.3 shown the attributes of new mixture.

Table 20.3 and Figure 20.3 show that the fresh characteristics diminish as the doses of recycled aggregates from construction demolition debris rise.

4.2. Crushing Strength

Table 20.4 shows the crushing strength at 7, 28, and 90 days for mixtures SC1 through SC8, or for different natural aggregate replacements using CD waste figure. This illustrates how the values of crushing strength vary with age for various mixes at 7, 28, and 90 days [8].

Table 20.6 and Figure 20.9 show that replaceable fine aggregate results in a decrease in crushing strength, while replaceable coarse aggregate results in an increase for SC5–SC8.

4.3. Split Tensile Strength

The split tensile strength values of mixes SC1–SC8, or for various replacements of natural aggregates with C&D waste [8], are displayed in Table 20.5 at seven, twenty-eight and ninety days respectively. Figure 20.7 illustrates the variation in cylinder splitting tensile strength with age for various mixes at seven, twenty-eight and ninety days respectively.

In this case observed that decrease in the cylinder splitting tensile strength in SC2 – SC4 for C&D waste fine replacement and started increase in cylinder splitting tensile strength from SC5 – SC8 which is replacement of C&D coarse aggregates.

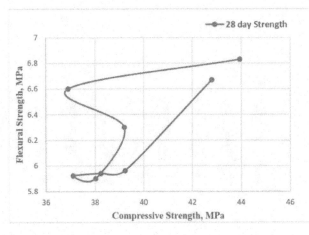

Figure 20.11: Plot of bending strength Vs crushing strength of SC.

Figure 20.12: Compression Test.

Figure 20.13: Propagation of crack on prism.

4.4. Bending Strength

The bending strength values for mixes SC1–SC8, or for various replacements with C&D

waste, are displayed in the table at 7, 28, and 90 days [8]. Figure 20.9 illustrates how the bending strength varies with age for various mixes at seven, twenty-eight and ninety days respectively.

It was noted that the bending strength of the mixes SC2 through SC4, which replace C&D with FA, decreased, while the mixes SC5 through SC7, which replaced CA, increased.

5. Conclusions

a. The Combined mix achieved 97%, 97.5%, 95.4% of conventional concrete's compression strength for the curing period of 7, 28, 90 days correspondingly.
b. The Combined mix achieved 90.6%, 93.8%, 92% of conventional concrete's split tensile strength for the curing period of 7, 28 and 90 days correspondingly.
c. The combined mix achieved 94.36%, 97.64% and 95.3% of conventional concrete's flexural strength at the end of the 7, 28 and 90 days correspondingly.
d. Higher replacement levels of Combined Mix have more noticeable impact on fresh and hardened characteristics. This behavior influenced by the source, shape, texture, porosity.

Acknowledgement

Authors acknowledge the support of VNRVJIET and other authors who have contributed to this field in this study.

References

[1] Bahrami, Nasrollah, et al. (2020). "Optimum recycled concrete aggregate and microsilica content in self- compacting concrete: Rheological, mechanical and microstructural properties." *J. Build. Eng.*, 31, 101361.
[2] Júnior, Nilson S. Amorim, et al. (2019). "Concrete containing recycled aggregates: Estimated lifetime using chloride migration test." *Constr. Build. Mater.*, 222, 108–118.
[3] Naredra, Kumar B. (2017). "Development of high strength self compacting concrete using quartz sand as an alternative of natural river sand." *Indian Concr J* 91.4, 43–50.
[4] Singh, N. and Singh, S. P. (2018). "Evaluating the performance of self compacting concretes made with recycled coarse and fine aggregates using non- destructive testing techniques." *Constr. Build. Mater.* 181, 73–84.
[5] Sangoju, Bhaskar, et al. (2015). "Use of portland pozzolana cement to enhance the service life of reinforced concrete exposed to chloride attack." *J. Mater. Civ. Eng.* 27.11, 04015031.
[6] Nedeljković, Marija, et al. (2021). "Use of fine recycled concrete aggregates in concrete: A critical review." *J. Build. Eng.* 38, 102196.
[7] Surendar, M., G. Beulah Gnana Ananthi, M. Sharaniya, M. S. Deepak, and T. V. Soundarya. (2021). "Mechanical properties of concrete with recycled aggregate and M– sand." *Mater. Today: Proc.* 44, 1723–1730.
[8] Kumar, B. N., Abhilash, B., Naveen Kumar, C. H., and Pavan, S. (2021). Effect of Nano Materials on Performance Characteristics of High Strength Self Compacting Concrete, *Nanonext*, 2, 26–35.
[9] N. K. Boppana, Kumar, G. V. (2020). Effects of Fly Ash, Slag, and Nano-Silica Combination on Corrosion Induced Crack in Reinforced Self Compacting Concrete.
[10] Apoorva, M. and Boppana, N. K. (2022). "Development of standard and high strength concretes using sustainable & recycled materials." *Mater. Today: Proc.*, 71, 202–208.
[11] Cartuxo, F., et al. (2015). "Rheological behaviour of concrete made with fine recycled concrete aggregates–Influence of the superplasticizer." *Construction and Building Materials* 89, 36–47.
[12] Ahmad, S. (2003). "Reinforcement corrosion in concrete structures, its monitoring and service life prediction—A review." *Cement and concrete composites* 25.4–5, 459–471.
[13] Rajamallu, C., Chandrasekhar Reddy, T., and Arunakanthi, E. (2021). "Service life prediction of self compacted concretes with respect to chloride ion penetration." *Mater. Today: Proc.*, 46, 677–681.

21 Proactive restoration: an evaluation of structural health monitoring-driven rehabilitation for reinforced concrete structures

D. Harun Kumar, B. Narendra Kumar[a], E. Gopichand, and N. Karthik

Department of Civil Engineering, VNR Vignana Jyothi Institute of Engineering & Technology, Hyderabad, India

Abstract

Structural health monitoring (SHM) significantly improves the protection of reinforced concrete (RC) structural elements' longevity and integrity and integrity. This work focuses on the application of SHM approaches to precisely assess the state and performance of The RC structures over the course of their operational lives. The study concentrates on several important elements, such as non-destructive testing (NDT) techniques, the deployment of sensors including strain gauges, accelerometers, and temperature sensors, and the following analysis of data gathered. in this work, the field of structural health monitoring concrete (RC) structural elements is closely linked to advanced rehabilitation techniques such as guniting, stitching, grouting, and jacketing. with the use of sensors and data processing, SHM forms the basis for proactive monitoring and early detection of deterioration in reinforced concrete structures. In order to systematically address and remedy structural flaws, innovative rehabilitation procedures like grouting—which involves filling gaps with cementitious material—jacketing, which involves external reinforcing, stitching, which involves crack reinforcement, and guniting, which involves pneumatically spraying concrete are also being researched. Empirical case studies verifying the mutually beneficial link between SHM and rehabilitation, showing how their cooperation guarantees prompt intervention and prolongs the service life of RC structures. This comprehensive strategy increases the sustainability and resilience of these structures, promising a more robust and long-lasting built environment.

Keywords: Structural health monitoring, proactive monitoring, structural resilience, deterioration identification

1. Introduction

Present civil infrastructure cannot function properly without reinforced concrete structure which provide stability and support for a vast array of important works such as bridges and buildings. During construction and after construction the structure is always exposure to the harsh environment and atmospheric conditions and increasing loading demands seriously compromising their life span and performance Deterioration of reinforced concrete components can lead to costly repairs, lowered serviceability, and safety hazards if it is not properly identified. And addressed. To resolve this problems scientist come with Structure health monitoring techniques to examine the structural behaviour for every minute to minimise damage and risk. As the construction material costs increases consistently due to inflation and lowering of availability of natural resources or increased in demand all the investors are interested to repair the damage by early detection to

[a]narendrakumar_b@vnrvjiet.in

DOI: 10.1201/9781003545941-21

minimise the repair works and to minimise the damage risk so that they can increase the structure life span and make structure more durable, and less expensive to maintain over the course of their whole life cycle by giving early notice of potential problems[1].

SHM consists of proper monitoring strategies, sensors, data processing techniques, and methodologies, along with their integration with state-of-the-art technology. This SHM provides information about the changes in load bearing capacity, changes in structural elements i.e. changes in length, cracks, thermal elongation, changes in chemical activity, voids detection, leakages, electricity etc. these are information continuously transferred to the technician to evaluate the structural behaviour and condition[2]. This study aims to provide a thorough evaluation of the application of SHM for structural elements made of reinforced concrete, emphasising the vital role that this technology plays in maintaining the security and resilience of civil infrastructure[3]. We applied this techniques in a existing structure and fixed sensors at required locations and connected to computer device so that we can examine structural behaviour day by day as it is the building age too old so that we can obtain the changes easily and we identified some cracked portions in walls and slabs to fix these issues with guniting and stitching methods and by core sampling techniques we examined the change in chemical composition changes and depth

of carbocation done, and mineral present in sample all repair work of the structure is done by rehabilitation techniques as shown in Figure 21.1.

1.1. *Benefits of Structural Health Monitoring (SHM)*

- Sensors and monitoring devices employed by the SHM systems allows early detection of changes in the strength, structural elements length due to thermal expansion, bearing capacity by noticing every minute changes in structural behaviour.
- Safety of structure is judged by the continuous monitorisation It has the ability to find problems that are difficult to see or are concealed yet could endanger structural integrity[4].
- SHM can help extend the service life of structures by allowing for better understanding of their Realtime performance, which helps in optimising maintenance schedules and making informed decisions about repairs and replacements.

SHM data can be used to validate and refine design assumptions, leading to more accurate predictions of structural behaviour. This information can inform future construction projects, contributing to safer and more efficient designs

2. Technologies Used to Detect Structural Member

- Visual inspection
- Non-Destructive Testing (NDT)
- Load Testing
- Structural Analysis
- Core Sampling and Laboratory Testing
- Monitoring Systems (Structural health monitoring use sensors and instrumentation

2.1. *Visual Inspection*

- Visual inspection is one of the most fundamental and widely used techniques for

Figure 21.1: Structural rehabilitation of reinforced concrete structures.

detecting failures in beams, slabs, and columns of structural elements. It entails a thorough inspection of the surface and parts of the structure by skilled people to spot any readily apparent signs of trouble, damage, or degradation [5].

2.2. NDT Testing

• Several non-destructive and partially destructive procedures for the analysis of the concrete structure are available to ascertain the cause of the current building's deterioration. These non-destructive testing (NDT) processes may be categorised into five fundamental categories: chemical, strength, durability, performance, and integrity tests[6]. These tests allow us to determine the concrete's in-situ strength and quality, which allows us to pinpoint exactly where the structure has been damaged and what is causing it to deteriorate as shown in Figure 21.2.

2.3. Rebound Hammer Test

This is the most rapid method to assess the hardness of concrete, which is represented by the rebound number. Concrete has a high rebound factor when its strength is strong. A spring-controlled mass bounces back when the plunger of the rebound hammer crashes on the surface of concrete, and the amount of

this rebound is based on how firm the concrete is on the surface [7]. The concrete's surface hardness determines the rebound number, which is further connected to the material's compressive strength. The most recent rebound hammer does not require angle correction. According to Indian Standard IS: 13311 Part-21992, the average rebound hammer as shown in Figure 21.3 value varies depending on the concrete quality.

The ultrasonic pulse velocity (UPV) Test method as shown in Figure 21.4. is typically used to analyse concrete deterioration, determine homogeneity of concrete, and determine the extent of cracks and honeycombing [8]. The foundation of this test is the calculation

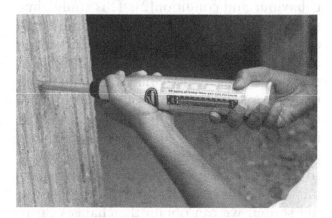

Figure 21.3: Testing concrete with a rebound hammer 2.4 ultrasonic pulse velocity (UPV).

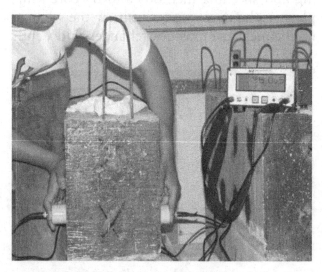

Figure 21.4: Testing the ultrasonic pulse velocity of a specimen.

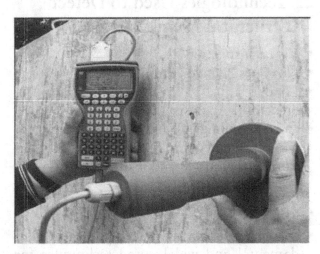

Figure 21.2: Testing of NDT on a concrete wall.

of the transit time for a known distance after an electroacoustic pulse has propagated through a concrete channel. The elastic modulus of the concrete is primarily responsible for UPV. The basic requirements for concrete quality according to Indian Standard IS: 13311 Part-11992.

Each specimen underwent six surface wave measurements as part of the weekly impact echo testing.

2.5. Concrete Carbonation

Concrete contains microscopic pores that are entirely occupied with liquid with a pH as high as 12.5 and which permeates the entire substance. As a result, the concrete has an alkaline character. Concrete becomes carbonated when $Ca(OH)_2$ reacts with ambient CO_2 and changes into $CaCO_3$.

The pore water's pH is reduced by this reaction by up to 8.5. As time goes on, carbonation begins in the concrete's outermost layer and moves deeper into the mass as CO_2 diffuses in from the surface. Steel is vulnerable to corrosion damage if the carbonation depth in the concrete approached that of the steel. By doing a carbonation test, we can determine the concrete's carbonation depth. Stage-by-stage drilling of a hole with phenolphthalein solution applied over it allows researchers to map the carbonation's route.

We stop drilling as soon as the concrete turns pink, and the hole's depth is then determined as shown in Figure 21.5.

2.6. Core Sampling and Laboratory Testing

To evaluate the condition and probable failure modes of structural elements, core sampling and laboratory testing are crucial techniques used in structural engineering. With the aid of these techniques, engineers may directly inspect the structure's material qualities and features, which helps them spot flaws or warning signals of impending failure as shown Figures 21.6. and 21.7.

Typically, a rotary cutting tool with diamond bits is used to cut concrete cores. This produces a cylindrical specimen with irregular, parallel, square ends and, on rare occasions, embedded reinforcement. With particular focus on compaction, aggregate distribution, the presence of steel, etc., the cores are visually described and documented. After good hydration with water at room temperature

Figure 21.6: Extraction of core sample.

Figure 21.5: Spraying a phenolphthalein indicator solution on alkali-activated concrete specimens to evaluate their level of carbonation.

Figure 21.7: Sample testing.

then specimen is covered with molten sulphur to make the two ends parallel and at a straight angle, at end core be tested under compression test in a moist atmosphere.

There are many uses for core samples in research and engineering. They can be used quantify the density and strength of the material, and to identify the depth of carbonation. To examine the Scanning electron microscopy to identify the mineral present in the core and cracks and crack portion in the samples and to examine the interfacial transition zone in the cracked portions and how chemical bonding occurs.

Using petrographic analysis is a method used to extract information from core samples regarding the properties of rocks and minerals. The ASHTO chloride Permeability Test used to ensure the permeability of chloride ions, is also made easier by these samples. Fundamentally, core samples serve as adaptable instruments that facilitate various inquiries, expanding our comprehension of various types of materials and structures.

3. Rehabilitation Technologies on Reinforced Concrete Structural Members

3.1. *Grouting*

3.1.1. Resin Injection

The Grouting method is also known as resin injection it used in the reinforcement and restoration of underground tunnels and pipelines. Resin is specifically used as a successful repair method for complicated cracks seen in reinforced concrete constructions. Epoxy coatings, in particular, provide a wide range of compelling benefits in concrete applications. These include outstanding chemical resistance, robustness against the impacts of hot tire markings, a professional-looking surface, cost-effectiveness, ease of cleaning, solid longevity, the capacity to illuminate dark locations, and a wide variety of potential uses. Epoxy resins are also extremely useful when coupled with concrete, acting as mortars, sealants, mortars, coatings, repair agents, adhesives, and modifiers for cement-based combinations, improving the overall performance and capabilities of concrete materials. Epoxy resins are applied to electronic components as coatings and adhesives as shown in Figure 21.8.

3.2. *Jacketing*

A technique named jacketing is used to reinforce reinforced concrete columns that have degraded as a result of unfavourable weather or poor maintenance. In order to increase the column's strength, this technique includes wrapping it in more material. There are a number of noteworthy advantages, including less work because no foundation strengthening is necessary, increased shear strength, improved concrete confinement in circular columns, low weight increase, and time savings during construction. Even underwater, jacketing can be used to repair damaged columns, piers, and piles. It can be used to reinforce and protect concrete, steel, and wood parts from further harm. After structural improvements, this approach is used to increase bearing load capacity or to repair damaged members' structural integrity. Jacketing is a corrective method used to reinforce damaged reinforced concrete columns as shown in Figure 21.9. This method is necessary for column degeneration, which is

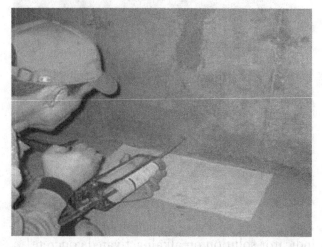

Figure 21.8: Pressure injection grouting.

Figure 21.9: Jacketing and collars for beams and columns for strengthening.

usually brought on by harsh weather or poor maintenance. The column's structural integrity is strengthened by encasing it in extra material. Because jacketing eliminates the need for foundation strengthening, it has several advantages: it increases shear strength, improves confinement of circular concrete columns, adds less weight, and speeds up construction by cutting down on curing times.

3.3. Guniting

Using a spray cannon to apply mortar or concrete is known as guniting. Usually, it is employed to finish certain rehabilitation projects like constructing retaining walls, swimming pools, tunnels, and concrete restoration in addition to stabilising slopes. A pressure tank, admixture dispenser, guniting gun, concrete guniting machine, nozzle, and hose are among the equipment used in the process. Its exceptional impermeability, flexibility for projects with possible work halts, strong compressive strength (between 56 and 70 N/mm^2 at 28 days), and speedy repairs are only a few of its prized attributes. Uniting is a one of construction technique to speed up the work and minimise time in construction by using mechanical devices used in construction of water tanks, retaining walls and foundation works, it offers the strength between

56 and 70 N/mm^2 at 28 days. Guniting used to fill the void areas with mortar using specialised canon. Its main uses are to guarantee slope stability and perform specialist rehabilitation tasks like retaining walls, tunnel construction, demoulding swimming pools as shown in Figure 21.10.

3.4. Stitching

To maintain the structural integrity of the concrete and stop movement between slabs, stitching technique used to fix cracks in RCC structures. In this technique initially cracked portions of slabs or walls are maintained its elevation or slope by using the scaffolding then holes are drilled along the direction of

Figure 21.10: Blue imperial engineers guniting.

Figure 21.11: Using stitching method repairing the active cracks.

the cracked portion by maintaining some gap for good joint strength, then using u-bar linking the cracked portions. For good appearance drilling to be done by covering the length of bar to the surface level of wall then cover it with mortar to protect from atmospheric conditions and for good appearance as shown in Figure 21.11.

3.5. *Polymer Impregnated Concrete*

A technique called polymer impregnated concrete (PIC) involves fusing polymer material with concrete to improve the structure's characteristics. The introduction of polymer monomers, drying, polymerisation, and curing are the four stages of this process. The end product is an exceptional composite material, as opposed to regular concrete. PIC is especially advantageous because of its enhanced resistance to freezing and thawing, reduced water permeability, and heightened protection against chemicals like as sulphates and chlorides. In addition to ensuring longevity and safety in maritime environments, pipelines, and bridges, it is used in a wide variety of buildings. Among its many advantages are PIC's greater impact resistance, compressive strength, durability, and reduced permeability as shown in Figure 21.12. This specific type of concrete provides a durable answer for difficult circumstances and performs better when it comes to sustainability issues.

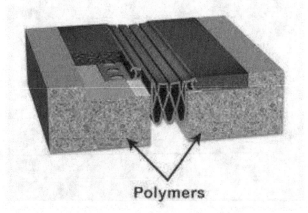

Figure 21.12: Polymer impregnated concrete.

4. Significance

This study has significance the importance of structural health monitoring on assessing the durability and safety of RCC structures. structural health monitoring (SHM) techniques can be used to detect the changes in structural behaviour i.e. bending strength, load bearing capacity, beam deflection, cracks and thermal expansion etc. this allowing for quick rehabilitation and minimising risks. The resistance of these structures is improved by the use of cutting-edge rehabilitation procedures. The outcomes of this research are essentials for improving maintenance strategies, minimising life-cycle costs, and ensuring the long-term viability of vital infrastructure.

5. Conclusion

As a result, this study emphasises how important structural health monitoring (SHM) is to preserving the longevity and integrity of reinforced concrete constructions. Early identification of any SHM problems allows for prompt resolution and reduces life-cycle costs. This is achieved by identifying even the smallest behavioural changes. By combining state-of-the-art restoration methods with SHM, structural flaws may be corrected holistically, increasing sustainability and resilience. The infrastructure is more dependable when SHM is implemented, notwithstanding problems with sensor location and data security.

Acknowledgement

The authors acknowledge the support of VNRVJIET and other authors who have contributed to this field in this study.

References

[1] Kumar, B. N., et al., (2021). Effect of nano materials on performance characteristics of high strength self compacting concrete. *Assian Res. Assoc.*, 2, 26–35.

[2] Boppana, N. K., & Kumar, G. V. Effects of Fly Ash, Slag, and Nano-Silica Combination on Corrosion Induced Crack in Reinforced Self Compacting Concrete. *Slag, and Nano-Silica Combination on Corrosion Induced Crack in Reinforced Self Compacting Concrete(2020)*.

[3] Naderpour, H., Nagai, K., Fakharian, P., and Haji, M. (2019). Innovative models for prediction of compressive strength of FRP-confined circular reinforced concrete columns using soft computing methods. *Compos. Struct.*, 215, 69–84.

[4] Otoom, O. F., et al. (2022). Flexural behaviour of circular reinforced concrete columns strengthened by glass fibre reinforced polymer wrapping system. In *Structures*. Vol. 38. Elsevier.

[5] Li, R. W., Zhang, N., and Wu, H. (2022). Effectiveness of CFRP shear-strengthening on vehicular impact resistance of double-column RC bridge pier. *Eng. Struct.*, 266, 114604.

[6] Mai, A. D., Sheikh, M. N., and Hadi, M. N. (2018). Investigation on the behaviour of partial wrapping in comparison with full wrapping of square RC columns under different loading conditions. *Constr. Build. Mater.*, 168, 153–168.

[7] Vijayan, D. S., et al. (2021). Experimental investigation on the ecofriendly external wrapping of glass fiber reinforced polymer in concrete columns. *Adv. Mater. Sci. Eng.*, 2021, 1–12.

[8] Mungale, A. N. M. and Joshi, D. A. Construction techniques for retrofitting by Jac-Neting of RCC member(2016).

22 An Advanced Ballistic Military Helmet Incorporating Kevlar129/Kenaf/Sisal Fiber/PALF-Fiber Orientation

B. Murali[1,a], B. Silambarasan[2], K. Sivasakthi Balan[3], M. Aiswarya[4], B. S. Navaneeth[5], and K. Karthik[1]

[1]Department of Mechanical Engineering, Vel Tech Rangarajan Dr. Sagunthala R&D Institute of Science and Technology, Chennai, India
[2]Department of Mechanical Engineering, Anjailai Ammal Mahalingam Engineering College, Kovilvenni, Thiruvarur, India
[3]Department of Mechanical Engineering, Sri Sairam College of Engineering, Bengaluru, India
[4]Department of Electronics and Communication Engineering, Karpagam Institute of Technology, Coimbatore, India
[5]Department of Aeronautical Engineering, Nehru Institute of Technology, Coimbatore, India

Abstract

This development is in the field of ballistic protective caps and especially connects with a better strategy for making ballistic head protectors using composite material produced using regular strands and polymers. The examination work is for the most part centered on creating and investigation a Normal-Fibre Situated Progressed Ballistic Protective cap at different circumstances. During the Help Hours, a significant number of our warriors lost their life because of higher openness from Ad libbed Dangerous Gadgets and the impact of shock waves. The effect of these shock waves causes Horrible Cerebrum Injury even without the event of ballistic effect. Half-breed Progressed Battle Protective cap are utilised for safe-monitoring the Military Men from supporting TBIs and different sorts of mind wounds because of gruff and ballistic conflict time. This cap packs the radiation/acoustic concealment during unequivocal trial and error and impact moderation. These head protectors have been used to give insurance against shrapnel and ballistic dangers. Previously, the shot infiltrates the cushioning, these materials scatter the shockwave energy. The blend of materials used for this manufacture work are Kevlar-129, Kenaf, Sisal Fiber and PALF (Pineapple Leaf Fiber). This development additionally focuses on the examination of froths like PU (Polyurethane) froths utilised as the protective cap cushioning materials. Modular Examination are finished for these cap materials to find the Regular Vibration Reaction and Damping Co-productive. This protective cap is fabricated with the accompanying properties like lightweight, high strength fiber and weighs just around 0.5 kg max. These are the fundamental attributes to be thought of, as per the NIJ and IS principles standards. This head protector offers the simplicity of development and great perceivability with next to no neck strain for armed force faculty. During this exploration, numerous different tests are gone through, for example, Electrical Conductivity test, and Warm Conductivity test, Imperviousness to fire test and Modular Examination.

Keywords: Natural-fibre oriented advanced military helmet, Kenaf, Kevlar-129, PALF

1. Introduction

Battle Cap used to safeguard the officer's head against shrapnel and ballistic risks during a fighting. This ballistic Battle Cap is considered to reduce lethal risks and terrible head wounds [1,2]. The head protector advancement integrates kevlar composite liner inside, Polyurethane Froth Cushioning Material and Regular Fiber as an external

[a]bmprojectss@gmail.com

DOI: 10.1201/9781003545941-22

shell followed its capacity for process span improvement [3]. High level Battle Head protector (ACH) was investigated by the utilisation of the FEM system for entertainment. It showed that the limit was better than the standard kind of helmet [4]. This ACH is viewed as our country's biggest acquirement as far as safeguarding the Military Workforce during their life time administration.

There are various past writing concentrates on the improvement of regular fiber situated progressed ballistic military caps. A portion of these investigations have zeroed in on the mechanical properties of normal filaments, like their solidarity, firmness, and strength. Different investigations have zeroed in on the advancement of assembling processes for regular fiber composites [5].

One review, distributed in the diary "Materials and Plan" in 2018, explored the mechanical properties of flax fiber composites. The investigation discovered that flax fiber composites had practically identical mechanical properties to Kevlar composites, however they were lighter in weight [6].

The review fostered an interaction for meshing normal strands into texture and afterward impregnating the texture with a polymer pitch. The investigation discovered that the subsequent composites had great mechanical properties and were lightweight [7].

These are only a couple of instances of the past writing concentrates on the improvement of normal fiber situated progressed ballistic military protective caps [8]. These examinations have shown that regular filaments can possibly be utilised in ballistic caps. Notwithstanding, there is even more examination that should be finished to foster regular fiber composites that have similar degree of execution as Kevlar composites.

Here are a portion of the critical discoveries from these past writing studies:

- Normal filaments, like flax, hemp, and jute, can be utilised to areas of strength for make lightweight composites that are reasonable for use in ballistic protective caps.

- The mechanical properties of normal fiber composites can be improved by upgrading the fiber direction and by utilising an elite presentation polymer tar.

- There are various assembling processes that can be utilised to deliver normal fiber composites, including winding around, sewing, and pultrusion [9].

- The expense of regular fiber composites is as yet higher than the expense of Kevlar composites, however this cost is supposed to descend as the innovation develops.

Generally, the past writing concentrates on the improvement of normal fiber arranged progressed ballistic military protective caps have shown that these head protectors can possibly be lighter, more agreeable, and more maintainable than conventional Kevlar protective caps. In any case, there is even more exploration that should be finished to foster these head protectors that have similar degree of execution as Kevlar caps [10].

2. Methodology

The moment development is portrayed all the more completely hereinafter concerning the going with drawings or potentially photos, in which at least one praiseworthy encapsulations of the creation are shown. The Figure.1 shows Flowchart of research work.

This development may, nonetheless, be epitomised in a wide range of structures and ought not be understood as restricted to the encapsulations put forward thus; rather, these exemplifications are given so this revelation will be usable, empowering, and complete. Likewise, the specific courses of action unveiled are intended to be illustrative just and not restricting regarding the extent of the creation. In addition, numerous epitomes, like transformations, varieties, adjustments, and comparable plans, will be certainly unveiled by the encapsulations portrayed thus and fall inside the extent of the present invention. Figure shows the Material creation philosophies which are utilised to deliver proficient

Unbeatable Caps relying on the danger type and speed of the shot. First and foremost, the Unrefined substances of Option in contrast to Kevlar (RMAK) were bought from the Market, the elective material were Pineapple fiber (Uni-directional), Kenaf fiber, Sisal fiber these materials are handled for cleaning process with NaOH arrangement, the components of the battle protective cap were set apart in the material (600 mm × 600mm) (420 mm × 420 mm) and the materials were handled for stacking. The thickness, weight of every material, aspects were noted.

The Sandwich technique for stacking process were as Pineapple fiber (3 layers), Kenaf Fiber (2 layers), Sisal fiber (2 layers) (PF-KF-SF PF-KF-SF) each and Hot-embellishment Machine were warmed up to 150°C in the span of 3 hours, applied wax inside the shape and put the stacked materials applying with General Epoxy Sap (GER) on each layer, set in to shape. The creation were utilised by pouring GER (250 ml) subsequent to setting into the shape, then, at that point, the hot squeezing worked at 150° temperature with 270 bar strain for 10 minutes. The head protectors have delivered utilising a Gaseous tension Check (APG), Besides, handled for testing reviews in research facility. Figure 22.2

shown the stacking sequence of sandwich composite and Figures 22.3 to 22.10 shows various materials used for fabrication of the composite this.

2.1. Experimental Works on Military Helmet Shell Fabrication, Performance Measures of Ballistic Helmets

An infantry trooper conveys all his/her gear. The obligations of such a warrior are genuinely requesting, and any expansion to the weight conveyed produces significant debilitation to the perseverance of the trooper. In this way, weight is an essential thought in planning any new cap framework.

2.2. Ballistic Helmet Measures and Consolidations

The Figures 22.11 to 22.13 Ballistic effect of a helmetand Modal Analysis KIT. Ballistic security is an essential thought in the handling of another protective cap. The ballistic presentation of a composite head protector relies upon the material utilised, cap thickness, and creation technique. A trade off frequently must be made between the weight permitted and the ballistic security prerequisites. In view of the utilisation of ad libbed dangerous gadgets (IEDs) in metropolitan fighting, an infantry trooper is presented to shoot occasions with a rising recurrence as per ballistic

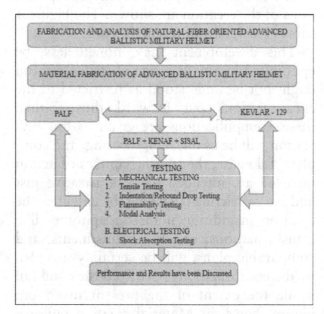

Figure 22.1: Flowchart of this research work.

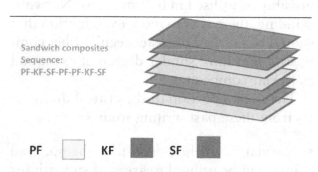

Figure 22.2: Schematic view of sandwich sequence of natural fibre.

Figure 22.3: PALF Mat Form.

Figure 22.5: PALF Stack Form.

Figure 22.4: Sisal Fiber Stack Form.

Figure 22.6: Manufacturing Process.

principles for a battle protective cap are legitimate for new and arising battling conditions.

2.3. Area of Focus of Mass

The ideal area for any weight on the head is on the straight line associating the focal point of mass (CM) of the head and the CM of the body. Any change in the weight balance on the head from the regular CM of the head will bring about stressing and weakness of neck muscles. It will likewise prevent the body balance during different developments like running, squatting, running, or strolling, due to solid facilities required. The power applied on the skull base is the speed increase duplicated by the mass of the helmeted head (in the event of effect). The size of the focal point of gravity (CG) offset force is corresponding to the CG offset distance, speed increase, and protective cap/head mass. Consequently, The CG and the offset distance should be littlest in limit.

2.4. Head Protector Measuring and Fit

Current caps are intended to give substantially more than simply ballistic insurance. On the off chance that the attack of the cap were not happy, the protective cap client would be hesitant to wear it. The attack of the defensive head gear accordingly influences the exhibition of the warrior. Attack of a thing relies upon the anthropometry. Generally, there have been two different ways for deciding the size and fit: (a) beginning with an essential size and utilising grade rules to anticipate higher sizes, and (b) anthropometric measuring.

2.5. Hot Embellishment Machine

The unrefined substance have been bought from the nearby market and pre-handled with the fundamental handling of cleaning, estimation, aspect computation of cap size and as per the size the unrefined substance

Figure 22.7: Kevlar - 129.

Figure 22.8: PALF.

Figure 22.9: KENAF.

Figure 22.10: PALF + KENAF + SISAL.

has been stamped and cutted to the ideal stacking arrangement. Besides, these materials have been handled for essential examination like properties of material thickness, layer, weight of natural substance, aspect, etc. The Epoxy Gum have been taken in an estimating cup and weighted 250g and afterward it is applied in each layer of the protective cap and stacked in a request for differentarrangement of material, for example, Kevlar-129, Kenaf, Sisal Fiber and PALF (Pineapple Leaf Fiber) these materials are organised in a consecutive request separately

2.6. *Modular Examination*

Modular Examination are performed for Battle Cap and fixed-end limit state of pillar utilising drive recurrence reaction test. An upsetting power was given to compelled cap shell through a motivation hammer went with a power transducer office (Dytron model 5800B3), and the vibration reaction was

estimated utilising piezoelectric accelerometer (Dytron model 3055B2). Information securing with FFT analyzer (Dytron model Photon 200) associated with piezoelectric accelerometer gets the reaction signal is recorded by a PC. The Modular Investigation of vibration information tentatively gained through drive test was created by application programming RT-Star, which is additionally handled for deciding first regular recurrence, progressive reverberation sets, and its comparing damping factor.

3. Results and Discussion

3.1. *Experimental Procedure & Testing Methods*

3.1.1. Tensile Properties

The composite example I (Palf fiber), II (Kenaf fiber) and III (Palf+Kenaf+Sisal) were assessed for the pliable properties.

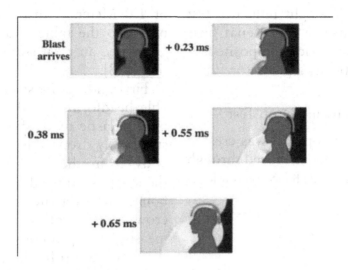

Figure 22.11: Ballistic effect of a helmet.

Figure 22.12: Hot Mould Pressing.

Figure 22.13: Modal Analysis KIT.

The outcomes outlines that in example III strains (Palf+Kenaf+Sisal) showed critical rigidity (93 Mpa) when contrasted with Palf and Kenaf fiber tests separately. Figures 22.14 and 22.15 shows tensile property of different composite specimen The outcomes outlines that the sandwich composites of Palf+Kenaf+Sisal showed higher tractable properties and assumes a crucial part in situated progressed ballistic cap structure. Additionally, Ramnath et al. [11] showed that the manufacture and assessment of jute and abaca strands half and half with glass fiber composites showed multiple times higher elasticity when contrasted with banana/sisal composite.

3.1.2. Impact Test

The effect gadget might be instrumented to assess the speed of the impactor at a given point before influence, to such an extent that the effect speed might be assessed. By and large, such frameworks utilise a twofold pronged banner, which impede a light bar between a photograph diode producer and identifier [12]. The main edges of the banner prongs are commonly isolated by 4.0 to 10.0 mm [0.135 to 0.450 in.].

The Table 22.1 shows Impact properites of sandwich composites, The banner prongs will be situated with the end goal that speed estimation is finished between 4 to 6 mm [0.12 to 0.35 in.] in an upward direction over the outer layer of the example. The effect speed is determined utilising the deliberate time the light bar is discouraged by every prong, as well as the time that an effect force is first distinguished. Speed estimation will be exact to inside 10 mm/s [0.35 in./s]. Table shows the outcome for drop weight influence test for Palf ballistic protective cap, Kenaf fiber ballistic cap and the sandwich composite Palf + Kenaf + Sisal ballistic head protector.

3.1.3. Indentation/Rebound Drop Test

A Space/Bounce back Drop Test was created to genuinely show force being moved through a plumb sway of weight 2.0 KGS from a level

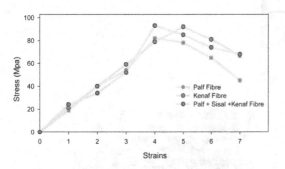

Figure 22.14: Tensile property of different composite specimen.

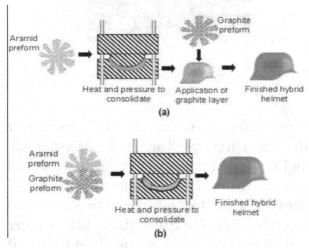

Figure 22.15: Schematic Diagram Experimental Testing Setup of a Helmet.

of 1.0 Meter. A displaying compound was put onto the table and was streamlined to guarantee an even surface for exact estimations [13].

First and foremost mount the cap in suitable headform, put the cap with headform on an elastic matt of 1 meter x 1 meter to stay away from the harm of plate/plumb sway from the level while testing. When the sway is dropped from 1 meter level, the bounce raises a ruckus around town of the crown of the protective cap. After the testing measure the profundity of entrance. Mark the entrance profundity, and further continued for examination. A video was taken of every preliminary to guarantee right estimation of the bounce back level. The space drop test mechanical assembly. Force versus time estimation of various composite fiber test results shows that the Palf + Kenaf + Sisal fiber composite ballistic head protector has higher power influence as long as 14 minutes of time taken when contrasted with the Palf ballistic cap and Kenaf Ballistic cap individually. Figure 22.19 Bounce back Drop trial of examination between Power versus season of (A) Palf Ballistic protective cap, (B) Kenaf Ballistic cap and (C) Palf+Kenaf+Sisal Ballistic Cap[14–18]. The Figure 22.16 shows Indentation/Rebound Drop Test and Figures 22.17 and 22.18: Peak Deformation (mm) for Palf, Kenaf, and Palf+Kenaf+Sisal Ballistic Helmets.

Top deformity examination of three distinct composite ballistic cap showed that the composite Palf+Kenaf+Sisal showed less deformity (10.56 mm) when contrasted with

Table 22.1: shows Impact properites of sandwich composites

Ballistic helmet	Impact energy (J)	Peak force (kN)	Peak deformation (mm)	Absorbed energy (J)	Energy to peak deformation (J)
Palf	4.75	0.63	18.67	3.12	3.15
Kenaf	4.89	0.62	13.67	4.53	4.43
Palf+ Kenaf +Sisal	4.95	0.68	15.67	5.67	4.89

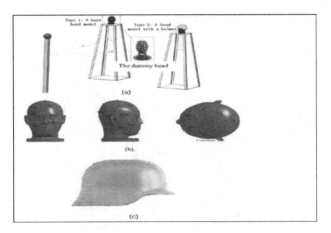

Figure 22.16: Indentation/Rebound Drop Test.

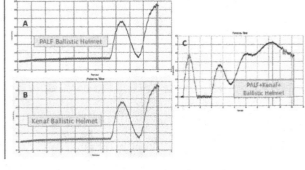

Figure 22.17: Indentation/Rebound Drop Test.

the Palf (18.67 mm) and Kenaf (16.54 mm) ballistic head protectors separately.

Notwithstanding, the retained energy (J) rate expanded in the Palf+Kenaf+Sisal Ballistic head protectors (4.87 J) when contrasted with the Palf (3.120 J) and Kenaf (3.430 J) separately. Despite the fact that Palf and Kenaf show more deformity after the effect, it shows less ingested energy after the effect. The outcomes charges that the Palf+Kenaf+Sisal Ballistic Protective caps properties are better which is not so much disfigurement but rather more ingested energy than the Palf and Kenaf head protectors. The deformity components and energy retention capacity of Palf, Kenaf, and Palf+Kenaf+Sisal Ballistic Head protectors are tentatively contemplated determined to foster an extensive constitutive regulation to be executed for genuine application and furthermore for influence examination for future exploration. Below Figure 22.19 shows Absorbed Energy (J) for Palf, Kenaf and, Palf+Kenaf+Sisal ballistic helmets.

Drop-weight influence test conveys a superior method for getting the retained energy during an effect test. Drop-weight influence tests were performed on a protective cap influence test rig which best means to connect the cap on the effect occasion. To decline the effect power of the head protector, material that utilised for the shell of the cap assume an indispensable part since the higher it ingested

energy from the effect occasion, less disfigurement will happen to the cap.

3.1.5. Shock Absorption Test

The Shock Retention test hardware comprise of a wooden striker, one striker lifting handle and a striker discharge switch. A heap which would move how much shock to the computerised load pointer. As soon the striker raises a ruckus around town, genuine unit and it is switched over completely to kgf by duplicating into 9.8 gravity. Eliminate the cap from the wooden head structure by raising the striker. The Figure 22.20 shows the Graph plot represents the average Head Injury criterion (HIC).

The outcomes showed that Head Injury model (HIC) was better in the Palf+Kenaf+Sisal fiber ballistic protective caps in Front (1156 HIC), Top (1350 HIC), and Back (1126 HIC) separately and it is altogether unique with Palf and Kenaf fiber ballistic caps.

3.1.6. Flammability Test

The combustibility test have been occurred to check whether the cap burst into flames or whether the cap retains higher fire, in such example there is a plate in the arrangement with fire spout that is utilised to control the fire range. The fire passes on all sides of the cap similarly for 5–10 seconds and fire will be

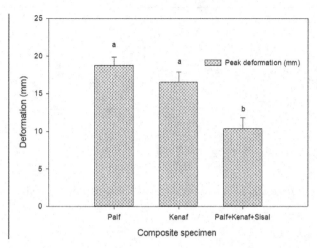

Figure 22.18: Peak Deformation (mm) for Palf, Kenaf, and Palf+Kenaf+Sisal Ballistic Helmets.

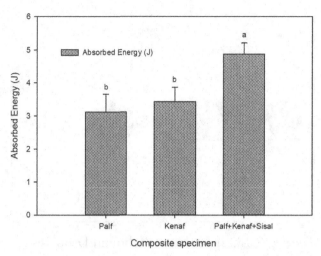

Figure 22.19: Absorbed Energy (J) for Palf, Kenaf and, Palf+Kenaf+Sisal ballistic helmets.

turned off and pivoted. At the point when the fire get off, the outer layer of the head protector will be examined for additional cycles. The combustibility measure shows that the intensity transition (kW/m²) and p-HRR (top Intensity Delivery rate) (kW/m²) of the three different normal composite ballistic caps. Below the Figure 22.20 shows Graph plot represents the flammability assay of Heat flux and peak Heat release rate (p-HRR).

The outcomes showed that both the intensity motion rate and p-HRR rate was altogether lesser in the Palf+Kenaf+Sisal fiber ballistic caps Intensity (Transition 33.45 kW/m2 and pinnacle Intensity Delivery rate 313.45 kW/m²) when contrasted with the Palf and Kenaf ballistic head protectors and the outcomes conveys that the sandwich composites Palf+Kenaf+Sisal shows higher fire opposition.

4. Conclusion

In this venture work, a Characteristic Fiber Situated Progressed Ballistic Military Cap have been created utilising Regular Filaments, for example, Pineapple Leaf Fiber (PALF), Kenaf Fiber, Sisal Fiber as a substitution of Kevlar Material, Different model have been performed under various stacking of

Figure 22.20: Graph plot represents the average Head Injury criterion (HIC) recorded from impacts against the three different types of natural fibre.

materials. It is drawn-out to close in regards to the tension in the cerebrum, since the difference in pressure is little between the different protective cap shell setups. Different injury components can likewise be sorted where the non-contact influences resulted dormancy prompted strain in the mind, while the contact influences give a confined space of the cap to the skull. By directing the tractable test on rectangular examples It is presumed that PALF+ Sisal Fiber + Kenaf Fiber had shown improved results contrasted with different materials. Combustibility Tests

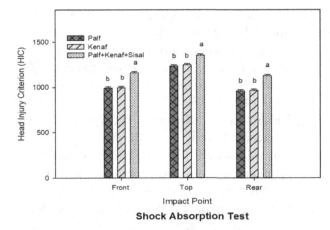

Figure 22.21: Graph plot represents the flammability assay of Heat flux and peak Heat release rate (p-HRR) across the three different bio-composite ballistic helmets.

have been led for every one of the Normal filaments material blends. The outcomes had demonstrated that these materials had passed for certain varieties in it. This is because of the fire obstruction force level of every single materials chose in this analysis. During this examination, the head protector handled through different states of Static Testing and dynamic testing followed according to the standards of Guard guidelines. The Unadulterated PALF material came about with high damping co-effective (15.54%) up toupper = 92.887 Hz perceptibility, gives superb damping factor best result similar to PALF+Kenaf+Sisal Fiber composite material came about in 8.16% fupper= 136.374 Hz of damping coefficient noted. From this time forward, as indicated by these arrangement of Tests led, it is infer that the recently proposed set of materials showed two times the presentation of existing battle head protector materials (Kevlar - 129) utilised. It is guaranteed that these materials are protected towards any shot and bomb impacts (Impacting explosives, Hand projectiles, IED and can be fabricated easily.

Acknowledgement

The authors express their sincere gratitude to the Department of Mechanical Engineering, Vel Tech Rangarajan Dr. Sagunthala R&D Institute of Science and Technology, Chennai, for providing the research facilities and necessary support throughout the entire phase of this research work.

References

[1] Murali, B. et al. (2014). Chemical treatment on hemp/polymer composites. *J. Chem. Pharmaceutical Res.*, 6(9), 419–423.

[2] Chandramohan, D. et al. (2014). Machining of composites - A review. *Acad. J. Manufacturing Eng.*, 12(3), 67–71.

[3] Rajesh, S. et al. Increasing combusting resistance for Hybrid composites. *Int. J. Appl. Eng. Res.*, 9(20), 6979–6985.

[4] Chandramohan, D., et al. (2013). Impact test on natural fiber reinforced polymer composite materials. *Carbon Sci. Technol.*, 5(3), 314–320.

[5] Rajesh, S., et al. Study of machining parameters on natural fiber particle reinforced polymer composite material. *Acad. J. Manufacturing Eng.*, 12(3), 72–77.

[6] Senthilathiban, A., et al. (2014). Effects Of chemical treatment on jute fiber reinforced composites. *Int. J. Appl. Chem.*, 10(2), 153–162.

[7] Murali, B., et al. (2014). Mechanical behavior of chemically treated jute/polymer composites. *Carbon – Sci. Tech.*, 6(1), 330–335.

[8] Chandramohan, D. and Ravikumar L. (2019). Free vibrational analysis of cortical / hard cancellous bone by using of FEA. *Mater. Today Proc.*, 16 (Part 2), 744–749.

[9] Ravi Kumar, L., et al. (2018). Experimental investigation of mechanical properties of GFRP reinforced with coir and flax. *Int. J. Mech. Eng. Technol.*, 9(1034–1042), 1034–1042.

[10] Murali, B., et al. (2019). Mechanical properties of Boehmeria nivea reinforced polymer composite. *Mater. Today Proc.*, 16, 883–888.

[11] Chandramohan, D., Murali, B., Vasantha-Srinivasan, P. and Dinesh Kumar, S., 2019. Mechanical, moisture absorption, and abrasion resistance properties of

bamboo–jute–glass fiber composites. Journal of Bio-and Tribo-Corrosion, 5(3), 66.

[12] Dinesh Kumar, S., et al. (2020). Mechanical and thermal properties of jute/aloevera hybrid natural fiber reinforced composites. *AIP Conference Proceedings*, 2283.

[13] Murali, B., et al. (2022). Multi-potency of bast fibers (flax, hemp and jute) as composite materials and their mechanical properties: review. *Mater. Today Proc.*, 62, 1839–1843.

[14] Murali Banu, Vijaya Ramnath Bindu Madhavan, Dhanashekar Manickam, Chandramohan Devarajan. (2021). Experimental Investigation on stacking Sequence of Kevlar and Natural Fibres/Epoxy Polymer Composites. *Polimeros: Ciencia e Tecnologia*, 31(1), 1–9.

[15] Banu Murali, Bindu Madhavan Vijaya Ramnath, Devaraj Rajamani, Emad Abouel Nasr, Antonello Astarita, Hussein Mohamed (2022). Experimental Investigations on Dry Sliding Wear Behaviour of Kevlar and Natural Fiber-Reinforced Hybrid Composites through an RSM–GRA Hybrid Approach. *Materials*, 15(3), 1–16.

[16] Murali, B., Yogesh, P., Karthickeyan, N. K., Chandramohan, D. (2022). Multi-potency of botanicals (flax, hemp and jute fibers) as composite materials and their medicinal properties: a review. *Mater. Today: Proc.*, 62, 1839–1843.

[17] Karthik, K. and Manimaran, A. (2020). Wear behaviour of ceramic particle reinforced hybrid polymer matrix composites. *Int. J. Ambient Energy*, 41(14), 1608–1612.

[18] Ganesh, R., Karthik, K., Manimaran, A., and Saleem, M. (2017). Vibration damping characteristics of cantilever beam using piezoelectric actuator. Int J Mech Eng Technol, 8(6), pp. 212–221.

[19] Karthik, K., Ganesh, R., Ramesh, T. (2017). Experimental investigation of hybrid polymer matrix composite for free vibration test. *Int. J. Mech. Eng. Technol.*, 8(8), 910–918.

[20] Ahmed, I., Renish, R.R., Karthik, K. and Karthik, M., 2017. Experimental investigation of polymer matrix composite for heat distortion temperature test. Int. J. Mech. Eng. Technol, 8(8), pp. 520–528.

[21] Murali, B. and Nagarani, J. (2013). Design and fabrication of construction helmet by using hybrid composite material. In *2013 International Conference on Energy Efficient Technologies for Sustainability* (pp. 145–147). IEEE.

[22] Krishnasamy Karthik, Jayavelu Udaya Prakash, Joseph Selvi Binoj, Bright Brailson Mansingh (2022). Effect of stacking sequence and silicon carbide nanoparticles on properties of carbon/glass/Kevlar fiber reinforced hybrid polymer composites. *Polymer Composites*, 43(9), 6096–6105. https://doi.org/10.1002/pc.26912.

[23] Fayaz, H., Karthik, K., Christiyan, K. G., Arun Kumar, M., Sivakumar, A., Kaliappan, S., Mohamed, M., Ram Subbiah, Simon Yishak. (2022). An investigation on the activation energy and thermal degradation of biocomposites of Jute/Bagasse/Coir/Nano TiO2/Epoxy–Reinforced Polyaramid Fibers. *J. Nanomat.*, Hindawi, 2022, ArticleID 3758212, 5. https://doi.org/10.1155/2022/3758212.

[24] Karthik, K., Rajamani, D., Venkatesan, E.P., Shajahan, M.I., Rajhi, A.A., Aabid, A., Baig, M., Saleh, B. (2023). Experimental Investigation of the Mechanical Properties of Carbon/Basalt/SiC Nanoparticle/Polyester Hybrid Composite Materials. *Crystals*, 13(3), 415.

[25] Ramesh, V., Karthik, K., Cep, R. and Elangovan, M., (2023). Influence of stacking sequence on mechanical properties of basalt/ramie biodegradable hybrid polymer composites. *Polymers*, 15(4), 985.

[26] barose Juliyana, S., Udaya Prakash, J., Čep, R. and Karthik, K. (2023). Multi-Objective Optimization of Machining Parameters for Drilling LM5/ZrO2 Composites Using Grey Relational Analysis. *Materials*, 16(10), 3615.

[27] Siva Kumar, M., Rajamani, D., El-Sherbeeny, A.M., Balasubramanian, E., Karthik, K., Hussein, H.M.A. and Astarita, A. (2022). Intelligent Modeling and Multi-Response Optimization of AWJC on Fiber Intermetallic Laminates through a Hybrid ANFIS-Salp Swarm Algorithm. *Materials*, 15(20), 7216.

23 Intricacies in machining of titanium grade 5 alloy: a review

K. N. Mohandas[1], N. R. Acharya[1], and Nagahanumaiah[2]

[1]Faculty, Department of Mechanical Engineering, Ramaiah Institute of Technology, Bangalore, India
[2]Central Manufacturing Technology Institute, Bangalore, India

Abstract

Machining is a form of subtractive manufacturing process resulting in the desired shape and size of the final work part. Titanium Grade 5 or Ti-6Al-4V is widely adopted in the aerospace and medical industries due to its excellent properties. The main rationale for engineers to choose titanium grade 5 is its relative low mass for a given strength and its high temperature resistance. It is mainly used in the front portions of aircraft engine. Also, it is one of the most difficult materials to machine. Titanium components are machined in forged condition and requires removal of significant percentage of the weight of the work piece. Its poor machinability can be associated with its' characteristics like high strength at elevated temperature, stronger chemical reactivity with tool materials, low specific heat, and low modulus of elasticity. This article reports the various findings on conventional machining of Ti6Al4V and highlights the solutions proposed to overcome the difficulties in Titanium machining.

Keywords: Machining, titanium grade 5, aerospace, modulus of elasticity

1. Introduction

Pure Titanium comprises mainly of α- phase at normal temperatures and subsequent translation to β phase occurs at around 880°C. Vanadium (V) is an alloying addition to decrease this temperature and aluminium (Al) is added for further stabilisation. Ti–6Al–4V or Ti grade-5 comprises 6 wt% of Al and 4 wt% V. [1]

Titanium alloys have been extensively adopted in the fabrication of aircraft fuselages and gas turbines owing to their outstanding properties such as excellent strength-to-weight ratio and higher corrosion resistance. It has been reported that titanium alloys form around 11% of fuselage weight of the Boeing 777. Also, the usage of titanium alloys is more evident in fighter aircrafts. Moreover, titanium alloys have wide range of applications in the automotive, chemical, and medical sectors. Titanium alloys are considered to be "difficult-to-machine" materials. The capability of these alloys to maintain their strength at elevated temperatures along with greater work hardening property result in increased machining forces. The machining techniques for Ti-6Al-4V alloy differs from machining of any conventional metals and alloys. According to Chichili D. R. et al. [2] the difficulties during fabrication of aerospace grade alloys is due to its natural behaviour to the action of cutting tool. High strength at elevated temperature, lower thermal conductance, and low elasticity modulus, greater strain reinforcement, and higher chemical affinity make significant contributions in the machining process of Titanium grade 5 alloy. Cutting tools available show chemical affinity to Ti-6Al-4V which causes the subsequent chemical wear. Adding to this, chips that are formed adhere readily to the

[a]mohandaskn@msrit.edu

DOI: 10.1201/9781003545941-23

cutting tool surface as built -up edge (BUE). Carbide tools are extensively used to cut Ti-6Al-4V.

2. Complexities in Machining Titanium Alloys

2.1. Tool Wear

The cutting tool is subjected to continuous wear during machining of titanium alloy. High pressure and temperature, more friction lead to an environment where damage to the cutting tools is more. The below sections highlight the adoption of various tool materials and its wearing out during cutting of titanium alloy.

According to Campbell [3] certain sub-grades of high speed steels (HSS) tool are generally used for cutting titanium alloys.

Although carbides are prone to rapid chipping while intermittent machining it can cause significant increase in stock removal rates. Ceramic cutting tools are rarely used while shaping titanium alloys because of its low fracture resistance, meagre heat conductance, and greater chemical affinity to titanium alloy.

Research works across the globe involved various grades of cutting tool, such as carbide based tool, TiN tool, aluminium oxide variant of ceramic tool, TiC coated cutting tool, CVD coated tools, binder less CBN, diamond based cutting tools. Major amount of wear due by chemical interaction between the cutting tool and component occurs with TiN and the TiC coated tools. CBN tools exhibit large amounts of wear at the flank and rake face. The carbide, natural and sintered diamond tools exhibit good performance while cutting aerospace titanium alloy. Also, binder less CBN delivers excellent results in machining titanium alloys.

Natural form of diamond tool underwent lesser amount of wear post 30-minutes cutting at low speeds (100 m/min) as well as at higher (198 m/min wet) cutting speeds [4,5]. Also, the tool wear was pronounced at cutting speed of 300 m/min. Polycrystalline cubic diamond (PCD) tool performs well during finishing operation of TiAl6V4. The operational life was found to be more with better surface characteristics and closer part dimensional accuracies.

Diffusion process, considered to be one of the main problems when machining titanium grade 5 alloy, involves dispersing of elements and exhibition of greater chemical affinity [6,7]. This process of material loss is more pronounced at higher surface cutting velocities or when the tool–chip interface is subjected to high temperatures. Elements of Ti grade 5 alloy have greater tendency to weld onto the surface of tungsten carbide (Co binder) tools during cutting at elevated temperatures. These alter the surface composition and influences the functioning of milling tool.

At lower cutting speeds, the tool wear is attributed to mechanical fatigue and pre-existing micro defects present in it [8]. Tool material constituents are easily pulled out from the deeply cracked areas and they slide on a definite interface leading to abrasion wear. Abrasion is considered to be the common wear mechanism when cutting with carbide tools at lower speeds.

2.2. High Heat Stress

It is generally dependent on cutting velocity, uncut chip thickness, thermal conductance, and specific energy of workpiece [9,10]. This interdependency raises the temperature even at low cutting speeds. Major portion of the heat energy is transferred to the cutting tool due to Titanium Grade 5 low levels of heat conduction.

Also, the heat stress becomes higher and diffusion and adhesion wear modes are enhanced. The elevated cutting temperatures deteriorates tool life, surface finish and subsequent dimensional accuracy. Among the machining parameters, the effect of cutting speed is the most important on cutting temperature. Keeping the cutting temperature to its minimum is desirable for improved machinability since the chemical reactivity of titanium workpiece with the cutting tool can be reduced [11,12].

2.3. Elastic Recovery

According to Schaal et al., in turning process the height of ploughed material is dependent on nose radius as shown in Figure 23.1 [13].

Very Young's modulus render titanium alloys its high elasticity which causes excessive workpiece bouncing back upon removal of cutting tool. The workpiece tends to bounce away from the cutting tool which leads to reduced clearance angle at the flank region, increased rubbing, chatter, and increased ploughing forces.

In this article it has been reported that due to elastic recovery there is significant effect on surface quality, tool wear and process forces.

2.4. Pressure on Cutting Edge of Tool

Because of higher loads on the cutting edge of the tool the titanium work area under the tool is reduced in contact and attains low plasticity.

Consequently, higher mechanical stresses arise around the cutting edge of tool. This stress amplifies as the cutting speed increases.

2.5. Presence of Residual Stress Before Machining

Aerospace products are fabricated from workpieces that are cast. When these cast workpieces are formed the temperature gradient during cooling and solidification sets up residual stresses. Upon machining where material removal takes place, the residual stresses are released and the machined workpieces show distortion in their size and shape. This distortion affects the dimensional accuracy of the machined components [14–16].

3. Solutions Reported for Efficient Machining of Ti-6Al-4V

In most of the methods to cut Ti-6Al-4V, more machining time is taken and it is not so economical. The various techniques to enhance machinability of Titanium grade-5 are stated here:

3.1. Using VibrationAanalysis Technique

This technique reduces chatter, and repetitive loads at higher cutting velocity. Typical setup will help in reducing vibrations, increase cutting speed and depth of cut leading to improved material removal rate [17].

3.2. Thermal Enhanced Machining

The workpiece is heated before the machining to soften it. Variation in chip thickness is eliminated. The cutting pressure load can be decreased by almost 50% [18].

3.3. Coolant Delivery

Coolant must penetrate tool-chip and tool-work interfaces to be more effective. Also heat conduction characteristics between the lubricant material and the ploughing region influences heat removal. High pressure coolant

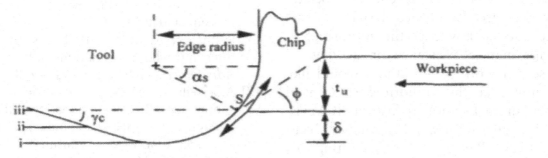

Figure 23.1: Influence of nose radius during material removal in turning process.

delivery results in formation of small sized chips, increased tool life. Also, it has been reported that machinability can be improved by cooling the workpiece or by introducing cryogenic material to the cutting region [19].

3.4. Hybrid Machining

This process simultaneously involves pre-heating the component and cryogenic treatment of cutting tool. This softens the component and reduces the cutting tool degradation [20,21].

4. Conclusion

The main difficulties in machining titanium are higher tool wear and chatter due to continuous change in chip thickness, higher amounts of heat accumulation, and greater pressure loads. These affect the material removal rate and raise the machining time. Majority of cutting tool materials have greater chemical affinity to titanium alloys under different machining condition. Carbides, binder less CBN, and diamond tools are considered to be appropriate for cutting titanium grade-5 alloy. Also, workpiece bounce back affects the dimensions of machined components. Delivery of coolant at increased pressure is the preferred technique to improve machinability of titanium alloys. Also there lies a greater scope in thermally enhanced machining and hybrid machining.

Acknowledgement

This work is a portion of the on-going research project number: TAR/2022/000118 sanctioned by SERB under Teachers Associateship for Research Excellence (TARE) scheme. The authors acknowledge the authorities of SERB for financial assistance for carrying out the research work. The authors would like to acknowledge the authorities of Central Manufacturing Technology Institute (CMTI), Bangalore and Ramaiah Institute of Technology (RIT), Bangalore for facilitating this research under the funding.

References

[1] Donachie Jr., M.J. (1988). Titanium: A Technical Guide, ASM International, Material Park, OH.

[2] Chichili, D. R., Ramesh, K. T., and Hemker, K. J. (1998). The high-strain-rate response of alpha-titanium: Experiments, deformation mechanisms and modelling. *Acta. Mater.*, 46, 1025–43.

[3] Campbell, F. C. (2006). Manufacturing Technology for Aerospace Structural Materials, 1st edition, Elsevier Ltd.

[4] Takeyama, H., Murakoshi, A., Motonishi, S., and Narutaki, N. (1983). Study on machining of titanium alloys. *Ann. CIRP.*, 32(1), 65–69.

[5] Sutter, G., Faure, L., Molinari, A., Ranc, N., and Pina, V. (2003). An experimental technique for the measurement of temperature fields for the orthogonal cutting in high speed machining.

[6] Takeyama, H., Murakoshi, A., Motonishi, S. and Narutaki, N. (1983). Study on machining of titanium alloys. *Ann. CIRP.*, 32 (1), 65–69.

[7] Ahmadi, K., and Ismail, F. (2012). Stability lobes in milling including process damping and utilizing multi-frequency and semi-discretization methods. *Int. J. Mac. Tools. Manuf.*, 54, 46–54.

[8] Komanduri, R., and Hou, Z. -B. (2002). On the thermoplastic shear instability in the machining of titanium alloy (Ti–6Al–4V). *Metall. Mater. Trans.*, 33A, 2995–3010.

[9] Abele, E., and Fröhlich, B. (2008). High speed milling of titanium alloys. *Adv. Prod. Eng. Manag.*, 3, 131–140.

[10] Schaal, N., Kuster, F., and Wegener, K. (2015). Springback in metal cutting with high cutting speeds, 15th CIRP. Conference on Modelling of Machining Operations, Procedia CIRP 31, 24–28.

[11] Pramanik, A. (2014). Problems and solutions in machining of titanium alloys. *Int. J. Adv. Manuf. Technol.*, 70, 919–928.

[12] Albertelli, P., Mussi, V., Ravasio, C., and Monno, M. (2012). An experimental investigation of the effects of spindle speed variation on tool wear in turning. *Procedia. CIRP.*, 4, 29–34.

[13] Albertelli, P., Musletti, S., Leonesio, M., Bianchi, G. and Monno, M. (2012). Spindle speed variation in turning: technological effectiveness and applicability to real industrial cases. *Int. J. Adv. Manuf. Technol.*, 62, 59–67.

[14] Altıntas, Y., Budak, E., and Engin, S. (1999). Analytical stability prediction and design of variable pitch cutters. *J. Manuf. Sci. Eng.*, 121(2), 173–178.

[15] Bäker, M., Rösler, J., and Siemers, C. (2002). A finite element model of high-speed metal cutting with adiabatic shearing. *Comput. Struct.*, 80(5), 495–513.

[16] Barry, J., Byrne, G., and Lennon, D. (2001). Observations on chip formation and acoustic emission in machining Ti–6Al–4V alloy. *Int. J. Mac. Tools. Manuf.*, 41, 1055–1070.

[17] Bediaga, I., Zatarain, M., Muno, a J., and Lizarralde, R. (2011). Application of continuous spindle speed variation for chatter avoidance in roughing milling. Proceedings of the Institution of Mechanical Engineers Part B. *J. Eng. Manuf.*, 225, 631–640.

[18] Bermingham, M. J., Palanisamy, S., Kent, D., and Dargusch, M. S. (2012). A comparison of cryogenic and high-pressure emulsion cooling technologies on tool life and chip morphology in Ti–6Al–4V cutting. *J. Mater. Process. Technol.*, 212, 752–765.

[19] Bouzakis, K. -D., Michailidis, N., Skordaris, G., Kombogiannis, S., and Hadjiyiannis, S., et al. (2002). Effect of the cutting-edge radius and its manufacturing procedure on the milling performance of PVD coated cemented carbide inserts. *Ann. CIRP.*, 51(1), 61–64.

[20] Barry, J. and Byrne, G. (2001). Study on acoustic emission in machining hardened steels. Part 1: acoustic emission during sawtooth chip formation. Proceedings of the Institution of Mechanical Engineers Part B. *J. Eng. Manuf.*, 215, 1549–1559.

[21] Bareggi, A., O'donnell, G.E., and Torrance, A. (2008). Modelling and experimental analysis of high-speed air jets used during metal cutting as a cooling technique. *Proc. Third CIRP Int. Conf. High Perf. Cutting*, 1, 337–346.

24 Experimental analysis on waste tyres for safest replacement in concrete composites

S. Muthukumarasamy[1,a], M. Siva[2], K. Sivasakthi Balan[3], K. Rajkumar[4], Rakesh Kumar Pandey[5], and A. Mohan[1]

[1]Department of Aeronautical Engineering, Vel Tech Rangarajan Dr. Sagunthala R&D Institute of Science and Technology, Chennai, India
[2]Department of Civil Engineering, Easwari Engineering College, Chennai, India
[3]Department of Mechanical Engineering, Sri Sairam College of Engineering, Bengaluru, India
[4]Department of Civil Engineering, ICCS College of Engineering and Management, Thrissur, India
[5]Department of Civil Engineering, MATS University, Raipur, India

Abstract

The principal objective of the current examination is to assess the potential outcomes of utilising scrap elastic as an incomplete substitution of sand and elastic chips as a fractional swap for blue metal. In this study, crumb rubber is used in place of 10%, 20%, and 30% of the typical fine aggregate (sand), and rubber chips are used in place of 10%, 20%, while 30% of the traditional coarse aggregate (blue metal). The replacements 10(5+5)%, 20(10+10)%, and 30(15+15)% have also been performed in combination. They made these samples with various ratios of rubber replacement as well as a control mix with no rubber. To check the fresh concrete workability, they performed a slump test. After allowing the concrete to cure for 7 and 28 days, they tested the hardened concrete samples. The tests included measuring the compressive strength, flexural strength, split tensile strength, and ultrasonic pulse velocity. The goal was to investigate the effects of partially replacing the coarse aggregates with rubber particles derived from recycled tires on the properties of the hardened concrete. The different rubber replacement ratios were compared against the control mix without rubber to evaluate changes in strength and other properties.

Keywords: Waste tyres, concrete mold, tensile strength, flexural test

1. Introduction

The exploratory examination of waste tires for the most secure substitution in substantial composites is a generally new field of examination [1]. Be that as it may, various examinations have been directed to explore the impacts of supplanting fine or coarse totals with squander tire elastic (WTR) on the properties of cement. The consequences of these investigations have demonstrated the way that the substitution of WTR can fundamentally affect the new and solidified properties of cement [2]. By and large, the supplanting of WTR with fine totals brings about a diminishing in the usefulness of cement, while the supplanting of WTR with coarse totals brings about an expansion in the functionality [3]. Be that as it may, the substitution of WTR can likewise work on the solidness and effect obstruction of cement. The substitution of 6% of fine total with WTR brought about the ideal properties of cement regarding compressive strength, flexural strength, and effect opposition [4]. The investigation likewise discovered that the substitution of WTR didn't altogether influence the

[a]muthukumarasamys@veltech.edu.in

DOI: 10.1201/9781003545941-24

functionality of cement. The substitution of 10% of coarse total with WTR brought about the ideal properties of cement concerning compressive strength and flexural strength [5]. The investigation likewise discovered that the substitution of WTR brought about a diminishing in the water retention and shrinkage of cement [6]. The consequences of these examinations recommend that the substitution of WTR can be a practical choice for working on the properties of cement. Notwithstanding, it is vital to take note of that the ideal measure of WTR to supplant relies upon the particular properties of the substantial being created [7].

Here are a portion of the variables that ought to be thought about while deciding the ideal measure of WTR to supplant in concrete:

- The sort of WTR being utilised.
- The properties of the fine or coarse totals being supplanted.
- The ideal properties of the substantial.
- The expense of WTR and the expense of the ordinary materials.

In general, the exploratory examination of waste tires for the most secure substitution in substantial composites is a promising field of examination. With additional examination, it is feasible to foster more maintainable and eco-accommodating substantial composites that are made with squander tire elastic [8].

Here are a portion of the advantages of involving waste tires in concrete:

- It can assist with diminishing the natural effect of waste tires.
- It can work on the solidness and effect obstruction of cement.
- It can diminish the expense of cement.

Be that as it may, there are additionally a few difficulties related with involving waste tires in concrete, for example,

- The need to appropriately deal with the waste tires to eliminate impurities.

- The potential for the waste tires to influence the usefulness of cement.
- The need to lead further examination to upgrade how much waste tires that can be utilised in concrete.

Regardless of these difficulties, the utilisation of waste tires in concrete is a promising area of examination with the possibility to make substantial more maintainable and eco-accommodating [9].

2. Experimental Investigation

The impact of different rubber types and contents on the characteristics of concrete were investigated [10]. M20 mix was employed in accordance with ACI Committee 211. 1-91 procedure. Concrete was hand mixed with the following specifications: 0.5 water content, 1:1.73, and 3.12 mix ratio. To prevent calcium hydroxide from concrete from leaching during testing, the specimens were demolded at least 24 hours after casting and wet-cured by fully submerging them in lime-saturated tap water for 7 or 28 days [11].

3. Materials and Methods

Two types of rubber are used in this study as shown in Figures 24.1 and 24.2.

1. Crumb rubber (1 to 4 mm) in place of sand &
2. Rubber chips (10mm to 20 mm) in place of Coarse Aggregate.

3.1. Specific Gravity Analysis

When formulating concrete mixes, engineers must consider the specific gravity of the particles being utilised. The specific gravity number indicates the relative density of the aggregate material compared to water. By determining the specific gravity of each constituent of concrete, we can accurately transform their masses into volumes when they are in solid form. This conversion is essential for

Figure 24.1: Crumb rubber.

Figure 24.2: Rubber Chips.

Type: Locally available angular coarse aggregate (Conforming to IS 383: 1970) and results shown in Figures 24.5 and 24.6.

Maximum Size: 20 mm; Minimum Size: 12.5 mm; Total weight taken = 2000 grams

Sieve analysis of Crumb Rubber as shown in Figure 24.7.

Total weight taken = 250 grams

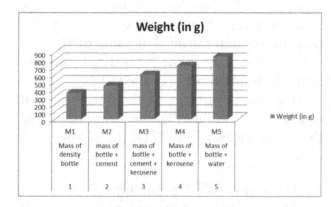

Figure 24.3: Specific Gravity Analysis Cement.

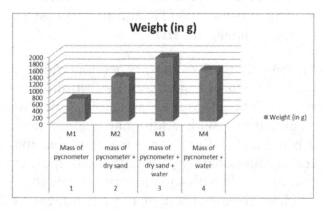

Figure 24.4: Specific Gravity Analysis Sand.

determining the theoretical yield of the concrete mixture per unit volume. The specific gravity of aggregates is particularly crucial when dealing with lightweight or heavyweight concrete mixtures. Figures 24.3 and 24.4 demonstrate the influence of aggregate density on the mix calculations for these particular types of concretes. application. The specific gravities facilitate the conversion of the weights of the combination constituents into their volumetric composition. Sieve analysis of Coarse Aggregate

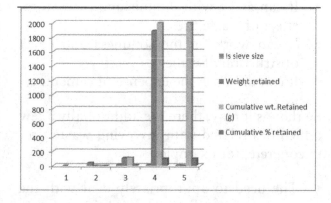

Figure 24.5: 20mm coarse aggregate.

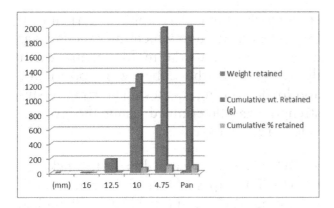

Figure 24.6: 12.5 mm Coarse aggregate.

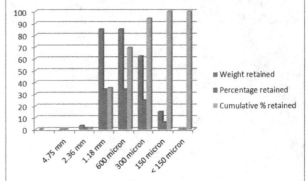

Figure 24.7: Sieve analysis of Crumb Rubber.

Table 24.1: Mix Design for the various batches

	N	F10	F20	F30	C10	C20	C30	B10	B20	B30
W/c	0.5	0.5	0.5	0.5	0.5	0.5	0.5	0.5	0.5	0.5
Rubber type	-	Crumb	Crumb	Crumb	Chips	Chips	Chips	Both	Both	Both
Rubber % (agg. Repl.)	0	10	20	30	10	20	30	10	20	30
Water, kg/m^3	185	185	185	185	185	185	185	185	185	185
Cement, kg/m^3	370	370	370	370	370	370	370	370	370	370
Sand, kg/m^3	642	577	513	449	642	642	642	610	577	546
Gravel, kg/m^3	1158	1158	1158	1158	1042	926	810	1100	1042	984
Crumb Rubber, kg/m^3	0	21.5	43	64.5	0	0	0	10.75	21.5	32.25
Rubber Chips, kg/m^3	0	0	0	0	26.2	52.4	78.6	13.1	26.2	39.3

4. Casting

Figure 24.8: Casting of Moulds.

Figure 24.9: Wet Concrete in Moulds.

Figure 24.10: Curing.

5. Curing

After 24 + 0.5 hours, all of the cast specimens were demolded and put in the curing tank for 7 or 28 days. The samples were then taken for testing, including the flexure test, split tensile strength test, and compression test. The outcomes were contrasted and examined. Figure 24.10 displays samples maintained for curing.

5.1. *Tests Procedure on Fresh and Hardened Concrete*

5.1.1. Slump test

A slump test was done to survey and look at the consistency that showed the simplicity of stream of newly blended concrete. A higher downturn suggested better consistency and usefulness. Droop test is the most normally utilised strategy for estimating consistency of cement. The device for leading the rut test basically comprises of a metallic shape as a frustum of a cone having the interior aspects as under

    Bottom diameter : 20 cm
    Top Diameter   : 10 cm
    Height   : 30 cm

5.1.2. Compressive Strength Test

For the compression test, the cubes were placed on their sides in the testing machine. This oriented the cubes such that the compressive force was applied perpendicular to the direction in which the concrete was originally cast into the molds. Figure 24.11 illustrates how the cubes were positioned and loaded during this test. The compressive strength was determined using the average value of the three cubes.

- 3D shapes are weighed prior to stacking.
- The 3D shapes are set in the compressive testing machine and the heaps are applied steadily.
- The advanced machine shows how much burden applied on the presentation. Care ought to be taken that the heap is applied at a uniform rate.
- The heap arrives at a greatest point and starts lessening from there on. The greatest burden is a definitive burden and that perusing is to be noted.

5.1.3. Split Tensile Test

- The cylinders are positioned horizontally in the compression testing equipment.
- Load is gradually imparted.

The greatest load applied to the specimen during the test is recorded. The load is increased until the specimen fails at a constant pace as shown in Figure 24.12.

Figure 24.11: Shows the compression test being carried out to find the compressive strength of the specimen.

Figure 24.12: Split tensile test on cylinder.

5.1.4. Flexural Strength Test

Prior to completing the flexural strength examination, the researchers made the necessary preparations for the concrete beam specimens. The regions on the surface of the beam where the supporting rolling elements and loading roll would come into contact were thoroughly examined and cleaned with great care. All loose particles of sand or other particulates were eliminated from these contact sites.

The specimen is then positioned in the machine so that the load is applied along two lines that are 20 cm apart to the uppermost surface that was cast in the mold.

The greatest load applied to the specimen during the test is recorded. The load is increased until the specimen fails. A beam is tested for flexural strength in Figure 24.12.

5.1.5. Ultrasonic Pulse Velocity

The ultrasonic heartbeat speed V, addressing the spread speed of longitudinal pressure wave beats through concrete, was determined. A higher V worth meant cement of better quality.

This is a non damaging test completed to track down the reviewing nature of cement. The ultrasonic heartbeat speed strategy includes the estimation of the hour of movement of electrically created mechanical heartbeats through the substantial. There are three different ways of estimating beat speed through the substantial. They are

1. Direct Transmission
2. Indirect Transmission
3. Surface Transmission

In this undertaking the surface transmission strategy is followed. The transmitter and recipient are situated on the contrary essences of the substantial and the beat wave is sent.

The perusing is noted from the advanced presentation meter. The recommended beat speeds for concrete are referenced in Table 24.2 and Figure 24.13 shows the UPV test being led on example.

5.1.6. Determination of hardened density

The solidified thickness was resolved with the end goal that a lower thickness esteem suggested solidified cement of lighter weight. The examples are weighted in Computerised weighting machine following 7 and 28 days relieving period as shown in Figure 24.15.

6. Analysis of Empirical Findings and Discussion

6.1. *New Properties of Concrete (Slump)*

On substitution of totals by elastic, a decrease in droop esteem was noticed. The decrease in

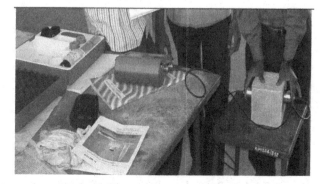

Figure 24.13: Ultrasonic pulse velocity test.

droop expanded with the expansion in volume of elastic totals. The decrease values are plotted is displayed in Figure 24.15.

6.2. *Effect of Rubber Type*

From Figure 24.16, it is plainly noticed that substantial with scrap elastic instead of sand has a higher downturn esteem than concrete with elastic chips substitution of a similar rate. From this it is seen that substantial with scrap elastic has preferred functionality over concrete with elastic chips.

6.3. *Hardened density*

From Figures 24.17 and 24.18, it is seen that there is a drop in solidified thickness with expansion in elastic substance. In the Figure FA signifies Fine Total substitution and CA means coarse total substitution. The drop

in solidified thickness can be credited to the lesser unit weight of scrap endlessly elastic chips contrasted with that of sand and rock. It is likewise seen that there is a slight expansion in solidified thickness with restoring age from 7 days to 28 days. It is seen that the solidified thickness of piece elastic (Fine total) filled concrete is higher than the elastic chips (coarse total) filled concrete.

The drop in solidified thickness concerning the control blend for piece elastic filled concrete was around 2%, 3% and 8% for 10%, 20% and 30% substitution separately. Essentially, the drop in solidified thickness regarding the control combination for elastic chips filled concrete and joined substitution was roughly 5%, 9%, 13% and 3%, 5%, 8% for 10%, 20%, 30% substitution separately.

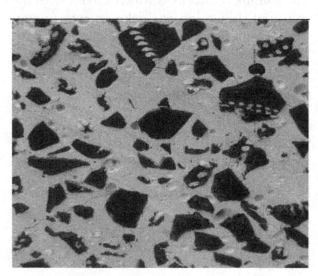

Figure 24.14: Hardened Concrete.

Figure 24.15: Partial replacement of Aggregates.

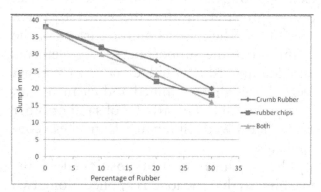

Figure 24.16: Variation of slump against rubber volume.

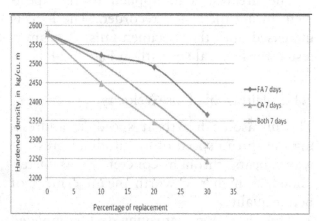

Figure 24.17: Hardened density for 7 days curing with increase in rubber content and change in rubber type.

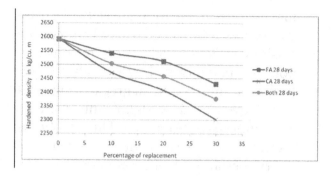

Figure 24.18: Hardened density for different 28 days curing with increase in rubber content and change in rubber type.

6.4. Compressive Strength

Figure 24.19 shows the variety in pressure strength with expansion in level of substitution. The adjustment of solidarity for various kinds of elastic and different relieving ages can likewise be seen from a similar Figure 24.20. The diminishing in compressive strength, in fine total supplanted concrete, taken toward the finish of 28 days were around 31%, 46% and 55% for 10%, 20% and 30% substitution separately. While, the lessening in compressive strength

In coarse total supplanted concrete and joined substitution taken toward the finish of 28 days were around 39%, 60%, 67% and 30%, 42%, 63% for 10%, 20%, 30% substitution separately. It is to be noticed that for a similar volume of substitution, the decrease of solidarity in piece elastic filled concrete is lesser than elastic chips filled concrete. This shows that morsel elastic filled concrete has better strength for a similar level of substitution. From the Figure 24.21 it is likewise noticed that, between the various kinds of elastic, the distinction in strength expanded with the relieving age. During the 7 days tests, the qualities are nearer to one another for scrap endlessly elastic chips substitution.

6.5. Split Tensile Strength

From Figures 24.20 and 24.21, it very well may be noticed that the rigidity likewise decreses with expansion in level of elastic aggregates. The rates of decline in rigidity for 10%, 20% and 30% of piece elastic substitution are 36.7%, 46.6% and 50% separately. Correspondingly the rates of diminishing in rigidity for 10%, 20%, 30% of elastic chips and consolidated substitution are 16%, 31.6%, 41% and 15%, 27%, 35% separately.

Not at all like the compressive strength, is the rigidity for fine and coarse total supplanted concrete higher than fine total substituted concrete for a similar level of elastic. The drop in rigidity from the control blend is only 10% for 10% consolidated substitution. However, it is additionally noticed that similarly as with expansion in level of coarse elastic chips, the drop is higher and proceeds

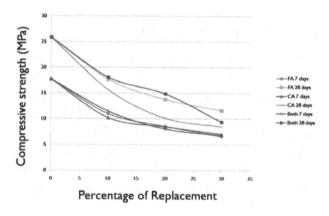

Figure 24.19: Compression strength for different curing ages with increase in rubber content and change in rubber type.

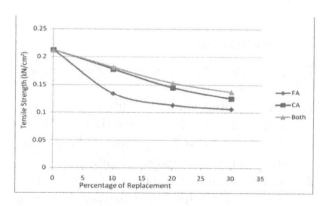

Figure 24.20: Tensile strength for different types of rubber with increase in rubber content.

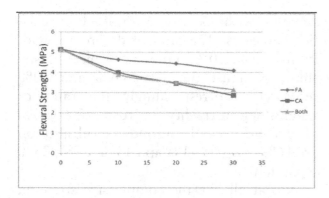

Figure 24.21: Variation of flexural strength for different types of rubber with increase in rubber content.

with a similar rate. However, on account of scrap elastic after an unexpected fall, the decrease is progressive and the distinction in upsides of rigidity is exceptionally close for both cases.

6.6. *Flexural Strength*

The elastic substance expands the flexural strength of substantial reductions. The decline from control combination was viewed as at 9.9%, 13.77% and 20.59% for 10%, 20% and 30% of morsel elastic. The lessening was at 22%, 32.8%, 44.2% and 24%, 32%, 38% for 10%, 20%, 30% of elastic chips filled concrete and consolidated substitution individually.

Like the compressive strength, the flexural strength was viewed as higher for morsel elastic supplanted concrete in contrast with elastic chips-filled concrete.

7. Conclusion

The research yields the following deductions:

1. Lower slump values in rubber-filled concrete indicate less workability. When compared to using crumb rubber, using rubber chips created less slump.
2. For crumb rubber and combination replacement of 30%, hardened density decreased by up to 8%, and for rubber chips replacement of 30%, by 12.5%. This demonstrates

that there is a significant weight decrease and rubber filled concrete can be used successfully in light weight concrete constructions like prefabricated wall panels.
3. Both fine and coarse aggregate substitution experienced a significant decrease in compressive strength. This could be ascribed to a weak connection between the rubber and the other concrete constituents. These kinds of concrete can be utilised for things like paving blocks where a lower compressive strength is adequate.
4. The ultrasonic pulse velocity test revealed that the concrete's quality rating is "Good" according to IS:13311-part 1.
5. For different replacements, there is a loss in flexural strength ranging from 10% to 25%.

Acknowledgement

The authors express their sincere gratitude to the Department of Mechanical Engineering, Vel Tech Rangarajan Dr. Sagunthala R&D Institute of Science and Technology, Chennai, for providing the research facilities and necessary support throughout this research work.

References

[1] Skripkiūnas, G., Grinys, A. and Černius, B., 2007. Deformation properties of concrete with rubber waste additives. Materials science [Medžiagotyra], 13(3), 219–223.
[2] Bakri1, M. M. A., Adam, S. N. F., Abu Bakar, M. D., and Leong, K. W, Comparison of rubber as aggregate and rubber as filler in concrete. *1st International Conference On Sustainable Materials* 2007_ICoMS 2007 9–11 June 2007, Penang.
[3] Lepech, M.D., Li, V.C., Robertson, R.E. and Keoleian, G.A., 2008. Design of green engineered cementitious composites for improved sustainability. ACI Materials Journal, 105(6), pp.567–575.
[4] Senthil Kumaran, G, Mushule, N., and Lakshmipathy, M. (2008). "A review on construction technologies that enables environmental protection: rubberized concrete.

Am. J. Eng. Appl. Sci., 1(1), 40–44. ISSN 1941-7020.

[5] Amirkhanian, S. N. and Arnold, L. C. A feasibility study of the use of waste tyres in asphaltic concrete mixtures. Report No. FHWASC-92-04 2001.

[6] Eldin, N. N and Senuci, A. B. (2002). Experts Join panels to guide industry and asphalt rubber technology transfer. Advisory Committee-RPA, Annual Meeting.

[7] Banu, M., Madhavan, V. R. B., Manickam, D., Devarajan, C. (2021). Experimental investigation on stacking sequence of kevlar and natural fibres/epoxy polymer composites. *Polimeros: Ciencia e Tecnologia,* 31(1), 1–9.

[8] Murali, B., Ramnath, B. M. V., Rajamani, D., Nasr, E. A., Astarita, A., Mohamed, H. (2022). Experimental investigations on dry sliding wear behaviour of kevlar and natural fiber-reinforced hybrid composites through an RSM–GRA hybrid approach. *Materials,* 15(3), 1–16.

[9] Murali, B., Yogesh, P., Karthickeyan, N. K., and Chandramohan, D. (2022). Multi-potency of botanicals (flax, hemp and jute fibers) as composite materials and their medicinal properties: a review. *Mater. Today Proc.,* 62, 1839–1843.

[10] Karthik, K. and Manimaran, A. (2020). Wear behaviour of ceramic particle reinforced hybrid polymer matrix composites. *Int. J. Ambient Energy,* 41(14), 1608–1612.

[11] Ganesh, R., Karthik, K., Manimaran, A. and Saleem, M. (2017). Vibration damping characteristics of cantilever beam using piezoelectric actuator. Int J Mech Eng Technol, 8(6), 212–221.

[12] Karthik, K., Ganesh, R., and Ramesh, T. (2017). Experimental investigation of hybrid polymer matrix composite for free vibration test. *Int. J. Mech. Eng. Technol.,* 8(8), 910–918.

[13] Irfan Ahmed, I., Rohith Renish, R., Karthik, K., Karthick, M. (2017). Experimental investigation of polymer matrix composite for heat distortion temperature test. *Int. J. Mech. Eng. Technol.,* 8(8).

[14] Murali, B. and Nagarani, J. (2013, April). Design and fabrication of construction helmet by using hybrid composite material. In *2013 International Conference on Energy Efficient Technologies for Sustainability,* IEEE, 145–147.

[15] Karthik, K., Prakash, J.U., Binoj, J.S. and Mansingh, B.B., 2022. Effect of stacking sequence and silicon carbide nanoparticles on properties of carbon/glass/Kevlar fiber reinforced hybrid polymer composites. Polymer Composites, 43(9), pp.6096–6105. https://doi.org/10.1002/pc.26912.

[16] Fayaz, H., Karthik, K., Christiyan, K. G., Arun Kumar, M., Sivakumar, A., Kaliappan, S., Mohamed, M., Subbiah, R., and Yishak, S. (2022). An investigation on the activation energy and thermal degradation of biocomposites of jute/bagasse/coir/nano TiO2/epoxy-reinforced polyaramid fibers. *J. Nanomater.,* 2022, 5. https://doi.org/10.1155/2022/3758212.

[17] Karthik, K., Rajamani, D., Venkatesan, E.P., Shajahan, M.I., Rajhi, A.A., Aabid, A., Baig, M. and Saleh, B. (2023). Experimental investigation of the mechanical properties of carbon/basalt/SiC nanoparticle/polyester hybrid composite materials. *Crystals,* 13(3), 415.

[18] Ramesh, V., Karthik, K., Cep, R. and Elangovan, M. (2023). Influence of stacking sequence on mechanical properties of basalt/ramie biodegradable hybrid polymer composites. *Polymers,* 15(4), 985; T Sathish, et al., A facile synthesis of Ag/ZnO nanocomposites prepared via novel green mediated route for catalytic activity. *Appl. Phys. A* 127 (9), 1–9.

[19] Fairburn, B. and Larson, J. (2001). Experience with asphalt rubber concrete – an overview and future direction. *National Seminar on Asphalt Rubber,* Cansas City, Missouri, 417–431.

[20] Joe, P. E. and Chandler, A. Z. (1992). Asphalt – rubber system in road rehabilitation, *Board of Asphalt Rubber Pavements,* Technical Committee.

25 Exploring the SARS-CoV-2 NSP12 interaction interface: a computational drug repurposing approach for COVID-19

S. Aarthy[1], S. G. Bhavya[1], Manjunath Dammalli[2], M. Adithya[2], and M. Keerthana[1]

[1]Department of Biotechnology, M S Ramaiah Institute of Technology, Bangalore, India
[2]Department of Biotechnology, Siddaganga Institute of Technology, Tumakuru, India

Abstract

The recent devastation caused by the SARS-CoV-2 virus through the spread of COVID-19 pandemic, has become a major concern for all the countries at global level. The rapid spread is due to the underlying characteristic of the SARS-CoV-2 Nsp12 protein, which is the key component of the viral replication and transmission mechanism. This Nsp12 exists in the form of a complex with its co- factors Nsp8 and Nsp7, which are found to be essential in primase activity and the replicative nature of the virus. As the virus spreads to different geographical locations, it constantly changes the sequence of its proteins by introducing mutations in its genome that enables it to survive better in the host. Rapid mutation renders this inhibition activity ineffective. Repurposing of existing FDA-drugs against the current SARS-CoV-2 is highly encouraged in order to narrow down the broad spectrum of drugs. In this research, we have focused on identifying potential antiviral compounds that are regarded as FDA-approved by targeting the interfacial domain (i.e. Nsp12-Nsp8) of the Wild and mutant RNA-Dependent RNA-Polymerases (RdRp) complex. Initially in our study, the sequence level analysis was interpreted to identify P323L as potential mutant and later structure-based approach was employed. In this study, we employed site-specific molecular docking technique to narrow down from 15 investigational ligands the 2 potential antiviral hits, which is further been subjected to dynamic simulations for 50 ns and analysed for RMSD, RMSF, RG, SS and MM-PBSA for stability studies with respect to Nsp12-Nsp8 interfacial region of wild type and mutant type. Saquinavir (wildtype: -9.3 kcal/mol, mutant type: -9.2kcal/mol) and Tegobuvir (wildtype: -9.2 kcal/mol, mutant type: -9.0 kcal/mol) was observed to be potential hits exhibiting high binding affinity w.r.t wild and mutant complexes. Further, our data analysis from MDS in terms of RMSD, RMSF, RG, Secondary Structure and MM-PBSA suggested that comparatively saquinavir was more significant than tegobuvir. Thus we recommend for further in-vivo biological evaluation that saquinavir could be a potential antiviral against wild as well as mutant SARS-CoV-2 RdRp.

Keywords: RNA-dependant RNA-polymerases (RdRp), SARS-CoV-2 interface domain COVID-19, conservational analysis, molecular docking dynamic simulations

1. Introduction

Coronaviruses (CoV) constitute a diverse group of RNA viruses which are predominantly responsible for causing pulmonary and respiratory ailments upon manifestation on its animal and human hosts. These corona viruses come under the family designated as Coronaviridae and basically they are categorised into four sub-groups namely the gamma-corona viruses, beta-corona viruses, alpha-corona viruses and the delta-corona

[a]bhavyasg@msrit.edu

DOI: 10.1201/9781003545941-25

viruses [1,2]. An important feature of coronaviruses is their classification as positive-sense RNA viruses. So far, seven species of corona viruses are claimed to induce infections in humans. Of these, the recent devastation caused by the SARS-CoV-2 viral through the spread of COVID-19 pandemic. As of May 2022, the pandemic has caused a worldwide infectivity count of around 52.3Cr and a mortality toll of approximately 62.7L.

At the 5'- Terminal of the SARS-CoV-2 genome, is present a ORF1a/1b gene which codes for a particular genre of proteins known as the 1a/1b polyproteins. These polyproteins possess the tendency which enables them to undergo proteomic lysis and generate several further non-structural proteins. The single-stranded RNA genome belonging to the SARS-CoV-2 virus has been determined to be round about 29.8 kb length, with 14 ORFs (Open Reading Frames) producing about 29 types of proteins, involving the structural proteins, non-structural proteins as well as the accessory proteins wherein each of them contribute unique property of function throughout the life cycle of the SARS-CoV-2 virus. The construction of SARS-CoV-2 comprises of sixteen Non-Structural Proteins (NSP) and about four structural proteins namely Membrane, Envelope, Nucleocapsid and Spike proteins [3]. A pivotal protein component of SARS-CoV-2 among these is the Non-Structural Protein-12 (i.e) Nsp12 that is also referred to as the RNA-dependant RNA-polymerases (RdRp) protein region. This is due to the underlying characteristic of the SARS-CoV-2 Nsp12 protein, which is identified as the key component of the replication and transmission mechanism [4,5]. This Nsp12 is crucial in determining the survivability factor of the virus inside the body of the host and also it's procreating ability.

Nsp12 is found to exist in the form of a complex along with its co-factors which includes, Nsp8 and Nsp7. This Nsp12-Nsp8-Nsp7 complex can either depend on the primers or be independent of them for its replicative activity. The template-primer RNA which is found to be present along with the Nsp12 complex is a prime point of target for many drugs for the inhibition activity. The complementary nucleotides that an RNA primer draws from the surrounding nucleoplasm aid in the construction of a new complementary strand of m-RNA. The functional activity played by these cofactors, Nsp7 and Nsp8 mainly encompasses the stability attributed towards the main element, Nsp12 for its replicative activity. These cofactors play a crucial role through its primase activity. The characteristic feature of this primase activity is the synthesis and construction of new RNA nucleotides that accounts in the fabrication of RNA primer template. Thus, these cofactors are anticipated to have an influence on the polymerase activity subsequently, it contributes to the replication and transmission phenomenon of the RdRp complex. Henceforth, it is an important target of study due to its prime role in the replication and transmission mechanism.

It is highly important to identify a potential drug against the manifestation caused at a very fast rate by SARS-CoV-2 virus. In such a scenario, re-purposing the existing FDA-drugs against the current SARS- CoV-2 is highly encouraged in order to narrow down the broad spectrum of drugs [6–9]. The significance of molecular dynamics (MD) simulations in molecular biology and drug discovery has significantly grown in recent years. These simulations enable the depiction of protein and biomolecule behaviors with meticulous atomic precision and at exceptionally fine temporal resolutions. Therefore, In silico approaches are carried out in order to identify an efficient antiviral drug compound which acts against the SARS-CoV-2 virus. This is considered to be time- saving and a cost-beneficial approach.

In this study, we have focused on identifying potential antiviral compounds that are regarded as FDA-approved by targeting the interfacial domain (i.e. Nsp12-Nsp8) of the RdRp complex through virtual screening and protein-ligand docking strategies

[10–12]. The stability studies aid in further refining which is carried out through Molecular Dynamic Simulations (MDS). Targeting the RdRp/Nsp12 inhibition of SARS-CoV-2 protein is also expected to address the broad classification of RdRp/Nsp12 in other related viruses such as SARS-CoV, MERS-CoV, ZIKV, DENV, HCV etc.

2. Materials and Methodology

2.1. *Protein Retrieval and Target Preparation*

The target RdRp protein of SARS-CoV-2, identified with PDB ID: 7L1F was obtained from the RCSB databank [26]. The crystal structure retrieved from the PDB serves as the wild type starting structure for the study (Figure 25.1). PyMOL suite by Schrodinger Inc was used to obtain the Mutant Type (MT) of target protein. P323L mutation was induced at the specific NSP12-NSP8 interface region (PDB ID: 7L1F) using PyMOL suite by Schrodinger Inc, and the prepared mutant structure was used as the starting mutant type structure for the study. All the necessary parameters were followed to prepare both wild and mutant type protein using the Auto Dock Tool-1.5.6 suite [12].

2.2. *Ligand Preparation*

The ligand molecules under investigation were retrieved from PubChem Database. The PubChem database is a carefully curated collection of chemical substances that are readily available commercially and created specifically for virtual screening. The list of 15 ligand molecules which are under investigation is shown in Table 25.1. The ligand molecule structures, obtained in SDF File formats, were converted to PDB File format using the Open Babel GUI suite. Additionally, AutoDock Tools-1.5 was employed to process the ligand, and the resulting ligands were then saved in PDBQT format. In order for a conformation to be regarded near to a living system, energy minimisation helps to attain a configuration with reduced free energy values. We have performed energy minimisation using Avogadro. The force field was fixed as MMFF94. The geometry organisation comprised of about 300 steps

Table 25.1: Catalogue of Ligands under Investigation

S.No	Ligand	PubChem ID	Molecular Formula
1	Beclabuvir	49773361	$C_{36}H_{45}N_5O_5S$
2	Cidofovir	60613	$C_8H_{14}N_3O_6P$
3	Dasabuvir	56640146	$C_{26}H_{27}N_3O_5S$
4	Dolutegravir	54726191	$C_{20}H_{19}F_2N_3O_5$
5	Entecavir	135398508	$C_{12}H_{15}N_5O_3$
6	Etravirine	193962	$C_{20}H_{15}BrN_6O$
7	Famciclovir	3324	$C_{14}H_{19}N_5O_4$
8	Nilotinib	644241	$C_{28}H_{22}F_3N_7O$
9	Peramivir	154234	$C_{15}H_{28}N_4O_4$
10	Raltegravir	54671008	$C_{20}H_{21}FN_6O_5$
11	Saquinavir	441243	$C_{38}H_{50}N_6O_5$
12	Sofosbuvir	45375808	$C_{22}H_{29}FN_3O_9P$
13	Stavudine	18283	$C_{10}H_{12}N_2O_4$
14	Tegobuvir	23649154	$C_{25}H_{14}F_7N_5$
15	Tenofovir	464205	$C_9H_{14}N_5O_4P$

with Steepest Descent Algorithm and energy convergence at 10e-7. After optimisation, the ligand is saved in suitable .pdb format and used for further studies [15].

2.3. Molecular Docking and Binding Pocket Predictions

Molecular docking was carried out in the interaction interface of SARS-CoV-2 RdRp protein (PDB ID: 7L1F), in specific interfacial region Nsp12-Nsp8 along with all the chains (A, B, C) in the structure. The likely active site of the target protein SARS-CoV-2 RNA-dependent RNA polymerase (7L1F) was determined by selecting the residues on the NSP12-NSP8 interface region as the binding site region. The active pockets at the NSP12-8 interfacial region were recorded (shown in Table 25.2) and represented in Figure 25.1. The ADT was used to prepare the complex nsp7–nsp8–nsp12 and then saved in PDBQT. The grid box size for x, y, and z was set at 176.286, 145.975 and 182.049Å, respectively. The distance between the grid points was 0.375 Å, with the grid center positioned at 60, 60, and 60 Å along the x, y, and z axes, respectively. All default settings were employed, except for the exhaustiveness value, which was fixed at 8 and the modes of binding generated was about 9 to increase accuracy. The above mentioned parameters were recorded and using the information retrieved from setting the grid box, the configuration file (conf.txt) was created. This file is used as input for running site-specific docking through "Auto Dock Vina" using specific commands [16,17]. The protein-ligand docked complex was visualised at 2-D level using LigPlot+ and 3D level using "PYMOL".

2.4. Protein Conservational Domain Analysis

The Nsp-12 region of SARS-CoV-2 possesses some functionally essential areas which are conserved in due course. These were predicted

through the ConSurf online server. This tool helps in carrying out the conservation analysis for our target protein (PDB ID: 7L1F) by evaluating the degree of conserved residues of amino acids based on its similarities [18]. The residual range for each motif of Nsp-12 is mentioned in the Figure 25.2.

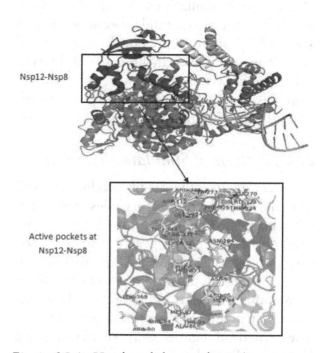

Figure 25.1: Hardened density for 7 days curing with increase in rubber content and change in rubber type.

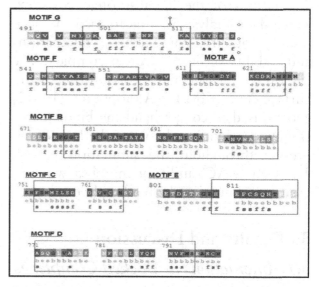

Figure 25.2: Conserved protein domains of the RdRp protein obtained from consurf profile.

2.5. *Molecular Dynamics Simulation*

Molecular dynamics (MD) simulation system inputs were created using Charmm-gui, while GROMACS 2020.6 was utilised to conduct MD simulations of the RdRp protein and protein-ligand complexes, employing the chosen CHARMM36m force field. Overall, MD simulations lasting 50 ns were conducted for two ligands in complex with the RdRp protein, as previously documented [15]. The complexes include (1) Wild type RdRp-Saquinavir (2) Mutant type RdRp-Saquinavir (3) Wild type RdRp-Tegobuvir and (4) Mutant type RdRp-Tegobuvir.

2.6. *Analysis of Simulated Trajectories*

After completing the 50 ns production MD run, the resulting trajectory files were analyzed using GROMACS utilities such as 'gmx rms', 'gmx rmsf', 'gmx gyrate' and 'gmx do_dssp' to assess Root Mean Square Fluctuations (RMSF), Radius of Gyration (Rg), Root Mean Square Deviation (RMSD) and the secondary structure of the protein, respectively. The structural figures were observed using VMD and PyMol software, while the graphs and plots were produced using the Xmgrace tool. This data analysis was conducted to assess the compactness, rigidity, stability and fluctuations of the ligand-protein complex. These data analysis was performed to check for the rigidity, compactness, fluctuations and stability of ligand-protein complex. The Molecular Mechanics Poisson-Boltzmann Surface Area (MM-PBSA) approach [19–22] was utilised to comprehend the binding free energy (ΔG binding) between the ligand and protein throughout the simulation period. The GROMACS utility g_mmpbsa was utilised for computing the binding free energy.

3. Results and Discussion

3.1. *Interactions in Docked Complexes*

The 15 selected ligands underwent molecular docking studies to evaluate their binding patterns with the RdRp protein (PDB:7L1f) (Figure 25.1). The binding energy of ligands with RdRp is given as the results of Auto-Dock software. The docking results for top 5 ligands is provided in Table 25.2.

From the results it is evident that, out of 15 selected compounds Saquinavir and Tegobuvir have more binding affinity with the receptor. The docking poses of Saquinavir and Tegobuvir for wild and mutant type are shown in Figure 25.2. Saquinavir is a FDA-approved antiviral that works by inhibiting the HIV protease which is an enzyme required for viral maturation and reproduction. The binding scores of Nsp12-Nsp8 interface region of wild type and mutant type were obtained to be –9.3 kcal/mol and –8.6 kcal/mol respectively. For the wild type RdRp-saquinavir complex, Saquinavir make 3 hydrogen bonds with Arg249 and Arg349 (2) and 14 hydrophobic bonds with Pro323, Phe326, Glu350, Thr394, Cys395, Phe396, Arg457, Asn459, Leu460, Pro461, Asn628, Val675, Lys676 and Pro677 [23–27]. Regarding mutant

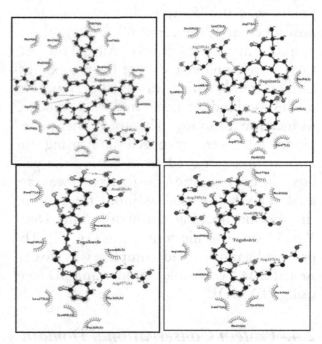

Figure 25.3: Docking interaction images at Nsp12-Nsp8 Interface for (A) Wild Protein with Saquinavir (B) Mutant Protein with Saquinavir (C) Wild Protein with Tegobuvir (D) Mutant Protein with Tegobuvir.

Table 25.2: Binding affinities of NSP12-8 interface protein-ligand complexes

Ligands	Mutant/Wild type	Score kcal/mol	H-bond	Hydrophobic bond
Saquinavir	Wild	-9.3	Arg249, Arg349(2)	Pro323, Phe326, Glu350, Thr394, Cys395, Phe396, Arg457, Asn459, Leu460, Pro461, Asn628, Val675, Lys676, Pro677
	Mutant	-8.6	Arg249, Asn459	Phe165, Pro169, Leu172, Arg173, Thr394, Cys395, Pro461, Arg457, Tyr458, Leu460, Pro677
Tegobuvir	Wild	-9.3	Asn628, Arg457	Phe165, Pro169, Leu172, Arg249, Tyr458, Leu460, Pro461, Pro677
	Mutant	-9.7	Asn628, Arg349	Phe165, Pro169, Leu172, Arg249, Tyr458, Asn459, Leu460, Pro461, Pro677
Beclabuvir	Wild	-8.2	Thr394, Ser397	Thr319, Phe321, Pro323, Arg349, Cys395, Phe396, Tyr456, Asn459, Leu460, Pro461, Asn628, Pro677
	Mutant	-8.0	Ser397	Leu122(B), Tyr149(B), Ser151(B), Ser255, Tyr265, Lys267, Trp268, Leu270, Pro322, Leu323, Thr324, Leu389, Cys395, Phe396
Cidofovir	Wild	-5.9	Ala74(B)Ala490, Val493, Tyr516, Asp517, Met519, Tyr521,	Gln69(B), Gln73(B), Ser520
	Mutant	-6.1	Tyr456 Thr556, Asp623, Arg624, Thr680, Ser682	Arg553, Arg555, Ser681
Dasabuvir	Wild	-7.2	-	Phe415, Phe440, Phe441, Phe442, Ala443, Leu837, Phe843
	Mutant	-7.8	-	Tyr149, Tyr265, Ile266, Trp268, Phe321, Pro322, Leu323, Thr324, Phe396

type RdRp-Saquinavir complex, the hydrogen bonds with Arg249 and Asn459 and the hydrophobic interactions with Leu172, Arg173, Thr394, Cys395, Pro461, Arg457, Tyr458, Leu460 and Pro677 are observed.

A potent non-nucleoside inhibitor called tegobuvir binds to the HCV RdRp and launches the inhibition process. The binding scores of Nsp12-Nsp8 interface region of wild type and mutant type were obtained to be –9.3 kcal/mol and –9.7 kcal/mol respectively. For wild type RdRp-Tegobuvir complex, Tegobuvir make 2 hydrogen bonds with Asn628 and Arg457 and 8 hydrophobic bonds with Phe165, Pro169, Leu172, Arg249, Tyr458, Leu460, Pro461 and Pro677. Regarding mutant type RdRp- Tegobuvir complex, we observed 2 hydrogen bonds with Asn628 and Arg349 and 7 hydrophobic interactions with Phe165, Pro169, Leu172, Arg249, Tyr458, Asn459, Leu460, Pro461 and Pro677 (Figure 25.3). To gain insights into the dynamic evolution of ligands in relation to RdRp, we conducted molecular dynamics simulation technique [28,29]. To proceed with molecular dynamics simulation, we selected Saquinavir and Tegobuvir ligands based on Molecular docking results.

3.2. *Data Analysis from Dynamic Simulations*

A molecular dynamic simulation process for about 50ns was performed for the Wildtype as well as the Mutant type complexes involving saquinavir and tegobuvir using CHARMM-GUI and GROMACS platforms. The stabilisation of proteins and protein-ligand combinations was investigated by employing dynamic itineraries, such as RMSD, RMSF, Rg, MM-PBSA and secondary structure prediction [7].

The RMSD values of wild type RdRp-Saquinavir complex show stability around 0.4 to 0.5 nm. In fact, in the first 10 ns of the dynamic simulation, RMSD values of the RdRpprotein increase slowly from 0.3 to 0.4nm for the initial 10ns and shows stability around 0.5nm for the next 50ns (Figure 25.4). The RMSD values of mutant type RdRp-Saquinavir complex show stability around 0.2 to 0.3 nm. For the initial 10ns the values of the RdRp protein increases upto 0.3nm and remain stable at 0.3nm for the next 50ns. It is observed that Saquinavir mutant complex is more stable when compared to the wild type complex [30–34].

The RMSD value of wild type RdRp-Tegobuvir complex show stability around 0.3 to 0.4 nm. During the initial 10 ns, the RMSD values exhibit a gradual increase from 0.2 to 0.35nm and for the next 50ns the values exhibit stability between 0.35 and 0.4nm. RMSD values of mutant type RdRp-Tegobuvir complex increase slowly from 0.2 to 0.25nm and show stability at 0.25 nm for the entire simulation. Tegobuvir mutant complex is more stable when compared to the wild type complex (Figure 25.4).

The RMSF graph of RdRp-Saquinavir complex is shown in Figure 25.5. Comparative study of the RMSF profiles for wild-type and mutant-type protein-ligand complexes revealed variations at the respective catalytic sites in the range of 0–0.3 nm. While certain regions within the protein-ligand complexes exhibit notable mobility, with fluctuations ranging from 0.4 nm to 0.8 nm, they are not regarded as significant for this study. Our primary emphasis is on the dynamic behavior of the catalytic sites within the protein-ligand complex [3,35–37]. However, at our mutational position P323L, we can observe a fluctuation in peak in comparison to that of the wild type.

A comparison of the RMSF profiles for protein-ligand complexes of the wild-type and mutant types RdRp-Tegobuvir complex is shown in Figure 25.5. As the primary focus of this study is on the dynamic behaviour of protein-ligand complex active site, some regions in complexes have shown increased

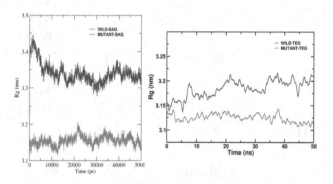

Figure 25.4: (A) RMSD of wild type RdRp-Saquinavir complex and (B) RMSD of mutant type RdRp-Tegobuvir complex.

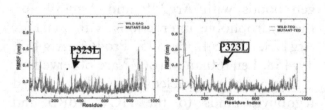

Figure 25.5: (A) RMSF of wild type RdRp-Saquinavir complex and (B) RMSF of mutant type RdRp-Tegobuvir complex.

mobility with fluctuations ranging between 0.1 nm and 0.9 nm. However, these regions are not considered to be significant for this study. However, we observe a change in peak at our mutational location P323L compared to the wild type. The graph for Rg is shown in Figure 25.6. Based on the compactness of protein structure with respect to folding and expanding, the Rg establishes the optimum of conformations in a complete system. From the results of Rg, we can analyse that the wild type complex of saquinavir exhibited greater variation in the value of Rg. At the beginning the value of Rg for wild type complex initiated from 3.4 nm and reached 3.35 nm undergoing variations. From this range, the value of Rg showed less oscillation, and stable equilibria were seen from 30ns for up to 50 ns. In the 50 ns dynamic simulation, the RdRp mutant complex with saquinavir persisted in a stable equilibrium with Rg values in the range of 3.15 nm. These findings show that, in contrast to mutant type complexes, wild type complexes have higher Rg values. It is evident that saquinavir, when it interacts with the RdRp complex, affects the organisation of protein which leads to conformational variations with respect to increased value of Rg.

Based on the compactness of protein structure with respect to folding and expanding, the Rg establishes the optimum of conformations in a complete system. From the RdRp-Tegobuvir complex results of Rg, (Figure 25.6) we can analyse that the wild type complex of tegobuvir exhibited greater variation in the value of Rg. At the beginning the value

of Rg for wild type complex initiated from 3.15 nm and reached 3.2 nm undergoing variations. From this range, the value of Rg showed less oscillation, and stable equilibria were seen from 35 ns for up to 50 ns. In the 50ns dynamic simulation programme, the RdRp mutant complex with tegobuvir persisted in a stable equilibrium with Rg values in the range of 3.8 nm. These findings show that, in contrast to wild type complexes, mutant complexes have higher Rg values. It is evident that tegobuvir, when it interacts with the RdRp complex, affects the organisation of protein which leads to conformational variations with respect to increased value of Rg.

After conducting a secondary structure study, it was shown that both the wild-type and mutant forms of the RdRp protein were more susceptible to structural variations when exposed to the potential ligands under investigation. Considering the wild type of RdRp - Saquinavir complex as shown in Figure 25.7, within the region of 845–850 amino acid residues from an 8.75 ns time interval of dynamic simulation, a secondary structural rearrangement with the transition of A-helix to turns was observed. With in range of 14 ns, another secondary structural transition from bend to coil was seen in the amino acid residues 645–655. A secondary structural shift from turns to A-helix was noted within the range of 17ns present between the residues

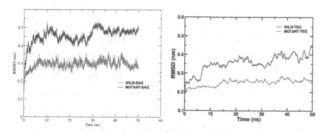

Figure 25.6: (A) Rg of wild type RdRp-Saquinavir complex and (B) Rg of mutant type RdRp-Tegobuvir complex.

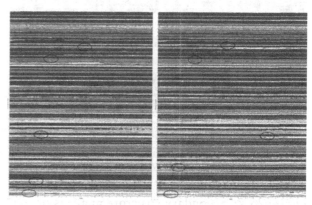

Figure 25.7: Secondary structure analysis of (A) Wild type RdRp-Saquinavir complex (B) Mutant type RdRp-Saquinavir complex.

100–110 and another shift from bend to coil within the range of 18ns between the residues 915–920. Yet another change was identified between the residues 75–80 from coil to turns in the range of 20ns. Within the range of 25ns, a change from turns to bend was observed for the residual range 790–795. A change from turns to coil was recorded in the range of 29ns between the residues 25–35.

Furthermore, a change from A-helix to turns was seen in the range of 30ns between the residues 820–825. Considering the mutant type of RdRp – Saquinavir complex as showin in Figure 25.7, within the region comprising 5–15 residues, a structural change from B-sheets to coil was noted in the range of 10ns. Another conformational variation was recorded in the range of 8ns from A-helix to turns within the residual range 20–30. We also observed a change from A-helix to turns in the range of 14ns between the residues 145–155 and one more from bend to coil in the range of 11ns between the residues 320–325. Further, between the residues 845–855 in the range of 25ns, a change from turns to A-helix was noticed. Thus, from the observations of wild type and mutant type secondary structures it is understood that the complex of RdRp with Saquinavir contributed towards many changes in secondary structural level.

Considering the wild type of RdRp - Tegobuvir complex as shown in Figure 25.8,

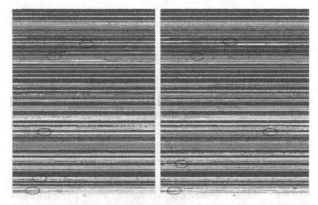

Figure 25.7: Secondary structure analysis of (A) Wild type RdRp-Saquinavir complex (B) Mutant type RdRp-Saquinavir complex.

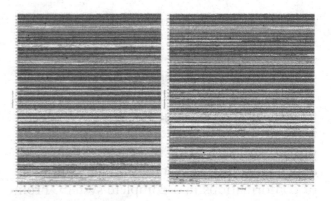

Figure 25.8: Secondary structure analysis of (A) Wild type RdRp-Saquinavir complex (B) Mutant type RdRp-Saquinavir complex.

within the region of 730–750 amino acid residues and 15–35 ns time interval of dynamic simulation, a secondary structural rearrangement with the transition of bends to turns was observed. Within range of 35 ns onwards, another secondary structural transition from coil to bend was seen in the amino acid residues 700–710. A secondary structural shift from sheets to coil was noted within the range of 15ns present between the residues 300–320 and another shift from A-helix to turns within the range of 20ns between the residues 65–85. Furthermore, a change from sheets to coil was seen in the range of 10ns between the residues 1–20. Considering the mutant type of RdRp – Tegobuvir complex as showin in Figure 25.8, within the region comprising 810–820 residues, a structural change from bend to turns was noted in the range of 25ns. Another conformational variation was recorded in the range of 20–25ns from A-helix to bends within the residual range 790–800. We also observed a change from turns to sheets in the range of 7.5–12.5 ns between the residues 740–745 and one more from sheets to turns in the range of 35ns between the residues 360–380. A secondary structural shift from sheets to coil was noted within the range of 40 ns present between the residues 320–340 and another shift from A-helix to turns within the range of 1.5–7.5

ns between the residues 150–160. Further, between the residues 1–20 in the range of 5ns, a change from sheets to coil was noticed. Thus, from the observations of wild type and mutant type secondary structures it is understood that the complex of RdRp with Tegobuvir contributed towards many changes in secondary structural level [38–40].

To assess the binding affinity of Saquinavir and Tegobuvir towards RdRp, we investigated the relative binding strength within the active center pocket. The MMPBSA method was employed for the binding strength comparison of wild and mutant type RdRp with respect to Saquinavir and Tegobuvir (Figure 25.9) In this analysis, the contribution of each residue to the total interaction energy was calculated throughout the stable simulation trajectory. The binding energy of wild type RdRp-Saquinavir was found to be –134.194 kJ/mol, while the mutant type RdRp-Saquinavir possesses a binding energy of -110.333 kJ/mol as shown in Figure 25.9. This implies that both wild and mutant type RdRp hold the Saquinavir and Tegobuvir ligands for the entire simulation. In summary, through a combination of molecular docking, MD simulation and MM-PBSA, we investigated the potential of Saquinavir and egobuvir ligands to inhibit RdRp. The secondary structure, RMSF and Rg values provide definitive evidence of the ligands' impact on altering the conformational state of RdRp.

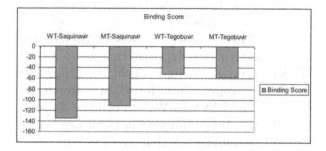

Figure 25.9: Binding free energy of wild and mutant type RdRp complex.

4. Conclusion

RNA-dependant RNA-polymerase (RdRp) is a crucial therapeutic target due to its involvement in RNA genome replication and the absence of a functioning host counterpart for this protein. For a very long time, numerous academic institutions and pharmaceutical firms have been conducting research on the development of efficient RdRp inhibitors to stop viral replication. Researchers are looking into potential therapeutic strategies to solve the issue and manage the viral infection due to the present rate of infection and the high pathogenic rate of SARS-CoV-2 around the world. In this regard, structure-based drug design employing drug re-purposing could be used as a robust strategy to investigate potential disease-treating strategies. Designing novel compounds that target the SARS-CoV-2 RdRp may be made possible by the novel inhibitors which are reported. The research reported here is an initial investigation into the possibility of using FDA-approved pharmaceuticals in combination to find a treatment that could potentially target the RdRp of the SARS-CoV-2 virus which is responsible for causing the COVID-19. The prevailing problem in the society now is that the prescribed treatment drugs does not address the evolving mutant varieties leading to newer waves of COVID-19.

Herein, we have investigated the potential function of the antivirals – Saquinavir and Tegobuvir against the RdRp (Nsp12) of SARS-CoV-2. Both the antivirals were observed to induce a significant change to an extent in the functionality of protein in both wild as well as mutant type infection. Comparatively, we recommend Saquinavir to be a more potential antiviral against SARS-CoV-2 RdRp based on our observations from RMSD, RMSF, RG, SASA & MM-PBSA. However, further extensive biological evaluation is required for the selection of potential combination therapy to deal with the current pandemic circumstances.

References

[1] Akinlalu, A. O., Chamundi, A., Yakumbur, D.T., Afolayan, F. I. D., Duru, I. A., Arowosegbe, M.A. and Enejoh, O. A., (2021). Repurposing FDA-approved drugs against multiple proteins of SARS-CoV-2: an in silico study. *Sci. Afr.*, 13, e00845.

[2] Dash, S., Aydin, Y., and Stephens, C. M. (2019). Hepatitis C Virus NS5B RNA-dependent RNA polymerase inhibitor: an integral part of HCV antiviral therapy. In *Viral Polymerases*. Academic Press. pp. 211–235.

[3] Tian, L., Qiang, T., Liang, C., Ren, X., Jia, M., Zhang, J., Li, J., Wan, M., YuWen, X., Li, H., and Cao, W. (2021). RNA-dependent RNA polymerase (RdRp) inhibitors: the current landscape and repurposing for the COVID-19 pandemic. *Eur. J. Med. Chem.*, 213, 113201.

[4] Dereeper, A., Guignon, V., Blanc, G., Audic, S., Buffet, S., Chevenet, F., Dufayard, J. F., Guindon, S., Lefort, V., Lescot, M., and Claverie, J. M. (2008). Phylogeny. fr: robust phylogenetic analysis for the non-specialist. *Nucleic Acids Res.*, 36(suppl_2), W465–W469.

[5] Elfiky, A. A., Azzam, E. B., and Shafaa, M. W. (2022). The anti-HCV, Sofosbuvir, versus the anti-EBOV Remdesivir against SARS-CoV-2 RNA dependent RNA polymerase in silico. *Mol. Diversity*, 26(1), 171–181.

[6] Elkarhat, Z., Charoute, H., Elkhattabi, L., Barakat, A., and Rouba, H., (2022). Potential inhibitors of SARS-cov-2 RNA dependent RNA polymerase protein: molecular docking, molecular dynamics simulations and MM-PBSA analyses. *J. Biomol. Struct. Dyn.*, 40(1), 361–374.

[7] Eweas, A. F., Alhossary, A. A., and Abdel-Moneim, A. S. (2021). Molecular docking reveals ivermectin and remdesivir as potential repurposed drugs against SARS-CoV-2. *Front. Microbiol.*, 11(592908), 3602.

[8] Gangadharappa, B. S., Sharath, R., Revanasiddappa, P. D., Chandramohan, V., Balasubramaniam, M., and Vardhineni, T. P. (2019). Structural insights of metallo-beta-lactamase revealed an effective way of inhibition of enzyme by natural inhibitors. *J. Biomol. Struct. Dyn.*, 38(14), 1–15. https://doi.org/10.1080/07391102.2019.1667265.

[9] Gordon, C. J., Tchesnokov, E. P., Woolner, E., Perry, J. K., Feng, J. Y., Porter, D. P., and Götte, M., (2020). Remdesivir is a direct-acting antiviral that inhibits RNA-dependent RNA polymerase from severe acute respiratory syndrome coronavirus 2 with high potency. *J. Biol. Chem.*, 295(20), 6785–6797.

[10] Guindon, S., Dufayard, J. F., Lefort, V., Anisimova, M., Hordijk, W., and Gascuel, O. (2010). New algorithms and methods to estimate maximum-likelihood phylogenies: assessing the performance of PhyML 3.0. *Syst. Biol.*, 59(3), 307–321.

[11] Hanwell, M. D., Curtis, D. E., Lonie, D. C., Vandermeersch, T., Zurek, E., and Hutchison, G. R. (2012). Avogadro: an advanced semantic chemical editor, visualization, and analysis platform. J. Cheminf., 4(1), 1–17.

[12] Hasöksüz, M., Kilic, S., and Saraç, F. (2020). Coronaviruses and sars-cov-2. *Turk. J. Med. Sci.*, 50(9), 549–556.

[13] Jack, P. K. B., Tyler, L. D., David, W. T., Kenneth, A. J. (2021). Remdesivir is a delayed translocation inhibitor of SARS-CoV-2 replication. *Mol. Cell*, 81(7), 1548–1552.

[14] Hosseini, M., Chen, W., Xiao, D., and Wang, C. (2021). Computational molecular docking and virtual screening revealed promising SARS-CoV-2 drugs. *Precis. Clin. Med.*, 4(1), 1–16.

[15] Jo, S., Kim, T., Iyer, V. G., and Im, W. (2008). CHARMM-GUI: a web-based graphical user interface for CHARMM. *J. Comput. Chem.*, 29(11), 1859–1865.

[16] Jockusch, S., Tao, C., Li, X., Chien, M., Kumar, S., Morozova, I., Kalachikov, S., Russo, J. J., and Ju, J. (2020). Sofosbuvir terminated RNA is more resistant to SARS-CoV-2 proofreader than RNA terminated by Remdesivir. *Sci. Rep.*, 10(1), 1–9.

[17] Kulkarni, S. A. and Ingale, K. (2022). *In Silico* Approaches for Drug Repurposing for SARS-CoV-2 Infection. *R. Soc. Chem.*, 1–80. https://books.rsc.org/books/edited-volume/948/chapter/755439/.

[18] Kumar, T. A. (2013). CFSSP: Chou and Fasman secondary structure prediction server. *Wide Spectrum*, 1(9), 15–19.

[19] Lobanov, M. Y., Bogatyreva, N. S., and Galzitskaya, O. V. (2008). Radius of gyration as an indicator of protein structure compactness. *Mol. Biol.*, 42(4), 623–628.

[20] MIu, L., Bogatyreva, N. S., and Galzitskaia, O.V. (2008). Radius of gyration is indicator of compactness of protein structure. *Molekuliarnaia Biologiia*, 42(4), 701–706.

[21] Modrow, S., Falke, D., Truyen, U., and Schätzl, H. (2013). Viruses with single-stranded, positive-sense RNA genomes. *Mol. Virol.*, 437–520.

[22] Nagar, P. R., Gajjar, N. D., and Dhameliya, T. M. (2021). In search of SARS CoV-2 replication inhibitors: Virtual screening, molecular dynamics simulations and ADMET analysis. *J. Mol. Struct.*, 1246, 131190.

[23] Nie, Q., Li, X., Chen, W., Liu, D., Chen, Y., Li, H., Li, D., Tian, M., Tan, W., and Zai, J. (2020). Phylogenetic and phylodynamic analyses of SARS-CoV-2. *Virus Res.*, 287, 198098.

[24] O'Boyle, N. M., Banck, M., James, C. A., Morley, C., Vandermeersch, T., and Hutchison, G. R. (2011). Open babel: an open chemical toolbox. *J. Cheminf.*, 3(1), 1–14.

[25] Pachetti, M., Marini, B., Benedetti, F., Giudici, F., Mauro, E., Storici, P., Masciovecchio, C., Angeletti, S., Ciccozzi, M., Gallo, R. C., and Zella, D. (2020). Emerging SARS-CoV-2 mutation hot spots include a novel RNA-dependent-RNA polymerase variant. *J. Transl. Med.*, 18(1), 1–9.

[26] Qiao, Z., Zhang, H., Ji, H. F., and Chen, Q. (2020). Computational view toward the inhibition of SARS-CoV-2 spike glycoprotein and the 3CL protease. *Computation (Basel)*, 8(2), 53.

[27] Raghav, S., Ghosh, A., Turuk, J., Kumar, S., Jha, A., Madhulika, S., Priyadarshini, M., Biswas, V. K., Shyamli, P. S., Singh, B., and Singh, N. (2020). Analysis of Indian SARS-CoV-2 genomes reveals prevalence of D614G mutation in spike protein predicting an increase in interaction with TMPRSS2 and virus infectivity. *Front. Microbiol.*, 11, 594928.

[28] Reva, B. A., Finkelstein, A. V., and Skolnick, J. (1998). What is the probability of a chance prediction of a protein structure with an rmsd of 6 Å?. *Fold. Des.*, 3(2), 141–147.

[29] Rodrigues, C. H., Pires, D. E., and Ascher, D. B. (2018). DynaMut: predicting the impact of mutations on protein conformation, flexibility and stability. *Nucl. Acids Res.*, 46(W1), W350–W355.

[30] Ruan, Z., Liu, C., Guo, Y., He, Z., Huang, X., Jia, X., and Yang, T. (2021). SARS-CoV-2 and SARS-CoV: Virtual screening of potential inhibitors targeting RNA-dependent RNA polymerase activity (NSP12). *J. Med. Virol.*, 93(1), 389–400.

[31] Šali, A. and Blundell, T. L. (1993). Comparative protein modelling by satisfaction of spatial restraints. *J. Mol. Biol.*, 234(3), 779–815.

[32] Satarker, S. and Nampoothiri, M. (2020). Structural proteins in severe acute respiratory syndrome coronavirus-2. *Arch. Med. Res.*, 51(6), 482–491.

[33] Panda, S. P., Atmakuri, L. R., and Guntupalli, C. (2021). In silico identification of potential inhibitors from Cinnamon against main protease and spike glycoprotein of SARS CoV-2. *J. Biomol. Struct. Dyn.*, 39(13), 4618–4632. https://doi.org/10.1080/07391102.2020.1779129.

[34] Stevens, L. J., Pruijssers, A. J., Lee, H. W., Gordon, C. J., Tchesnokov, E. P., Gribble, J., George, A. S., Hughes, T. M., Lu, X., Li, J., and Perry, J. K. (2022). Mutations in the SARS-CoV-2 RNA dependent RNA polymerase confer resistance to remdesivir by distinct mechanisms. *Sci. Transl. Med.*, 14(656), eabo0718.

[35] Trott, O. and Olson, A. J. (2010). AutoDock Vina: improving the speed and accuracy of docking with a new scoring function, efficient optimization, and multithreading. *J. Comput. Chem.*, 31(2), 455–461.

[36] V'kovski, P., Kratzel, A., Steiner, S., Stalder, H., and Thiel, V. (2021). Coronavirus biology and replication: implications for SARS-CoV-2. *Nat. Rev. Microbiol.*, 19(3), 155–170.

[37] Vella, S. and Floridia, M. (1998). Saquinavir. Clinical pharmacology and efficacy. *Clin. Pharm.*, 34(3), 189–201.

[38] Wallace, A. C., Laskowski, R. A., and Thornton, J. M. (1995). LIGPLOT: a program to generate schematic diagrams of

protein-ligand interactions. *Protein Eng. Des. Sel.*, 8(2), 127–134.

[39] Wu, C., Liu, Y., Yang, Y., Zhang, P., Zhong, W., Wang, Y., Wang, Q., Xu, Y., Li, M., Li, X., and Zheng, M. (2020). Analysis of therapeutic targets for SARS-CoV-2 and discovery of potential drugs by computational methods. *Acta Pharm. Sin. B*, 10(5), 766–788.

[40] Yin, W., Mao, C., Luan, X., Shen, D.D., Shen, Q., Su, H., Wang, X., Zhou, F., Zhao, W., Gao, M., and Chang, S. (2020). Structural basis for inhibition of the RNA-dependent RNA polymerase from SARS-CoV-2 by remdesivir. *Science*, 368(6498), 1499–1504.

Abbreviation Index

SARS-CoV-2: Severe Acute Respiratory Syndrome Coronavirus 2

NSP12: Non-Structural Protein 12

COVID-19: Coronavirus Disease 2019

FDA: Food and Drug Administration

RdRp: RNA-Dependent RNA-Polymerase

RMSD: Root Mean Square Deviation

RMSF: Root Mean Square Fluctuation

Rg: Radius of Gyration

SS: Secondary Structure

MM-PBSA: Molecular Mechanics Poisson-Boltzmann Surface Area

MDS: Molecular Dynamics Simulation

ORF: Open Reading Frame

MT: Mutant Type

PDB: Protein Data Bank

SDF: Structured Data File

PDBQT: Protein Data Bank, Partial Charge (Q) and Atom Type (T)

MMFF94: Merck Molecular Force Field 94

CHARMM-GUI: Chemistry at Harvard Molecular Mechanics – Graphical User Interface

GROMACS: Groningen Machine for Chemical Simulation

26 Packed bed column studies on denitrification of water using immobilised modified kaolin

Bhagyeshwari D.[a], Chalageri, Rajeswari M. Kulkarni, Archna N. Anju Maria Mathew, Ishani Agrawal, Neha P. S. Rao, and Sandeep P. Raman

Department of Chemical Engineering, Ramaiah Institute of Technology, Bengaluru, India

Abstract

Nitrate contamination is a major source of groundwater and surface water pollution. Industrial effluents, municipal, agricultural runoff which mostly comprise chemicals, fertilisers and pharmaceuticals are sources of nitrate contamination. Nitrate may be dangerous to human health and the environment. Due to such risks, the WHO has prescribed 50 mg/L nitrate level in water. The present study focuses on the denitrification performance of modified kaolin in a column. To investigate the impact of the operating parameters, adsorption studies were conducted for bed height (2.5 cm to 10 cm) and flow rate (5 mL/min to 20 mL/min) using immobilised beads of modified kaolin as an adsorbent. The dynamic behaviour of the column was predicted using Bohart-Adams (BDST) and Thomas model. The modified kaolin demonstrated the highest adsorption capacity at 108.3 mg/g kaolin. Thomas model rate constant at 5 mL/min flow rate was $K_{Th} = 0.076$ mL/mg min. BDST model estimated values were $N_0 = 47.2$ mg/L and Ka= 0.0406 L/mg min. It can be concluded that modified kaolin shows high nitrate removal potential as a low-cost adsorbent for the continuous system.

Keywords: Denitrification, modified kaolin, packed bed adsorption, breakthrough analysis, thomas model

1. Introduction

In both emerging and industrialised countries, there has been an increase in the utilisation of groundwater resources. The majority of the world's freshwater supplies derive from groundwater [1]. Despite being underground, groundwater can become contaminated through streams, lakes, reservoirs, wells, and other groundwater sources which has impaired the quality of water [2].

Increased industrial and agricultural activity in recent years has produced hazardous contaminants such as metal ions, inorganic anions, and synthetic organic chemicals raising public concern about the quality of groundwater [3–6]. Due to its great solubility in water, nitrate is the most pervasive groundwater contaminant in the globe [7]. Groundwater nitrate pollution has been linked in large part to the use of high doses of fertilisers for agriculture [8]. Nitrate contamination of groundwater is from septic systems and agricultural fertilisers which is a concern to the environment and public health, thus causing methemoglobinemia, thyroid, kidney ailments etc., [9–12]. Nitrate in high concentrations can promote algae growth thus stimulating eutrophication in water bodies [13].

USEPA and other environmental regulatory organisations have established approximately

[a]bhagyadc01@gmail.com

DOI: 10.1201/9781003545941-26

50 mg/L of NO^{3-} limit in drinking water [14,15]. In drinking water nitrate content can be roughly 45 mg/L for adults and 15 mg/L for newborns [16,17].

Therefore, there is a need for the development of innovative, scalable, technically and financially viable effective options for treating nitrate in groundwater. The different methods to achieve the desirable nitrate contamination for the denitrification process include reverse osmosis, ion exchange and electrodialysis [18–22]. In comparison to other processes, adsorption technology using inexpensive adsorbents has been demonstrated to be effective in eliminating nitrate ions. Because of its practicality, simplicity, and ease of use, the adsorption technique is typically regarded as being superior in the treatment of water. This technique can also get rid of a number of different organic and inorganic water pollutants.

In the present study, adsorption was the chosen mode of operation which is a technique that involves the removal of nitrate by adhering it to the surface of a substance [23]. The adsorbent kaolin is abundant in nature and it exhibits excellent physico-chemical stability, and structural and surface properties [24,25]. Surface modification of kaolin using acid treatment was done to enhance the active sites and the adsorbent capacity. Batch study was already conducted with modified kaolin and reports on the effects of nitrate removal have been made regarding influence of pH (2 to 7), time of contact (5to 240 min) and modified kaolin dosage (1–30 g/L at 27± 2°C [25]. Based on the batch study, the optimal range was chosen for continuous nitrate removal using modified kaolin in a packed bed column. The Yoon-Nelson, Adams-Bohart, Bed Depth Service Time, and Thomas models are frequently used for the investigation of column data. The most popular models for examining the bed height and flow rate behavior of a continuous column are the Bed Depth Service Time and Thomas models, which are employed in the current study.

2. Materials and Methods

2.1. Modified Kaolin

Pure kaolin procured from Nice Chemicals was used to produce modified kaolin. The surface of the kaolin was altered by activating it with sulfuric acid to enhance the active sites to increase the adsorbent capacity. Acid activation was done by soaking 1 g kaolin with 5 mL dilute sulphuric acid (18% wt/wt) for a period of 24 h. After that, the supernatants were filtered many times. Distilled water was used to wash away the residue. For four hours, the altered kaolin was kept at 100°C in a hot air oven. and the prepared modified kaolin was stored in an air-tight container. The procedure to prepare modified kaolin was adopted from our previous study in a batch system [25].

2.2. Immobilised Modified Kaolin

The modified kaolin was used in immobilised form because the powdered form of kaolin was either getting washed off from the column or forming a wet cake at the bottom of the column. Since kaolin swells easily, the thick slurry blocks liquid flow in the packed column. To make the immobilised beads, 3 g of sodium alginate were dissolved in 100 mL of distilled water. The mixture was placed on a magnetic stirrer at 30°C for about 45 min at 500 rpm speed. In 100 mL of distilled water, 2.2 g of calcium chloride (dihydrate) was dissolved and chilled for 30 min in a refrigerator. Sodium alginate and modified kaolin were mixed in 1:1 proportion until a thick consistency free from lumps was obtained. The mixed paste was dropped in the chilled calcium chloride solution using a dropper to get beads of uniform shape and size [26,27]. Figure 26.1 (a) and (b) show the image of modified kaolin before and after immobilisation.

2.3. Nitrate Solution

For a 100 mg/L standard nitrate solution, 0.163 g of potassium nitrate were dissolved in 1000 mL of distilled water. 0.1N HCl was added to the prepared solution to bring its pH down to 2. Maximum nitrate uptake was observed at pH 2 with modified kaolin by Kulkarni et al. [25] in a batch system.

Kulkarni et al. [25] has reported higher nitrate removal rate at lower pH values of 2 and this was a result of the anionic nitrate ions' electrostatic interaction with the cationic groups on modified kaolin. Hence in column study nitrate solution pH was maintained at pH 2.

2.4. Analysis of the Nitrate Content in the Sample

A calibration graph was prepared by measuring absorbance for different known strengths (20,40,60,80 and 100 mg/L) of nitrate ion solution in a Shimadzu UV-spectrophotometer. At regular intervals, 10 milliliters of the effluent were taken out of the column and heated on the hot plate until all the samples were evaporated. To the evaporated sample, 2mL of phenol disulphonic acid and 10 mL of ammonia were added dropwise (1:5). The contents were diluted with distilled water up to the mark of 100 mL. The sample was analysed for nitrate ion concentration in effluent at 410 nm with UV-VIS spectrophotometer [28,29].

2.5. Column Studies

Removal of nitrate using modified kaolin was studied in a packed bed column that was fabricated to allow nitrate solution to enter the column at a known flow rate. Packing of ceramic beads and glass wool were filled in column for uniform distribution of nitrate solution and to increase the contact time between modified kaolin and nitrate ion. In the middle, immobilised modified kaolin adsorbent was mixed with beads and glass wool.

The setup consists of a feed jar containing 100 mg/L nitrate solution connected to a peristaltic pump. Nitrate solution enters the adsorption column of diameter 3 cm and height of 40 cm at a fixed flow rate in the downward direction. A three-neck connector was fixed onto the packed bed column where the sprinkler was attached to ensure uniform solution distribution in a column. Denitrified solution at exit was analyzed using spectrophotometer. The study explores the impact of two crucial parameters bed height and flow rate of nitrate removal at 27 ± 2°C temperature and pH 2 with influent nitrate concentration 100 mg/L (Figure 26.2).

2.6. Equations Applied

Percentage nitrate removal and nitrate uptake are calculated using Eqs.1 and 2 respectively.

$$\% \text{ Nitrate removal} = \frac{C_{adsorbed}}{C_{total}} \times 100 \tag{1}$$

$$Q_e = \frac{C_{adsorbed}}{W_{adsorbent}} \tag{2}$$

Figure 26.1: (a) Modified kaolin (b) Immobilised modified kaolin beads.

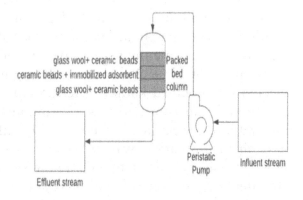

Figure 26.2: Packed bed column employed for denitrification.

$W_{adsorbent}$ - mass of adsorbent in column (g). $C_{adsorbed}$ and C_{total} are mass of nitrate ions adsorbed and the total mass of ions of nitrate (mg) sent to the column till exhaustion time T_e (min) is reached. Te is the exhaustion time where effluent nitrate concentration is equal to 90% influent nitrate concentration (90 mg/L). $C_{adsorbed}$ value was calculated using the region above the effluent nitrate concentration curve multiplied by flow rate. C_{total} is calculated by Eq.3.

$$C_{total} = C_i V_0 T_e \qquad (3)$$

Where C_i, V_0 and T_e are influent nitrate concentration, volumetric flow rate and exhaustion time respectively.

Breakthrough time (T_b), time to reach 10% of influent nitrate concentration at exit.

$$Total\ volume\ of\ nitrate\ solution\ V_e = V_0 T_e \qquad (4)$$

$$breakthrough\ volume\ V_b = V_0 T_b \qquad (5)$$

Length of unused bed (LUB)which is required for design of adsorber can be determined from Eq.6. Z is bed height (cm) used for experimentation. Scale up can be done based on LUB to find required packing height.

$$LUB = Z \left(\frac{T_e - T_b}{T_e} \right) \qquad (6)$$

3. Results and Discussions

3.1. *Effect of Bed Height*

Denitrification using modified kaolin was investigated with a change in bed height at an influent flow rate 10 mL/min with C_i= 100 mg/L nitrate ion maintained at an ambient temperature of 27 ± 2°C. The bed heights selected for modified kaolin were 2.5 cm to 10 cm. Nitrate solution was continuously pumped into the packed bed column. 10 mL of denitrified solution was collected at the outlet at specific time intervals and residue left after evaporating the collected solution was tested for nitrate concentration.

The breakthrough curve plot is depicted in Figure 26.3, which shows the trend of nitrate concentration variation with time. Notably, at a modified kaolin bed height of 10 cm, nitrate removal percentage, nitrate uptake, and the extent of the mass transfer zone ($\Delta T = T_e - T_b$) reached their peak values. In particular, increasing the immobilised bed height from 2.5 to 10 cm resulted in nitrate uptake increase from 88.83 to 105.53 mg/g, and a rise in nitrate removal from 59.20% to 83.15% which indicates a direct correlation between bed height and nitrate removal. Table 26.1 shows nitrate uptake, % removal, breakthrough time and bed exhaustion time. Bed exhaustion time was 1350 min for 2.5cm bed height compared to 4950 min at 10 cm bed height. The findings showed that, due to the availability of more nitrate adsorption sites, the percentage of nitrate removed increased as kaolin bed height increased [30,31]. A reduced breakthrough time is seen at lower bed depths due to lower number of active sites. Modified kaolin being a good adsorbent has shown a delayed breakthrough, high removal efficiency, sharp breakthrough curve and high nitrate uptake. Nitrate removal with modified kaolin (83.15%) was found to be higher in continuous system than other low-cost adsorbents such as activated carbon-74% (Wang et al. 2024), hazelnut shells-62% (Vilardi et al. 2020) and chitosan/alumina composite-59.82% (Goli et al. 2017) [32–34].

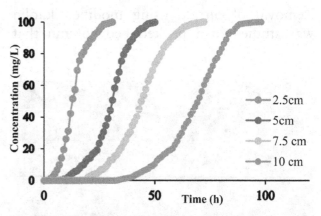

Figure 26.3: Effect of kaolin bed height on nitrate removal.

Table 26.1: Calculated parameters for effect of bed height on nitrate removal (C_i =100 mg/L and V_0=10mL/min)

Z (cm)	T_b (min)	T_e (min)	ΔT (min)	V_e (mL)	V_b (mL)	Cadsorbed (mg)	Ctotal (mg)	% nitrate removal	LUB (cm)	Qe (mg/g)
2.5	420	1350	930	13500	4200	799.5	1350	59.20	1.72	88.83
5	1050	2430	1380	24300	10500	1700	2430	69.95	2.83	94.44
7.5	1770	3510	1740	35100	17700	2569.5	3510	73.20	3.71	102.78
10	2970	4950	1980	49500	29700	4116	4950	83.15	4.00	105.53

Table 26.2: Calculated parameters for the effect of flowrate on nitrate removal (Ci =100 mg/L and Z=5cm)

V_o (mL/min)	T_b (min)	T_e (min)	ΔT (min)	V_e (mL)	V_b (mL)	$C_{adsorbed}$ (mg)	C_{total} (mg)	% removal	LUB (cm)	Q_e (mg/g)
5	2160	5340	3180	53400	21600	1950	2670	73.03	2.97	108.30
10	1050	2430	1380	24300	10500	1700	2430	69.95	2.83	94.44
20	510	1020	510	10200	5100	1530	2460	62.19	2.60	85.00

3.2. *Effect of Flow Rate*

Denitrification using modified kaolin was observed as a function of flowrate for C_i= 100 mg/L nitrate concentration and a fixed 5 cm bed height maintained at an ambient temperature of 27 ± 2°C. Nitrate solution was continuously pumped at different flow rates 5, 10, and 20 mL/min into the packed bed column. Periodically, 10 mL of effluent was collected and analyzed.

Figure 26.4 shows the trend of the concentration-time data of the collected effluent sample. Table 26.2 represents all parameters obtained at diverse flow rates. All the breakthrough curves at different flow rates showed the typical S-shaped curve. Nitrate uptake was found to decrease from 108.30 to 85 mg/g and nitrate removal percentage was decreased from 73.03% to 62.19% for flow rate from 5mL/min to 20 mL/min. It is significant to observe that the bed exhaustion time was notably low at 20 mL/min (1020 min) compared to 5 mL/min (5340 min).

The highest nitrate removal percentage 73.03%, occurred at 5 mL/min flow rate. This observation highlights that adsorption process is primarily governed by intra-particle mass

transfer, validating the mechanism behind the efficient nitrate removal. Overall, these results offer valued insights into the relationship of flow rates, nitrate uptake, removal efficiency and bed exhaustion kinetics.

The high nitrate removal was due to the increase in contact time at lower flow rate. The column functioned satisfactorily even with the 5 mL/min flow rate. Early breakthrough was noticed as flow rates increased. The column breakthrough time was decreased through an increase in flow rates from 5 to 20 mL/min. The inability of nitrate to diffuse

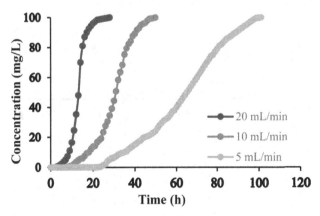

Figure 26.4: Effect of flowrate on nitrate removal.

into adsorbent pores may have been due to reduced residence time and high flow rates, preventing nitrate interaction and causing column exit before equilibrium is reached [30,35].

3.3. Thomas Model

Thomas Model is a kinetic model that is frequently and extensively used to forecast column operation performance [31]. Adsorption rate constant and adsorbent uptake can be determined using the Thomas model (Eq. 7).

$$\ln\left(\frac{C_i}{C_t} - 1\right) = \frac{K_{Th}q_0W_{adsorbent}}{V_0} - K_{Th}C_it \qquad (7)$$

Where K_{Th} = Thomas rate constant, q_0 = maximum concentration of adsorbate, Wadsorbent =adsorbent mass, V_0 = flow rate, t = time. C_i and C_t are influent nitrate concentration and concentration at time t respectively (mg/L). Figure 26.5 shows the Thomas model linear plot and Table 26.3 presents the results of K_{Th}.

Table 26.3: Calculated Thomas model parameters

C_i (mg/L)	Z (cm)	V_O (mL/min)	K_{Th} (mL/mg min)
100	5	5	0.076
100	5	10	0.065
100	5	20	0.064

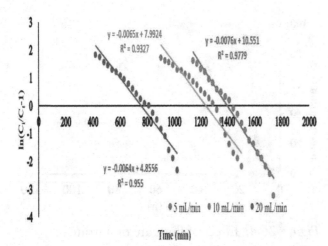

Figure 26.5: Thomas Model fit for various flow rates.

3.4. Bed Depth Service Time (BDST)

The relationship between modified kaolin bed height and breakthrough time is explained by BDST model for concentrations and other breakthrough curve adsorption parameters [36]. Without conducting further tests, the process may be scaled up for various bed heights, flow variations and concentration using the BDST model constant. Eq. 8 shows the BDST Model.

$$t_b = \left(\frac{Z\,N_0}{F\,C_i}\right) - \frac{\left(\ln\left(\frac{C_i}{C_b} - 1\right)\right)}{K_a C_i} \qquad (8)$$

Where K_a = kinetic constant, N_O, C_i, C_b are saturation concentration, influent nitrate concentration and breakthrough concentration, t=time, Z = immobilised bed height, F = linear velocity. Figure 26.6 shows the BDST model linear plot and Table 26.4 presents the results of K_a and N_o.

4. Conclusion

The study demonstrated the use of immobilised modified kaolin beads in a packed bed

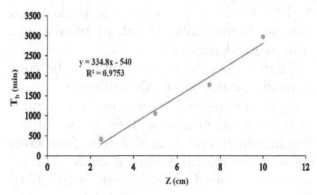

Figure 26.6: A plot of BSDT model.

Table 26.4: Calculated BDST model parameters

C_i (mg/L)	V_o (mL/min)	Z (cm)	T_b (min)	N_o (mg/L)	K_a (L/mg min)
100	10	2.5	420		
100	10	5	1050	47.2	0.0406
100	10	7.5	1770		
100	10	10	2970		

column as a potent clay adsorbent for nitrate removal. The effect of flowrate (5 mL/min to 20 mL/min) and the influence of bed height (2.5 cm to 10 cm) on denitrification efficiency were studied at room temperature. High nitrate removal was facilitated by a 10 cm bed height and 5 mL/min lower flow rate. Experimental data obtained was used to calculate mass transfer parameters. The data showed high agreeability with the mathematical models selected and thus the constants obtained can be used for industrial scale-up. Thomas Model constants obtained at 5 cm bed height and 5, 10, 20 mL/min flow rate were 0.076, 0.065, 0.064 L/mg min respectively. The BDST Model constants obtained were N_o = 47 mg/L and K_a = 0.0406 L/mg min. The regression coefficients obtained were high and thus we can conclude that the data obtained fit the selected mathematical models.

Acknowledgement

The authors are thankful to MSRIT alumni association for providing financial resource.

References

[1] Margat, J. and Van der Gun, J. (2013). Groundwater Around the World: A Geographic Synopsis. CRC Press.

[2] Kapoor, A. and Viraraghavan, T. (1997). Nitrate removal from drinking water. *J. Environ. Eng.*, 123(4), 371–380.

[3] Billen, G., Garnier, J., and Lassaletta, L. (2013). The nitrogen cascade from agricultural soils to the sea: modelling nitrogen transfers at regional watershed and global scales. *Philos. Trans. R. Soc. B*, 368(1621), 1–13.

[4] Kyllmar, K., Forsberg, L. S., Andersson, S., and Mårtensson, K. (2014). Small agricultural monitoring catchments in Sweden representing environmental impact. *Agric. Ecosyst. Environ.*, 198, 25–35.

[5] Bergquist, A. M., Choe, J. K., Strathmann, T. J., and Werth, C. J. (2016). Evaluation of a hybrid ion exchange-catalyst treatment technology for nitrate removal from drinking water. *Water Res.*, 96, 177–187.

[6] da Silva, J. F., de Carvalho, A. M., Rein, T. A., Coser, T. R., Júnior, W. Q. R., Vieira, D. L., and Coomes, D. A. (2017). Nitrous oxide emissions from sugarcane fields in the Brazilian Cerrado. *Agric. Ecosyst. Environ.*, 246, 55–65.

[7] Hayatsu, M., Tago, K., and Saito, M. (2008). Various players in the nitrogen cycle: diversity and functions of the microorganisms involved in nitrification and denitrification. *Soil Sci. Plant Nutr.*, 54(1), 33–45.

[8] Abbasi, M. K. and Adams, W. A. (2000). Gaseous N emission during simultaneous nitrification–denitrification associated with mineral N fertilization to a grassland soil under field conditions. *Soil Biol. Biochem.*, 32(8–9), 1251–1259.

[9] Bloomfield, R. A., Welsch, C. W., Garner, G. B., and Muhrer, M. E. (1961). Effect of dietary nitrate on thyroid function. *Science*, 134(3491), 1690–1690.

[10] Avery, A. A. (1999). Infantile methemoglobinemia: reexamining the role of drinking water nitrates. *Environ. Health Perspect.*, 107(7), 583.

[11] Archna., Sharma, S. K., and Sobti, R. C. (2012). Nitrate removal from ground water: a review. *E-J. Chem.*, 9(4), 1667–1675.

[12] Adimalla, N. and Li, P. (2019). Occurrence, health risks, and geochemical mechanisms of fluoride and nitrate in groundwater of the rock-dominant semi-arid region, Telangana State, India. *Hum. Ecol. Risk Assess.: Int. J.*, 25(1–2), 81–103.

[13] Haller, L., Tonolla, M., Zopfi, J., Peduzzi, R., Wildi, W., and Poté, J. (2011). Composition of bacterial and archaeal communities in freshwater sediments with different contamination levels (Lake Geneva, Switzerland). *Water Res.*, 45(3), 1213–1228.

[14] United States Environmental Protection Agency: https://www.epa.gov/.

[15] World Health Organization: https://www.who.int/.

[16] Kom, K. P., Gurugnanam, B., and Bairavi, S. (2022). Non-carcinogenic health risk assessment of nitrate and fluoride contamination in the groundwater of Noyyal basin, India. *Geod. Geodyn.*, 13(6), 619.

[17] Lin, L., St Clair, S., Gamble, G. D., Crowther, C. A., Dixon, L., Bloomfield, F. H., and Harding, J. E. (2023). Nitrate contamination

in drinking water and adverse reproductive and birth outcomes: a systematic review and meta-analysis. *Sci. Rep.*, 13(1), 563.

[18] Bhatnagar, A. and Sillanpää, M. (2011). A review of emerging adsorbents for nitrate removal from water. *Chem. Eng. J..*, 168(2), 493.

[19] Cyplik, P., Marecik, R., Piotrowska-Cyplik, A., Olejnik, A., Drożdżyńska, A., and Chrzanowski, Ł. (2012). Biological denitrification of high nitrate processing wastewaters from explosives production plant. *Water, Air, Soil Pollut.*, 223, 1791–1800.

[20] Belkada, F. D., Kitous, O., Drouiche, N., Aoudj, S., Bouchelaghem, O., Abdi, N., and Mameri, N. (2018). Electrodialysis for fluoride and nitrate removal from synthesized photovoltaic industry wastewater. *Sep. Purif. Technol.*, 204, 108–115.

[21] Huno, S. K., Rene, E. R., van Hullebusch, E. D., and Annachhatre, A. P. (2018). Nitrate removal from groundwater: a review of natural and engineered processes. *J. Water Supply: Res. Technol. AQUA*, 67(8), 885.

[22] Scholes, R. C., Vega, M. A., Sharp, J. O., and Sedlak, D. L. (2021). Nitrate removal from reverse osmosis concentrate in pilot-scale open-water unit process wetlands. *Environ. Sci.: Water Res.*, 7(3), 650–661.

[23] Hu, Q., Chen, N., Feng, C., and Hu, W. (2015). Nitrate adsorption from aqueous solution using granular chitosan-Fe_{3+} complex. *Appl. Surf. Sci.*, 347, 1–9.

[24] Katal, R., Baei, M. S., Rahmati, H. T., and Esfandian, H. (2012). Kinetic, isotherm and thermodynamic study of nitrate adsorption from aqueous solution using modified rice husk. *J. Ind. Eng. Chem.*, 18(1), 295–302.

[25] Kulkarni, R. M., Chalageri, B. D., Narula, A., and Sachindran, A. (2022). Denitrification performance of kaolin and modified kaolin for the treatment of nitrate contaminated water: isotherm and kinetic studies. *Nanotechnol. Environ. Eng.*, 7(2), 405–413.

[26] Eltaweil, A. S., Omer, A. M., El-Aqapa, H. G., Gaber, N. M., Attia, N. F., El-Subruiti, G. M., and Abd El-Monaem, E. M. (2021). Chitosan based adsorbents for the removal of phosphate and nitrate: a critical review. *Carbohydr. Polym.*, 274, 118671.

[27] Khan, M. N., Chowdhury, M., and Rahman, M. M. (2021). Biobased amphoteric aerogel derived from amine-modified clay-enriched chitosan/alginate for adsorption of organic dyes and chromium (VI) ions from aqueous solution. *Mater. Today Sustainability*, 13, 100077.

[28] Taras M. J. (1950). Phenoldisulfonic acid method of determining nitrate in water. Photometric study. *Anal. Chem.*, 22(8), 1020–1022.

[29] Srivastava S., Nahar A. S., Brighu U., and Gupta A. B. (2019). Comparative study of three methods for the analysis of nitrate nitrogen in synthetic water and wastewater samples. *Int. J. Environ. Anal. Chem.*, 99(12):1164–1185.

[30] Jahangiri-Rad, M., Jamshidi, A., Rafiee, M., and Nabizadeh, R. (2014). Adsorption performance of packed bed column for nitrate removal using PAN-oxime-nano Fe_2O_3. *Environ. Health Sci. Eng.*, 12, 90–94.

[31] Salman Tabrizi, N., and Yavari, M. (2020). Fixed bed study of nitrate removal from water by protonated cross-linked chitosan supported by biomass-derived carbon particles. *J. Environ. Sci. Health, Part A*, 55(7), 777–787.

[32] Wang, J., Amano, Y., and Machida, M. (2024). Nitrate removal from aqueous solution by glucose-based carbonaceous adsorbent: batch and fixed-bed column adsorption studies. *Colloids Surf. A*, 686, 133296.

[33] Vilardi, G., Bubbico, R., Di Palma, L., and Verdone, N. (2020). Nitrate green removal by fixed-bed columns packed with waste biomass: modelling and friction parameter estimation. *Chem. Eng. Res. Des.*, 154, 250–261.

[34] Golie, W. M. and Upadhyayula, S. (2017). An investigation on biosorption of nitrate from water by chitosan based organic-inorganic hybrid biocomposites. *Int. J. Biol. Macromol.*, 97, 489–502.

[35] Liu, L., Ji, M., and Wang, F. (2018). Adsorption of nitrate onto $ZnCl_2$-modified coconut granular activated carbon: kinetics, characteristics, and adsorption dynamics. *Adv. Mater. Sci. Eng.*, 2018, 1–12..

[36] Thirunavukkarasu, A., Nithya, R., and Sivashankar, R. (2021). Continuous fixed-bed biosorption process: a review. *Chem. Eng. J. Adv.*, 8, 100188.

27 A study on theoretical and experimental investigations on hybrid composites subjected to low velocity impact tests

R. S. Shilpa[a] and R. Jyothilakshmi

Department of Mechanical Engineering, M. S. Ramaiah Institute of Technology, Bangalore, India

Abstract

This is a research progress on the reaction of the hybrid laminates exposed to low velocity impact loads. In the present context many researchers have investigated on various types of hybrid composites such as glass fibers, woven flax fibers, carbon fibers, Fibre metal laminates reinforcement done on epoxy resins, epoxy vinyl resins etc. Different types of fabrications techniques were also followed such as vacuum bag technique, vacuum infusion process and various alternating stacking arrangement. After fabrication the laminates were subjected to low velocity instrumental falling weight impact tests. After the conduction of test the damages and delamination on the contact forces, stress and the deformation of the structures were studied. 3D finite element analysis, micro computed tomography (μ - ct) techniques, digital image correlations were used to find the behavior of laminates on the impacted and non-impacted areas and also inter and intra laminar damages were studied. Various mathematical models like Fiber failure models and delamination models were also developed in order to accurately simulate the impact cases on composites laminates. The data obtained in various studies are discussed to analyze the parameters like peak tension stress, Fibre breakages, matric cracking, delamination, at various energy levels and energy absorption capabilities. Results obtained by analyzing the above said parameters studied were compiled and analyzed to establish the comparative evaluation about Fibre breakage, delamination and matric cracking between above said composite materials and methods involved.

Keywords: Hybrid laminates, reinforcement, delamination, laminar damages

1. Introduction

Indeed! Natural examples like wood represent the core concept of composite materials. Wood, as a prime instance, embodies a composite structure comprising cellulose strands held together by lignin. These materials coexist without merging, preserving their distinct properties. Composite materials, whether found in nature or artificially created, result from the fusion of two or more materials with vastly different traits. These materials retain their individual characteristics while synergistically interacting to impart unique and enhanced properties to the composite. In wood, for instance, cellulose fibers contribute strength, while lignin acts as a bonding agent, granting the composite material durability, flexibility, and resilience that surpass the properties of its individual components. This collaborative synergy among dissimilar materials defines composite materials, allowing them to attain remarkable mechanical, structural, and functional properties.

Certainly! Impact classification usually categorises velocities into low, medium, and high segments, with speeds under 10 m/s falling into the low velocity range. Aircraft typically encounter impact damage in areas like the tips, tail (due to runway contact), near doors, and within cargo compartments.

[a]shilpars1519@gmail.com

DOI: 10.1201/9781003545941-27

Common sources of impact damage involve tool drops during assembly, hail impacts, mishandling, bird strikes from airborne debris, and collisions with objects.

Understanding the initiation and spread of damage within underlying components, termed "Barely Visible Impact Damage" (BVID), is a crucial concern within aviation industries. Low-velocity impacts often result in internal defects, causing a significant reduction in local strength, accompanied by matrix cracks and fiber or layer separation, as observed in various research studies.

The primary challenge with BVID lies in its concealed nature, making these damages challenging to detect from the surface of the affected structure. Therefore, early identification of such damages, along with understanding their inception and propagation, becomes imperative. Employing diverse methodologies to predict the remaining lifespan of a structural component under dynamic loading conditions is essential in addressing this issue.

GFRP (Glass fiber reinforced polymer) and CFRP (carbon fiber reinforced polymer) are widely acknowledged composite materials, particularly esteemed in automotive and aviation industries for their myriad benefits. Yet, these composites encounter impact loads during their operational lifespan, primarily causing a decrease in stiffness and the composite's design strength. Numerous variables significantly influence the properties of composite laminates subjected to low-velocity impacts, encompassing impact shape, velocity, mass, fiber and resin types, and laminate layup configurations.

In the study by Arachchige and Ghasemnejad [1], focus centered on identifying post-impact damages and thickness fluctuations in curved composite panels. They employed ply models to discern damages in curved plates, considering stiffness alterations within a semi-analytical spring-mass model [2–5]. Lee et al. [6] utilised finite elemental analysis and the Hertzian contact hypothesis to investigate the dynamic behaviour of low-velocity impacts between elastic property of the curved shell structures and spheres. Similarly,

Chandrashekhara et al. [7] employed a modified Hertzian contact law to examine the nonlinear impact responses of cylindrical composite laminates and doubly curved shells.

However, composite laminates remain vulnerable to impacts from foreign objects, like low velocity debries and the dropped tools during maintenance encountered during operations. Consequently, impact damages commonly arise during routine activities or maintenance may be required due to the absence of reinforcement through the thickness and the presence of weak interfaces in composite materials [8–10].

Over the last two decades, researchers have dedicated efforts to exploring the shape memory alloys (SMAs) integration of into composite structures to fortify their resilience against low-velocity impacts. These impacts typically result in critical damage, notably matrix cracking and delamination, severely compromising the composite materials mechanical strength and longevity. The allure of SMAs stems from their unique ability to revert to predetermined shapes under specific stimuli, presenting a potential solution to enhance composite impact resistance. Strategically embedding SMAs within laminates aims to counteract damage mechanisms like matrix cracking, fiber breakage, and delamination. Leveraging the shape memory effect and mechanical properties of SMAs holds promise in effectively absorbing and dispersing impact energy, fortifying composite structures and extending their operational lifespan, with implications across diverse industries from aerospace to automotive engineering. [11–25]. Due to the exceptional characteristics exhibited by shape memory alloys (SMAs), including their high damping limit, super-elasticity and SME (shape memory effect) and, many researchers have investigated how these alloys react when incorporated into composites. Lei et al. [11] performed both experimental and mathematical analyses focusing on the overall behavior of shape memory alloy hybrid composites subjected to quasi-static loading, by considering the factors such as effects of weak interface and the propagation of damages. Aurrekoetxea et al. [12] The analysis was done to find the influence of super elastic shape memory alloy wires on the behavior of carbon-reinforced poly(butylene terephthalate) composites. Their findings indicated a positive influence of SMAs on the maximum absorbed energy, thereby enhancing the influence of impact load on the composite blend. Furthermore, Taheri et al. [13] conducted a study involving the characterisation of a shape memory alloy hybrid...

2. Materials and Manufacturing Process

1. Super elastic wires, along with unidirectional glass fibers of varying layer thicknesses and surface densities, have been employed to reinforce epoxy resins in conjunction with shape memory alloys.
2. Woven flax fiber orientation with biaxial stitched non crimp fabrics with vinyl ester resins.
3. Carbon and flax fibers with epoxy prepregs, with two stacking sequences like fiber carbon fiber (FCF) and carbon fiber carbon (CFC).
4. Fiber metal laminates with aluminum structures with alternating stacking arrangement of FRP layers
5. Fabrications was done by using vacuum assisted resin injection, vacuum bag technique, vacuum infusion techniques were adopted to produce hybrid specimens with multi-layer laminates.

2.1. Tests Conducted

1. Low-velocity impact experiments were carried out at room temperature employing drop weight impact testing apparatus. These tests involved utilising a falling weight impact using a hemispherical tip with differing diameters.
2. Some researchers conducted four-point bending test on impacted and non-impacted surfaces of the specimens.

2.2. Methods of Analysis

1. Scanning Electron Microscopy (SEM) was employed to observe the macro structure in the vicinity of the damaged region, while the analysis focused on scrutinising the morphology of micro-damage.
2. The characterisation of impact damage utilised micro computed tomography (μ - ct) techniques, while Finite Element Analysis (FEM) was employed to replicate the impact test.
3. Numerical analysis was cross-referenced with experimental outcomes and imaging, confirming the accurate identification of delamination positions within the composite laminates resulting from the impact.

3. Result and Discussions

Laminates with SMAs with different thickness has some breakages in the fiber and merely delamination region and on the backside no breakages of fibers were noticed.

As illustrated in the Table 27.1, the laminate lacking SMAs displays the smallest peak force, measuring at 6.1012 kN. Conversely, SMAs inserted laminates at a position 15/16ths into the specimen's thickness exhibits the highest peak force, reaching 6.9370 kN. For laminates incorporating SMAs positioned at 1/8, 1/2, and 14/16ths thicknesses of the specimen, the peak forces are documented at 6.8290 kN, 6.4666 kN, and 6.3241 kN, respectively. This suggests that the inclusion of one layer of SMAs at a 15/16 thickness position leads to a 13.70% enhancement in peak force compared to the SMA-free laminate. Additionally, the laminates featuring SMAs placed at 1/8, 1/2, and 14/16ths thickness positions demonstrate peak force improvements of 11.93%, 5.99%, and 3.65%, respectively.

The peak tension stress emerges in the rear region when concentrated forces strike the central section of the laminates. Consequently, this tension stress disperses as matrix cracks and stress waves, propagating through the thickness, originating from the backside and moving towards the impact side.

Table 27.1: The peak force and the corresponding parameters of different laminates with one layer of SMAs

The code of laminates	Peak force (kN)	Displacement (mm)	Energy (J)	Velocity slow down (%)
Ply mock 1	6.1012	7.4650	25.1551	53.4839
Ply mode II	6.8290	7.8381	28.7238	69.1575
Ply mode III	6.4666	7.6828	27.2288	61.8073
Ply mode IV	6.3241	7.7232	26.6954	59.3399
Ply mode V	6.9370	7.8036	28.6716	68.0312

Two layers Laminates with SMAs inserted shows much better improvement at the peak force than the one-layer SMAs inserted as shown in Figure 27.2

According to the Figure 27.3 during the impact event of the initial loading, there is a noticeable similarity in the slope of the curve until the maximum force is attained. However, at approximately 2.7 mm displacement, the first instances of delamination within the composite plate become evident, marking a discernible alteration in the curve's trajectory. This alteration indicates a transition in the material's behavior, signifying structural damage induced by the impact. Delamination, characterised by the separation of layers within the composite, signifies a critical point where the impact force surpasses the material's capacity to endure it, resulting in internal failure. This shift in behavior serves as a pivotal indicator for evaluating the material's response to impact loading and its resilience in practical scenarios. This small changes in

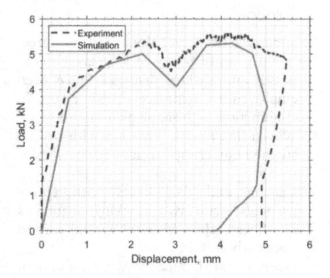

Figure 27.2: Load Vs displacement.

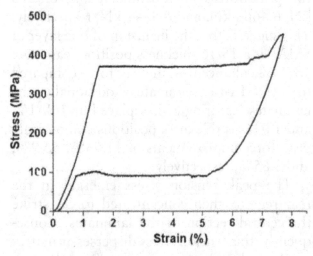

Figure 27.1: Tensile stress–strain curve of shape memory superelastic wire.

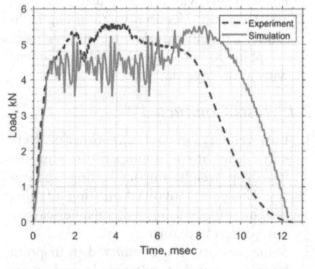

Figure 27.3: Load Vs Time.

the load is presented. Composite laminates, comprising flax, glass, and hybridised glass-flax, underwent low-velocity damage assessment at two distinct energy levels, with investigations conducted through experimental and numerical methods. Utilising Micro-CT, the study assessed damage mechanisms. Findings showcased significant enhancements in impact-related characteristics—like load capacity and absorbed energy—in flax-based natural fiber composites when integrated with glass fiber as hybrid reinforcement. Experiments conducted at 25 J and 50 J reaffirmed that while flax specimens absorbed high energy during impacts, they generally exhibited higher impact damage at less energy levels compared to glass-flax counterparts. The finite element model closely replicated impact scenarios, aligning well with experimental results, establishing itself as a reliable predictor of material behavior under low-velocity impact loads.

X-ray micro-CT analysis revealed extensive matrix cracks and debonding at the fiber/matrix interface in glass-flax specimens at both energy levels. The hybridisation of flax-reinforced natural fiber composites demonstrated notably superior impact performance, displaying increased resistance to perforation and penetration while maintaining lower environmental impact compared to non-hybridised glass fiber laminates. Properly designed flax-glass hybrid laminates strike a balance between performance and environmental advantages. Moreover, stacking flax fiber laminates with carbon fiber ones to create a hybrid composite revealed environmental benefits at end-of-life and other advantages.

According to the data obtained from experimentations, the quantifyable results obtained from the FE model developed to replicate the mass-drop experiments with a 25 J impact energy exhibit a strong resemblance as shown in the Figure 27.4. The FE model predicts a total displacement of the 6x6 mm² central area of the specimen to be 10.12 mm. Notably, this value closely matches the

real maximum vertical deflection observed on the specimen, which was measured at 11 mm. This close correspondence between the numerical predictions and experimental observations underscores the accuracy and reliability of the FE model in simulating the impact event and forecasting the structural response of the specimen.

Microtomography confirmed carbon fiber laminates' higher sensitivity to impacts compared to hybrids. Among hybrid configurations, Flax-Carbon Fiber (FCF) specimens outperformed Carbon Fiber-Carbon Fiber (CFC) ones in mechanical properties and impact absorption, displaying BVID thresholds solely at higher energy levels (30 J) while sustaining robust load-bearing capabilities.

The comparison between the experimental data and the Finite Element (FE) model, as depicted in Figure 27.5, illustrates the capabilities of the developed model. It is evident from this analysis that the FE model closely aligns with the experimental results, particularly during the initial loading phase of the phenomenon, extending up to 0.84 milliseconds. This correlation underscores the accuracy and reliability of the FE model in capturing the behavior of the phenomenon within this specified time interval. Such close agreement between the model and experimental data indicates the effectiveness of the

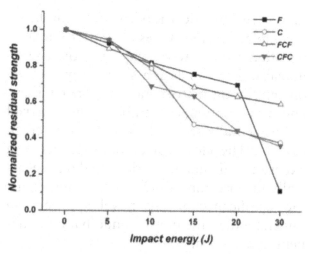

Figure 27.4: Behaviour on impact energyImpact energy.

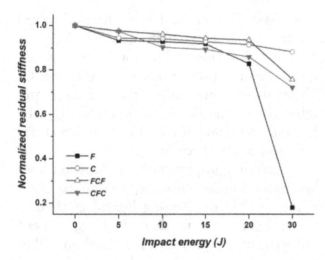

Figure 27.5: Behaviour on impact energyImpact energy.

FE model for simulating and predicting the phenomenon, suggesting its potential for further analysis and prediction beyond the initial loading stage.

The application of Digital Image Correlation (DIC) in tensile testing demonstrated its efficacy in detecting similar damage evolution, notably cracks in carbon plies and delamination, across both hybrid configurations. However, these damages emerged earlier in CFC specimens, affirming the beneficial impact of the stacking order comprising flax skins and a carbon core.

4. Mathematical Models

Elaborate numerical models were employed to simulate various types of impact damages resulting from low-velocity and low-energy impacts. The simulation specifically focused on replicating matrix cracking, fiber failure, and delamination, chosen based on observed experimentally seen debris obstructing crack closure. The close correspondence between experimental and numerical findings, especially in post-impact deformation, highlights the significance of this novel approach in understanding the physics underlying permanent indentation.

Accurately portraying the impact case of a composite laminate with unidirectional reinforcement required identification of only a few material parameters, with the majority sourced from traditional experimental tests detailed in existing literature. Using the same model for simulating both impact damage and residual mechanical properties post-impact is crucial for numerically optimising composite structure designs with improved impact damage tolerance.

The research investigated the dynamic response given by the Fiber Metal Laminates (FMLs) subjected to low-velocity impacts using analytical methods based on the Fourier series and the first-order shear deformation theory. Employing a two-degree-of-freedom system comprising springs and masses simulated the interaction between the plate and the impactor. Experimental outcomes highlighted that integrating aluminum sheets within fiber-reinforced plastic plates significantly improved overall performance against impacts, ensuring more uniform and swift dampening of the impact effect. Furthermore, the study emphasised the influential role of parameters like layer sequence, mass and velocity at constant impact energy, and the plate's aspect ratio, all crucial factors shaping the FML's dynamic response.

5. Conclusion

The review paper synthesised and analyzed results from diverse studies to establish a comparative evaluation among different composite materials and methods utilised. It delved into discussing the correlation between the energy of low-velocity impact and resultant damages, while also presenting the immediate properties of composites post-impact. The compilation and analysis of findings from various studies aimed to provide a comprehensive understanding of the relationship between different composite materials, the methodologies used, and the impact-related outcomes are as follows.

1. The inclusion of a laminate featuring a single layer of Shape Memory Alloys

(SMAs) placed at the 15/16 thickness of the specimen yields significant alterations compared to the SMA-free laminate. Specifically, there is a 13.70% enhancement in peak force and an 8.41% reduction in maximum energy. These findings indicate that integrating SMAs into the laminate structure has the potential to enhance its peak force performance while concurrently diminishing the maximum energy requirements.

2. With reference to the flax and vinyl composite laminates Upon comparison, it's apparent that the developed Finite Element (FE) model correlates well with the experimental results during the initial loading stage of the phenomenon, up to 0.84 milliseconds. However, as the damage initiation takes place within the components, the model starts to deviate from the experimental data.

3. The outcomes propose that multiple variables, including layer arrangement, mass, velocity at constant impact energy, and aspect ratio of the plate, play a significant role in influencing the dynamic response of Fiber Metal Laminates (FMLs).

References

[1] Arachchige, B. and Ghasemnejad, H. (2017). Post impact analysis of damaged variable-stiffness curved composite plates. *Compos. Struct.* 166, 12–21. https://doi.org/10.1016/j.compstruct.2017.01.018.

[2] Seifoori, S., Izadi, R., and Yazdinezhad, A. R. (2019). Impact damage detection for small- and large-mass impact on CFRP and GFRP composite laminate with different striker geometry using experimental, analytical and FE methods. *Acta Mech.*, 230(12), 4417–4433. https://doi.org/10.1007/s00707-019-02506-8.

[3] Kim, S. J., Goo, N. S., and Kim, T. W. (1997). The effect of curvature on the dynamic response and impact-induced damage in composite laminates. *Compos. Sci. Technol.*, 57(7), 763–773. https://doi.org/10.1016/S0266-3538(97)80015-2.

[4] Hu, C., Yang, B., Xuan, F.-Z., Yan, J., and Xiang, Y. (2020). Damage orientation and depth effect on the guided wave propagation behavior in 30CrMo steel curved plates. *Sensors*, 20(3), 849.

[5] Saad, M. and Ouinas, D. (2018). Damage analysis of the stratified curved plate. *J. Mater. Eng. Struct.*, 5, 223–236.

[6] Lee, D. I. and Kwak, B. M. (1993). An analysis of low-velocity impact of spheres on elastic curve dshell structures. *Int. J. Solids Struct.* 30(21), 2879–2893. https://doi.org/10.1016/0020-7683(93)90201-H.

[7] Chandrashekhara, K. and Schroeder, T. (1995). Nonlinear impact analysis of laminated cylindrical and doubly curved shells. *J. Compos. Mater.* 29(16), 2160–2179. https://doi.org/10.1177/002199839502901604.

[8] Sevkat, E., Liaw, B., and Delale, F. (2013). Drop-weight impact response of hybrid composites impacted by impactor of various geometries. *Mater. Des.* 52, 67–77.

[9] De Moura, M. and Marques, A. T. (2002). Prediction of low velocity impact damage in carbon-epoxy laminates. *Compos. A Appl. Sci. Manuf.*, 33(3), 361–368.

[10] Davies, G. A. O. and Zhang, X. (1995). Impact damage prediction in carbon composite structures. *Int. J. Impact Eng.*, 16(1), 149–170.

[11] Lei, H. S., Wang, Z. Q., Tong, L. Y., Zhou, B., and Fu, J. (2013). Experimental and numerical investigation on the macroscopic mechanical behavior of shape memory alloy hybrid composite with weak interface. *Compos. Struct.*, 101, 301–312.

[12] Aurrekoetxea, J., Zurbitu, J., de Mendibil, I. O., Agirregomezkorta, A., Sánchez-Soto, M., and Sarrionandia, M. (2011). Effect of superelastic shape memory alloy wires on the impact behavior of carbon fiber reinforced in situ polymerized poly (butylene terephthalate) composites. *Mater. Lett.*, 65(5), 863–865.

[13] Taheri-Behrooz, F., Taheri, F., and Hosseinzadeh, R. (2011). Characterization of a shape memory alloy hybrid composite plate subject to static loading. *Mater. Des.*, 32(5), 2923–2933.

[14] Raghavan, J., Bartkiewicz, T., Boyko, S., Kupriyanov, M., Rajapakse, N., and Yu, B.

(2010). Damping, tensile, and impact properties of superelastic shape memory alloy (SMA) fiber-reinforced polymer composites. *Compos. B Eng.*, 41(3), 214–222.

[15] Kang, K. W. and Kim, J. K. (2009). Effect of shape memory alloy on impact damage behavior and residual properties of glass/epoxy laminates under low temperature. *Compos. Struct.*, 88(3), 455–460.

[16] Pappadà, S., Gren, P., Tatar, K., Gustafson, T., Rametta, R., Rossini, E., et al. (2009). Mechanical and vibration characteristics of laminated composite plates embedding shape memory alloy superelastic wires. *J. Mater. Eng. Perform.* 18(5–6), 531–537.

[17] Zhou, G. and Lloyd, P. (2009). Design, manufacture and evaluation of bending behaviour of composite beams embedded with SMA wires. *Compos. Sci. Technol.*, 69(13), 2034–2041.

[18] Zhang, R., Ni, Q. Q., Natsuki, T., and Iwamoto, M. (2007). Mechanical properties of composites filled with SMA particles and short fibers. *Compos. Struct.*, 79(1), 90–96.

[19] Lau, K. T., Ling, H. Y., and Zhou, L. M. (2004). Low velocity impact on shape memory alloy stitched composite plates. *Smart Mater. Struct.*, 13(2), 364–370.

[20] Pappadà, S., Rametta, R., Toia, L., Coda, A., Fumagalli, L., and Maffezzoli, A. (2009). Embedding of superelastic SMA wires into composite structures: evaluation of impact properties. *J. Mater. Eng. Perform.*, 18(5–6), 522–530.

[21] Khalili, S. M. R., Shokuhfar, A., Malekzadeh, K., and Ghasemi, F. A. (2007). Low-velocity impact response of active thin-walled hybrid composite structures embedded with SMA wires. *Thin-Walled Struct.*, 45(9), 799–808.

[22] Roh, J. H. and Kim, J. H. (2002). Hybrid smart composite plate under low velocity impact. *Compos. Struct.*, 56(2):175–182.

[23] Tsoi, K. A., Stalmans, R., Schrooten, J., Wevers, M., and Mai, Y. W. (2003). Impact damage behaviour of shape memory alloy composites. *Mater. Sci. Eng. A*, 342(1), 207–215.

[24] Paine, J. S. N. and Rogers, C. A. (1994). The response of SMA hybrid composite materials to low velocity impact. *J. Intell. Mater. Syst. Struct.*, 5(4):530–535.

[25] Mili, F. and Necib, B. (2001). Impact behavior of cross-ply laminated plated composite plates under low velocities. *Compos. Struct.*, 51(3):237–244.

[26] Taheri-Behrooz, F., Taheri, F., and Hosseinzadeh, R. (2011). Characterization of a shape memory alloy hybrid composite plate subject to static loading. *Mater. Des.*, 32(5), 2923–2933.

[27] Sarasini, F., Tirillo, J., D'Altilia, S., Valente, T., Santulli, C., Touchard, F., Chocinski-Arnault, L., Mellier, D., Lampani, L., and Gaudenzi, P. (2016). Damage tolerance of carbon/flax hybrid composites subjected to low velocity impact. *Compos. Part B: Eng.*, 91, 144–153.

[28] Payeganeh, G. H., Ghasemi, F. A., and Malekzadeh, K. (2010). Dynamic response of fiber–metal laminates (FMLs) subjected to low-velocity impact. *Thin-Walled Struct.*, 48, 62–72.

[29] Sun, M., Wang, Z., Yang, B., and Sun, X. (2017). Experimental investigation of GF/epoxy laminates with different SMAs positions subjected to low-velocity impact. *Compos. Struct.*, 171, 170–184.

28 Recent progress in multifunctional transition metal di-chalcogenides and their applications

V. S. Kavyashree[1], D. L. Shruthi[2], and G. N. Anil Kumar[1,a]

[1]Department of Physics, M S Ramaiah Institute of Technology, Bangalore, India
[2]Department of Physics, Sai Vidya Institute of Technology, Bangalore, India

Abstract

This review highlights recent advances in the field of transition-metal dichalcogenides. The transition metal dichalcogenides are two-dimensional layered materials of the type MX_2, where M is transition metal and X is chalcogen. The nanocomposites of these materials can either be metals or semiconductors with direct or indirect band gaps, depending on their stoichiometry. The two-dimensional crystal structure of these materials has emerged as a complementary material to graphene. Hence, these materials find a variety of applications in next-generation electronics. This review mainly focuses on transitional metal dichalcogenides, synthesis methods, structural and morphology features with their applications in various areas of research such as optoelectronics, Nanoelectronics, sensors, energy storage, etc. Finally, the review also summarizes the multifunctional applications of the versatile MX_2 and its composites in lubricants, in lithium-ion batteries, as a catalyst for water-splitting hydrogen evolution reactions, oxygen evolution reactions, and various biomedical and environmental applications.

Keywords: Di-chalcogenides, transition metals, nanocomposites, two dimensional materials

1. Introduction

Transition Metal Dichalcogenides (TMDs) are layered materials with formula MX_2 where M is the transition metal [Mo, W, Nb, Hf, V, Ta, Ti, Zr, Pt, Pd, etc] and X is the chalcogen atom [S, Se, Te and Po] [Figure 28.1]. Each layer in the X-M-X layered structure is attracted by the Van der Waals force of attraction. The word chalcogen is derived from the Greek word khalkós and genes meaning copper born. The compounds MoS_2 (Molybdenite) and WS_2 (Tunstenite) are naturally occurring minerals with hexagonal structures and exhibit excellent friction-reducing capabilities due to their weak Van der Waals interlayer interaction making them a good lubricant [1]. TMDs are semiconductors with a finite bandgap unlike graphene [2]. This property of tunable bandgaps [3] in TMDs has drawn great attention in the field of research and is widely used in electronics, optoelectronics, biomedical, sensors, energy devices, and many more. TMDs were found to be a suitable type of semiconductor having a layered structure like graphene. The TMDs were used in many applications including biosensing due to their chemical and physical properties. Hence TMDs can be used as a replacement for graphene as graphene has zero bandgap which restricts its application in many potential fields. Several methods are to synthesise the layer –few layers of TMDs by a different approach. By using both liquid exfoliation and mechanical exfoliation transform the bulk crystal into few layers of

[a]anilgn@msrit.edu

DOI: 10.1201/9781003545941-28

TMDs. The method of preparation is simple to operate and produces the high-quality material. In 2015, Chiya et al. reported on a wide range of applications of TMDs in oxygen reduction, hydrogen gas evolution etc. In 2004 A.K. Geim and K.S. Novoselov for the first time successfully exfoliated graphene using the micromechanical exfoliation method which is now commonly known as the "Scotch tape" method [4]. Xiong et al. [5] show the fabrication of TMD films in a continuous, large scale by using the Chemical Vapour deposition technique. The structural, mechanical, electronic, and optical properties of TMDCs are changed by the preparation techniques including mechanical and CVD methods. Chio et al. [6] show the when bulk material converts to monolayer the 2D TMDCs show the transition from indirect bandgap to direct bandgap which enhancement in the optical and electrical properties. The unique electrical and optical properties with direct band-gap are used for an ideal optoelectronic device for better light utilisation [7]. TMDC nanomaterials are deposited onto flexible substrates which shows mechanical strength and is well adapted to the human body, this shows the characteristic of implantable biosensor devices [8,9]. TMDCs show excellent performance with good semiconducting and metallic electrical conductivity [10], large specific surface area, and tunable bandgap [11].

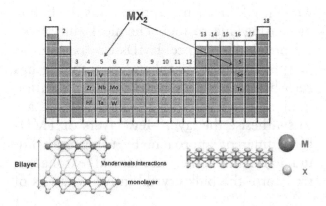

Figure 28.1: Layered transition metal dichalogenides atomic structure.

This review summarises the outstanding capabilities of TMDCs with their significant optical, electrical, mechanical, and structural properties. Here this review also emphasises synthesis techniques which include top-down and bottom-up approaches. The different TMDCs modification methods with advantages and disadvantages are also discussed. Finally, the application of TMDCs is presented which makes them excellent materials compared to other conventional materials.

2. Synthesis of 2D Materials

2D materials have gained greater attention in recent years due to their unique properties and remarkable applications in various fields such as sensors, optoelectronic devices, energy devices, electronic devices, flexible gadgets, catalysis, biomedical devices, and many more. The synthesis of 2D materials is classified under top-down and bottom-up approaches [12]. In the top-down approach, a bulk solid 2D material is broken down to form desired nanostructures at controlled processes with preferable properties. In the bottom-up approach, the 2D material is synthesised by a reaction of atomic or molecular precursors under optimal temperature and pressure for a defined period. The top-down method involves, Liquid-phase, chemical, mechanical, electrochemical, and ultrasonic exfoliations whereas, the bottom-up method includes chemical vapor deposition (CVD), Atomic layer deposition (ALD), hydrothermal or solvothermal method, pulsed laser deposition, epitaxial growth, and microwave-assisted method. Compared to a top-down approach, bottom-up approaches are efficient in producing large quantities of 2D material. The commonly used methods for the development of 2D materials are CVD, mechanical exfoliation (scotch tape method), hydrothermal synthesis, liquid-phase exfoliation and physical vapor deposition (PVD).

2.1. Chemical Vapor Deposition (CVD)

Chemical vapor deposition (CVD) is an implicit method of synthesis of high-quality TMDs at a lost cost. In CVD methods, TMDs are synthesised by heating the transition metal oxide and a pure chalcogen precursor which is placed on a SiO_2/Si substrate in a furnace. It is maintained at a high temperature ranging from 650 to 1000°C [Table 28.1]. The nucleation of the first layer of TMD is hindered by the reverse carrier gas flow from the substrate to the precursor source [13]. The common route of synthesis by CVD is through a single-step TMDs formation by decomposition of the precursors under an inert environment, where the presence of H_2 at low temperature avoids oxidation and directly converts the precursors to TMDs [14]. Another route is through selenisation/sulfurisation of pre-deposited metal oxide films on a suitable substrate. The layer-by-layer deposition process controls the number of layers being deposited on the substrate and a high-quality TMD with desirable properties are synthesised [15]. Morphological properties of TMDs are an important criterion that ensures the high quality and performance of the TMD. Different morphologies of the single-layered TMDs can be synthesised, like stars, triangles, hexagons, pentagons, flower-like, etc. by the CVD method [16]. The TMDs grown using the CVD method have found a wide range of applications in vast areas such as electronics, optoelectronics, spintronic and ferroelectronics [17].

2.2. Hydrothermal/Solvothermal

Hydrothermal is the widely reported method for TMD synthesis, which involves the heating of suitable precursors and solvents at a very high temperature and for a defined period. It involves centrifugation, filtration, and washing of the produced TMDs for further analysis [18]. The hydrothermal approach for the synthesis of TMDs dates back to the year 2000, when Fan et al., successfully synthesised $MoSe_2$ and MoS_2 nanocrystallites [19] for the first time using Na_2MoO_4 and $Na_2S_2O_3$/ Na_2SeSO_3 as the precursors of the transition metals and chalcogenides respectively [Table 28.2]. The TMDs synthesised by the hydrothermal process have a wide variety of potential applications as it can be modified based on size, morphology, and phase. Some of the applications are catalysis [20], electronics [21], and biomedical [22] by the usage of different precursors and reactive temperature and time.

2.3. Chemical Exfoliation

Chemical exfoliation is a solvent-based exfoliation technique that involves intercalation, which involves the separation of the layers of the material through a chemical reaction which helps in the synthesis of large quantities and a high-quality TMDs [23]. This technique involves the intercalation of ions between the layers which induces an expansion force between the layers to separate it [24]. It is then ultrasonicated to yield an atomically thin TMD layer rapidly with high quality. Ultrasonication generates pressure waves that facilitate the intercalating compound to infiltrate easily between the chalcogenide layers [Table28.3]. It also reduces the time required for intercalation. Intercalation with lithium(Li) ions is one the most effective methods of synthesising TMDs with remarkable electrical and mechanical properties, used in highly efficient solar cells and FETs [25].

2.4. Liquid Phase Exfoliation (LPE)

LPE involves the dissolving of TMDs in a suitable solvent to exfoliate them into thin layers and the thickness of the exfoliated flakes can be controlled using chemical, thermal or mechanical methods[Table 28.4]. It is one of the efficient, cost-effective and simple techniques to synthesise TMDs with exceptional properties [26]. The TMDs synthesised using LPE technique have been used in a wide area of applications like sensors [27], electronic devices [28], biomedical [29] and catalysts [30,31].

2.5. *Mechanical Exfoliation*

TMDs are exfoliated by breaking the Van der Waals force between the adjacent layers by mechanical forces. This technique is commonly known as the "Scotch-tape" method as the layers of TMDs are exfoliated by pressing the adhesive tape against the bulk TMD and then repeatedly peeling it off which helps in obtaining single or several layers of TMDs [32]. This is then transferred onto the substrate surface for further characterisation.

The TMDs synthesised are of high quality, have minimal defects and the crystal size ranges from few nanometers to micrometers. Mechanically exfoliated TMDs are used in the fabrication of electronic devices like FET [33], detectors, etc.

3. Different Methods to Synthesise Various TMDCs with Its Applications

3.1. *Different Methods for the Synthesis of MoS$_2$ and Its Applications*

Table 28.1: Methods for synthesis, substrate, characterization and application of MoS$_2$ [27,30,33–37]

Method of synthesis	Substrate/ Precursor	Characterisation	Applications
CVD	Silicon with 285 nm SiO$_2$ layer	HRTEM, Raman, AFM	FET, catalysis, sensors; The MoS$_2$ nanosheets exhibits enhanced field emission properties with good emission stability and low turn-on electric field [34].
CVD	Silicon with 90 nm SiO$_2$ layer	SEM, Raman, XRD	Back-gated FET devices : The FET device fabricated on the synthesised MoS$_2$ shows a n-type behaviour with field-effect mobility to be 1.46 cm^{2}V^{-1}s^{-1} and carrier concentration of 4.5×10^{12} cm^{-2} [35].
Liquid Phase Exfoliation (LPE)	MoS$_2$ powder, Sodium cholate (NaC)	UV-Visible absorption, AFM, XRD, TEM, Raman	Gas sensors: Response optimisations was investigated for thin-film thickness and temperature. The thinner films (150nm) shows higher response to NO$_2$ gas compared to films of 300nm thickness. Also, thin-film response decreases and recovery time increases with increasing temperature [27].
LPE	MoS$_2$ powder with particle size ~ 6 µm	UV-Visible spectroscopy, TEM, SEM, Raman, PL spectroscopy	Photocatalysis: The synthesised MoS$_2$ was of 2–7 layers of nanosheets. It is useful in cleaning of waste water using sunlight. The ultrathin nanosheets acts as an excellent catalyst to enhance the photocatalytic activity by suppressing the recombination of photo-excited holes and electrons [30].

(continued)

Table 28.1: continued

Method of synthesis	Substrate/ Precursor	Characterisation	Applications
Hydrothermal	Molybdenum oxide (MoO_3), Potassium thiocyanate (KSCN)	XRD, TEM, Raman, FESEM, XPS	Fast absorbent for the removal of BPA: The wrapped nanosheets with the highest specific surface area and biggest pore volume were made with a 3:1 M S/Mo ratio and heated to 240°C for 37 hours to demonstrate the best adsorption capability. It was discovered that the highest adsorption capacity was 39.03 mg/g and the adsorption rate constant was 0.0053 g/(mg/min) [36].
Hydrothermal	Ammonium molybdate tetrahydrate, L-cysteine	XRD, FESEM, TEM	Photocatalytic degradation of Methylene orange (MO) and Rhodamine blue (PhB) Antifungal activity. The dyes degraded in the presence of sunlight due to these nanosheets' broad surface area (6.46 m2g-1) and small bandgap (1.79 eV) [37].
Mechanical exfoliation	Top layer: Si with 300 nm SiO_2 layer Bottom layer: Highly doped n-type Si	PL, Raman, AFM	FET: The fabricated FET was found to have on/off high mobility of 24.26 cm²/Vs better current ratio ~10^5, at V_{DS}=10 V and saturated current 1.28×10^3 µA when V_G=10 V, which is consistent with those of the devices previously reported [33].

3.2. Different Methods of Synthesis of WS_2 with Applications

Table 28.2: Methods for synthesis, substrate, characterization and application of WS_2 [30,31,38–41]

Method of synthesis	Substrate/ Precursor	Characterisation	Applications
CVD	SiO_2	Optical microscopy, AFM, SEM, PL, Raman	Triangular monolayers of WS_2 are formed with 35 µm size. PL spectra confirms a significant single sharp peak at 1.97 eV, corresponding to the direct band gap emission of the monolayer WS_2 [38].
CVD	SiO_2/Si	SEM, AFM, Raman, Energy dispersive X-Ray spectrometer (EDX)	Nanoelectronics and Optoelectronics: The thickness of 4 layers of WS_2 was found to be ~2.6 nm. The thickness and size of the nanoplates depends on the temperature and the growth time affects the quantity of the sample. The electrostatic properties exhibit uniform surface potential and charge distributions [39].

(continued)

Table 28.2: continued

Method of synthesis	Substrate/ Precursor	Characterisation	Applications
Liquid Phase Exfoliation (LPE)	Bulk WS_2, N-Methyl pyrrolidone	SEM, TEM, Raman, XRD, TGA, DSC	Catalytic thermal decomposition: Compared to bulk WS_2, the few layered sheets of WS_2 showed better catalytic decomposition of TKX-50 due to its ability to enhance proton transfer and more exposed active sites [40].
LPE	WS_2 powder, Liquid Nitrogen	FESEM, AFM, TEM, XRD, XPS	Catalysis: The extracted WS_2 nanosheets were of thickness 1.5nm. It was observed to have an onset potential of 65mV with remarkable stability even after 1000 cycles [31].
Hydrothermal	Sodium tungstate dehydrate, L-cysteine	TEM, HRTEM, XPS, UV-Vis spectra, PL	Sensors: The quantum dots produced were of 4–7 nm. These QDs shows the capability in H_2O_2 sensing as the H_2O_2 causes partial oxidation of QDs which leads to oxidation induced quenching [41].

3.3. Synthesis Methods of MoSe₂ with Applications

Table 28.3: Methods for synthesis, substrate, characterization and application of $MoSe_2$ [28,42–44]

Method of synthesis	Substrate/ Precursor	Characterisation	Applications
CBV (Chemical Bath Vaporisation)	Substrate: SiO_2/Si and Au/Si Precursor: Ammonium molybdate, Selenium dioxide	XRD, XPS, Raman, SEM, PL, AFM, TEM	Catalyst, FET: The $MoSe_2$ films exhibits a low potential of 88mV and highest current density of 0.845 mA.cm^{-2}. Its stability for long term HER performance was tested to be excellent. The FETs shows lesser performance as compared to those synthesised by CVD method due to its vertical alignment. Post annealing or doping can help to enhance the performance [42].
CVD	Silicon with 285 nm SiO_2 layer	TEM, SAED, AFM, FESEM, Raman, XPS, Optical microscopy	Photodetectors: 6nm thick $MoSe_2$ layers were selected for the fabrication of photodetector. When LED light source was illuminated, the current significantly increases and is strongly dependent on the power of the LED. The photoresponsivity was found to be 1.26 AW^{-1} for a bias voltage of 6V which was higher than that of a monolayer $MoSe_2$ [43].
Liquid Phase Exfoliation (LPE)	Bulk $MoSe_2$, 230 nm SiO_2	HRTEM, FESEM, XRD, Raman, UV-Vis Spectroscopy	FET device: Bandgap was found to be 1.55eV. Linear relation between drain current and drain voltage is observed up to -0.5V which indicates a good ohmic contact between the Au electrode and $MoSe_2$ channel material. The on/off current ratio was estimated to be 10^2 [28].

(continued)

Table 28.3: continued

Method of synthesis	Substrate/ Precursor	Characterisation	Applications
Hydrothermal	Selenium powder, Sodium molybdate dehydrate	XRD, SEM, HRTEM, EDS, FTIR, PL	Catalytic activity: $MoSe_2$/montmorillonite composite nanosheets with a spacing of 0.65 nm are synthesised which have high adsorption properties and photocatalytic activity due to the synergy of $MoSe_2$ and MMT [44].

3.4. Different Methods to Synthesis of WSe₂ and Its Applications

Table 28.4: Methods for synthesis, substrate, characterization and application of WSe_2 [45–47]

Method of synthesis	Substrate/ Precursor	Characterisation	Applications
CVD	SiO_2/Si	AFM, PL, Raman	Back gated transistors, MOSFETs : The monolayer WSe_2 was found to have a thickness of 0.54 nm. The average carrier mobility was found to be 24.8 cm^2/V-s. The threshold voltage of back gated WSe_2 transistor increases as the temperature increases from 35K to 300K [45].
CVD	WO_3 and Se powder	AFM, TEM, SAED, EDS, PL, Raman	Photonics, excellent saturable absorbers (SA): The lattice constant was found to be 0.33 nm and the microfiber of WSe_2 integrated exhibits a large modulation depth of 54.5%. Whereas, large area WSe_2 generates stable pulses with duration of 477fs (at 1.5 μm) and 1.18ps (at 2 μm). Hence crystalline, large area WSe_2 is suitable for ultrafast photonic application [46].
Hydrothermal	Se, Na_2WO_4, $NaBH_4$	XRD, XPS, HRTEM, SEM	Gas sensors: The work function of Au is 5.1 eV which is greater than that of WSe_2 (3.61 eV) which leads to energy transfer from WSe_2 to Au NPs resulting in the potential barrier which increases the resistance of the gas sensors. At low concentration for isoamylol. It has short response/ recovery time and great stability [47].

4. Conclusion

In summary, we reviewed the TMDCs properties, the synthesis routes, and various applications with their structure. Different synthesis methods are addressed with bottom-up and top down approaches for the preparation of TMDCs. Chemical exfoliation method and mechanical methods in top-down and chemical vapour deposition method, liquid phase exfoliation has been addressed in detail. Studies show that good-quality TMDCs are obtained by the CVD technique and it is the most effective method of synthesis. On the other hand, here we also present the applications and benefits of TMDCs with different materials are discussed in detail. The large surface area with the semiconducting properties and unique structure shows great applications in various fields. TMDCs are used as a nano enzyme which provides the place for targeted drug delivery for therapy of cancer treatment. Excellent elasticity and strong mechanical strength with various substrates used as a biosensor. Overall TMDCs are used in X-ray computed tomography, magnetic Resonance imaging (MRI), includ-ing fluorescence, bioimaging etc. TMDCs have acquired interest in solid-state research with most of the applications and associated unique properties with Vander waal's bonding between the layers.

References

[1] Polcar, T. and Cavaleiro, A. (2011). Review on self-lubricant transition metal dichalcogenide nanocomposite coatings alloyed with carbon. *Surf. Coat. Technol.*, 206(4), 686–695.

[2] Nandee, R., Chowdhury, M. A., and Rana, M. (2022). Results in engineering, bandgap formation of 2D material in graphene: future prospect and challenges. *Results Eng.*, 15, 100474.

[3] Tang, H., Neupane, B., and Neupane, S. (2022). Tunable band gaps and optical absorption properties of bent MoS$_2$ nanoribbons. *Sci. Rep.*, 12, 3008.

[4] Novoselov, K. S. (2004). Electric field effect in atomically thin carbon films. *Science*, 306, 666–669.

[5] Xiong, L., Wang, K., Li, D., Luo, X., Weng, J., Liu, Z., and Zhang, H. (2020). Research progress on the preparations, characterizations and applications of large scale 2D transition metal dichalcogenides films. *FlatChem*, 21(3).

[6] Choi, W., Choudhary, N., Han, G. H., Park, J., Akinwande, D., and Lee, Y. H. (2017). Recent development of two-dimensional transition metal dichalcogenides and their applications. *Mater. Today*, 20, 116–130.

[7] Monga, D., Sharma, S., Shetti, N., Basu, S., Kakarla, R. R., and Aminabhavi, T. (2021). Advances in transition metal dichalcogenide-based two-dimensional nanomaterials. *Mater. Today Chem.*, 19, 100399.

[8] Sarkar, D., Liu, W., Xie, X., Anselmo, A. C., Mitragotri, S., and Banerjee, K. (2014). MoS2 field-effect transistor for next-generation label-free biosensors. *ACS Nano*, 8(4), 3992–4003.

[9] Choi, C., Lee, Y., Cho, K. W., Koo, J. H., and Kim, D. H. (2019). Wearable and implantable soft bioelectronics using two-dimensional materials. *Acc. Chem. Res.*, 52(1), 73–81.

[10] Kim, H. I., Yim, D., Jeon, S. J., Kang, T. W., Hwang, I. J., Lee, S., Yang, J. K., Ju, J. M., So, Y., and Kim, J. H. (2020). Modulation of oligonucleotide-binding dynamics on WS$_2$ nanosheet interfaces for detection of Alzheimer's disease biomarkers. Biosens. *Bioelectron.*, 165, 112401.

[11] Zhang, X., Teng, S. Y., Loy, A. C. M., How, B. S., Leong, W. D., and Tao, X. (2020). Transition metal dichalcogenides for the application of pollution reduction: a review. *Nanomaterials*, 10, 1012.

[12] Brent, J. R., Savjani, N., O'Brien, P. (2017). Synthetic approaches to two-dimensional transition metal dichalcogenide nanosheets. *Prog. Mater. Sci.*, 89, 411–478.

[13] Wang, J., Li, T., Wang, Q., Wang, W., Shi, R., Wang, N., Amini, A., and Cheng, C. (2020). Controlled growth of atomically thin transition metal dichalcogenides via chemical vapor deposition method. *Mater. Today Adv.*, 8, 100098.

[14] Brito, J. L., Ilija, M., and Hernández, P. (1995). Thermal and reductive decomposition of ammonium thiomolybdates. *Thermochim. Acta*, 256 (2), 325–338.

[15] Wang, X., Feng, H., Wu, Y., and Jiao, L. (2013). Controlled synthesis of highly crystalline MoS2 flakes by chemical vapor deposition. *J. Am. Chem. Soc.*, 135(14), 5304–5307.

[16] Zhang, G., Wang, J., Wu, Z., Shi, R., Ouyang, W., Amini, A., Chandrashekar, B. N., Wang, N., and Cheng, C. (2017). Shape-dependent defect structures of monolayer MoS$_2$ crystals grown by chemical vapor deposition. *ACS Appl. Mater. Interfaces*, 9(1), 763–770.

[17] Zeng, H., Wen, Y., and Yin, L. (2023). Recent developments in CVD growth and applications of 2D transition metal dichalcogenides. *Front. Phys.*, 18, 53603.

[18] Berwal, P., Rani, S., Sihag, S., Singh, P., Dahiya, R., Kumar, A., Sanger, A., Mishra, A. K., and Kumar, V. (2024). Hydrothermal synthesis of MoS2 with tunable band gap for future nano-electronic devices. *Inorg. Chem. Commun.*, 159, 2024, 111833.

[19] Fan, R., Chen, X., and Chen, Z. (2000). A novel route to obtain molybdenum dichalcogenides by hydrothermal reaction. *Chem. Lett.*, 29(8), 920–921.

[20] Xie, J., Zhang, H., Li, S., Wang, R., Sun, X., Zhou, M., Zhou, J., Lou, X. W., and Xie, Y. (2013). Defect-rich MoS$_2$ ultrathin nanosheets with additional active edge sites for enhanced electrocatalytic hydrogen evolution. *Adv. Mater.*, 25, 5807–5813.

[21] Urban, F., Passacantando, M., Giubileo, F., Iemmo, L., and Di Bartolomeo, A. (2018). Transport and field emission properties of MoS$_2$ bilayers. *Nanomaterials*, 8(3), 151.

[22] Agarwal, V. and Chatterjee, K. (2018). Recent advances in the field of transition metal dichalcogenides for biomedical applications. Nanoscale, 10(35), 16365–16397.

[23] Pariari, D. and Sarma, D. D. (2020). Nature and origin of unusual properties in chemically exfoliated 2D MoS$_2$. *APL Mater.*, 8(4), 040909.

[24] Zhang, Q., Mei, L., Cao, X., Tang, Y., and Zeng, Z. (2020). Intercalation and exfoliation chemistries of transition metal dichalcogenides. *J. Mater. Chem. A, R. Soc. Chem.*, 8(31), 15417–15444.

[25] Lebogang, M., Fru, J. N., Kyesmen, P. I., Diale, M., and Nombona, N. (2020). Electrically enhanced transition metal dichalcogenides as charge transport layers in metallophthalocyanine-based solar cells. *Front. Chem.*, 8, 612418.

[26] Islam, M. A., Serles, P., and Kumral, B. (2002). Exfoliation mechanisms of 2D materials and their applications. *Appl. Phys. Rev*, 9(4), 041301.

[27] Ni, P., Dieng, M., Vanel, J.-C., Florea, I., Bouanis, F. Z., and Yassar, A. (2023). Liquid shear exfoliation of MoS$_2$: preparation, characterization, and NO$_2$-sensing properties. *Nanomaterials*, 13, 2502.

[28] Sharma R., Dawar A., and Ojha S. (2023). A thrifty Liquid-Phase Exfoliation (LPE) of MoSe$_2$ and WSe$_2$ nanosheets as channel materials for FET application. *J. Electron. Mater.*, 52, 2819–2830.

[29] Nguyen, E.P., Daeneke, T., Zhuiykov, S., and Kalantar-zadeh, K. (2016). Liquid exfoliation of layered tranition metal dichalcogenides for biological applications. *Curr. Protoc. Chem. Biol.*, 8, 97–108.

[30] Sahoo, D., Kumar, B., and Sinha, J. (2020). Cost effective liquid phase exfoliation of MoS$_2$ nanosheets and photocatalytic activity for wastewater treatment enforced by visible light. *Sci. Rep.*, 10, 10759.

[31] Qin, Y. Q. (2020). Ultrathin exfoliated WS$_2$ nanosheets in low-boiling-point solvents for high-efficiency hydrogen evolution reaction. *IOP Conf. Ser., Mater. Sci. Eng.*, 770, 012079.

[32] Li, Y. (2022). Recent progress on the mechanical exfoliation of 2D transition metal dichalcogenides. *Mater. Res. Express*, 9(12), 122001.

[33] Yu, Z., Xiong, C., Hao, Z., Shaozu, H., Guohong, Z., Meifang, Z., Wei, Q., Zhaohua, W., Xiaowei, H., and Jun, W. (2011). Facile exfoliation for high-quality molybdenum disulfide nanoflakes and Relevant Field-Effect Transistors Developed with Thermal Treatment. *Front. Chem.*, 9, 650901.

[34] Li, H., Wu, H., and Yuan, S. (2016). Synthesis and characterization of vertically standing MoS$_2$ nanosheets. *Sci. Rep.*, 6, 21171.

[35] Mathew, S., Reiprich, J., Narasimha, S., Abedin, S., Kurtash, V., Thiele, S., Hähnlein, B., Scheler, T., Flock, D., and Jacobs, H. O. (2023). Three-dimensional MoS$_2$ nanosheet structures: CVD synthesis, characterization, and electrical properties. *Crystals*, 13, 448.

[36] Luo, L., Shi, M., Zhao, S., Tan, W., Lin, X., Wang, H., and Jiang, F. (2019). Hydrothermal synthesis of MoS2 with controllable morphologies and its adsorption properties for bisphenol A. *J. Saudi Chem. Soc.*, 23(6), 762–773.

[37] Mengist, M., RamaDevi, D., Belachew, N., and Basavaiah, K. (2021). Hydrothermal green synthesis of MoS2 nanosheets for pollution abatement and antifungal applications. *RSC Adv.*, 11(40), 24536–24542.

[38] Asgary, S., Ramezani, A., and Nejad, Z. (2022). Characterization of high quality, monolayer WS2 domains via chemical vapor

deposition technique. *Appl. Phys. A*, 128(2), 139.

[39] Fan, Y., Hao, G., Luo, S., Qi, X., Li, H., Ren, L., and Zhong, J. (2014). Synthesis, characterization and electrostatic properties of WS_2 nanostructures. *AIP Adv.*, 4(5), 057105.

[40] Shang, Y., Yang, W., and Xu, Y. (2019). Preparation of few-layered WS2 and its thermal catalysis for dihydroxylammonium-5,5-bistetrazole-1,1-diolate. *J. Nanomater.*, 2019, 7458645.

[41] Hang, D. R., Sun, D. Y., and Chen, C. H. (2019). Facile bottom-up preparation of WS_2-based water-soluble quantum dots as luminescent probes for hydrogen peroxide and glucose. *Nanoscale Res Lett.*, 14, 271.

[42] Vikraman, D., Hussain, S., Akbar, K., Adaikalam, K., Lee, S. H., Chun, S.-H., Jung, J., Kim, H.-S., and Park, H. J. (2018). Facile synthesis of molybdenum diselenide layers for high-performance hydrogen evolution electrocatalysts. *ACS Omega*, 3(5), 5799–5807.

[43] Wu, Z. (2018). Large-area synthesis and photoelectric properties of few-layer MoSe2 on molybdenum foils. *Nanotechnology*, 29(12),125605.

[44] Li, X. and Peng, K. (2018). $MoSe_2$/montmorillonite composite nanosheets: hydrothermal synthesis, structural characteristics, and enhanced photocatalytic activity. *Minerals*, 8, 268.

[45] Yao, Z., Liu, J., and Xu, K. (2018). Material synthesis and device aspects of monolayer tungsten diselenide. *Sci. Rep.*, 8, 5221.

[46] Yin, J., Li, J., Chen, H., Wang, J., Yan, P., Liu, M., Liu, W., Lu, W., Xu, Z., Zhang, W., Wang, J., Sun, Z., and Ruan, S. (2017). Large-area highly crystalline WSe2 atomic layers for ultrafast pulsed lasers. *Opt. Express*, 25, 30020–30031.

[47] Zhang, X., Tan, Q., Wang, Q., Yang, P., and Liu, Y. (2022). Enhanced gas sensitivity of Au-decorated flowery WSe_2 nanostructures. *Nanomaterials*, 12, 4221.

29 Fabrication and evaluation of composites with an aluminium metal matrix for wire electrical discharge machining parameters

S. Jayavelu[1,a], J. Udayaprakash[1], S. Ganesan[2], C. Rajkumar[1], N. Kumaran[3], and Prem G. Roshan[1]

[1]Department of Mechanical Engineering, Vel Tech Rangarajan Dr. Sagunthala R&D Institute of Science and Technology, Chennai, India
[2]Department of Aeronautical Engineering, Vel Tech Rangarajan Dr. Sagunthala R&D Institute of Science and Technology, Chennai, India
[3]Department of Physics, Vel Tech Rangarajan Dr. Sagunthala R&D Institute of Science and Technology, Chennai, India

Abstract

The article reports on machinability ofAluminium Hybrid Composites utilising wire electrical discharge machining.In this study, stir casting is utilised to create 5%, 10%, and 15% zirconium silicate (ZrSiO4) and boron carbide (B4C). using aluminium metal matrix composites. By varying the Wire Feed, Reinforcement, Gap Voltage, Pulse on Time, and Pulse off Time, the cutting experiments were carried out. Also examined how these modifications affected the property of Material Removal Rate (MRR), Surface Roughness value (R), Time is provided in minutes, and Kerf (Width of Cut). Regression equations were created in order to decide which variables in MMCs were important. Model sufficiency in MMCs with varying responses was assessed using the ANOVA test results. The results demonstrated that the ratio of B4C to ZrSiO4 ceramic particles affects the machinability and surface quality properties of Al2024. Both brittle and ductile fracture modes are indicated by the hybrid composite's ploughing effect and cavities inside its machined surface.

Keywords: MRR, kerf width, ANOVA, boron carbide, aluminium hybrid composites

1. Introduction

Metal matrix composite (MMC) may meets the requirements of today's cutting-edge engineering applications. MMCs are typically composites of two ceramics and a metal basis which exhibit improved mechanical qualities such membrane strength, balanced bending, and stiffness. Compared to MMCs with a single reinforcement, this material is lighter and has superior fatigue resistance, damping, and radiation shielding. Among the many MMCs, aluminium MMCs are in high demand because of their improved mechanical qualities, decreased weight, and low production costs have used in variety of applications, which includes aerospace, automobile, and marine. B4C is a material crucial for many fields which is alsomajorly used in tank's armour and radiantsof nuclear absorbers becauseof its excellent ballistic qualities in situations where the base material is an aluminium metal matrix. Another form of hard reinforcement utilised in MMCs is ZrSiO4,

[a]drjayavelusundaram@gmail.com

DOI: 10.1201/9781003545941-29

which has the great properties such as extraordinary hardness, strength, and wear resistance. Additionally, it preserves chemical reactions and high temperatures. Although the materials are much homogenous and anisotropic by nature, the inclusion inhardest reinforcement causes the aluminium metal matrix composites to become much more brittle as well as hard. In order to enhance customise qualities including stiffness, strength, wear resistance, thermal expansion, friction, and thermal conductivity metals are reinforced with fibres or other particles. MMCs, for instance, are more frequently seen in the automotive sector. These materials use silicon carbide fibres to strengthen a matrix made of a metal, such aluminium. The need for MMCs with high specific stiffness and almost low coefficient of thermal expansion has grown in the space age. MMCs typically include silicon carbide or graphite fibres or particles that strengthen a low-density metal, like aluminium or magnesium. MMCs, in comparison to metals, also have several drawbacks. The most significant of these is the higher fabrication cost for high-performance.

1.1. *Experimental Procedure*

The aluminium alloy Al2024 is the matrix material utilised in creation of Metal Matrix Composite (MMCs).Chemical compositions of Al2024 alloy is represented in weight percentages. The manufacturing process involves the stir casting technique, which combines Al2024 with reinforcements such as B4C (64 μm) and ZrSiO4 (44 μm).Various compositions of reinforcements were prepared through the stir casting process, including combinations like 5wt% B4C and 5wt% ZrSiO4, 10wt% B4C and 5wt% ZrSiO4, and 15% B4C and 5% ZrSiO4. Initially, the unreinforced Al6063 was heated and melted at 750 °C using an electrically fired crucible furnace. The reinforcements, B4C and ZrSiO4, underwent preheating for one hour in separate electrical furnaces at temperatures up to 800 °C to remove moisture and any absorbed hydroxide or gas. Additionally, 20 g of magnesium particles were applied to enhance the wetability of the surface. Following thispreheating processB4C and ZrSiO4 were added to the molten metal and agitated for 15 minutes at 300 rpm.

2. Literature Review

Based on Material Materials called MMCs provide the custom property combinations needed for a variety of engineering applications. Hybrid composites are more homogenous than conventional composites and have better attributes as a result of combining two or more reinforcing materials with different characteristics at the molecular level in a continuous metal matrix phase. The evaluation of machining research of hybrid MMCs use both standard and non-conventional machining techniques, including drilling, turning, and milling as well such as laser, wire-cut EDM, and electrical discharge machining (EDM).In 2019, Yuvaraj et al.In this project, a hybrid metal matrix reinforced with B4C and Utilising an abrasive aqua jet (AAJ), ZrSiO4 particles in the ratio of 5wt% B4C and 5wt% ZrSiO4 aremachined. The Response surface approach was used to conduct experiments with a central composite design by varying the traverse rate, abrasive flowing rate, and aqua jet pressure were noted. Response surface graphs have been utilised to analyse the experiment's outcomes, which were acquired using a range of abrasive kinds and mesh sizes. The striation length and its frequency's were used to assess the striation effect on the bottom-machined surfaces. The composite kerf wall cuts produced by AAJ machining were

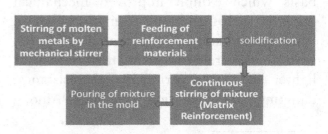

Figure 29.1: Processofstircasting.

examined for surface topography and morphology. Wear track and impurities formed in the metal surface of the machined surface, which was exhibiting the inherent qualities of AAJ. The experimental results showed that a highest material removal rate, a lower kerf taper angle, and a reduction in surface roughness were achieved by increasing the abrasive flow rate (400.0 g/min), traverse rate (30 mm/min), and aqua jet pressure (300.0 MPa)..For the production of 3D complicated shape and geometry of alloys or composites, a non-traditional electro thermal machining process called WEDM is utilised. With sharp edges and a variety of sizes or hardnesses, wire EDM can precisely create objects that are difficult to make using conventional manufacturing techniques. Process variables and performance indicators are numerous. This machining process has been examined and explored by numerous researchers. In this work, recent research trends in wire electric discharge machining are reviewed.Few of the various optimisation techniques that may be used on materials such as alloys, superalloys, and MMCs. This review of the literature lists the allowable process variables and their impact on several performance indicators under standard or cryogenic tool electrode conditions, as well as their ranges in the machining of various materials.

3. Stir Casting

The inception of MMC stir casting dates back to 1968. Typically, mechanical stirring is employed to effectively disperse powder-form reinforcing phases into molten metals. Prior to introducing reinforcing particles into the vortex created by a mechanical stirrer, the metal undergoes heating in a crucible. The mechanical stirrer consists of a revolving blade composed of materials with higher melting points than the liquid metal. Subsequently, the fluid has poured into a prepared mold and left to solidify. The final step involves sintering and compacting within a hollow

die.To establish a winder in the material grid for enhanced consolidating support through the mix projecting technique, a mechanical stirrer is utilised. This innovation revolves around the integration of a supporting material with flowing liquid metal via a mechanical stirrer to generate composite materials. Subsequently, the fluid metal lattice composite material is poured into a mold utilising the standard projection process. This interaction between the components influences the characteristics and microstructure of the composite in various ways. Wire Cut EDM samples of Al2024+5%B4C+5%ZrSiO4 shown in Figure 29.2, Al2024+10%B4C+5%ZrSiO4 shown in Figure 29.3, and Al2024+15% B4C+5%ZrSiO4 shown in Figure 29.4.

4. Results and Discussion

The parametric Taguchi method was employed to carry out the WEDM experiments. In this part, spcific WEDM parameter process on the chosen quality attributes - MRR - are discussed. Experimental data were used to calculate the response characteristics' S/N ratio for every variable at various levels. Plots are made depicting process variables' primary effects on the S/N data. Analysing the parametric influences in the characteristics response is done using the response graphs (also known as the main effects plot). By reviewing the ANOVA tables

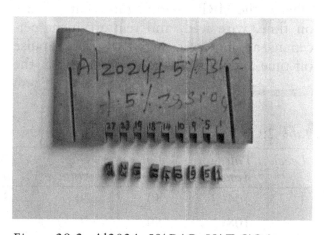

Figure 29.2: Al2024+5%B4C+5%ZrSiO4.

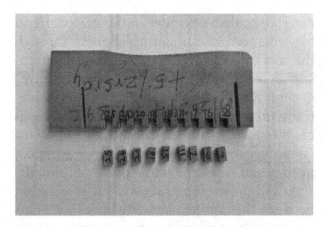

Figure 29.3: Al2024+10%B4C+5%ZrSiO4.

Figure 29.4: Al2024+15%B4C+5%ZrSiO4.

MRR falls. This is because a greater MRR result from the discharging energy increasing as the Pulse on Time increases. A larger MRR results from more discharges occurring during a stipulated period as the pulse off time lowers. Lower MRR is the result of the average discharge gap widening with an increase in gap voltage.

Through the analysis of the ANOVA tables and response graphs. Table 29.2 provides the experimental result for MRR and MRR S/N ratios. The 27 rows of the L27 orthogonal array, with thirteen columns and three levels, correspond to the number of experiments.

4.1. Schedule of Optimal Levels

The parameterprocess for the WEDM ispresented in Table 29.1 for the different composites, compiled based on various literature sources. Experimental results for WEDM were tabulated in Table 29.2.

The rankings show how important each aspect was in relation to the outcome. According to delta values and ranks values, the factors that have the largest impact on MRR are pulses on-time, pulses off-time, voltage gap, reinforcement, and wire feed values, in that order in the Table 29.3 Response Table. Given that MRR is a "higher is better" type of quality characteristic, demonstrates that the WEDM process's greatest MRR is produced by the Pulse on-Time on third level, Pulse off-Time of first level, voltage gap on first level, feed wire on third level, and third level of reinforcement.

The significance of the factors process in MRR was investigated using the Analysis of

and response graphs, it is possible to determine the best settings for the process varies in the terms of the mean responsive characteristics. Experiments were carried out utilising L27 OA to examine how process factors affected the MRR.Demonstrates that as pulse on time, wire feed, and reinforcement percentage are increased, the MRR rises; as pulse off time and gap voltage were increased, the

Table 29.1: Process parameter of WEDM

Levels	Pulse-on Time(μs)	Pulse-off Time(μs)	Gap Voltage(V)	Wire feed(ml/min)	Reinforcement(%)
1	104	40	20	3	Al2024+5%B4C+5%ZrSiO4
2	108	50	30	6	Al2024+10%B4C+5%ZrSiO4
3	110	60	40	9	Al2024+15%B4C+5%ZrSiO4

Table 29.2: Experimental results for WEDM

Ex. No.	A Pulseon-Time (μs)	B Pulseoff-Time (μs)	C Voltage Gap (V)	D Wire Feed (m/min)	E Reinforcement (wt%)	Time in min	MRR (mm³/min)	Kerf (mm)
1	104	40	20	3	Al2024+5%B4C+5%ZrSiO4	3:50	119.4	3.98
5	104	50	30	9	Al2024+5%B4C+5%ZrSiO4	4:54	88.8	3.95
9	104	60	40	6	Al2024+5%B4C+5%ZrSiO4	9:45	39.3	3.93
10	108	40	20	3	Al2024+5%B4C+5%ZrSiO4	3:59	118.5	3.95
14	108	50	30	9	Al2024+5%B4C+5%ZrSiO4	4:47	89.1	3.96
18	108	60	40	6	Al2024+5%B4C+5%ZrSiO4	7:20	50.9	3.96
19	110	50	20	3	Al2024+5%B4C+5%ZrSiO4	3:47	118.5	3.95
23	110	55	30	9	Al2024+5%B4C+5%ZrSiO4	4:16	89.5	3.98
27	110	60	40	6	Al2024+5%B4C+5%ZrSiO4	6:04	59.5	3.97
2	104	40	30	6	Al2024+10%B4C+5%ZrSiO4	5:14	71.2	3.96
6	104	50	40	3	Al2024+10%B4C+5%ZrSiO4	6:32	59.5	3.97
7	104	60	20	9	Al2024+10%B4C+5%ZrSiO4	7:00	51.1	3.98
11	108	40	30	6	Al2024+10%B4C+5%ZrSiO4	4:32	89.3	3.97
15	108	50	40	3	Al2024+10%B4C+5%ZrSiO4	5:45	71.2	3.96
16	108	60	30	9	Al2024+10%B4C+5%ZrSiO4	4:16	88.8	3.95
20	110	50	30	6	Al2024+10%B4C+5%ZrSiO4	3:43	119.1	3.97
24	110	55	40	3	Al2024+10%B4C+5%ZrSiO4	5:18	71.6	3.98
25	110	60	20	9	Al2024+10%B4C+5%ZrSiO4	3:35	118.5	3.95
3	104	40	40	9	Al2024+15%B4C+5%ZrSiO4	7:40	93.8	3.98
4	104	50	20	6	Al2024+15%B4C+5%ZrSiO4	6:56	109.1	3.97
8	104	60	30	3	Al2024+15%B4C+5%ZrSiO4	8:20	81.4	3.95
12	108	40	40	9	Al2024+15%B4C+5%ZrSiO4	6:43	108.6	3.95
13	108	50	20	6	Al2024+15%B4C+5%ZrSiO4	6:16	108	3.93
17	108	60	30	3	Al2024+15%B4C+5%ZrSiO4	8:35	81.4	3.95
21	110	50	40	9	Al2024+15%B4C+5%ZrSiO4	5:52	130.6	3.96
22	110	55	20	6	Al2024+15%B4C+5%ZrSiO4	5:18	131.3	3.98
26	110	60	30	3	Al2024+15%B4C+5%ZrSiO4	6:40	108.9	3.96

Variance (ANOVA) method. If the p value is >than 0.050, the parameters is considered significant. ANOVA Table 29.4 indicates that the following variables are significant: Reinforcement Percentage, Voltage Gap, Pulse On-Time, and Pulse Off Time. The wire feed, the interactions between the Pulse on-Time and the voltage gap, and the Pulse on-Time and the feed wire were all merged into one error term.

5. Conclusion

In this work, WEDM was used to conduct machinability investigations on various aluminium metal matrix composites. The findings showed that the machinability of Al2024 is influenced by the ratio of B4C with combination of ZrSiO4 Ceramic particles. The Cavities and the ploughing effect on the surface of machined hybrid composites indicate both ductile and brittle fracture modes. Using WEDM, machinability tests on several aluminium metal matrix composites were performed on this study analysis. By altering the proportions of B4C (5%, 10%, 15%), 5% ZrSiO4, and Aluminium Al2024, metal matrix composites were created. In the current work, performance and surface properties were examined. The optimal parameter combinations for achieving minimal Removal of materialprocess ie Material Removal Rates (MRR) were determined through Taguchi optimisation method.

The optimal input parameters for various conditions are as follows:

- Pulse on time: 40.17 µs for Al2014 + 15wt %B4C + 5wt %ZrSiO4
- Pulse off time: 40.51 µs for Al2014 + 15wt %B4C + 5wt %ZrSiO4
- Gap Voltage: 40.32 V for Al2014 + 10wt %B4C + 5wt %ZrSiO4
- Wire feed: 39.34 m/min for Al2014 + 15wt %B4C + 5wt %ZrSiO4
- Reinforcements: Al2014 + 15wt %B4C + 5wt %ZrSiO4 at 40.38 wt%

Table 29.3: Responses table

Level	Pulse-on	Pulse-off	Gap Voltage	Wire feed	Reinforcement
1	37.5	40.51	40.32	39.03	38.11
2	38.8	38.94	39.08	38.1	37.98
3	40.17	37.03	37.07	39.34	40.38
Delta	2.67	3.47	3.25	1.25	2.4
Rank	3	1	2	5	4

Table 29.4: ANOVA

Variation of Source	D.O.F	Sum of square	Mean value of squares	F0	p	Contribution%
A	2.	33.003	17.002	12.44	0.002	16.05
B	2	54.471	27.236	21.17	0	27.31
C	2	48.467	24.233	18.84	0	24.30
E	2	32.794	16.397	12.75	0.002	16.44
AB	4	8.789	2.197	1.71	0.224	4.41
AE	4	10.034	2.508	1.95	0.179	5.03
Error (Pooled)	10	12.864	1.286			6.45
Total	26	199.422				100.00

References

[1] Asthana, R. (1998). Reinforced cast metals: Part I Solidification microstructure *Journal of Materials Science*, 33, 1679–1698.

[2] Calhoun, R. B., Dunand, D. C., Kelley, A., and Zweben, C. (2000). Comprehensive composite materials. *Metal Matrix Compos.*, 3, 27–59.

[3] Clyne, T. W. (2001). Metal matrix composites: matrices and processing. In *Encyclopaedia of Materials: Science and Technology*, Volume1.

[4] Clyne, T. W. and Withers, P. J. (1993).*An Introduction to Metal Matrix Composites*. Cambridge University Press.

[5] Lloyd, D. J. (1994). Particle reinforced aluminium and magnesium matrix composites. *Int. Mater.Rev.*, 39(1), 1–23.

[6] Maruyama, B. (1998). Progress and promise in aluminum metal matrix composites. *AMPTIAC Newsl.*, 2(3).

[7] Surappa, M. K. and Rohatgi, P. K. (1981). Preparation and properties of cast aluminium-ceramic particle composites.*J.Mater. Sci.*, 16, 983–993.

[8] Nicholls, C. J., Boswell, B., Davies, I. J., and Islam, M. N. (2017). Review of machining metal matrix composites. *Int. J. Adv. Manuf. Technol.*, 90, 2429–2441.

[9] Ibrahim, I. A., Mohamed, F. A., and Lavernia, E. J. (1991). Particulate reinforced metal matrix composites—a review. *J.Mater. Sc.*, 26, 1137–1156.

[10] Davim, J. P. (2002). Diamond tool performance in machining metal–matrix composites. *J. Mater. Process. Technol.*, 128(1–3), 100–105.

[11] Rai, R. N., Datta, G. L., Chakraborty, M., and Chattopadhyay, A. B. (2006). A study on the machinability behaviour of Al–TiC composite prepared by in situ technique. *Mater. Sci. Eng. A*, 428(1–2), 34–40.

[12] Torralba, J. D., Da Costa, C. E., and Velasco, F. (2003). P/M aluminum matrix composites: an overview. *J. Mater. Process. Technol.*, 133(1–2), 203–206.

[13] Chawla, K. K. and Chawla, K. K. (1998). *Metal Matrix Composites*. Springer, New York, 164–211.

[14] Lin, J. T., Bhattacharyya, D., and Ferguson, W. G. (1998). Chip formation in the machining of SiC-particle-reinforced aluminium-matrix composites. *Compos. Sci. Technol.*, 58(2), 285–291.

[15] Gururaja, S., Ramulu, M., and Pedersen, W. (2013). Machining ofMMCs: a review. *Machining Scienceand Technology 17, no.1.*, 41–73.

30 Effect of strain rates on short glass fiber reinforced thermoplastics

S. Jayavelu[1,a], SaralaRuby[2], S. Ganesan[1], Prem G. Roshan[1], C. Rajkumar[1], and N. Kumaran[3]

[1]Department of Mechanical Engineering, Vel Tech Rangarajan Dr.Sagunthala R&D Institute of Science and Technology, Chennai, India
[2]Department of Physics, Vel Tech Rangarajan Dr.Sagunthala R&D Institute of Science and Technology, Chennai, India
[3]Department of Mathematics, Vel Tech Rangarajan Dr.Sagunthala R&D Institute of Science and Technology, Chennai, India

Abstract

In recent times, the utilisation of composite materials has significantly surged across various applications, owing to their exceptional properties. However, many of these structures are subjected to dynamic loading environments, necessitating a thorough understanding of their mechanical properties under medium to high rates of loading.In our current investigation, we focus on thermoplastic polypropylene reinforced with short glass fibers, augmented with grafted maleic anhydride as a coupling agent. To facilitate this study, we have prepared reinforcements in three distinct ratios: 10%, 30%, and 50%, employing a twin-screw extrusion process. Subsequently, these reinforcements are utilised in an injection moulding process to fabricate the requisite specimens.The primary objective of our study is to conduct testing across varying strain rates in the quasi-static region. Specifically, we aim to assess tensile properties within the range of 10-4 to 10-1 S-1 and flexural properties within the range of 10-5 to 10-2 S-1. To achieve this, we maintain the desired strain rates by adjusting the crosshead speed of the Universal Testing Machine relative to the gauge length for tensile testing and the span length for flexural testing.By meticulously examining the mechanical behaviour of this composites in different loading conditions, we seek in gain insights into their performance characteristics, thus facilitating their optimisedutilisation in practical applications.

Keywords: Thermoplastic polypropylene, short glass fibre, twin screw extrusion process, high strain rate

1. Introduction

In recent years, short glass fiber reinforced thermoplastics (SGFRTPs) have gained significant attention due to their widespread use in various engineering applications, ranging from automotive components to consumer goods. These materials offer a unique combination of strength, stiffness, and impact resistance, making them desirable for lightweight and durable structures. However, their mechanical properties can be significantly influenced by the rate at which they experience deformation, commonly known as strain rate.

Understanding effect ofrates in strain on the mechanical behaviour of SGFRTPs is crucial for optimising their performance and ensuring their suitability for specific applications. The mechanical response of these materials under different strain rates is influenced by several factors, including the viscoelastic nature of the polymer matrix, the orientation and aspect ratio of the glass fibers, and the interfacial bonding between the fibers and the matrix.

[a]drjayavelusundaram@gmail.com

DOI: 10.1201/9781003545941-30

Several studies have investigated the impact of strain rates on SGFRTPs, aiming to elucidate the underlying deformation mechanisms and characterise their mechanical properties under dynamic loading conditions. These investigations have revealed complex relationships between strain rates and key mechanical parameters such as tensile strength, modulus of elasticity, fracture toughness, and energy absorption capacity.

The behaviour of SGFRTPs under varying strain rates is particularly important in applications where dynamic loading conditions are prevalent, such as impact and crashworthiness scenarios in automotive and aerospace industries. Understanding how these materials respond to rapid deformations can help engineers design more robust and resilient structures, ultimately enhancing safety and performance.

Fitoussi et al. [1] delved into the effects of strain on discontinuous glass fiber reinforced ethylene-propylene copolymer (EPC) matrix composites commonly employed in automotive contexts. Their investigation revealed a profound relationship between microscopic observations of deformation and damage mechanisms and the resulting macroscopic mechanical properties. Tensile tests conducted at elevated strain rates, reaching up to 200 S-1, unveiled a highly strain rate-dependent behavior, with damage primarily attributed to the viscous nature of the EPC matrix, as evident in scanning electron microscopy (SEM) analyses.Similarly [2], explored the mechanical enhancements achieved through heat treatment of glass fiber reinforced unsaturated plastic (GFRP) composites. Their study, conducted via hand layup fabrication followed by heat treatment at temperatures up to 150°C, demonstrated notable improvements in mechanical properties. The ultimate tensile strength exhibited sensitivity to heat treatment temperatures, with optimal performance observed at lower temperatures. However, elongation properties displayed a

contrasting behaviour, showcasing higher values in untreated materials compared to heat-treated counterparts.Furthermore, Martinez et al. [2] investigated the strain rate dependency of polypropylene glass fiber woven composites through a series of tension, shear, and compression tests. Their findings highlighted a significant influence of strain rates on the strength and stiffness of the composite, with dynamic tests revealing distinct softening behaviour.Studies by Sun et al. [3] and McKownand Cantwell[4] further explored the mechanical behaviour of fiber composites under varying strain rates, revealing insights into stress-relaxation phenomena and strain-rate sensitivity, respectively. Additionally, investigations by Zhang et al. [5], Schobig et al. (2006), Eriksen et al. [6], Xie et al. [7], Elanchezhian et al. [8], Al-Mosawe et al. [9], Singh [10], and Zhou and Mallick[11] have contributed valuable insights into the influence of factors such as temperature, coupling agents, and fiber orientation on the mechanical properties of composite materials.

The primary objective of this research is to examine how strain rates affect the stress-strain behavior of thermoplastic composite materials under both tensile and compressive loading conditions.

2. Methods and Materials

Twin screw extruder is used to prepare the polypropylene with C-glass fiber reinforcement at three fiber volume percentages i.e.,10%, 30% and 50%. Specimens required for testing is prepared by injection moulding process.Glass fiber has relatively high strength with low young's modulus and brittle. Glass fiber is used to reinforce the polymer matrices.Polypropylenehomopolymer (Repol) used as a base material with melt flow rate of 15g/10min. Maleic anhydride modified polypropylene (Bondyram 1001) used as a couplingagent. Polypropylene homopolymer had the following characteristics.

2.1. *Extrusion and MouldingProcess*

The twin screw extrusion process is used to produce a homogeneous glass fiber reinforced composite material by mixing three raw materials. This continuous manufacturing process is called as compounding process. Based on the reinforcement percentages the thermoplastic, fiberglass and coupling agent is conveyed and fed into the extruder by separate feeding system. The three compositions with 10%, 30% and 50% reinforcements are listed in Table 30.1.

Material 1 – Polypropylene
Material 2 – Coupling agent
Material 3 – Glass fiber

1.5% coupling agent is used constant for all the three compositions, base material and fiber reinforcement material is only changed. Each composition consumed 10kg of material as given below in Table 30.2.

Initially material 1 and material 2 is blended by mixer for 3 minutes and then conveyed to main feeder storage bin. The glass fiber is loaded manually in the side feeder for fillers. When all the materials are in the hopper storage bin of the twin-screw extruder, the preblended material is fed into the extrusion hopper by single screw feeder. The material is gradually melted by the mechanical energy of the rotating screws in which heaters are attached along the barrel length. When the material starts melting the reinforcement material is fed through the side feeder. The fiber generally have higher melting temperature. Feeding the reinforcement material in the main hopper with polymer starts blocking the temperature. To avoid these the fiber is always fed after the polymer is melted. The molten polymer is forced into a die, which forms the shape after cooling.

The preblended material from the hopper is fed by gravimetric force into the barrel of the extruder. The material enters the feed throat and meets screw. The screw rotates at 120rpm, forces the material towards the front of the heated barrel. The temperature required is set and is controlled by PID controller. PID stands for Proportional-Integral-Derivative. It is a control algorithm widely used in electronics and engineering for regulating systems to achieve desired performance. The barrel heaters are divided into three zones and the temperatures are set in such a way that gradual increase from the rear to the front. By this way of maintaining the heat, the material melts and flows gradually towards the front and the degradation of polymer is minimised. Additional heat is also generated inside the barrel between the screw and the material by friction. At the front portion the molten material leaves through a die. The die is designed to create back pressure inside the barrel, which helps is uniform melting and proper dispersion. The molten

Table 30.1: Mixing ratio details

Material	Composition A	Composition B	Composition C
Material 1 (%)	88.50	68.50	48.50
Material 2 (%)	1.5	1.5	1.5
Material 3 (%)	10	30	50

Table 30.2: Formulation details

Material	Composition A	Composition B	Composition C
Material 1 (kg)	8.85	6.85	4.85
Material 2 (kg)	0.15	0.15	0.15
Material 3 (kg)	1.00	3.00	5.00

material flowing through a die at the front is cooled using water bath. The material is pulled out of the die and called as the extrudate. The extrudate is cut to desired shape and size using roller cutter and collected in a separate bin. Three percentages of reinforcement are done by changing the feeding value of the preblended material and the reinforcement material.

2.2. *Injection Moulding Process*

For the manufacturing of plastic products injection moulding is widely used. In this process products of any shape and size can be moulded with ease. In this process the plastic granules are melted inside the barrel, which is then injected into the mould cavity through nozzle. After injection into the mould cavity the part is cooled and then ejected from the cavity.

The composite used in this work is in the form of granules and fed to the barrel containing single screw through hopper by gravimetric feed. The barrel is heated by means of electric heater surrounded on the outer layer. The barrel is divided into six zones and each zone is maintained at different temperatures in such a way that the temperature from the last zone is in increasing order to nozzle. This is followed to avoid the decomposition of the material inside the barrel. The moulding process is divided into four stages: Clamping, Injection, Holding, Cooling and Ejection. The mould consists of two halves, one half is fixed to the injection unit that is nozzle and the other half is fixed to the variable drive which slides on the guides by toggle mechanism. This is operated by means of hydraulic mechanism using variable drive servo motor, which moves the mould towards the other half and closed airtight. The granules melted inside the barrel is injected through nozzle and by means of pressure it is fed into the mould cavity. The required volume of the material to fill the mould cavity is called as the shot volume. The injection time is fixed using the pressure of injection and position of

the screw inside the barrel. After injecting the molten material into the mould cavity, the holding pressure is maintained at the nozzle, which tightly packs the material in the mould cavity.

3. Testing of Mechanical Properties

3.1. *Tensile Strength*

For mechanical properties characterisation to study the effect of strain rates of the thermoplastic composite, following tests are carried out in this project at different strain rates listed below. Before characterisation to ensure the fibre content of the composite material, ash content test is performed.

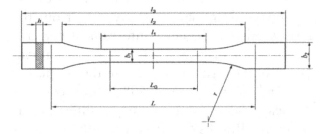

Figure 30.1: ISO 527 tensile specimen Type1A.

Table 30.3: Tensile specimen dimensions as per ISO 527

Dimensions (mm)	
Length (l3)	170.0
Narrow parallel-sided portion Length (l1)	80±2.0
Radius (r)	24±1.0
Distance between broad parallel-sided portions (l2)	109.3±3.2
Width at both ends (b2)	20±0.20
Width at narrow portion (b1)	10±0.20
Thickness Preferred(h)	4±0.20
Length of Gauge (L0)	50±0.50
Initial distance between grips (L)	115±1.0

The shape of the tensile specimen as per ISO 527[13] is shown in Figure 30.1 and dimensions are listed in Table 30.3.

The strain rates are maintained using the crosshead speed as listed in Table 30.4. The strain rate is calculated by a fixed gauge length of 50 mm and the strain of the specimen is measured using DAK Advanced Video Extensometer which captures continuous images when the specimen is under load and deformation is measured by the actual distance between the gauge lengths as shown in Figure 30.1. Five specimens tested at each strain rates and total 35 specimens are prepared.

With increasing strain rate an increase in tensile strength is clearly seen and it shows that the polymer is clearly strain rate dependence. With increase in reinforcements the strength of the polymer composite also increases. With increase in strain rate the deformation of the composite decreases is clearly seen from the results. Also, at higher reinforcement percentage a major decrease in the deformation can be seen. Deformation rate of thermoplastic material with 30% and 50% reinforcements as nearly same and no major difference is observed. Results of tensile strength and elongation at break is listed Table 30.5.

3.2. Flexural Properties

DAK Series 7200 Universal Testing Machine of capacity of 1kN load cellused to measure

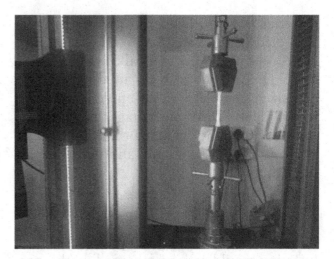

Figure 30.2: Specimen under test in tensile testing machine.

the flexural properties as per ISO 178 [12] as shown in Figure 30.3 and the strain rates are maintained using the crosshead speed as listed in Table 30.5.

The strain rate is calculated by a fixed span length of 64mm and the strain of the specimen measured using displacement of crosshead. Five specimens tested at each strain rates and 35 specimens are prepared.

Figure 30.4 shows the flexural strength of each fiber percentage reinforcement at seven different strain rates. With increasing strain rate an increase in flexural strength is clearly seen and it shows that the polymer is clearly strainrate dependence in compression. With increase in reinforcements the strength of the polymer composite also increases.

The tendency of the material to resist bending decreases with increase in the strain rate in all the fiber reinforcements. In the 30% and 50% reinforcement the flexural modulus is nearly same, and no major difference is observed whereas in the 10% reinforcement there is a linear drop in the modulus. Results of flexural properties is listed Table 30.7.

4. Conclusion

This research presents findings on the stress-strain behavior of glass fiber reinforced thermoplastic polypropylene composites under

Table 30.4: Tensile strain rate

Gauge Length (mm)	Crosshead Speed (mm/min)	Strain Rate (S^{-1})
50	1	3.3×10^{-4}
50	5	1.7×10^{-3}
50	50	1.7×10^{-2}
50	250	8.3×10^{-2}
50	450	1.5×10^{-1}
50	650	2.2×10^{-1}
50	800	2.7×10^{-1}

Table 30.5: Tensile property test results

Reinforcement Percentage	Strain Rate (S-1)	Tensile Strength at Break (MPa)	Elongation at Break (%)
	3.3×10^{-4}	31.932	12.381
	1.7×10^{-3}	34.529	9.018
	1.7×10^{-2}	38.328	8.365
10% GF	8.3×10^{-2}	40.988	7.647
	1.5×10^{-1}	42.081	6.802
	2.2×10^{-1}	42.669	6.009
	2.7×10^{-1}	42.749	4.390
	3.3×10^{-4}	70.156	3.043
	1.7×10^{-3}	74.045	2.996
	1.7×10^{-2}	80.851	2.876
30% GF	8.3×10^{-2}	84.805	2.558
	1.5×10^{-1}	89.865	2.447
	2.2×10^{-1}	89.692	1.911
	2.7×10^{-1}	90.826	1.580
	3.3×10^{-4}	90.532	2.690
	1.7×10^{-3}	93.406	2.677
	1.7×10^{-2}	103.349	2.572
50% GF	8.3×10^{-2}	115.214	2.560
	1.5×10^{-1}	118.022	2.183
	2.2×10^{-1}	120.876	1.392
	2.7×10^{-1}	119.779	1.035

Table 30.6: Flexural strain rate

Gauge Length (mm)	Crosshead Speed (mm/min)	Strain Rate (S-1)
64	1	1.6×10^{-5}
64	5	8.1×10^{-5}
64	50	8.1×10^{-4}
64	250	4.1×10^{-3}
64	450	7.3×10^{-3}
64	650	1.1×10^{-2}
64	800	1.3×10^{-2}

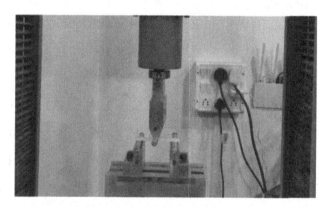

Figure 30.4: Flexural specimen under test.

varying strain rates, specifically ranging from low to medium levels (10-4 to 10-1 for tensile and 10-5 to 10-2 for flexural loading). The study investigates three different fiber reinforcement configurations, with glass fiber contents of 10%, 30%, and 50% by weight.

The experimental results clearly demonstrate that the mechanical properties of the composite materials are influenced by strain rate dependency within the investigated range. Notably, an increase in strain rate from 10-4 to 10-1, corresponding to the quasi-static region, leads to an increase in tensile strength alongside a decrease in elongation at break. This trend suggests that higher strain rates induce more rapid and limited deformation in the thermoplastic composite material.

Similarly, in compression testing, an increase in strain rate from 10-5 to 10-2

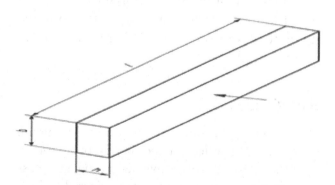

Figure 30.3: ISO 178 flexural specimen.

Table 30.7: Flexural Property test results

Reinforcement Percentage	Strain Rate (S^{-1})	Flexural Strength (MPa)	Flexural Modulus (MPa)
	1.63×10^{-5}	50.991	2069.51
	8.14×10^{-5}	56.104	2209.87
	8.14×10^{-4}	62.520	2410.91
10% GF	4.07×10^{-3}	67.363	1536.79
	7.32×10^{-3}	68.216	969.44
	1.06×10^{-2}	71.421	851.58
	1.30×10^{-2}	71.408	604.93
30% GF	1.63×10^{-5}	117.685	7745.64
	8.14×10^{-5}	124.752	7512.91
	8.14×10^{-4}	134.497	7502.74
	4.07×10^{-3}	144.978	5181.54
30% GF	7.32×10^{-3}	149.406	3245.93
	1.06×10^{-2}	151.090	2490.60
	1.30×10^{-2}	151.349	1797.03
	1.63×10^{-5}	145.861	6877.17
	8.14×10^{-5}	158.036	6582.73
	8.14×10^{-4}	172.029	6184.02
50% GF	4.07×10^{-3}	185.151	5234.70
	7.32×10^{-3}	189.638	3359.61
	1.06×10^{-2}	191.094	1937.82
	1.30×10^{-2}	189.236	1569.61

results in enhanced flexural strength. However, the flexural modulus decreases with higher strain rates, indicating a reduced ability of the material to resist deformation under increased strain rates.

Overall, the findings highlight a positive correlation between strength and strain rate in both tension and compression within the quasi-static region. Moreover, the observed increase in strength aligns with the performance of certain engineering plastics such as Polyamide and Acetal polymers, suggesting the potential for thermoplastic composite materials to serve as viable alternatives in various applications.

References

[1] Fitoussi, J., Bocquet, M., and Meraghni, F. (2013). Effect of the matrix behavior on the damage of ethylene–propylene glass fiber reinforced composite subjected to high

strain rate tension. *Compos. Part B: Eng.*, 45(1), 1181–1191.

[2] Elahi, A. F., Hossain, M. M., Afrin, S., and Khan, M. A. (2014). Study on the mechanical properties of glass fiber reinforced polyester composites.*Int. Conf. on Mechanical Industrial and Energy Engineering*, Volume ICMIEE-PI.(pp. 1–5).

[3] Babu, T. J., Khaja, M., and Hussain, G. (2017). A study of mechanical properties of HDPE reinforced with lime stone composite material. *Int. J. Sci. Eng. Dev. Res.*, Volume 2, Issue 1.

[4] McKown, S. and Cantwell, W. J. (2007). Investigation of strain-rate effects in self-reinforced polypropylenc composites. *J. Compos. Mater.*, 41(20), 2457–2470.

[5] Zhang, D., Guo, J., and Zhang, K. (2015). Effects of compatilizers on mechanical and dynamic mechanical properties of polypropylene–long glass fiber composites. *J. Thermoplast. Compos. Mater.*, 28(5), 643–655.

[6] Eriksen, R. N. W. (2014). High strain rate characterisation of composite materials. *DTU Mechanical Engineering.*

[7] Xie, H. Q., Zhang, S., and Xie, D. (2005). An efficient way to improve the mechanical properties of polypropylene/short glass fiber composites. *J. Appl. Polym. Sci.*, 96(4), 1414–1420.

[8] Elanchezhian, C., Ramnath, B. V., and Hemalatha, J. (2014). Mechanical behaviour of glass and carbon fibre reinforced composites at varying strain rates and temperatures. *Procedia Mater., Sci.*, 6, 1405–1418.

[9] Al-Mosawe, A., Al-Mahaidi, R., and Zhao, X. L. (2017). Engineering properties of CFRP laminate under high strain rates. *Compos. Struct.*, 180, 9–15.

[10] Singh, V. (2018). *Literature Survey of Strain Rate Effects on Composites.* Swerea SICOMP AB, Sweden.

[11] Zhou, Y. and Mallick, P. K. (2005). A nonlinear damage model for the tensile behavior of an injection molded short E-glass fiber reinforced polyamide-6, 6. *Mater. Sci. Eng.: A*, 393(1–2), 303–309.

[12] ISO 178. (2019).*Determination of Flexural Properties.* International Standard for Plastics.

[13] ISO 527-2. (2012).*Determination of Tensile Properties.* International Standard for Plastics.

31 Hydrogen storage in MOFs: a machine learning approach

Yash Misra[1], Meeradevi[2,a], Sindhu Pranavi[1], B. J. Sowmya[3], Darshan Bankapure[2], and Spandana S. Mentha[1]

[1]Department of Chemical Engineering, Ramaiah Institute of Technology Bengaluru, Karnataka, India
[2]Department of AI & ML, Ramaiah Institute of Technology, Bengaluru, Karnataka, India
[3]Department of AI & DS, Ramaiah Institute of Technology, Bengaluru, Karnataka, India

Abstract

Energy generation, consumption, and energy systems must navigate a significant transformation in pursuit of the global energy shift to a carbon-neutral society. A substantial possibility emerges for hydrogen to speed up clean and renewable energy proliferation swiftly. Although safely storing a sufficient amount of hydrogen for transportation while maintaining a suitable volumetric density poses challenges with multiple technological obstacles. Metal-organic frameworks (MOFs), porous constructs crafted from modular molecular constituents, predominantly encompassing metal clusters and organic linkages, offer substantial potential owing to their attributes of minimal operating pressures, rapid reaction kinetics, reversibility, and elevated gravimetric densities, which are characteristic hallmarks of MOFs. Although exceptionally few MOFs have been experimentally characterised, their building block arrangement indicates a limitless array of potential materials, making it challenging to determine the best MOFs. This study employs machine-learning models to identify the hydrogen storage capacity of MOFs.. Hydrogen storage/capacity in MOFs is estimated by employing an effective machine-learning approach. The proposed model uses regression algorithms such as KNN, gradient, lasso, ridge, and Adaboost. The model performance is evaluated using RMSE, AUE, and accuracy values. KNN model performs better when compared to other models.

Keywords: Metal-organic frameworks, energy materials, regression, RMSE, clean energy transition

1. Introduction

The world is heading towards an energy crisis with this rapid increase in the consumption of conventional resources requiring researchers to find suitable clean alternatives. Being a versatile, abundant, and renewable source, hydrogen is the most promising replacement. When in use, hydrogen produces water vapor as the only byproduct making it a clean energy source. In terms of weight, Hydrogen contains roughly three times the energy content of gasoline [1]. It is practically feasible to produce and use hydrogen on a micro, macro, and mega scale [2]. Long-term energy storage

using hydrogen can enable excess renewable energy to be harvested and preserved during times of high generation and released during times of high demand or low generation. This helps with energy management, improves grid stability and dependability, and helps with supply and demand balancing difficulties. The usage of hydrogen is widespread, with applications in heating, industrial, and transportation.Substituting hydrogen for fossil fuels in these sectors has the potential to drastically decrease greenhouse gas emissions and diminish our dependence on non-renewable resources.

[a]meera_ak@msrit.edu

DOI: 10.1201/9781003545941-31

However, the absence of an efficient, cost-efficient, and secure onboard hydrogen storage system has been a barrier for hydrogen-driven energy future. The current storage methods include liquidnhydrogen tanks, compressed cryogenic compressed hydrogen, chemical hydrogen, nano porous adsorbents, hydrogen gas tanks, metal hydrides and storage materials. Recent advancements in hydrogen storage include various chemical storage methods like Organic Chemical Hydrides, synthesised hydrocarbons, imidazolium ionic liquids, ammonia, metal hydrides, Liquid Organic Hydrogen Carriers, complex metal hydrides, Liquid hydrogen peroxide, Liquid organic amides, and methanol, carbonite substances. Substances like hydrates, porous organic polymers, doped polymers, hypercrosslinked polymers, polymers with intrinsic microporosity, covalent organic frameworks (COFs), metallic organic frameworks (MOFs), etc. are some physical storage techniques [3].

Among them, MOF materials, in particular, have piqued the interest of researchers due to their comparatively simple and inexpensive production when compared to chemical vapor deposition and their significant storage levels (some of them are summarised in Table 31.1, and good thermal stability up to 400°C [4]. Metal-Organic Frameworks (MOFs) are a class of crystalline porous material made up of metal ions or clusters that are bonded together by carboxylate rigid organic ligands thus forming extended frameworks with low density (0.2–1 g/cm^3), good structural, thermal stability and a high specific surface area (500–6500 m^2/g) [5,6]. Outstanding resilience is provided by the strong metal-oxygen bond, which keeps the framework intact even after the solvent molecules utilised in the synthesis are eliminated [7]. Organic ligands and metal clusters are chosen for their size, shape, and chemical properties, which define the structure and properties of the final MOF. Consequently, MOFs find utility across a broad spectrum of applications, encompassing gas storage and separation, catalysis, drug delivery, and sensing. The unique properties of MOFs arise from their tuneable pore sizes and vast surface area, which allow them to selectively adsorb gases or molecules based on size, shape, and chemical affinity. Metal-organic frameworks (MOFs) can be subjected to functionalisation with additional chemical groups, a process that further augments their selectivity and functionality tailored

Table 31.1: Summary of different hydrogen storage levels by MOFs

MOF (Formula)	Pressure (bar)	Temperature (K)	Hydrogen Storage by WT%	Reference
MOF-5 [Zn4O(BDC)3]	50	77	4.7	[8]
MOF-74 [Zn2(dhtp)	1	77	1.75	[9]
TUDMOF-1 [Mo(BTC)2]	1	77	1.75	[10]
TUDMOF-2 [Mg3(NDC)3]	1	77	0.78	[11]
PCN-9 [H2[Co4O(tab)8/3]]	1	77	1.53	[12]
PCN-11 [Cu(sbtc)]	3.5	30	7.89	[13]
SNU-5 [Cu2(abtc)]	40-110	77	4.71	[14]
MOF-177 [Zn4O(BTB)2]	100	77	0.62	[15]
MOF-808 [Zr6O4(OH)4(BTC)2]	40	77	7.31	[16]
UMCM-150 [Cu3(bhtc)2]	45	77	5.7	[17]
FMOF-1 [Ag2[Ag4(trz)6]]	64	77	2.33	[18]

for specific applications. These materials are therefore ideal for storing hydrogen.

These actual Metal-Organic Frameworks (MOFs) [19] crystal structures are available in the Cambridge Structural Database (CSD). However, a significant portion of these structures may exhibit disorder, contain missing atoms, or possess limited porosity. Consequently, they are not immediately suitable for assessment through computer modelling [20].

Moreover, due to their ability to form clusters with many functional groups, metals, mixing metals, linkers etc the number of MOFs to be studied is increased beyond calculations, thus elevating the complexity of the entire MOFs database. There is an extensive database of the diverse range of Metal-organic frameworks (MOFs), such as those exhibiting high hydrogen storage capacity, elevated hydrogen adsorption enthalpy, extensive surface area, or exceptional stability, that have been synthesised specifically for hydrogen storage applications. These MOFs are characterised by their distinctive properties and attributes, which render them suitable for fulfilling the requirements of hydrogen storage applications. The key performance metrics such as gravimetric and volumetric capacity, operating conditions, and stability are used to evaluate MOFs for hydrogen. Moreover, there are various challenges associated with MOF-based hydrogen storage, including kinetics, reversibility, and MOF synthesis scalability. The wide range of available options presents a significant challenge in conducting comprehensive searches across all materials, even with high-throughput methods. Additionally, comparing screening investigations can be problematic due to variations in execution, such as differences in temperature/pressure settings or interatomic potentials utilised. Hence, there is a need to develop screening techniques that are both more efficient and reliable for evaluating the gas storage capacity of Metal-Organic Frameworks (MOFs) within existing and future databases. Thus, increasing the ease

to enhance MOF performance and increasing the utilisation prospects of MOFs in terms of commercialisation and integration into hydrogen storage systems.

Machine Learning is a potential sophisticated solution to enhance the MOFs screening procedure. Being an interdisciplinary field that integrates mathematics, statistics, and computer science machine learning (ML) is a precise and effective instrument for data handling and evaluation to provide cognitive analysis of dataset. ML model in a database is constructed based on certain features. To screen MOFs supervised learning ML model is being utilised. ML methods like neural networks and regression algorithms come under this category. Supervised learning involves training a model with a labelled dataset consisting of input feature (x) and a corresponding output label (y). The objective of the proposed model is learning relationship between data so that the model and generalise the unseen data. This model can effectively investigate the concealed rules and unique trends in study.

2. Insight into Various ML Models

The proposed work's performance is evaluated using a range of machine learning (ML) models. Regression models, in particular, are employed to identify correlations between continuous inputs and outputs based on example data. Subsequently, these models can predict novel inputs by extrapolating from the learned patterns.[21]. Regression analysis is used to predict a broad range of future values using the input data and past data. It aids in identifying a prediction connection between labels and data points. It is primarily used for tasks, which involve predicting a continuous numerical value. Several approaches for modelling regression have been devised. Adaboost, KNN, gradient, LASSO, and Ridge are frequently employed.

The K-nearest neighbours (KNN) algorithm, abbreviated as KNN, is a learning classifier that doesn't rely on predefined parameters. In this regression technique, it leverages data proximity to classify or predict the category of an individual data point. Instead of going through a formal training phase, it maintains a training dataset, which means all calculations take place when performing classification or predictions. This method is also referred to as instance-based or memory-based learning due to its strong reliance on memory for storing training data. Gradient descent, on the other hand, is an iterative optimisation technique employed to locate the minimum of a function. In this process, the algorithm progressively approaches the least-squares regression line by minimising the sum of squared errors over several iterations. Whereas, Adaptive Boosting, often known as AdaBoost, is a machine learning ensemble technique that may be used for various classification and regression applications. By integrating many weak or base learners (such as decision trees) into a strong learner, this supervised learning process is used to categorise data. The training dataset's examples are weighted by AdaBoost according to how accurately prior classifications were made.

LASSO and Ridge regression models are employed when multicollinearity is taken into consideration. Least Absolute Shrinkage and Selection (LASSO) takes advantage of shrinkage. Shrinkage is the process of reducing data values to a median or other middle value. The lasso approach encourages sparse, basic models having a small number of parameters. This regression method is particularly effective in cases where models exhibit significant multicollinearity or when specific processes within the model selection phase, such as parameter removal and variable selection, need to be automated. On the other hand, Ridge regression implements L2 regularisation. It is employed when multicollinearity is present, resulting in predicted values deviating from actual values, unbiased least-squares, and high variances. This method is advantageous when independent variables are highly correlated.

3. Screening of MOFs using ML

Machine learning (ML) has demonstrated commendable performance in classification and regression tasks related to identifying high-performance Metal-Organic Frameworks (MOFs). ML leverages computational results obtained from High-Throughput Computational Screening (HTCS) to extract valuable insights from a newly developed MOF database. By investigating the intricate connections between MOF components, structures, and performance, ML has proven its capacity to enhance the performance of MOFs beyond what HTCS calculations can achieve. This improvement is notably achieved by optimising the variety of MOFs and their constituent elements. For example, Fanourgakis et al. made a study on ML to forecast the absorptions of CO2 and CH4 capabilities resulting in efficient prediction [22]. This innovative approach allowed for a tenfold reduction in the size of the training dataset while maintaining a high level of predictive accuracy.

Li et al. explored the use of MOFs and COFs as adsorption materials for cascade heat pumps. Machine learning analysis revealed that the pore volume available is a critical factor that greatly influences cooling effectiveness [23]. In a recent study, we screened MOFs for CO2 capture at low concentrations, and ML algorithms identified the heat of adsorption as the primary MOF descriptor for adsorption selectivity. Notably, ML techniques have been widely employed in various applications, including CH4 storage, H2 storage, and CO2 capture and separation within the MOF screening process. These ML methods are typically integrated with HTCS studies, leading to the identification of MOF structures with superior performance in diverse applications. Furthermore, data mining techniques akin to ML are employed to explore the structure-function relationship within the acquired data, facilitating the design of novel MOFs [24].

4. Hydrogen Storage

According to recent studies, Metal-Organic Frameworks (MOFs) are experiencing growth, particularly in hydrogen (H_2) storage applications. Machine learning (ML) algorithms have emerged as integral tools in screening and selecting MOFs for this specific purpose.

Thornton et al. conducted screening of 850,000 nonporous materials for H2 storage, employing a combination of molecular simulation and ML [25]. This intricate process involved simulating the isotherm at pressures ranging from 1 to 100 bar and was divided into three phases. To train the model using NN initially 1000 materials are used for best predictive performance. Following this, the model was utilised to forecast the subsequent set of materials for simulation. The NN model evolved with each phase and yielded an impressive prediction accuracy (R^2 = 0.88). Through this NN model, an optimal descriptor indicator was identified, including a pore size of 10 Å, diameter of 0.5, and VSA of 5000 m^2 g^{-1}. The model undergoes analysis utilising a dataset comprising 850,000 materials, revealing that CoRE-MOFs demonstrate superior characteristics, surpassing zeolites in absorption capacity while boasting larger pore sizes. Additionally, hMOFs demonstrate notable adsorption capacity in terms of VSA, whereas porous polymer networks (PPNs) and covalent organic frameworks (COFs) exhibit excellence in terms of ϕ across both categories.

Bucior et al. introduced a novel descriptor called the energy histogram, derived from the mutual energy between MOFs and guest molecules [26]. They combined it with ML for efficient MOF screening. By coupling a LASSO model with the energy histogram, the regression model demonstrated impressive precision in forecasting gas adsorption, with discrepancies of under 3 g L^{-1} across diverse databases. Subsequently, the model was employed to anticipate the deliverable capacity of H_2 This groundbreaking work demonstrated the efficiency of machine learning (ML) methods, facilitating the swift exploration of numerous Metal-Organic Frameworks (MOFs) within MOF databases. The results were achieved over three orders of magnitude faster than traditional molecular simulations.

In Anderson et al.'s study, ML, in conjunction with molecular simulation, facilitated the highest achievable hydrogen storage capacity determination, reaching 62 g L^{-1} (adsorption heat 10 kJ mol^{-1}) under different thermodynamic conditions, a remarkable 138% improvement compared to room temperature and 100 bar conditions (26 g L^{-1}) [27]. Under varying pressure and temperature conditions, the maximum storage capacity of hydrogen only decreased by 3%, indicating a practical solution for designing and securing H_2 storage tanks.. The descriptors employed in the ANN model were easy to compute or acquire from existing literature, making it an invaluable instrument for assessing new materials, theoretically uncovering novel materials, and guiding experimental material synthesis.

5. Methodology

MOFs, identified as adaptable nanoporous materials, exhibit high surface areas and flexibility in pore size. Over 100,000 MOFs have been developed and created for an array of applications. In order to get beyond the limitations of tedious experimental synthesis, computational databases for prospective MOF structures have been constructed. Additionally, to the approximately 100,000 experimental structures now in existence, these databases effectively create and assess MOFs for various applications [28]. An early example is the database formulated by Wilmer et al., containing 137,000 Metal-Organic Frameworks (MOFs) characterised by a limited diversity of topologies, predominantly pcu. Our study focuses on a subset of MOFs from Northwestern University's database, known as ToBaCCo, which comprises approximately 13,500 hypothetical MOFs [29]. In our analysis, we investigate crucial crystallographic attributes, specifically LCD (largest cavity diameter), surface area, void fraction and PLD (pore limiting diameter), which serve as descriptors for the Metal-Organic Frameworks (MOFs) under study.

The ToBaCCo dataset utilised in this study is a product of a computer algorithm and program specifically designed for generating Metal-Organic Frameworks (MOFs) based on edge-transitive topological nets. It comprises 13,512 MOFs, encompassing 41 unique topologies. These MOFs were meticulously constructed and assessed using ToBaCCo, focusing on applications within energy-related domains such as gaseous fuel storage, xenon and krypton mixtures separation, and cryo-adsorbed hydrogen storage. This algorithm likely integrated existing MOF databases and computational methods to curate the MOFs within our dataset. In this research, the configuration of nodes and edges within the abstract network, guiding the positioning of Metal-Organic Framework (MOF) building blocks, is termed "topology" [30]. The researchers also investigated how topology affected MOF attributes like gas storage capacity and selectivity by using ToBaCCo to build MOFs with specific topologies suited for energy-related applications.

Segmenting the dataset into two subsets, one for training the model and the other for validation, is a common practice in data pre-processing. Many researchers advocate for a training/testing set ratio of 70/30 or 80/20 to facilitate model construction and evaluation. In this study, we adhere to an 80/20 split for our dataset, aiming to train a machine-learning model for forecasting hydrogen uptake in Metal-Organic Frameworks (MOFs). This partition assigns 80% of the data for training the model while reserving the remaining 20% for validation purposes.

We employed supervised learning methodologies and regression-based algorithms to train the models, considering the database contained linear and nonlinear data points as shown in Figure 31.1 and performance is analyzed using various ML models taking MOFs as input and trained using ML models and output is validated by analyzing performance as shown in Figure 31.2. Additionally, we adopted a methodical approach that included combining two regression models while investigating regression-based models and neural networks. The subsequent sections will comprehensively examine and discuss the findings derived from the analysis.

6. Results and Discussions

Neural networks and regression models have been used to screen MOFs for hydrogen uptake capacity. The performance of different machine learning (ML) models has been evaluated using three metrics: Area Under

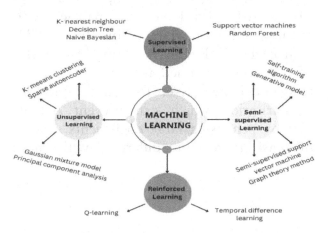

Figure 31.1: Classification of ML models.

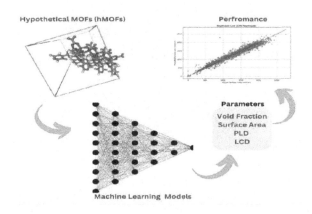

Figure 31.2: Proposed workflow of the model.

the Error Curve (AUE), Root Mean Squared Error (RMSE), and R-squared (R^2). A lower AUE (Absolute Unsigned Error) indicates better predictive accuracy. A lower RMSE (Root Mean Squared Error) indicates a model that better fits the data and a higher R-squared value indicates a model that explains a greater proportion of the variance in the dependent variable.

For regression problems, it's frequently the baseline model. The goal variable and the input characteristics are assumed to have a linear relationship. When the dataset is evaluated using this model, an R^2 of 0.66, RMSE of 273.36 and 211.95 AUE is obtained by which we can conclude that the data is not in linear relationship with hydrogen uptake. Figure 31.3 illustrates a rough relation between the training and testing data thus making this model ineffective.

As seen in Figure 31.4, Ridge and LASSO ML models also have discrepancies with training and testing data. A low R^2 of 0.692, high AUE and RMSE values of 190.95 and 253.36 were obtained respectively, making this model ineffective. This model includes regularisation which can prevent overfitting by penalising large coefficients. But in our case, there is no difference with linear

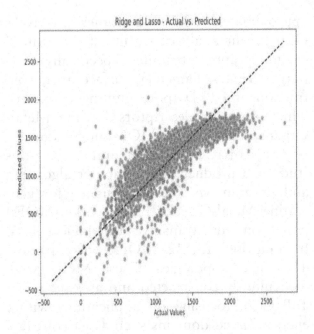

Figure 31.4: Ridge and Lasso model.

regression model proving there is no effect of regularisation.

LSTM model have decent values with R^2 of 0.95, RMSE of 107.27 and AUE of 78.90. Figure 31.5 shows not much deviation of testing of data from testing data. It is Good for capturing long-term dependencies in sequence data which is why it is not giving better results as our data is not sequential but it is better than the others above as it is focusing on the learning curves over epochs.

CNN and DNN have been also been employed and tested. From Figures 31.6 and 31.7 it can be deciphered that the model shows good agreement with training and testing data. They have R^2 of 0.953 and 0.955. ASME of 101.033 and 99.276 and AUE of 72.42 and 68.213 respectively. It is well known that CNN excels at collecting data's spatial hierarchies and make use of spatial links between features in the dataset whereas DNNs can recognise intricate patterns in data by adding additional layers and learns progressively over the large dataset. Thus, giving agreeable results.

As seen in Figure 31.8 CNN+LSTM model is evaluated and have an R^2 of 0.866, AMSE of 171.501 and AUE of 126.680. In general, these hybrid neural network designs are

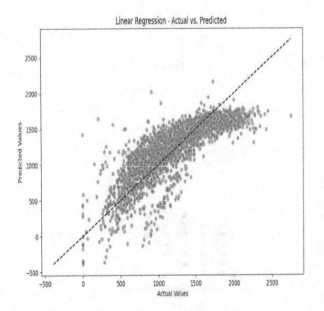

Figure 31.3: Linear Regression model.

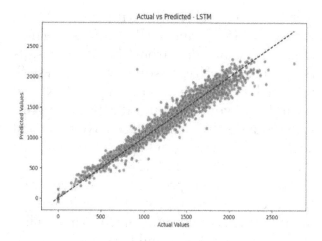

Figure 31.5: LSTM model.

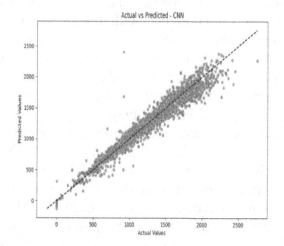

Figure 31.6: CNN model.

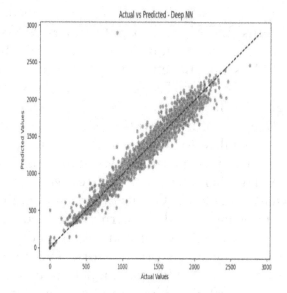

Figure 31.7: Deep CNN model.

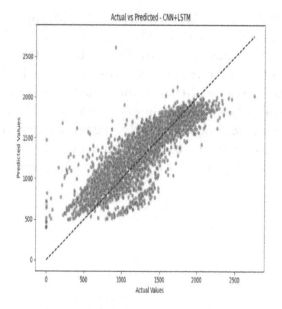

Figure 31.8: CNN + LSTM model.

able to extract temporal/sequence patterns (LSTM) and spatial features (CNN) from the input, however because of the LSTM's sequential effect, they perform worse than CNN and LSTM separately.

Gradient Boosting and KNN algorithms gave robust results with R² values of 0.96 each. They also have 96.50 and 94.16 AMSE values along with satisfactory low AUE values of 70.31 and 63.98. Figures 31.9 and 31.10 demonstrate a strong alignment between actual and predicted values. This alignment underscores the effectiveness of the boosting model's methodology, which constructs trees iteratively. With each new tree, errors from preceding trees are corrected, facilitating the model's ability to encapsulate the intricate relationships and non-linearities inherent within the dataset. While, KNN is non-parametric method that captures the data and plots the decision boundaries of the nearest neighbors where the distance between samples is a good indicator of their similarity.

As per the dataset, which contains both complicated non-linear patterns and linear patterns, is connected to predict the storage capabilities of hydrogen in MOFs. We use Stacking method to capture the whole data by using the combination of AdaBoost and

KNN as the Base-learners and selecting Linear regression as the Meta-learner to combine their predictions. AdaBoost facilitates the capture of intricate patterns, whereas KNN addresses local trends and Linear Regression establishes a global linear framework. Stacking of Adaboost and KNN regressors gave the best results for the taken values which can be seen in Figure 31.11. The testing and training

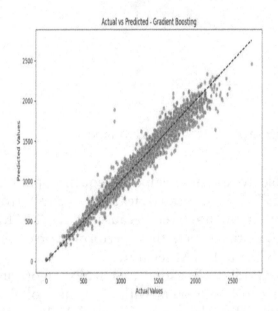

Figure 31.9: Gradient Boosting.

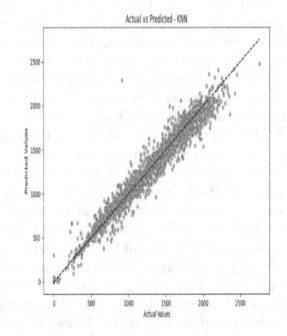

Figure 31.10: KNN model.

were in good agreement with R^2 of 0.976, AMSE of 85.13 and AUE of 55.23. Thus, it can be concluded from the given results that stacking of regressors gives the best fit followed by Gradient boosting, KNN and Neural Networks. While, the linear regression models were evaluated to have the worst accuracy, this is in good agreement with the findings of Ahmed et al. [28].

For regression problems, it's frequently the baseline model. The goal variable and the input characteristics are assumed to have a linear relationship. Ridge and Lasso includes regularisation which can prevent overfitting by penalising large coefficients. But in our case, there is no difference with linear regression model as there is no effect of regularization.

LSTM is good for capturing long-term dependencies in sequence data which is why it is not giving better results as our data is not sequential but it is better than the others above as it is focusing on the learning curves over epochs as shown in Figure 31.5. CNN excels at collecting data's spatial hierarchies and make use of spatial links between features in the dataset as shown in Figure 31.6.

CNN excels at collecting data's spatial hierarchies and make use of spatial links between features in the dataset as shown in Figure 31.7. These hybrid neural network designs are able to extract temporal/sequence patterns (LSTM) and spatial features (CNN) from the input, however because of the LSTM's sequential effect, they perform worse than CNN and LSTM separately as shown in Figure 31.8.

Deep Neural Networks (DNNs) possess the capability to discern intricate data patterns by incorporating additional layers, allowing them to learn progressively from extensive datasets. On the other hand, gradient-boosting models construct trees sequentially, with each new tree aimed at rectifying errors made by its predecessors. This iterative process enables the model to capture intricate relationships and non-linearities within the dataset, illustrated in Figure 31.9.

Meanwhile, K-Nearest Neighbors (KNN) operates as a non-parametric method, delineating data and decision boundaries based on the proximity of neighbouring samples. Figure 31.10 illustrates how the distance between samples serves as a reliable indicator of their similarity.

The dataset exhibits intricate non-linear and linear patterns and is utilised to forecast the hydrogen storage capacities within Metal-Organic Frameworks (MOFs). To capture the dataset comprehensively, we employ a technique known as Stacking, which involves amalgamating multiple classification or regression models via a meta-classifier or meta-regressor. Specifically, we leverage a combination of AdaBoost and KNN algorithms as the base learners, with linear regression serving as the meta-learner to merge their predictions effectively. Ada-Boost facilitates the capture of intricate patterns, whereas KNN addresses local trends and Linear Regression establishes a global linear framework as shown in Figure 31.11. AdaBoost reduces both bias and variance through its boosting cycles, whereas KNN has low bias but high variance and linear regression has a large bias and low variance, therefore these models create a model with improved generalisation capabilities by striking a balance between bias and variance.

Table 31.2 illustrates the different machine learning (ML) models employed for screening

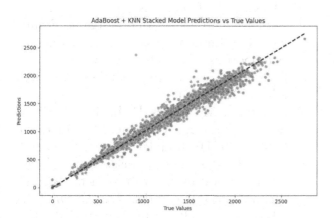

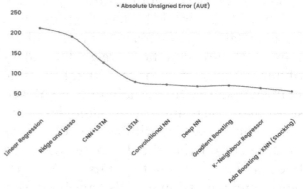

Figure 31.12: AUE values various ML model.

Figure 31.11: Proposed approach Adaboost integrated with KNN stacked model.

Table 31.2: RMSE, R^2 and AUE results of various ML models

Models	Mean square error	Root mean squared error (RMSE)	R-squared error	Absolute unsigned error (AUE)
Linear Regression	74726.03	273.36	0.66	211.95
Ridge and Lasso	63714.25	253.36	0.692	190.95
CNN+LSTM	29412.818	171.501	0.866	126.680
LSTM	11505.86	107.27	0.95	78.90
Convolutional NN	10207.586	101.033	0.953	72.42
Deep NN	9855.840	99.276	0.955	68.213
Gradient Boosting	9312.98	96.50	0.96	70.31
K-Neighbour Regressor	8866.28	94.16	0.96	63.98
Ada Boosting + KNN (Stacking)	7592.055	85.13	0.976	55.23

Metal-Organic Frameworks (MOFs) and the corresponding parameters obtained. Based on the results summarised in Table 31.2 and Figure 31.12, it becomes apparent that the stacking method yielded the lowest error values compared to other ML models.

6. Conclusion

Metal-organic frameworks (MOFs), porous constructs crafted from modular molecular constituents, predominantly encompassing metal clusters and organic linkages, offer substantial potential owing to their attributes of minimal operating pressures. This study employs machine-learning models that forecast the hydrogen capacity of MOFs for accelerating the identification approach. Machine learning (ML) has demonstrated commendable performance in classification and regression tasks related to identifying high-performance Metal-Organic Frameworks (MOFs). The model employed supervised learning methodologies and regression-based algorithms to train the models, considering the database contained linear and nonlinear data points. Neural networks and regression models have been used to screen MOFs for hydrogen uptake capacity. Three metrics namely- AUE, RMSE and R^2 have been used to evaluate the performance of the various ML models. We use Stacking to capture the whole data by using the combination of AdaBoost and KNN as the base learners and selecting Linear regression as the Meta-learner to combine their predictions.

References

[1] Suh, M. P, Park, H. J, Prasad, T. K., and Lim, D.-W. (2012). Hydrogen Storage in Metal-Organic Frameworks. *Chem. Rev*, 112(2), 782–835. https://doi.org/10.1021/cr200274s.

[2] Mazloomi, K., and Gomes, C. (2012). Hydrogen as an Energy Carrier: Prospects and Challenges. *Renew. Sustain. Energy Rev.*, 16(5), 3024–3033. https://doi.org/10.1016/j.rser.2012.02.028.

[3] Eberle, U., Felderhoff, M., and Schüth, F. (2009). Chemical and Physical Solutions for Hydrogen Storage. *Angew. Chem. Int. Ed.*, 48(36), 6608–6630. https://doi.org/10.1002/anie.200806293.

[4] Han, S. S., Deng, W.-Q., and Goddard, W. A. (2007). Improved Designs of Metal-Organic Frameworks for Hydrogen Storage. *Angew. Chem.*, 119(33), 6405–6408. https://doi.org/10.1002/ange.200700303.

[5] Eddaoudi, M, Li, H, Reineke, T, Fehr, M, Kelley, D, Groy, T. L, and Yaghi, O. M. (1999). Design and Synthesis of Metal-Carboxylate Frameworks with Permanent Microporosity. *Top. Catal.*, 9(1), 105–111. https://doi.org/10.1023/A:1019110622091.

[6] Meek, S. T., Greathouse, J. A., and Allendorf, M. D. (2011). Metal-Organic Frameworks: A Rapidly Growing Class of Versatile Nanoporous Materials. *Adv. Mater.*, 23(2), 249–267. https://doi.org/10.1002/adma.201002854.

[7] Eddaoudi, M., Moler, D. B., Li, H., Chen, B., Reineke, T. M., O'Keeffe, M., and Yaghi, O. M. (2001). Modular chemistry: secondary building units as a basis for the design of highly porous and robust metal–organic carboxylate frameworks. *Acc. Chem. Res.*, 34(4), 319– 330. https://doi.org/10.1021/ar000034b.

[8] Panella, B. Hirscher, M., Pütter, H., and Müller, U. (2006). Hydrogen Adsorption in Metal–Organic Frameworks: Cu-MOFs and Zn-MOFs Compared. *Adv. Funct. Mater.*, 16(4), 520– 524. https://doi.org/10.1002/adfm.200500561.

[9] Rowsell, J. L. C., and Yaghi, O. M. (2006). Effects of functionalization, catenation, and variation of the metal oxide and organic linking units on the low-pressure hydrogen adsorption properties of metal–organic frameworks. *J. Am. Chem. Soc.*, 128(4), 1304–1315. https://doi.org/10.1021/ja056639q.

[10] Kramer, M., Schwarz, U., and Kaskel. S. (2006). Synthesis and properties of the metal-organic framework Mo3(BTC)2 (TUDMOF-1). *J. Mater. Chem.*, 16(23), 2245. https://doi.org/10.1039/b601811d.

[11] Senkovska, I. and Kaskel, S. (2006). Solvent-induced pore-size adjustment in the metal-organic framework [Mg 3 (Ndc) 3 (Dmf) 4] (Ndc = Naphthalenedicarboxylate). *Eur. J. Inorg. Chem.*, 22, 4564–4569. https://doi.org/10.1002/ejic.200600635.

[12] Ma, S. and Zhou. H.-C. (2006). A metal-organic framework with entatic metal centers exhibiting high gas adsorption affinity. *J. Am. Chem. Soc.*, 128(36), 11734–11735. https://doi.org/10.1021/ja063538z.

[13] Wang, X.-S., Ma, S., Rauch, K., Simmons, J. M., Yuan, D., Wang, X., Yildirim, T., Cole, W. C., López, J. J., Meijere, A. D., and Zhou, H.-C. (2008). Metal–organic frameworks based on double-bond-coupled di-isophthalate linkers with high hydrogen and methane uptakes. *Chem. Mater.*, 20(9), 3145–3152. https://doi.org/10.1021/cm800403d.

[14] Xue, M., Zhu, G., Li, Y., Zhao, X., Jin, Z., Kang, E., and Qiu, S. (2008). Structure, hydrogen storage, and luminescence properties of three 3D metal–organic frameworks with NbO and PtS topologies. *Cryst. Growth Des.*, 8(7), 2478–2483. https://doi.org/10.1021/cg8001114.

[15] Li, Y., Yang, and R. T. (2007). Gas Adsorption and Storage in Metal–Organic Framework MOF- 177. *Langmuir*, 23(26), 12937–12944. https://doi.org/10.1021/la702466d.

[16] Xu, J., Liu, J., Li, Z., Wang, X., Xu, Y. Chen, S., and Wang, Z. (2019) Optimized synthesis of Zr(IV) metal organic frameworks (MOFs-808) for efficient hydrogen storage. *New J. Chem.*, 43(10), 4092–4099. https://doi.org/10.1039/C8NJ06362A.

[17] Wong-Foy, A. G., Lebel, O., and Matzger, A. J. (2007). Porous Crystal derived from a tricarboxylate linker with two distinct binding motifs. *J. Am. Chem. Soc.*, 129 (51), 15740–15741. https://doi.org/10.1021/ja0753952.

[18] Yang, C., Wang, X., and Omary, M. A. (2007) Fluorous metal–organic frameworks for high- density gas adsorption. *J. Am. Chem. Soc.*, 129(50), 15454–15455. https://doi.org/10.1021/ja0775265.

[19] Ahmed, A., Seth, S., Purewal, J., Wong-Foy, A. G., Veenstra, M., Matzger, A. J., and Siegel, D. J. (2019). Exceptional hydrogen storage achieved by screening nearly half a million metal-organic frameworks. *Nat. Commun.*, 10(1), 1568. https://doi.org/10.1038/s41467-019-09365-w.

[20] Moghadam, P. Z., Li, A., Wiggin, S. B., Tao, A., Maloney, A. G. P., Wood, P. A., Ward, S. C., and Fairen-Jimenez, D. (2017). Development of a cambridge structural database subset: a collection of metal–organic frameworks for past, present, and future. *Chem. Mater.*, 29(7), 2618–2625. https://doi.org/10.1021/acs.chemmater.7b00441.

[21] Stulp, F., and Sigaud, O. (2015) Many regression algorithms, one unified model: a review. *Neural Netw.*, 69, 60–79. https://doi.org/10.1016/j.neunet.2015.05.005.

[22] Fanourgakis, G. S., Gkagkas, K., Tylianakis, E., and Froudakis, G. E. (2020). A universal machine learning algorithm for large-scale screening of materials. *J. Am. Chem. Soc.*, 142 (8), 3814–3822. https://doi.org/10.1021/jacs.9b11084.

[23] Li, W., Xia, X., and Li, S. (2020) Screening of covalent–organic frameworks for adsorption heat pumps. *ACS Appl. Mater. Interfaces*, 12(2), 3265–3273. https://doi.org/10.1021/acsami.9b20837.

[24] Shi, Z., Yang, W., Deng, X., Cai, C., Yan, Y., Liang, H., Liu, Z., and Qiao, Z. (2020) Machine- learning-assisted high-throughput computational screening of high performance metal–organic frameworks. *Mol. Syst. Des. Eng.*, 5(4), 725–742. https://doi.org/10.1039/D0ME00005A.

[25] Thornton, A. W., Simon, C. M., Kim, J., Kwon, O., Deeg, K. S., Konstas, K., Pas, S. J., Hill, M. R., Winkler, D. A., Haranczyk, M., and Smit, B. (2017). Materials genome in action: identifying the performance limits of physical hydrogen storage. *Chem. Mater.*, 29 (7), 2844–2854. https://doi.org/10.1021/acs.chemmater.6b04933.

[26] Bucior, B. J., Bobbitt, N. S., Islamoglu, T., Goswami, S., Gopalan, A., Yildirim, T., Farha, O. K., Bagheri, N., and Snurr, R. Q. (2019). Energy-based descriptors to rapidly predict hydrogen storage in metal–organic frameworks. *Mol. Syst. Des. Eng.*, 4(1), 162–174. https://doi.org/10.1039/C8ME00050F.

[27] Anderson, G., Schweitzer, B., Anderson, R., and Gómez-Gualdrón, D. A. (2019). Attainable volumetric targets for adsorption-based

hydrogen storage in porous crystals: molecular simulation and machine learning. *J. Phys. Chem. C*, 123(1), 120–130. https://doi.org/10.1021/acs.jpcc.8b09420.

[28] Ahmed, A. and Siegel, D. J. (2021). Predicting hydrogen storage in MOFs via machine learning. *Patterns*, 2(7), 100291. https://doi.org/10.1016/j.patter.2021.100291.

[29] Wilmer, C. E. Leaf, M., Lee, C. Y., Farha, O. K., Hauser, B. G. Hupp, J. T., and Snurr, R. Q. (2012). Large-scale screening of hypothetical metal–organic frameworks. *Nat. Chem.*, (2), 83–89. https://doi.org/10.1038/nchem.1192.

[30] Colón, Y. J., Gómez-Gualdrón, D. A., and Snurr, R. Q. (2017). Topologically guided, automated construction of metal–organic frameworks and their evaluation for energy-related applications. *Cryst. Growth Des.*, 17(11), 5801–5810. https://doi.org/10.1021/acs.cgd.7b00848.

32 Minimization of armature shaft deflection in micro dc motor using sintered stainless steel collar followed by FEA

Dharmalingam G[1,a], N. Srinivasan[1], and Vishal Naranje[2]

[1]Department of Mechanical Engineering, Vel Tech Rangarajan Dr. Sagunthala R&D Institute of Science and Technology, Chennai, India
[2]Department of Mechanical Engineering, Amity University, Dubai International Academic City, Dubai, UAE

Abstract

Natural Frequency, as the name implies is the set of frequencies acting on an object when it vibrates naturally. The vibration acting on any object under the agitation of external forces then it is called Forced vibration. Agitation frequency occurs when the object is forced to vibrate at this frequency. At this agitation frequency, the resonance is generated when it matches with the system's natural frequency. The purpose of this research work is to increase the natural frequency in the area of the armature shaft in between the lamination stack and the copper commutator in a Micro DC Motor thus preventing the copper wire fracture when subjected to different mechanical and thermal stress. The scope of increasing the natural frequency of the assembly is carried out by introducing a stainless steel collar-ferrous diffusion alloyed (metal powder industries Federation (MPIF)-35-FD-0408-45 powder metallurgy part (sample 2)) in the required area that is in between the lamination stack and the Commutator thereby increasing the stiffness of shaft on that particular area thereby minimizing the shaft deflection. Taking the assembly's natural frequency out of the operating frequency improves the vibration stability of the component; it can take more vibrational load. The engine operating frequency in any automotive is carried in between the frequency ranges of 20 ~ 2000 Hz. It was confirmed that the parts produced through powder metallurgy part (MPIF-35- FD-0408-45) are more economical without compromising on the technical and functional requirements of the product than sample 1.

Keywords: Natural frequency, DC motors, resonance, powder metallurgy, collar-ferrous diffusion alloyed, armature shaft

1. Introduction

In the current motor industry, companies are struggling with field failures related to the armature coil cut which will happen during the field run and in the Engine level testing. By increasing the natural frequency of the system or by taking it away from the engine operating frequency these failures can be eliminated in a greater way. Taking the assembly's Natural frequency out of the operating frequency improves the vibration stability of the component or in other words, it can take more vibrational load. In design principle, any component's Natural frequency (or) Resonance frequency should be out of the operating frequency range in order to avoid failure.

Due to their low cost and compact design, small direct current (DC) motors are frequently used in toys, cars, and home appliances. They normally include housing, permanent magnets (stator), an armature (rotor), brushes, etc. Small DC motors come

[a]dharma21sona@gmail.com

DOI: 10.1201/9781003545941-32

in a variety of designs, each having a varied size and number of poles [1,2].

Despite the compact size and straightforward structural design of these motors, the causes of motor noise and vibration are often intricate and can be attributed to a wide range of factors, including the electro-magnetic forces between the permanent magnets and armature, brush switching, resonances in the housing, bearings, etc. [3].

The amount of noise emitted by brushless DC motors has been quantified, and in addition to simulation matching, resonance from structural and acoustical excitation has been demonstrated [4].

Bandar Al-Mangour et al. [5] powder metallurgy components are in high demand across a wide range of industries, such as biomedical aerospace, automotive, and chemical processing. This has generated a lot of interest in the research community. Knowing the right processing methods is essential to creating such important components and expanding the application of powder metallurgy technology, which has tremendous economic value in an expanding variety of applications [6–9].

By combining the compressibility of a pure iron powder with the lack of segregation of a prealloyed powder, diffusion-alloyed steel powders get beyond the unique constraints of both premixed and prealloyed powders. High-strength sintered powder metallurgy (P/M) products have been successfully made using diffusion-alloyed steel powders by single-press/single-sinter processing [10,11]. Diffusion-alloyed steel powders have the following benefits: maintaining the base iron's strong inherent compressibility, higher green strength, decreased chance of dusting and alloy segregation while transporting the powder mixture, uniform distribution of the alloying components, excellent sinterability [12].

1.1. Scope of the Research

The aim of this research is to increase the Natural Frequency in the Armature shaft in between the lamination stack and the copper commutator in a Micro DC Motor thus preventing the copper wire fracture when it is subjected to different mechanical and thermal stress. In design principle, any component's Natural frequency (or) Resonance frequency should be out of the operating frequency range in order to avoid Failure. This is the fundamental rule for any product design. The scope of increasing the Natural frequency of the assembly is carried out by introducing a stainless steel collar in the required area that is in between the lamination stack and the commutator there by increasing the stiffness of the shaft in that particular area thereby minimizing the shaft deflection taking the assembly's Natural frequency out of the operating frequency improves the vibration stability of the component or in other words it can take the more vibrational load and hence the life of the component will be increased further or the failure of the parts will be eliminated because of several failure modes.

This research will focus on the concept design, analysis and manufacturing of the proposed stainless steel collar. The manufacturing is carried out by both the conventional machining process and the Powder Metallurgy technique. From the analysis, it is evident that introducing the collar in the neck will increase or shift the system's natural frequency from 1446 Hz to 1629 Hz. The deflection of the armature shaft was performed with 50 g input at maximum frequency and resonant frequency respectively.

2. Materials and Methods

As the application of this DC motor is for the Automotive Engine management system, the selection of materials is the key to defining the function and durability of the product.

The research is carried out with two different materials for the proposed collar as described below. The collar placement is shown in Figure 32.1.

Sample1 – Japanese Industrial Standard (JIS) G 4303, SUS 416

Sample 2 – MPIF Standard, MPIF-35, FD-0408-45

2.1. Sample 1 – JIS G 4303, SUS 416

SUS 416 grade of material falls under the stainless steel which is procured from Atlas Steel Industries Private Limited in Himachal Pradesh. The highest machinability of all stainless steels is found in SUS 416, which is around 85% as machinable as free-machining carbon steel. The improvement in machinability is achieved, as with most other free-machining stainless steels, by adding sulfur, which results in the formation of manganese sulfide inclusions. This sulfur addition also reduces corrosion resistance, weldability, and formability. This steel was directly machined through the CNC-Lathe turning process.

2.2. Chemical Composition – SUS 416

The general chemical composition and mechanical properties of the SUS416 material grade are shown in Tables 32.1 and 32.2. This grade of stainless steel is majorly used in the automotive industry to withstand high

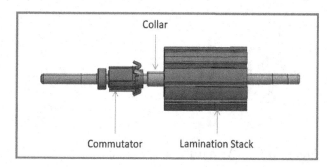

Figure 32.1: Armature assembly with rotor collar.

Table 32.1: Chemical composition of SUS 416

S.No	Elements	SUS 416 steel (wt. %)
1	Iron	Rest
2	Carbon	0.15
3	Silicon	1.0
4	Manganese	1.2
5	Molybdenum	0.6
6	Chromium	12~14

Table 32.2: Mechanical properties of SUS 416

S.No	Properties	SUS 416 steel
1	Density	7.80 g/cc
2	Yield stress	345 N/mm^2
3	Tensile Strength	540 N/mm^2
4	Hardness	166 HV

axial and radial loads where the part is press fitted to the shaft material in order to prevent deformation and to take high residual stresses.

2.3. Sample 2 – MPIF-35, FD-0408-45

The diffusion alloyed steel powder Fe-8Ni-1.5 Cu-0.5 Mo-0.2 Cr-0.7 C (%wt.) that complied with MPIF FD-0405 standard was employed as the foundation powder in this investigation. The diffusion alloyed steel powders were processed through the powder metallurgy (PM) route followed by vacuum hot pressing (VHP). The diffusion alloyed steel powders were poured into a graphite die and kept in a vacuum furnace. The powders were sintered at 1120 °C and holding for 28 minutes with a constant pressure level of 60 MPa was used in the study. The vacuum level was maintained at 10^{-3} Torr. The furnace cooling was performed for 45 minutes. The theoretical density of sample 2 is 7.93 g/cm^3. The sintered density (7.33g/cm^3) of the samples was measured using the Archimedes method.

2.4. Chemical composition – FD-0408-45

For this research, an equivalent material grade of the P/M technique is selected and the chemical composition and mechanical properties of the material grade are shown in Tables 32.3 and 32.4. The grade of material is selected to withstand high axial and radial load similar to that of the SUS 416 grade to prevent deformation and to take high residual stresses.

Table 32.3: Chemical composition of FD-0408-45

S.No	Elements	FD-0408-45 (wt. %)
1	Iron	Rest
2	Carbon	0.7
3	Nickel	8.0
4	Copper	1.5
5	Molybdenum	0.5
6	Chromium	0.2

Table 32.4: Mechanical properties of FD-0408-45

S.No	Properties	FD-0408-45
1	Density	7.23 g/cc
2	Yield stress	380 N/mm^2
3	Tensile Strength	610 N/mm^2
4	Hardness	178 HV

2.5. Concept Design

The concept design or the 3D modelling and the 2D drafting of the proposed assembly is carried out using the below software's like Creo Parametric – Version 7.0 for the 3D Modelling Auto CAD LT-2017 for the 2D drafting. The samples 1&2 materials are considered to withstand higher operating temperatures and high mechanical strength, so that a real motor can be conceptualized and produced.

2.6. Armature without Rotor Collar

The Armature model of a typical DC motor is shown in Figure 32.2 where there is no collar placed in between the Commutator and the Lamination stack.

2.7. Armature with Rotor Collar

The Armature model of a typical DC motor is shown in Figure 32.3 where the rotor collar is placed in between the Commutator and the Lamination stack.

2.8. Structural Analysis

The structural analysis for this proposed design of the introduction of the collar in the

DC motor armature is carried out using the software - ANSYS Workbench-2021-R1. The armature 3D model is fed into the ANSYS workbench to perform the resonance Analysis. Suitable materials are assigned to the respective parts. The constraints are defined as per the product requirement. The Bearing supports were defined as per the functional requirement. The mass of the copper wire and other irrelevant masses are taken as reference weights. The analysis input is shown in Figure 32.4.

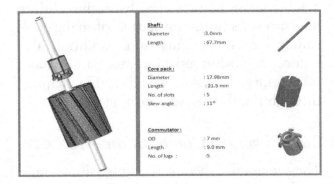

Figure 32.2: Armature without collar.

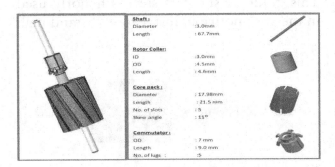

Figure 32.3: Armature with collar.

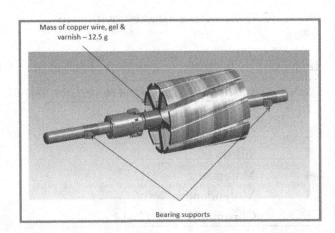

Figure 32.4: Analysis input for FEA simulation.

2.9. Manufacturing Process

As per the idea of the research the manufacturing is planned to process with two different manufacturing methods to study the differences in the manufacturing process and the effect of cost involved in doing.

Process 1 – Conventional CNC Machining & Grinding (Sample 1)

Process 2 – Powder Metallurgy Technique (Sample 2)

The manufacturing processes of both methods are well known and this research is not going to talk about the steps involved in the process but will define the process flow that is involved in both the techniques and is described in the below flow chart which is self-explanatory.

2.10. CNC Machining Process Flow for Collar

The process flow for the Armature collar using the CNC Turing machine is described below in Figure 32.5.

2.11. Powder Metallurgy Process Flow for Collar

The process flow for the Armature collar using the powder metallurgy techniques is described as below in Figure 32.6.

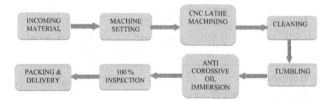

Figure 32.5: Rotor CNC turning process flow.

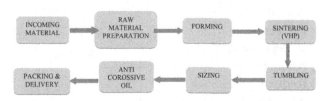

Figure 32.6: Rotor process flow with PM technique.

2.12. Collar Manufacturing

The collar manufactured through Casting and powder metallurgy techniques followed by the CNC-Lathe turning process to obtain the required shape is shown in Figure 32.7. The surface of the collar looks similar in both the process. Anti-corrosive oil is sprayed on the surface of the collar in order to prevent from oxidation and act as a lubricant during shaft pressing. Both the manufacturing process has the tumbling operation to remove unwanted burrs generated during the Turning and sintering. The PM technique involves the sintering process where the parts are fed into a furnace and heated at a temperature around 1120°C and it also involves the sizing operation which is required to meet the dimensional accuracy of the parts similar to the conventional turning operation.

2.13. Collar Assembly

The collar assembly into the motor armature is carried out with both the proposed manufacturing process. The initial push in load and the pull out load are achieved as per the design intent as shown in Figure 32.8.

2.14. Microstructural and Mechanical Properties Analysis

The microstructure of samples was prepared using mechanical polishing with the help of

Figure 32.7: Collar manufacturing a1) collar made by casting-CNC lathe (sample 1), a2) collar made by P/M (sample 2).

a double disc polishing machine (Chennai metco) and then etched with Villella's reagent for 10 seconds. The microstructures of the samples were characterized by optical micros-copy and scanning electron microscopy with X-ray energy dispersive spectroscopy (SEM-EDS). The hardness of the samples was meas-ured by Vickers hardness tester under a load of 30kgf for 10 seconds. The measured hard-ness values were taken at an average of eight readings for each condition.

3. Results and Discussion

The research is carried out with two different materials for the proposed collar as described below.

3.1. Finite Element Analysis (FEA) using Ansys Workbench

The analysis is carried out with Ansys work-bench version 2021-R1. In order to find the resonance the modal analysis is carried out with the structural physics environment. The model geometries were imported from CREO and materials were assigned as per the design. The cartesian coordinate system is consid-ered. Distributed mass structure is assigned for non-dimensional parts. Connections are defined for each part and all are bonded con-nections based on the assembly requirement. Meshing is done for the complete assembly.

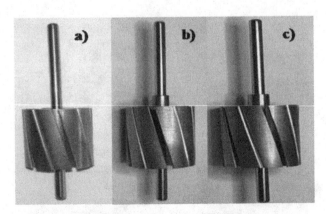

Figure 32.8: Rotor collar assembly a) assembly without collar, b) assembly with conventional collar, c) assembly with PM collar.

The tetrahedron mesh type is followed with patch confirming algorithm. The element size is 1.0mm. The number of nodes generated is 81523 and the number of elements gener-ated during the meshing is 45502. During the modal analysis, the mode shape requirements are restricted to 6 mode shapes. For the anal-ysis, cylindrical support is provided for both the bearing ends. The radial and tangential supports are fixed and the axial support is left free to simulate the real time conditions. From the modal analysis, the natural frequency of the armature assembly is 1446 Hz.

The simulation was run for two predefined conditions for the DC motor armature to cal-culate the deflection values as per below.

- 50 g Acceleration at maximum Frequency
- 50 g Acceleration at resonance Frequency

3.2. FEA Results of the Comparative Study of Armature with and without Rotor Collar

The resonance frequency analysis and the deflection of the armature shaft are carried out for both the proposed designs and the results are shown in Table 32.5.

From the analysis, it is evident that intro-ducing the Collar in the neck will increase or shift the system's natural frequency from

Table 32.5: Analysis results of the comparative study of armature with and without rotor collar

Items	Armature without rotor collar	Armature with rotor collar
Resonance frequency in Hz	1446	1629
Deflection at maximum frequency in microns	48	12
Deflection at maximum resonant frequency in microns	13	6.5

1446 Hz to 1629 Hz. The deflection of the Armature shaft with 50 g input at maximum frequency is reduced from 48 microns to 12 microns as shown in Figures 32.9 and 32.10. The deflection of the Armature shaft with 50 g input at the resonant frequency is reduced from 13 microns to 6.5 microns as shown in Figures 32.11 and 32.12.

3.3. *Evaluation of Microstructural Analysis*

Sample 1 collar made by casting route and machined through a CNC Lathe and its optical microstructure is shown in Figure 32.13. It was observed that sample 1 consisted of ferrite phase and Manganese – silicon inclusions (Mn-Si). Also, these phases and constituents were confirmed by SEM-EDS analysis as shown in Figure 32.14. The sample 2 collar was fabricated through powder metallurgy route and its optical microstructure is shown in Figure 32.15.

The optical microstructure of sample 2, which consists of a rich austenite phase marked by the white arrow, and a ferrite phase indicated by the red arrow. These phases and compositions were confirmed

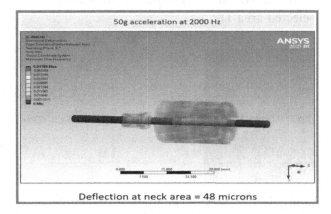

Figure 32.9: Deflection at maximum frequency armature without rotor collar.

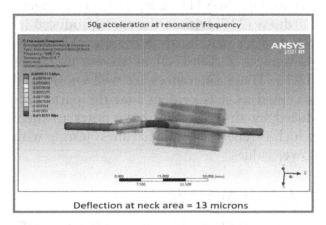

Figure 32.10: Deflection at resonance frequency armature without rotor collar.

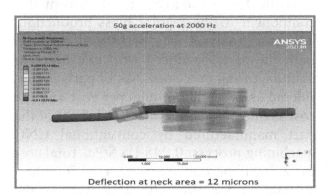

Figure 32.11: Deflection at maximum frequency with rotor collar.

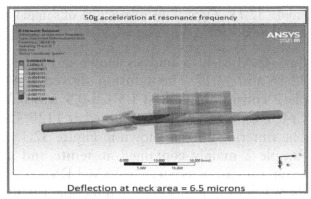

Figure 32.12: Deflection at resonance frequency with rotor collar.

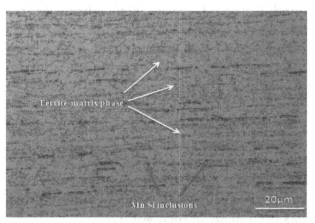

Figure 32.13: Optical microstructure of sample 1 at 500X.

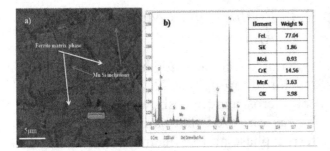

Figure 32.14: SEM-EDS analysis of sample 1 (selected area 1).

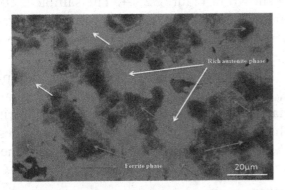

Figure 32.15: Optical microstructure of sample 2 at 500X.

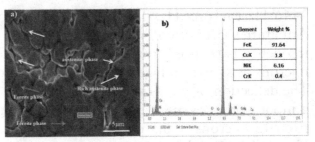

Figure 32.16: SEM-EDS analysis of sample 2 (selected area 1).

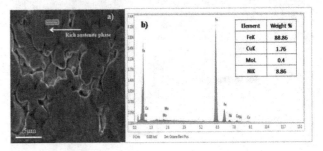

Figure 32.17: SEM-EDS analysis of sample 2 (selected area 2).

through SEM-EDS analysis shown in Figures 32.16 and 32.17. The diffusion alloyed sample 2 consists of Fe-8Ni-1.5 Cu-0.5 Mo-0.2 Cr-0.7 C (%wt.) which is sintered at 1120°C. The sintering cycle is shown in Figure 32.18. Sample 2 mainly contained austenite and a lesser amount of ferrite. The SEM-EDS analysis indicates that nickel (Ni) and copper (Cu) were uniformly distributed. This uniform distribution was obtained because copper could coat the iron powder uniformly and good wetting between them when sintered at 1120°C. Because of the high nickel, low carbon, and chromium content, there is no formation of bainite and martensite structure was observed. Moreover, with the addition of chromium (Cr) and molybdenum (Mo), the distribution of Ni was improved, also carbon was dissolved into nickel due to the bainite and martensite phase disappeared. But lower ferrite phase were observed in the study. Wu et al. [10–12] similar kind of observations has been made in their study.

The hardness testing results emphasize that the parts produced by the conventional rod machining (SUS416 sample 1) have higher hardness (217Hv)than the parts produced by the powder metallurgy technique (FD-0408-45 sample 2 (206Hv)) as shown in Figure 32.19. These differences are well known and accepted as it is evident that the parts produced by sintering will have pores to them which is reflected in the hardness testing with lower values.

3.4. Cost Analysis

As the design, analysis & component manufacturing is completed the next step is to perform the cost analysis to implement this component to the regular mass production. The cost analysis was performed between both the parts produced with two different manufacturing processes. No. of parts is considered as 2.7 Million units per year and the cost saving per year is calculated. Cost of the part manufactured by conventional CNC machining process: 0.048 US $/Pc. Total cost for 2.7 Million units/year: 0.048*2700000 = 129600 US $ = 10627200 INR. Cost of the part manufactured by Powder Metallurgy process: 0.025 US $/Pc. Total cost for 2.7

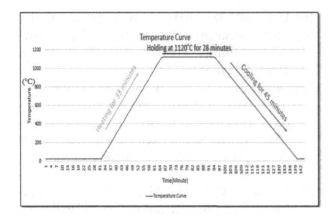

Figure 32.18: Sample 2 sintering cycle.

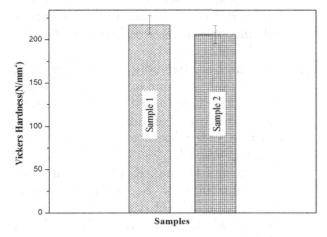

Figure 32.19: Vickers hardness of samples.

Million units/year: 0.048*2700000 = 67500 US $ = 5535000 INR. Total Savings per year = 10627200-5535000 = 5092200 INR = ~51 Lakhs INR.

4. Conclusions

The minimizing of the shaft deflection is carried out with the rotor collar and without the rotor collar through casting machined sample 1(JIS G 4303, SUS 416) and powder metallurgy sample 2(MPIF-35-FD-0408-45). The following findings were observed.

The effect of Vibration on small DC motors is studied and understanding of Natural Frequency and Resonant Frequency on DC motor Armatures was performed. Structural analysis (FEA) of the armature with Rotor collar and without rotor collar is completed using Ansys. The deflection at maximum frequency and Resonance frequency are compared between the armature with a rotor collar and without a rotor collar. From the Analysis, the Armature with a rotor collar gives better results than the armature without a rotor collar. From the analysis, it is evident that introducing the Collar in the neck will increase or shift the system's natural frequency from 1446 Hz to 1629 Hz.

- The Deflection of the Armature shaft with 50 g input at maximum frequency is reduced from 48 Microns to 12 Microns.
- The Deflection of the Armature shaft with 50 g input at resonant frequency is reduced from 13 Microns to 6.5 Microns.
- The proposed collar material is sample 2; it is cost cost-effective and produced through the Powder metallurgy route than the conventional CNC turning sample1.
- Based on the optical microscopy analysis, it was observed that sample 1 has a ferrite phase and Manganese – silicon inclusions (Mn-Si), which is also confirmed through SEM-EDS analysis. The analysis shows the parts are produced as per the designed intended material and confirms the standard material composition.
- Optical microscopy analysis observed that sample 2 consists of a rich austenite phase, and a lower ferrite phase, which is also confirmed through SEM-EDS analysis. The analysis shows the parts are produced as per the designed intended material and confirms the standard material composition. Moreover, with the addition of chromium (Cr) and molybdenum (Mo), the distributions of Ni were improved, also carbon was dissolved into nickel due to the bainite and martensite phase disappeared.
- Vickers hardness test is performed on the fabricated parts. The hardness testing results emphasize that the parts produced by the conventional rod machining (SUS 416 material-sample 1) have a little higher hardness than the parts produced by the powder metallurgy technique (FD-0408-45) sample2. These differences are well known and accepted as it is evident

that the parts produced by sintering have pores to them which is reflected in the hardness testing with little lower values.

- The sample 2 cost analysis is performed and the total savings per year is calculated as 6200 US $, Equivalent to 51 Lakhs INR approximately than the sample 1.
- Based on the research, it was confirmed that the parts produced through the powder metallurgy technique are most economical without compromising on the technical & functional requirements of the product.

References

[1] Hsiao, C.-Y., Yeh, S.-N., and Hwang, J.-C. (2011). A novel cogging torque simulation method for permanent-magnet synchronous machines. *Energies*, 4(12), 2166–2179.

[2] Espíndola-López, E., Gómez-Espinosa, A., Carrillo-Serrano, R. V., and Jáuregui-Correa, J. C. (2016). Fourier series learning control for torque ripple minimization in permanent magnet synchronous motors. *Appl. Sci.*, 6(9), 254.

[3] Cho, Y. T. (2018). Characterizing sources of small DC motor noise and vibration. *Micromachines*, 9(2), 84

[4] Lee, H. J., Chung, S. U., and Hwang, S. M. (2008). Noise source identification of a BLDC motor. *J. Mech. Sci. Technol.*, 22, 708–713.

[5] Bandar, A. M. (2015) "Powder metallurgy of stainless steel: State of the art, challenges and development." *Stainless Steel, A. Pramanik and AK Basak, Eds., Jubail, Saudi Arabia: Nova Science Publishers*, 37–80..

[6] Uzunsoy, D. (2010). Investigation of dry sliding wear properties of boron doped powder metallurgy 316L stainless steel. *Mater. Des.*, 31(8), 3896–3900.

[7] Salahinejad, E., Amini, R., and Hadianfard, M. (2010). Contribution of nitrogen concentration to compressive elastic modulus of 18Cr–12Mn–xN austenitic stainless steels developed by powder metallurgy. *Mater. Des.*, 31(4), 2241–2244.

[8] Hedberg, Y., Norell, M., Hedberg, J., Szakálos, P., Linhardt, P., and Odnevall Wallinder, I. (2013). Surface characterisation of fine inert gas and water atomised stainless steel 316L powders: formation of thermodynamically unstable surface oxide phases. *Powder Metall*, 56(2), 158–163.

[9] Ganesan, D., Sellamuthu, P., and Prashanth, K. G. (2020). Vacuum hot pressing of oxide dispersion strengthened ferritic stainless steels: effect of Al addition on the microstructure and properties. *J. Manuf. Mater. Process.*, 4(3), 93.

[10] Wu, M. and Hwang, K. (2006). Improved homogenization of Ni in sintered steels through the use of Cr-containing prealloyed powders. *Metall. Mater. Trans. A*, 37, 3577–3585.

[11] Wu, M., Hwang, K., and Huang, H. (2007). In-situ observations on the fracture mechanism of diffusion-alloyed Ni-containing powder metal steels and a proposed method for tensile strength improvement. *Metall. Mater. Trans. A*, 38, 1598–1607.

[12] Wu, M., Hwang, K., Huang, H., and Narasimhan, K. (2006). Improvements in microstructure homogenization and mechanical properties of diffusion-alloyed steel compact by the addition of Cr-containing powders. *Metall. Mater. Trans. A*, 37, 2559–2568.

33 Development of banana fiber-reinforced composite materials for small windmill blade

M. Venkatasudhahar[a], P. B. Senthilkumar, G. Dharmalingam, and G. S. Sibikannan

Department of Mechanical Engineering, Vel Tech Rangarajan Dr. Sagunthala R&D Institute of Science and Technology, Avadi, Chennai, Tamilnadu, India

Abstract

In our rapidly evolving world, growing concerns about pollution and the conservation of nonrenewable resources have urged scholars to pursue the creation of new substances and products based on sustainable principles. Constructing a windmill to power a home has become increasingly feasible; however, the expense associated with building traditional wind turbines remains significant. This experimental endeavor focuses on fabricating windmill blades using a natural fiber composite material comprised of banana fibers, epoxy resin, and hardener. The blades are crafted via an open die mold utilizing a hand lay-up method. Careful consideration is given to optimizing the blade's shape, size, and orientation. A horizontal axis windmill proves highly efficient even at low wind speeds, making it a viable option for energy generation. Moreover, this windmill design emphasizes lightweight construction to minimize costs, resulting in a final power output ranging from 5V to 12V suitable for domestic usage.

Keywords: Horizontal axis windmill blade, renewable energy, banana fiber, composite materials

1. Introduction

In order to accomplish anything in day-to-day life, energy is absolutely necessary and essential. Conventional fossil fuels, including oil, coal, and gas, are widely employed in the energy conversion process. The big problems behind non renewable energy sources unstable and global warming are harmful to the environment. Renewable energy sources are the best way to solve this problem. Solar, winds, tidal, and biogas power, among others, are widely available and environmentally friendly options for satisfying demand for energy. Wind power is the purest form of renewable energy found in electricity generation [1,2]. Air is a natural resource that can affect the environment. Wind power is becoming increasingly popular as a conventional energy source in nations like India due to its relative affordability in comparison with other renewable energy sources [3]. Energy from the wind is used to generate electricity by transferring the amount of kinetic energy in the air into a mechanical form. The function of a horizontal-axis wind turbine, also known as a HAWT, is to transform potential mechanical energy into usable kinetic energy [4].

Experimentation and analysis of high-power pumping systems that are not connected to the grid are being conducted with the purpose of improving the economy and lowering the demand placed on the grid. Additionally, research has proposed a wind-electric freshwater pumping system that is capable of operating independently from the electric

[a]venkatasudhahar@gmail.com

DOI: 10.1201/9781003545941-33

utility and is intended for use in rural areas to provide domestic and livestock water [5]. An analysis of the wind energy potential in coastal areas is provided, along with a proposal for an economically viable wind-powered storage system to store water and generate power via water turbines during peak loads [6]. However, this system is not very reliable because it requires a lot of extra space for water storage. The improvement of the airfoil structure has increased wind turbine blade reliability and decreased noise levels generated when the blades are operating [7].

Numerical modelling and optimization methods were required because researchers employed superior manufacturing techniques and more appropriate composite materials to create the windmill blade. The two areas of review are the vibrational response of the blades under aerodynamic loading and blade deflection and performance under aerodynamic loading [8,9].

With the help of composite materials, we make doors, windows, furniture, buildings, and bridges. Wind power convert kinetic energy into a simple energy method, similar to using wind turbines to generate electricity. It was the first source of natural energy used by humans. There are indications windmills found in Babylon and China to pump water, grind grain, and transport ships to the beginning of the industrial era. The economic efficiency of using windmills to generate electricity is highly dependent on the cost of installation required. It seems that these capital costs are still too high for this connection [10,11].

Banana fibre-reinforced composites are extensively used in technical industries like aviation, vehicles, space bikes, and shipbuilding because of their unique structure in comparison to more commonplace materials [12,13]. They have a high strength-to-weight ratio, a high corrosion resistance, a low thermal conductivity, a high fatigue resistance, a high chemical resistance, and a low density. In this study developed banana fiber-reinforced composite materials for small windmill blade.

2. Experiment Techniques and Materials

In the present study, banana fiber with a modulus of 27–32 GPa along with a density of 1.75 g/cm^3 and strength ranging from 392–677 MPa is employed, in combination with epoxy resin, to fabricate a composite laminate. A 900 GSM banana fiber shown in Figure 33.1. It is utilized for the composite laminate production. The epoxy resin (LY556) is combined with Teta hardener (HY951) sourced from Chennai, Tamil Nadu, as depicted in Figure 33.2. Teta hardener is chosen to enhance surface adhesion and reinforce the compound. Resin and hardener were mixed in the ratio 10:1 for obtaining a comprehensive matrix composition. The characteristics of both epoxy resin and banana fiber are outlined in Table 33.1 [12].

Table 33.1: Epoxy resin and banana fiber properties

S no.	Properties	Banana fiber	Epoxy resin
1	Density (g/cm^3)	1.75–2.00	1.2
2	Tensile strength (MPa)	392–677	50–60
3	Modulus of Elasticity (GPa)	27–32	4.4–4.6

Figure 33.1: Banana fiber mat.

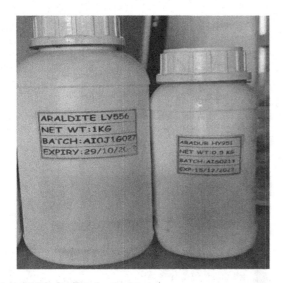

Figure 33.2: Resin & Hardener.

2.1. Details of Wind Turbine Blade and Fabrication

With excellent ejection pattern technology, blades are precisely made. In order to maximize wind power utilization, the generator might be started at a low speed. Windmill blade fabrication using natural fibers with ejection pattern technology involves a process that utilizes the hand layup method, as illustrated in Figures 33.3–33.5. The ejection pattern technology includes applying a wax and boron nitride spray on both sides of the mold cover in order to remove the part after completion of the curing process. To produce durable and sustainable blades by using the natural fibers are processed and formed into mats or sheets. After curing (24 to 48 hours) at room condition, the part removed from the mold, suited for polymer-based thermosetting compounds. Once the blade has cured, it undergoes finishing processes such as trimming and sanding to achieve the desired surface finish and aerodynamic profile. Finally, the completed wind turbine blades are assembled onto the turbine hub, ready for installation and use in wind farms.

No. of Blades: 3

Power Efficiency (η): $2\Omega T / \rho A V^3$

Power Efficiency (η): Power Generated by Wind Power / Windmill

Power generated by Windmill: ΩT

Wind Power (W): $\rho A V^3 / 2$

Where, Ω: Angular Velocity, T: Torque, ρ: Density of air, A: Area of Blade, V: Speed of the Wind, Angle of Blades: $360\bullet / 3$ (No. of Blades) = 120°

Small windmills have a different aerodynamic behavior than the high-performance wind turbines. The low performance of small windmills is due to laminar separation and, more frequently, on rotor blades due to the low number of Reynolds (Re) caused by lower wind speed and rotor power. Minimum numbers of Reynolds aerodynamic blades allow them to start at low wind speeds, increase initial torque, and thus improve the performance of all turbines. The construction of the windmill rotor will involve the rotation of the rotor and its components to achieve maximum power and efficiency. Voice rotation

Figure 33.3: Marking & Cutting of Banana Fibre.

Figure 33.4: Mould Preparation.

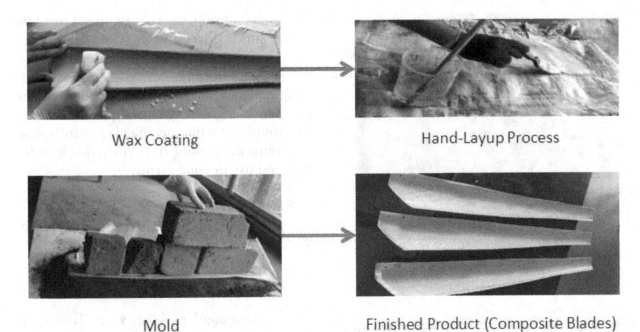

Figure 33.5: Process to create composite blades.

Figure 33.6: Structural design of project.

generate electricity. The horizontal-axis windmill structure as shown in Figure 33.6.

2.2. *Power generation at Different Wind Speeds*

Test procedure as follows. Measure the wind mill speed using tachometer, record the voltmeter reading for the corresponding blade speed, measure the voltage of the stored battery with a multi meter, and take the above reading every 5 minutes at regular intervals.

Formula for calculating blade velocity as following equation (2)

$$V= \pi \times 3.50/1000 \times 954/60 = 0.174 \text{ m/s} \qquad (1)$$

Types of experiments are conducted shape of the blade, size of the blade and number of

and chord length distribution are improved based on conserving angular pressure and aerodynamic force theory in the air sheet.

This windmill design is lightweight and has fewer mechanical connections, which reduces losses. Apart from this, we also

Table 33.2: Observed wind velocity and blade velocity

Sl.no	Time in min	Speed in RPM	Blade velocity in m/s	Minimal atmosphere velocity of air in m/s	Max. Atmosphere velocity of air in m/s
1	0	954	0.174	1.3	6.9
2	5	1200	0.219	1.3	6.8
3	10	1349	0.247	1.3	6.8
4	15	1099	0.201	1.2	6.7
5	20	1166	0.213	1.1	6.5

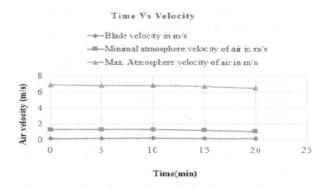

Figure 33.7: Time versus air velocity.

Table 33.3: Battery charge reading

Sl.no	Stages	Storage In volts	Battery capacity in volts	Battery charging in hours
1	1	0.20	12	20
2	2	0.28	12	14.28
3	3	0.36	12	11.11

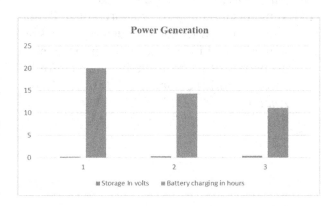

Figure 33.8: Power Generation.

blades. The shape experiments aim to optimal shape of a blade. Experiment carried out for different shapes of the blade at different wind speeds for optimal power generation. The size experiments aim to optimal size of a blade. Experiment carried out for different sizes of the blade at different wind speeds for optimal power generation. Number of blade experiments aim to optimal orientation of a blade on the hub. Experiment carried out for different orientation of the blade included angle of 180 degree (two blades) and 120 degree (three blades) at different wind speeds for optimal power generation.

The fully charged battery energy can be calculated by using a simplistic equation (2) shown below that is,

$$E = 7\ Ah \times 12\ V = 0.084\ kWh = 84\ WH \qquad (2)$$

The battery has a capacity of approximately 7Ah (Amp-hr). Hence the battery can produce an electrical current equal to 7 amps per hour with a battery voltage of 12 V.

3. Conclusions

The wind mill blade fabricated using natural fiber composite material (Banana and Epoxy resin with Hardener). All the wind mill blades are manufactured in open die mold with hand lay-up method. Optimized wind mill blade design such as shape, size and orientation. The various test conducted on the domestic wind mill shows a very positive outcome of the power. This project has been a great success and the equipment is fine working condition. Thus the objective of wind mill is achieved by some performance test like, speed versus velocity and speed versus voltage consumption. We found out the domestic wind mill is fulfill the home requirement such as light and fan the wind power is stored in a 12Vdc supply battery and is inverted in to 230V ac supply.

Declarations

In regards to the publication of this article, the authors state that there is no conflict of interest.

References

[1] Joseph, S., Sreekala, M. S., Oommen, Z., Koshy, P., and Thomas, S. (2002). A comparison of the mechanical properties of phenol formaldehyde composites reinforced with banana fibres and glass fibres. *Compos. Sci. Technol.*, 62 (14), 1857–1868.

[2] Scheurich, F. and Brown, R. E. (2013). Modelling the aerodynamics of vertical-axis wind turbines in unsteady wind conditions. *Wind Energy*, 16 (1), 91–107.

[3] Danao, L. A., Edwards, J., Eboibi, O., and Howell, R. (2014). A numerical investigation into the influence of unsteady wind on the performance and aerodynamics of a vertical axis wind turbine. *Appl. Energy*, 116, 111–124.

[4] Bausas, M. D. and Danao, L. A. M. (2015). The aerodynamics of a camber-bladed vertical axis wind turbine in unsteady wind. *Energy*, 93, 1155–1164.

[5] Danao, L.A. (2012). "The Influence of Unsteady Wind on the Performance and Aerodynamics of Vertical Axis Wind Turbines," University of Sheffield.

[6] Wekesa, D. W., Wang, C., Wei, Y., and Danao, L. A. M. (2014). Influence of operating conditions on unsteady wind performance of vertical axis wind turbines operating within a fluctuating free-stream: A numerical study. *J. Wind Eng. Ind. Aerodyn.*, 135, 76–89.

[7] Danao, L. A., Eboibi, O., and Howell, R. (2013). An experimental investigation into the influence of unsteady wind on the performance of a vertical axis wind turbine, *Appl. Energy*, 107, 403–411.

[8] Wekesa, D. W., Wang, C., Wei, Y., Kamau, J. N., and Danao, L. A. M. (2015). A numerical analysis of unsteady inflow wind for site specific vertical axis wind turbine: A case study for Marsabit and Garissa in Kenya, *Renewable Energy*, 76, 648–661.

[9] Rogowski, K. (2018). Numerical studies on two turbulence models and a laminar model for aerodynamics of a vertical-axis wind turbine. *J. Mech. Sci. Technol.*, 32, 2079–2088.

[10] Qin, N., Howell, R., Durrani, N., Hamada, K., and Smith, T. (2011). Unsteady flow simulation and dynamic stall behaviour of vertical axis wind turbine blades. *Wind Eng.*, 35(4), 511–527.

[11] Hariprasad, T., Dharmalingam, G., and Raj, P. P. (2013). Study of mechanical properties of banana-coir hybrid composite using experimental and fem techniques. *J. Mech. Eng. Sci.*, 4, 518–531.

[12] Venkatasudhahar, M., Ravichandran, A., and Dilipraja, N. (2022). Effect of stacking sequence on mechanical and moisture absorption properties of abaca-kenaf-carbon fiber reinforced hybrid composites. *J. Nat. Fiber*, 19(13), 7229–7240.

[13] Venkatasudhahar, M., Kishorekumar, P., and Dilip Raja, N. (2020). Influence of stacking sequence and fiber treatment on mechanical properties of carbon-jute-banana reinforced epoxy hybrid composites. *Int. J. Polym. Anal. Charact.*, 25(4), 238–251.

34 Corrosion resistant coating techniques to prevent silver corrosion in solar panels: a review

J. Hemanth Kumar

Department of Manufacturing Engineering, Vellore Institute of Technology, Vellore, India

Abstract

The demand for solar energy has been growing by leaps and bounds. Silver is an integral part of the solar panel as it has a high electrical conductivity and efficiently transfers the charge from one solar panel to another. Thereby maximizing the energy output of the solar cell. Silver is susceptible to corrosion when being exposed to an environment with chlorine gas, sulphide gas, oxides, moisture and other harmful pollutants. When the silver layer corrodes, the conductivity and efficiency of charge transfer in the solar drastically reduces. To prevent the silver layer corrosion, several new coating techniques have been developed. This review article encapsulates some of the latest work done in the domain of protective coatings to prevent the silver layer from being corroded. The paper also highlights the ways by which the various protective coating methods or techniques can be leveraged to prevent the corrosion of the silver layer in the solar panels.

Keywords: Solar panels, silver corrosion, corrosion resistant coatings, silver

1. Introduction

Silver is a noble metal. It has the lowest emissivity, highest reflectivity and highest conductivity [1,2]. Silver is used as paste to provide a conductive layer on the front and back of the silicon solar cells. Silver has a temperature coefficient of 0.0038/°C; enables the solar panel to generate the same amount of energy consistently irrespective of the ambient temperature [3]. When the sunlight is incident on the Photovoltaic (PV) cell, the energy is absorbed by the conductors and the electrons become mobile. The high conductivity of silver (6.30×10^7S/m) efficiently stores and transports the free electrons; thereby enhancing the energy output of the solar cells [4]. According to a study by the University of Kent, each solar panel approximately comprises of 20g of silver [5]. This implies that 8% of the world's total silver supply is only used for the solar panels [6].

Silver is susceptible to corrosion when being exposed to an environment with sulphides(H_2S), oxides and halogens (Cl_2 gas) forming silver sulphides (Ag_2S/AgS), silver oxides (AgO) and silver halides(AgCl) respectively [7–9] (Figure 34.1). When the humidity is higher silver tends to corrode faster and react with H_2S to form Ag_2S or AgS. Oxidising agents(ozone(O_3) and nitrogen dioxide (NO_2)) and Ultraviolet (UV) radiations would accelerate the rate of silver corrosion [7,8, 10–12]. The other atmospheric pollutants are equally responsible for the corrosion of silver. The formation of these silver sulphides, halides and oxides would hinder the performance of the solar panel. The forthcoming section highlights the various methods used by the researchers to prevent the corrosion of the silver layer. The

hemanthkumar.j2022@vitstudent.ac.in

DOI: 10.1201/9781003545941-34

SILVER CHLORIDE CLUSTERS

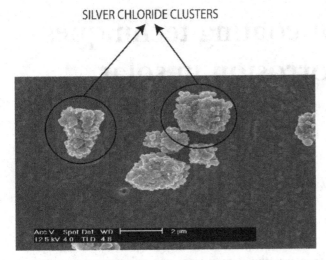

Figure 34.1: The SEM image of the Silver Chloride Clusters formed on the silver surface of the solar panel [13].

feasibility of leveraging these protective coating techniques to prevent the corrosion of the silver layer in the solar panels.

2. Protective Coating Techniques to Prevent Silver Corrosion in Solar Panels

Metal oxides, nitrides, fluorides and other inorganic compounds have been used as protective coatings to prevent the corrosion of the silver layer [14–17]. There are four main methods used to coat these inorganic compounds on the surface of the silver layer. They are Magnetron Sputtering (MS), Plasma Beam Sputtering (PBS), Chemical Vapor Deposition (CVD) and Atomic Layer Deposition (ALD) (Figure 34.2). In this section, the above-mentioned four-processes will be explained in brief and some of the application of these coating techniques to prevent silver corrosion will be highlighted. This section will also analyze whether each of the above-mentioned methods would be suitable to prevent the silver corrosion in the solar panels.

2.1. Magnetron Sputtering (MS)

Xu *et al.* [18] used Radio Frequency (RF) Magnetron Sputtering (MS) [19] to build

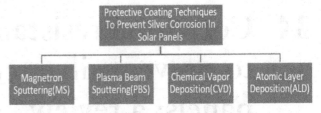

Figure 34.2: Protective Coating Techniques to Prevent Silver Corrosion in Solar Panels.

SiO$_2$ films on the silver layer's surface in the solar reflector. The 320 nm thick SiO$_2$ layer shielded the silver layer in the reflector from being damaged by oxygen/sulphur dioxide, the researchers note that the solar reflectivity (SR) was higher than the light reflectivity (LR) of a solar reflector with that layer [20]. When the SiO$_2$ layer thickness was increased beyond 320nm, it was discovered that the reflectivity was declining because it would be challenging for light to permeate the reflector's surface. The silver coating is vulnerable to corrosion if the protective layer thickness is less than 320nm. After a 456-hour aging test, the LR and SR of the SiO$_2$-coated solar reflector only decreased by 0.95% and 0.55%, respectively. Without significantly reducing the reflector's performance, the SiO$_2$ layer shields the silver layer not only from corrosion but also from abrasion. The MS-coated SiO$_2$ layer is also quite durable. Only small-scale or customised solar panel production units can use the MS. This cannot be scaled up for mass manufacturing since it is expensive and challenging to maintain a big vacuum chamber for producing a low pressure. With defects like columnar structures and discontinuities, PVD-type processes have an intrinsic residual porosity of the order of 1%, which increases the risk of the silver layer corroding (Figure 34.3) [22]. Therefore, it cannot be used to prevent the silver corrosion in solar panels.

2.2. Plasma Beam Sputtering (PBS)

PBS technique is best suited to produce corrosion and abrasion resistant coating to protect the silver layer from being corroded [23,24]. For example, Folgner *et al.* [25] coated SiNx

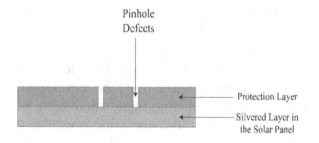

Figure 34.3: The Presence of Pin-Hole defects in the protective layer [21].

on the surface of silver with NiNx interlayer as an adhesive using the PBS technique. The strong adhesion between the silver layers and SiNx prevented the layer being affected by corrosive agents and abrasion. Maintaining a chamber with pressure as high as 1 to 100 mtorr is quite expensive and challenging. The rate of deposition of protective coating using PBS is very slow [26]. Therefore, it might not be suitable for creating corrosion resistant coatings for the silver layer in solar panels for a large scale despite its advantages.

2.3. Chemical Vapor Deposition (CVD)

CVD-based protective coatings have protected the silver layer from being corroded when being exposed to an acidic environment [27,28]. For example, Saad *et al.* [29] used PECVD to deposit SiO_x on the layer of silver in the presence of hexamethyldisiloxane (HMDSO) (or again TEOS) gas with oxygen gas. The protective layer prevented the corrosion of the silver layer when being exposed to H_2SO_4. Due to differences in the gas-to-particle conversion process' vapor pressure, nucleation, and growth rates, multi-component deposition utilising CVD is challenging. In the CVD process, chemical and operational safety are major concerns. The whole process takes a lot of time and energy. Depending on the price of the precursors, it may become expensive [30]. As a result of its poor visible-to-infrared selectivity, CVD is typically not utilised to coat protective coatings on the surface of silver or solar panels because doing so

would also impair visible transmission [31]. Despite the advantages of CVD, it cannot be used to create protective coatings on the silver layer in the solar panels as it might impair the working of the solar panel.

2.4 Atomic Layer Deposition

ALD can be used to produce excellent coatings on silver without any pinhole defects [32,33]. Figure 34.4 illustrates the steps involved in ALD. For example, Pusa *et al.* [34] coated the silver layer with alumina/titania monolayers using ALD. The silver layer with alumina/titania monolayer did nor tarnish even after being exposed to H_2S for 48h. The flawless coating of the alumina/titania monolayers with only 0.00028% of porosity is responsible for the silver's excellent resistance to H_2S. The ALD method can be utilised to create protective layers with a consistent thickness and no flaws. Researchers have been successful in protecting the silver-coated reflectors by using coatings produced using the ALD method [35] ALD uses less energy than the CVD and PVD techniques while still having greater layer deposition control.

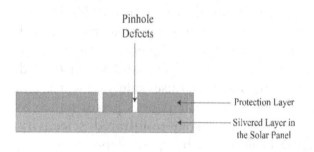

Figure 34.4: A Schematic Diagram Representing the Steps Involved in The ALD Process (a) The substrate surface is functionalised naturally or has been modified to do so. (b) The surface reaction is caused by pulsed precursor 1. (c) Extra precursor and reaction byproducts are expelled with an inert carrier gas. (d) Surface reaction caused by pulsed precursor 2. (e) An inert carrier gas is used to remove any surplus precursor and reaction byproducts. (f) Repeat steps (a) to (e) to obtain the specified material thickness [15].

The silver layer in solar panels can be safe-guarded against tarnishing or corrosion by employing coatings created using the ALD technique. The ALD process has to be more scalable, and new precursors and techniques must also be created. The effects of various operational factors, such as operating power, precursor concentration, and others must be investigated. Researchers must investigate the kinetics, mechanisms, and corrosion resistance behaviour of the ALD-based coatings. Using the ALD method, newer film materials other than oxides must be formed over the silver layer, and their reaction with corrosive reagents must be investigated[36].

3. Conclusion

This section of the review article discusses the potential of various coating techniques, including magnetron sputtering, plasma beam sputtering, chemical vapor deposition, and atomic layer deposition, to protect the silver layer in solar panels. Magnetron sputtering (MS) is known for producing corrosion and abrasion-resistant coatings for silver, but its intrinsic porosity (around 1%) leads to the formation of discontinuities like pinholes and columnar struts, making it unsuitable for preventing silver corrosion in solar panels. Both plasma beam sputtering and chemical vapor deposition are expensive and sophisticated. CVD, in particular, raises safety concerns and can reduce the solar panel's performance by decreasing its visible-to-infrared selectivity. CVD is slow and has poor deposition accuracy, making it unsuitable for protecting the silver layer.

On the other hand, atomic layer deposition (ALD) uses less energy than CVD and PVD techniques while providing greater layer deposition control. Coatings created with ALD can safeguard the silver layer against tarnishing or corrosion. ALD can deposit conformal and uniform films on any substrate, and free acid, silver, copper and lead concentrations on silver electrorefining electrolyte conductivityincluding superlattices, nanolaminates, and heterostructures, where conventional methods would fail. The porosity of ALD coatings is as low as 0.00028%.

To make ALD more scalable, new precursors and techniques must be developed. Researchers should investigate the effects of operational factors like operating power and precursor concentration on ALD coatings. The kinetics, mechanisms, and corrosion resistance behaviour of ALD-based coatings must also be studied. Additionally, researchers should explore the formation of new film materials over the silver layer using ALD and their reactions with corrosive reagents.

Acknowledgements

The author acknowledges the faculty members of the Department of Manufacturing in the School of Mechanical Engineering (SMEC) at Vellore Institute of Technology, Vellore for nurturing my interest in the domain of corrosion resistant coatings.

References

[1] Sheikh, D. (2016). Improved Silver Mirror Coating for Ground and Space-Based Astronomy. https://doi.org/10.1117/12.2234380.

[2] Boccas, M, Vucina, T., Araya, C., Vera, E., and Ahhee, C. (2006). Protected-silver coatings for the 8-m gemini telescope mirrors. *Thin Solid Films* 502(1), 275–80. https://doi.org/https://doi.org/10.1016/j.tsf.2005.07.295.

[3] Arif, A. T., Kalliomäki, T.,. Wilson, B. P., Aromaa, J., and Lundström, M. (2016). "Modelling the effect of temperature and free acid, silver, copper and lead concentrations on silver electrorefining electrolyte conductivity." *Hydrometallurgy*. https://doi.org/10.1016/j.hydromet.2016.09.006.

[4] Sharipov, M., Lee, Y., Han, J, and Lee, Y. I. (2021). Patterning microporous paper with highly conductive silver nanoparticles: via pvp-modified silver-organic complex ink for development of electric valves. *Mater. Adv.* https://doi.org/10.1039/d0ma00960a.

[5] Feger, F., Pavanini, N., and Radulescu, D. (2022). Welfare and redistribution in

residential electricity markets with solar power. *Rev. Econ.*, https://doi.org/10.1093/restud/rdac005.

[6] Kumar, A., and Akbar Hussain, D. M. (2018). A review paper on solar energy in India. *GJECS.*, https://doi.org/10.21058/gjecs.2018.31001.

[7] Graedel, T. E. (1992). Corrosion mechanisms for silver exposed to the atmosphere. *J. Electrochem. Soc.*, 139(7), 1963. https://doi.org/10.1149/1.2221162.

[8] Rice, D. W., Peterson, P., Rigby, E. B., Phipps, E. B., Cappell, R. J., and Tremoureux, R. (1981). Atmospheric Corrosion of Copper and Silver. *J. Electrochem. Soc*, 128 (2), 275. https://doi.org/10.1149/1.2127403.

[9] Lin, H., and Frankel, G. S. (2013). Accelerated atmospheric corrosion testing of Ag. *Corrosion.* https://doi.org/10.5006/0926.

[10] Pellicori, S. (1980). scattering defects in silver mirror coatings. *Appl. Opt.*, 19 (September), 3096–98. https://doi.org/10.1364/AO.19.003096.

[11] Wiesinger, R, Martina, I., Kleber, Ch., and Schreiner, M. (2013). Influence of relative humidity and ozone on atmospheric silver corrosion. *Corros. Sci.*, 77, 69–76. https://doi.org/https://doi.org/10.1016/j.corsci.2013.07.028.

[12] Liang, D., Allen, H. C., Frankel, G. S., Chen, Z. Y., Kelly, R. G., Wu, Y., and Wyslouzil, B. E. (2010). Effects of sodium chloride particles, ozone, UV, and relative humidity on atmospheric corrosion of silver. *J. Electrochem. Soc.*, 157(4), C146. https://doi.org/10.1149/1.3310812.

[13] Lin, H., Frankel, G. S., and Abbott, W. H. (2013). 'Analysis of Ag corrosion products'. *J. Electrochem. S.*, 160(8), C345. https://doi.org/10.1149/2.055308jes.

[14] Phillips, A., Miller, J., Bolte, M., Dupraw, B., Radovan, M., and Cowley, D. (2012). Progress in UCO's search for silver-based telescope mirror coatings. *Proc. SPIE - Inter. Soc. Optical Eng.*, 8450 (September). https://doi.org/10.1117/12.925502.

[15] Schneider, T., and Stupik, P. (2018). A UV-Enhanced Protected Silver Coating for the Gemini Telescopes. https://doi.org/10.1117/12.2313661.

[16] Phillips, A., Fryauf, D., Kobayashi, N., Bolte, M., Dupraw, B., Ratliff, C., Pfister, T.,

and Cowley, D. (2014). *Prog. New Tech. Protected-Silver Coatings.*, 9151. https://doi.org/10.1117/12.2055706.

[17] Phillips, A. C., Miller, J. S., Bolte, M. J., Dupraw, B., Radovan, M., and Cowley, D. J. (2012). Progress in UCO's search for silver-based telescope mirror coatings. In *Other. Conf.*, https://api.semanticscholar.org/CorpusID:120746829.

[18] Xu, Y. J, Cai, Q. W., Yang, X. X., Zuo, Y. Z., Song, H., Liu, Z. M., and Hang, Y. P. (2012). Preparation of novel SiO2 protected Ag thin films with high reflectivity by magnetron sputtering for solar front reflectors. *Sol. Energy Mater. Sol. Cells.*, 107, 316–21. https://doi.org/https://doi.org/10.1016/j.solmat.2012.07.002.

[19] Meyer, O. (1982).Sputtering by particle bombardment, Vol. I: physical sputtering of single-elements solids'. *J. Nuc. Mater.*, https://doi.org/10.1016/0022-3115(82)90441-x.

[20] Wang, Y H, Her, S C., Hsiao, C.N., and Chen, H. (2008). Optical and mechanical properties of a metallic layer. *J. Sci.Eng. Technol.*, 4 (January), 81–88.

[21] Schwinde, S., Schürmann, M., Jobst, P. J., Kaiser, N. and Tünnermann, A. (2015). Description of particle induced damage on protected silver coatings. *Appl. Opt.* 54 (16), 4966–71. https://doi.org/10.1364/AO.54.004966.

[22] Vacandio, F., Massiani, Y., Gergaud, P., and Thomas, O. (2000). Stress, porosity measurements and corrosion behaviour of AlN films deposited on steel substrates. *Thin Solid Films.*, https://doi.org/10.1016/S0040-6090(99)00763-4.

[23] Barrie, J. D., Fuqua, P. D.,. Folgner, K A., and Chu, C T. (2011). Control of stress in protected silver mirrors prepared by plasma beam sputtering. *Appl Opt.*, https://doi.org/10.1364/AO.50.00C135.

[24] Drake, R P. (2018). Properties of high-energy-density plasmas. In https://doi.org/10.1007/978-3-319-67711-8_3.

[25] Folgner, K. A., Chu, C. T., Lingley, Z. R., Kim, H. I., Yang, J. M., and. Barrie, J. D. (2017). Environmental durability of protected silver mirrors prepared by plasma beam sputtering. *Appl. Opt.*, https://doi.org/10.1364/ao.56.000c75.

[26] Gudmundsson, J. T., Anders, A. and Von Keudell, A. (2022). Foundations of physical vapor deposition with plasma assistance. *PSST.*, https://doi.org/10.1088/1361-6595/ac7f53.

[27] Carlsson, J. O., and. Martin, P.M. (2009). Chemical vapor deposition. In *Handbook of Deposition Technologies for Films and Coatings: Science, Applications and Technology.* https://doi.org/10.1016/B978-0-8155-2031-3.00007-7.

[28] Dobkin, D. M. and Zuraw, M. K. (2003). Principles of chemical vapor deposition. principles of chemical vapor deposition. https://doi.org/10.1007/978-94-017-0369-7.

[29] Tarazi, S. A., Volpe, L., Antonelli, L., Jafer, R., Batani, D., d'Esposito, A., and Vitobello, M. (2014). Deposition of SiOx layer by plasma-enhanced chemical vapor deposition for the protection of silver (Ag) surfaces. *Radiat. Eff.Defects in Solids.* https://doi.org/10.1080/10420150.2013.860972.

[30] Piszczek, P., and Radtke, A. (2018). Silver Nanoparticles fabricated using chemical vapor deposition and atomic layer deposition techniques: properties, applications and perspectives: review. In *Noble and Precious Metals - Properties, Nanoscale Effects and Applications.* https://doi.org/10.5772/intechopen.71571.

[31] Ebisawa, J., and Ando, E. (1998). Solar control coating on glass. *Curr. Opi. Solid State Mater. Sci.*, https://doi.org/10.1016/S1359-0286(98)80049-1.

[32] Johnson, R. W., Hultqvist, A., and. Bent, S. F. (2014). A brief review of atomic layer deposition: from fundamentals to applications. *Mater. Today.*, https://doi.org/10.1016/j.mattod.2014.04.026.

[33] George, S. M. (2010). Atomic Layer Deposition: An Overview. *Chem. Rev.*. https://doi.org/10.1021/cr900056b.

[34] Paussa, L., Guzman, L., Marin, E, Isomaki, N., and Fedrizzi, L. (2011). Protection of silver surfaces against tarnishing by means of alumina/titania-nanolayers. *Surf. Coat. Tech.*, https://doi.org/10.1016/j.surfcoat.2011.03.101.

[35] Fedel, M., Zanella, C., Rossi, S., and Deflorian, F. (2014). Corrosion protection of silver coated reflectors by atomic layer deposited Al2O3. *Sol. Energy.* https://doi.org/10.1016/j.solener.2013.11.038.

[36] Cremers, Véronique, Riikka L. Puurunen, and Jolien Dendooven.(2019). 'Conformality in Atomic Layer Deposition: Current Status Overview of Analysis and Modelling'. *Applied Physics Reviews.* https://doi.org/10.1063/1.5060967.

35 Review on role of solar energy, exploring into photovoltaic technology: various fabrication methods and characteristics of solar cells

Koteswara Rao Jammula, Abdul Azeez, Akshay Kumar,
G. M. Madhu, Jitin Bangera, and Ritesh H. Shetty[a]

Department of Chemical Engineering, MS Ramaiah Institute of Technology, Bangalore, India

Abstract

In the face of escalating energy demands and environmental concerns, the world stands at a pivotal crossroads. Fossil fuel reserves are depleting, while their ecological consequences loom large. Solar energy emerges as alternate sustainable energy promising an eco-conscious future. This comprehensive review embarks on an exploration of the pivotal role of solar energy, delving into photovoltaic (PV) technology and its transformative potential. The monocrystalline and polycrystalline cells, each with its merits, vie for prominence, while thin-film innovations such as CIGS, CdTe, and CZTS offer promising alternatives. Third-generation solar cells, including DSSCs, Quantum Dot, and Perovskite cells, push the boundaries of efficiency and sustainability. The review provides holistic information on types of photovoltaic cells, fabrication methodology, efficiency. The concise report on the work carried out by the various researchers are consolidated.

Keywords: Solar cells; photovoltaic; thin-film; fabrication methods

1. Introduction

In an age marked by high energy demands and ever-increasing environmental concerns, the global landscape stands at an unprecedented crossroads. The dependance on fossil fuels, the foundation of modern energy consumption, is approaching a crucial juncture [1]. Depletion begins over the finite reservoirs of oil, coal, and gas that have fueled progress for generations, while the ecological impacts of fossil fuel combustion has resulted in finding an alternate resource [2]. Amid this urgency renewable energy sources are an alternate, promising an energy-efficient and eco-conscious future [3,4].

Renewables hold the potential to navigate resource shortage and ecological unrest, with solar energy taking center stage in this transformative mission [5–7]. Solar energy, harnessed offers a boundless and permanent resource that could transform the global energy landscape [8,9]. This review aimed at comprehensive assessment of the importance of solar energy and photovoltaic (PV) technology, navigate the global energy landscape, and reveal the streams within the solar PV sector [10,11].

Solar cells emerge as the great sustainable alternative amidst escalating energy needs and environmental concerns [12]. The photovoltaic energy conversion devices hold the key to resolving the energy challenge [13]. As fossil fuel reliance faces a critical juncture, solar cells offer a sustainable path, harnessing the solar energy [14,15].

Solar cells convert light into energy, generating electron hole pairs through sunlight

[a]rithushetty071@gmail.com

DOI: 10.1201/9781003545941-35

absorption [16,17]. The p-n junction plays a vital role, facilitating charge separation that yields renewable electricity [11,18]. The journey of solar cells began in 1954 with the pioneering silicon solar cell, marking a path for scientific advancement and transformation and has the potential to redefine the global energy panorama [19].

Solar energy conversion through photovoltaic cells has seen remarkable progress [19]. Monocrystalline and polycrystalline solar cells are prominent candidates, with monocrystalline cells having high efficiency due to their single-crystal structure [20]. Polycrystalline cells offer cost advantages, but has less efficiency because of less space for flow of electrons compared to mono crystalline cell [20–22]. Thin-film technologies, such as Copper Indium Gallium Di selenide (CIGS) and Cadmium Telluride (CdTe), demonstrate promise, with Copper Zinc Tin Sulfide (CZTS) providing an alternative that utilises abundant elements [23,24]. Dye-Sensitised Solar Cells (DSSCs), Quantum Dot (QD) dye-sensitised solar cells, and Perovskite solar cells play a part in reshaping the renewable energy scene [25,26].

Nanotechnology's ability to manipulate materials at the nanoscale has unlocked new possibilities for solar cells. Nanostructured materials offer unique advantages, including flexible bandgaps, improved optical properties, and reduced charge carrier recombination, which are dependent on crystal sise and shape. These unique properties enable innovative approaches to light trapping and efficient collection, made nanomaterials highly adaptable in solar cell technology [27].

Different solar cell generations, covering traditional first-generation (silicon-based cells), subsequent second-generation (thin-film cells), and emerging third-generation technologies, experience advantages through the integration of nanomaterials. This integration extends to various solar cell types, including Dye-Sensitised Solar Cells (DSSCs), Hybrid Solar Cells (HSCs), Perovskite Solar Cells (PSCs), and Organic Solar Cells (OSCs),

fostering diversity in materials and applications [28]. Utilising nanomaterials in solar cell fabrication offers several advantages, such as increased surface area, compatibility with liquids, deep cell penetration, improved strength and ductility. Characterisation techniques are vital for understanding nanomaterial properties accurately [27,28].

The Government of India has implemented a range of initiatives, including the National Solar Mission (NSM) and a range of initiatives like the Solar Park Scheme, Viability Gap Funding (VGF), and the Central Public Sector Undertaking (CPSU) scheme. to promote solar power generation. These efforts reflect India's commitment to harnessing renewable energy sources and addressing climate change. Additionally, measures like utilising canal banks and rooftops for solar installations, encouraging foreign investment, and waiving transmission charges for solar power showcase India's dedication to fostering sustainability and reducing reliance on fossil fuels. Collectively, these initiatives aim to achieve renewable energy targets, bolster energy security and contribute to worldwide endeavors to alleviate climate change [29].

The present review deals with solar cell technology, from their history to fabrication methods, offering insights into their implications for a sustainable future. A brief study on nanotechnology in revolutionising solar cell technology, enhancing efficiency and cost-effectiveness across different generations of solar cells are explored.

2. Fabrication of Different Solar Cells

2.1. *First Generation Solar Cells*

2.1.1. Monocrystalline Solar Cells

Monocrystalline solar cells are created from single-crystal silicon wafers, which are highly regarded for their remarkable efficiency and purity [20]. The manufacturing process involves several steps, including epitaxial growth, thermal oxidation, texturing, emitter formation, and deposition of anti-reflection coatings [21]. These cells can be transferred to different substrates for cost-effective thin-film applications using specialised techniques [30]. The high cost of producing monocrystalline solar cells is attributed to the utilisation of premium materials and processes. This cost factor restricts their applicability in low-cost thin-film photovoltaics [31]. Enhancing the efficiency of monocrystalline solar cells involves integrating silicon nanowires or nanorods into their design. This augmentation amplifies surface area and light-trapping capabilities, facilitating improved electron movement and thereby enhancing overall performance. Furthermore, utilising sophisticated deposition methods like Plasma-Enhanced Chemical Vapor Deposition (PECVD) to achieve passivation and anti-reflection coatings provides enhanced control over material properties, ultimately resulting in improved efficiency [21].

2.1.2. Polycrystalline Solar Cells

Polycrystalline solar cells employ silicon wafers composed of multiple crystal grains, reducing material and manufacturing costs [22]. To enhance efficiency, the silicon undergoes treatment with phosphorus and aluminum, optimising its electrical properties for sunlight absorption and conversion. This process reduces defects and boosts electrical conductivity. Additionally, a double layer antireflective coating, featuring silicon nanoparticles, minimises light reflection, ensuring greater light absorption and electricity conversion. These measures synergise to optimise the overall efficiency of polycrystalline solar cells [31,32].

2.2. Second Generation/Thin film Solar Cells

2.2.1. Copper Indium Gallium Di selenide (CIGS) Thin-Film Solar Cells

The fabrication process of CIGS thin-film solar cells includes several crucial stages, starting with the preparation of a substrate using soda lime glass and a molybdenum layer [23]. The CIGS absorber layers are carefully grown through the deposition of precursor materials and precise temperature control [24,33]. Buffer and window layers are added to improve electron transport, and electrodes are deposited for efficient current collection. An anti-reflection coating is applied to maximise light absorption [34]. This comprehensive process, along with the specific ZnO/ Buffer nZnS/p-CIGS/Mo cell architecture and the utilisation of spin coating, Improving the combined efficiency of CIGS thin-film solar cells empowers them to efficiently convert sunlight into electrical energy.

2.2.2. CdTe Thin Film Solar Cells

The increased efficiency of CdTe solar cells is achieved through the use of innovative fabrication methods and materials, including Cadmium Stannate (CTO) transparent conductive oxide films, a ZnSnOx (ZTO) buffer layer, oxygenated nanocrystalline CdS: O window layers, and a modified device structure comprising CTO/ZTO/CdS/CdTe layers. [24,35]. Concurrently, novel manufacturing processes enhance $CdCl_2$ treatment and layer adhesion, collectively bolstering cell performance and stability [35,36]. These advancements operate through mechanisms that optimise charge carrier transport, improve layer interface quality, enhance light absorption and charge separation, and guarantee manufacturing integrity.

2.2.3. Copper Zinc Tin Sulfide (CZTS)

Thin Film Solar Cells Enhancing the efficiency of CZTS) thin-film solar cells involve a comprehensive technical strategy. This encompasses precise composition control, optimising Cu, Zn, Sn, S, and Se ratios for bandgap engineering and better charge carrier mobility. Achieving high-quality films through sputtering, CVD, and ink coating reduces defects and grain boundaries, enhancing charge transport and overall efficiency [24,37]. Incorporating Se in the composition aligns the bandgap with the solar spectrum, improving light absorption. Surface passivation and the integration of a back surface field (BSF) layer reduce recombination losses and enhance electrical properties. Ink formulation and spin coating ensure uniform, high-quality films, crucial for cost-efficiency and scalability. Controlled doping and alloying enhance electrical characteristics and boost charge separation and transport. Light-trapping structures and efficient contact layers increase light absorption and charge carrier extraction, respectively [38–40].

2.3.   *Third Generation Solar Cells*

2.3.1. Dye Sensitised Solar Cells (DSSC)

The Dye-Sensitised Solar Cell (DSSC) operates through a concise sequence, converting photons to electricity. Initially, incident light energises electrons in a photosensitiser dye, elevating them to higher energy levels. These energised electrons are promptly transferred into the conduction band of a nanoporous titanium dioxide (TiO2) electrode, inducing oxidation of the dye. Traveling through the TiO2 nanoparticles, these electrons generate an electric current as they progress towards the cell's rear contact. Simultaneously, at the counter electrode, electrons from the external circuit reduce triiodide ions (I3-) to iodide ions (I-), thus rejuvenating the dye's base state. This repetitive cycle persists, ensuring continual current production whenever sunlight is present. In essence, DSSCs offer a direct and effective method for capturing solar energy. [41]. To enhance Dye Sensitised Solar Cell (DSSC) performance, optimise the nanocrystalline titanium dioxide (TiO_2) photoanode and sensitising dyes, like Ru (II) complex HD-2 and SA-1, for efficient light absorption and electron collection [42,43]. Tailor counter electrode materials using precise deposition methods and control layer thickness

for improved electrocatalytic activity [44]. Employ advanced redox mediators and ionic electrolytes for enhanced electron transport. Use sealants like Surlyn and Bynel to protect against moisture and UV light, ensuring long-term stability. Apply anti-reflective coatings, optimised photoanode geometries, and alternative conductive materials for better photon capture and electron transport. Durability is enhanced through robust encapsulation techniques. Continuous monitoring and systematic experimentation are essential for achieving optimal DSSC efficiency and performance [43,45].

2.3.2. Quantum Dot Solar Cells

In order to boost the efficiency and overall performance of a Quantum Dot Solar Cell (QDSC), a multi-faceted approach can be employed [46]. Bandgap engineering is crucial, enabling tailored QDs to efficiently absorb photons across the solar spectrum [46]. Precise layer thickness control during QD active layer deposition using techniques like spin or dip coating ensures optimised light absorption [46]. Further improvements in charge transport efficiency can be achieved through ligand exchange treatments. Efficient charge extraction, minimising recombination losses, is facilitated by depositing back contacts (e.g., MoOx, Al, or Au) onto the QD layer. To reduce wasteful spin coating, alternative deposition techniques like doctor blading or inkjet printing should be explored. Enhancing electron and hole mobility, encapsulating and protecting the QDSC, and implementing tandem or multi-function configurations with complementary absorption materials are additional strategies to maximise overall performance [47]. Advanced electrode materials, improved light-trapping techniques, and environmental protection measures can further bolster the efficiency and competitiveness of QDSCs in the realm of renewable energy [47,48].

2.3.3. Perovskite Solar Cells (PSCs)

The Perovskite Solar Cell (PSC) functions through meticulously layered configurations to maximise its effectiveness. When exposed to sunlight, electrons in the perovskite layer move from the valence band to the conduction band, resulting in the perovskite being oxidised. To preserve charge balance, an adjacent hole transporting layer (HTL) accepts an electron from the perovskite, while the excited electron is injected into the electron transporting layer (ETL) and progresses towards the front contact. The movement of electrons and holes leads to the generation of an electric current. Precise alignment of energy levels across layers is imperative for optimal functionality. Ideally, the band gap of the perovskite should absorb visible light, typically ranging from 1.1 to 1.4 electron volts (eV), with its conduction band edge exceeding that of the electron transporter. Similarly, the valence band edge should be slightly lower than that of the hole conductors. Interfaces between layers facilitate the separation of electron-hole charges, with electrons and holes guided through selective conductor layers to maximise charge collection. This refined process ensures efficient harnessing of sunlight and elevates the overall performance of the device [49]. Enhancing the efficiency and performance of Perovskite Solar Cells (PSCs) is a complex procedure. Essential measures include refining solution-based film deposition techniques like spin-coating, blade coating, and slot-die coating, which require careful adjustment of parameters to ensure uniform film quality [50]. Crystallisation methods, including solvent engineering and vapor annealing, enhance perovskite film quality. Electron Transport Layers (ETL) like TiO_2, SnO_2, and ZnO require precise thickness control, while Hole Transport Layers (HTL) range from spiroOMeTAD to inorganic options like CuI and CuSCN [50–52]. Back electrodes formed with an inorganic buffer layer protect against ion bombardment. The deposition method choice (e.g.,

spray coating or inkjet printing) impacts uniformity. By optimising these steps, PSCs can achieve enhanced efficiency and performance [51].

3. Nanotechnology in Solar Cells

Nanotechnology is revolutionising solar cell technology, enhancing efficiency and cost-effectiveness across generations of solar cells. Solar cells play a vital role in harnessing renewable solar energy to address global energy challenges. Nanotechnology's ability to manipulate materials at the nanoscale has unlocked new possibilities for solar cells. Nanostructured materials offer unique advantages, including flexible bandgaps, improved optical properties, and Nanostructured materials offer versatile possibilities in solar cell technology due to their ability to minimise charge carrier recombination, a phenomenon influenced by their size and shape. This characteristic opens up innovative avenues for enhancing light trapping and optimising photo collection, contributing to increased efficiency in solar cells [51].

Nanomaterial integration proves advantageous for different solar cell generations, including initial silicon-based cells, subsequent thin-film cells, and emerging third-generation technologies. An array of solar cell varieties, including Dye Sensitised Solar Cells (DSSCs), Quantum Dot-Sensitised Solar Cells (QDSSCs), Hybrid Solar Cells, Perovskite Solar Cells (PSCs), and Organic Solar Cells, present a broad spectrum of materials and potential applications [52]. Various types of solar cells, there properties and fabrication methods are listed in Table 35.1.

Utilising nanomaterials in solar cell fabrication offers several advantages, such as increased surface area, compatibility with liquids, deep cell penetration, and improved strength and ductility. Challenges include safety concerns, potential job displacement, and security risks. Table 35.2 presents the performance of key solar cells from the first, second, and third generations.

Table 35.1: Solar cell types, properties and fabrication methods

Cell Type	Fabrication Method	Size	Cost	Fabrication Complexity	Material Purity	Scalability	Efficiency	Durability	Life Span	Band Gap (eV)	References
First Generation Solar cells — Monocrystalline	Czochralski method	Produce the same amount of power with significantly less volume.	Twice as costly as thin-film.	Complex	High	High	14%-17.5%	Long	Around 40 years	1.1	[23,48–49]
	Float-zone method			Complex	High						
Polycrystalline	Casting & slicing			Complex	High	Moderate	12%-14%	Long	Around 35 years	1.1	[23,49]
	Ribbon growth			Complex	High						
Second gen/ Thin Film Solar Cells	CIGS	Providing a broad spectrum of product designs, including those that are flexible, lightweight, and durable.	Costs 50 percent less than traditional silicon cells.	Moderate	Moderate	High	10%-12%	Moderate	Around 25 years	1.01–1.68	[23,49–51]
CdTe				Moderate	High	9%-11%	Moderate	Around	30 years	1.45	[23,49,52,53]
CZTS				High	High	10%	Moderate	Moderate		1.49–1.51	[23,49,54]
Third gen Solar cells — DSSC	Screen-printing & assembly		Moderate	Low	Moderate	Low	10%	Moderate	6 years	3-3.2	[23,49,55]
	Nano-crystalline TiO$_2$ photo anode		Moderate	Low	High						
QDSC	Quantum dot synthesis			High	High	Low	Varies	Short-Moderate	High	1.34	[23,49]
	Nanocrystal deposition			High	High						
PSC	Solution processing			Moderate	Moderate	Low	31%	Short-Moderate	Around 30 years	1.1-1.3	[23,49]

Table 35.2: Solar cell Technologies and performance

Cell Type	Cell Architecture/Nanomaterial used	Types of Coating Method	Voltage (Voc)	Highest Recorded Efficiency	Reference
First Generation Solar cells	Monocrystalline	Silicon Nanowires, Silicon Nanorods	Plasma Enhanced Chemical Vapor Deposition	48.7	19.33%	[58]
	Polycrystalline	Silicon Nanoparticles	Double Layer Antireflective Coating	46.3	17.27%	[58]
Second gen/Thin Film Solar Cells	CIGS	ZnO/Buffer n-ZnS/p-CIGS/Mo	Spin Coating	0.804	23.54%	[59,60]
	CdTe	Glass/ZnO: Al/n-CdS/p-CdTe/ZnTe	Sputtering, CSVT, Chemical Spray Pyrolysis, Electrodeposition	1.52	21.57%	[59,61]
	CZTS	$CZTS_{0.4}Se_{0.6}$ BSF Layer	Spin Coating	0.99	28.59%	[58,62]
	DSSC	Cell Type:2-cyanoacetani lide based organic dyes Dye: Ru(II) complex HD-2 with SA-1	Thermal Chemical Vapor deposition, Spin Coating, Spray Coating	0.68	8.02%	[59,63]
Third gen Solar cells	QDSC	ITO/SnO2: RCQs/ Cs0.05FA0.81PbI2.25Br0.45/ Spiro-OMeTAD/MoO$_3$/Au	Wasteful Spin Coating	1.14	22.77%	[59,64]
	PSC	FTO/TiO$_2$/Perovskite/(Me-PDA) Pb$_2$I$_6$/Spiro-OMETAD/Au	Spray Coating, Spin Coating, Inkjet printing method, Blade coating method, Slot die Coating method	1.13	22%	[56,65–66]

4. Conclusion

Photovoltaic technology will play pivotal role harnessing solar energy in a sustainable path. Brief review from the evolution of solar cells from the of single-crystal silicon to the latest innovations in perovskite technology, along with thin-film alternatives for harnessing the solar power is explained. The emergence of third-generation solar cells, with their innovative materials and fabrication techniques, holds the promise of even greater energy efficiency and sustainability. Dye-sensitised, quantum dot, and perovskite cells open new horizons in photovoltaic technology, pushing the boundaries of what is possible. The review concludes that photovoltaic cells offer brighter, cleaner, and more sustainable future.

References

[1] Rashedi, A., Khanam, T., and Jonkman, M. (2020). On reduced consumption of fossil fuels in 2020 and its consequences in global environment and exergy demand. *Energies* 13(22), 6048. https://doi.org/ 10.3390/en13226048.

[2] Höök, M., and Tang, X. (2013). Depletion of fossil fuels and anthropogenic climate change—A review. *Energy. Policy.*, 52, 797–809. https://doi.org/10.1016/j.enpol.2012.10.046.

[3] Choubey, P. C., Oudhia, A., and Dewangan, R. (2012). A review: Solar cell current scenario and future trends. *Recent. Res. Sci. Technol.*, 4(8).

[4] Jacobson, M. Z., and Delucchi, M. A. (2009). A path to sustainable energy by 2030. *Sci. Am.*, 301(5), 58–65. https://doi.org/10.1038/scientificamerican1109-58.

[5] Dincer, I., and Rosen, M. A. (1999). Energy, environment and sustainable development. *Appl. Energy.*, 64(1–4), 427–440. https://doi.org/10.1016/S0306-2619(99)00111-7.

[6] Wall, A. (2013, August). Advantages and disadvantages of solar energy. *Process. Ind. Forum.*, 7, 395–408).

[7] Kopp, G., and Lean, J. L. (2011). A new, lower value of total solar irradiance: Evidence and climate significance. *Geophys. Res. Lett.*, 38(1). https://doi.org/10.1029/2010GL045777.

[8] Foukal, P., Fröhlich, C., Spruit, H., and Wigley, T. M. L. (2006). Variations in solar luminosity and their effect on the Earth's climate. *Nature.*, 443(7108), 161–166. https://doi.org/10.1038/nature05072

[9] Mekhilef, S., Saidur, R., and Safari, A. (2011). A review on solar energy use in industries. *Renew. Sustain. Energy. Rev.*, 15(4), 1777–1790. https://doi.org/10.1016/j.rser.2010.12.018

[10] Tyagi, V. V., Rahim, N. A., Rahim, N. A., Jeyraj, A., and Selvaraj, L. (2013). Progress in solar PV technology: Research and achievement. *Renew. Sustain. Energy. Rev.*, 20,443–461. https://doi.org/10.1016/j.rser.2012.09.028

[11] Green, M., Dunlop, E., Hohl-Ebinger, J., Yoshita, M., Kopidakis, N., and Hao, X. (2021). Solar cell efficiency tables (version 57). Progress in photovoltaics: *Res. Appl.*, 29(1), 3–15. https://doi.org/10.1002/pip.3228

[12] IEA, U. (2020). Global energy review 2020. Ukraine. [Online] https://www. iea. org/countries/ukraine [Accessed: 2020-09-10], 810.

[13] "RENEWABLES 2021 GLOBAL STATUS REPORT." https://www.ren21.net/gsr2021

[14] Arias, P. A., Bellouin, N., Coppola, E., Jones, R. G., Krinner, G., Marotzke, J., ... and Zickfeld, K. (2023). Intergovernmental Panel on Climate Change (IPCC). Technical summary. *Climate Change 2021: The Physical Science Basis. Contribution of Working Group I to the Sixth Assessment Report of the Intergovernmental Panel on Climate Change*, 35–144. Cambridge University Press. https://doi.org/10.1017/9781009157896.002

[15] Lewis, N. S., and Nocera, D. G. (2006). Powering the planet: Chemical challenges in solar energy utilization. *Proc. Natl. Acad. Sci.*, 103(43), 15729–15735. https://doi.org/10.1073/pnas.0603395103

[16] Bisquert, J. (2017). *The physics of solar cells: perovskites, organics, and photovoltaic fundamentals.* CRC press. https://doi.org/10.1201/b22380.

[17] Shockley, W., and Queisser, H. (2018). Detailed balance limit of efficiency of p–n

junction solar cells. In *Renewable Energy*, 2_35—l2_54. Routledge. https://doi.org/10.1063/1.1736034

[18] "Solar Photovoltaic Technology Basics." https://www.nrel.gov/research/rephotovoltaics.html (accessed Aug. 18, 2023).

[19] Chapin, D. M., Fuller, C. S., and Pearson, G. L. (1954). A new silicon p-n junction photocell for converting solar radiation into electrical power. *J. Appl. Phys.*, 25(5), 676. https://doi.org/10.1063/1.1721711

[20] Han, K. M., Lee, H. D., Cho, J. S., Park, S. H., Yun, J. H., Yoon, K. H., and Yoo, J. S. (2012). Fabrication and characterization of monocrystalline-like silicon solar cells. *J. Korean. Phys.l Soc.*, 61, 1279–1282. https://doi.org/10.3938/jkps.61.1279

[21] Blakers, A., Zin, N., McIntosh, K. R., and Fong, K. (2013). High efficiency silicon solar cells. *Energy. Procedia.*, 33, 1–10. https://doi.org/10.1016/j.egypro.2013.05.033

[22] Sana, P., Salami, J. A. L. A. L., and Rohatgi, A. J. E. E. T. (1993). Fabrication and analysis of high-efficiency polycrystalline silicon solar cells. *IEEE Transactions on Electron. Devices.*, 40(8), 1461–1468. https://doi.org/10.1109/16.223706

[23] Sharma, S., Jain, K. K., and Sharma, A. (2015). Solar cells: in research and applications—a review. *Mater. Sci. Appl.*,6(12), 1145–1155. http://dx.doi.org/10.4236/msa.2015.612113

[24] Chopra, K. L., Paulson, P. D., and Dutta, V. (2004). Thin-film solar cells: an overview. Progress in Photovoltaics: *Res.Appl.*, 12(2–3), 69–92. https://doi.org/10.1002/pip.541

[25] Yan, J., and Saunders, B. R. (2014). Third-generation solar cells: a review and comparison of polymer: fullerene, hybrid polymer and perovskite solar cells. *Rsc.. Adv.*, 4(82), 43286–43314. https://doi.org/10.1039/C4RA07064J

[26] Rajendran, S., Naushad, M., Raju, K., and Boukherroub, R. (Eds.). (2019). *Emerging nanostructured materials for energy and environmental science,* Vol. 23, Berlin: Springer.

[27] Green, M. A., Ho-Baillie, A., and Snaith, H. J. (2014). The emergence of perovskite solar cells. *Nat. Photonics.*, 8(7), 506–514.

[28] Sethi, V. K., Pandey, M., and Shukla, M. P. (2011). Use of nanotechnology in solar PV cell. *Int. J. Chem. Eng.Appl,.* 2(2), 77.

[29] Raina, G., and Sinha, S. (2019). Outlook on the Indian scenario of solar energy strategies: Policies and challenges. *Energy. Strategy. Rev.*, 24, 331–341. https://doi.org/10.1016/j.esr.2019.04.005

[30] Chu, A. K., Wang, J. S., Tsai, Z. Y., and Lee, C. K. (2009). A simple and cost-effective approach for fabricating pyramids on crystalline silicon wafers. *Sol. Energy. Mater. Sol. Cells.*, 93(8), 1276–1280. https://doi.org/10.1016/j.solmat.2009.01.018

[31] Nijs, J. F., Szlufcik, J., Poortmans, J., Sivothhaman, S., and Mertens, R. P. (1999). Advanced manufacturing concepts for crystalline silicon solar cells. *IEEE Trans. Electron. Devices.*, 46(10), 1948–1969. https://doi.org/10.1109/16.791983

[32] Narayanan, S., Wenham, S. R., and Green, M. A. (1990). 17.8-percent efficiency polycrystalline silicon solar cells. *IEEE Trans. Electron Devices*, 37(2), 382–384. https://doi.org/10.1109/16.46370

[33] Ramanathan, K., Teeter, G., Keane, J. C., and Noufi, Y. R. (2005). Properties of high-efficiency CuInGaSe2 thin film solar cells. *Thin. Solid. Films.*, 480, 499–502. https://doi.org/10.1016/j.tsf.2004.11.050

[34] Song, H. K., Jeong, J. K., Kim, H. J., Kim, S. K., and Yoon, K. H. (2003). Fabrication of CuIn1– xGaxSe2 thin film solar cells by sputtering and selenization process. *Thin. Solid. Films.*, 435(1–2), 186–192. https://doi.org/10.1016/S0040-6090(03)00350-X

[35] Bonnet, D., and Meyers, P. (1998). Cadmium-telluride—Material for thin film solar cells. *J. Mater. Res.*, 13(10), 2740–2753.

[36] McCandless, B. E., and Dobson, K. D. (2004). Processing options for CdTe thin film solar cells. *Solar. Energy.*, 77(6), 839–856. https://doi.org/10.1016/j.solener.2004.04.012

[37] Katagiri, H. (2005). Cu2ZnSnS4 thin film solar cells. *Thin. Solid. Films.*,480, 426–432. https://doi.org/10.1016/j.tsf.2004.11.024

[38] Katagiri, H., Saitoh, K., Washio, T., Shinohara, H., Kurumadani, T., and Miyajima, S. (2001). Development of thin film solar cell based on Cu2ZnSnS4 thin films. *Solar. Energy. Mater. Solar. Cells.*, 65(1–4), 141–148. https://doi.org/10.1016/S0927-0248(00)00088-X

[39] Moriya, K., Tanaka, K., and Uchiki, H. (2007). Fabrication of Cu2ZnSnS4 thin-film solar cell prepared by pulsed laser deposition. *Jpn. J. Appl. Phys.*, 46(9R), 5780. https://doi.org/10.1143/JJAP.46.5780.

[40] Tang, D., Wang, Q., Liu, F., Zhao, L., Han, Z., Sun, K., ... and Liu, Y. (2013). An alternative route towards low-cost Cu2ZnSnS4 thin film solar cells. *Surf. Coat. Technol.*, 232, 53–59. https://doi.org/10.1016/j.surfcoat.2013.04.052

[41] Sharma, K., Sharma, V., and Sharma, S. S. (2018). Dye-sensitized solar cells: fundamentals and current status. *Nanoscale. Res. Lett.*, 13, 1–46. https://doi.org/10.1186/s11671-018-2760-6.

[42] Gong, J., Sumathy, K., Qiao, Q., and Zhou, Z. (2017). Review on dye-sensitized solar cells (DSSCs): Advanced techniques and research trends. *Renew. Sustain. Energy. Rev.*, 68, 234–246. https://doi.org/10.1016/j.rser.2016.09.097

[43] Richhariya, G., Kumar, A., Tekasakul, P., and Gupta, B. (2017). Natural dyes for dye sensitized solar cell: A review. *Renew. Sustain. Energy. Rev.*, 69, 705–718. https://doi.org/10.1016/j.rser.2016.11.198

[44] Yoo, K., Kim, J. Y., Lee, J. A., Kim, J. S., Lee, D. K., Kim, K., ... and Ko, M. J. (2015). Completely transparent conducting oxide-free and flexible dye-sensitized solar cells fabricated on plastic substrates. *ACS. Nano.*, 9(4), 3760–3771. https://doi.org/10.1021/acsnano.5b01346

[45] Sugathan, V., John, E., and Sudhakar, K. (2015). Recent improvements in dye sensitized solar cells: A review. *Renew. Sustain. Energy. Rev.*, 52, 54–64. https://doi.org/10.1016/j.rser.2015.07.076

[46] Chernomordik, B. D., Marshall, A. R., Pach, G. F., Luther, J. M., and Beard, M. C. (2017). Quantum dot solar cell fabrication protocols. *Chem. Mater.*, 29(1), 189–198. https://doi.org/10.1021/acs.chemmater.6b02939

[47] Zhang, X., Santra, P. K., Tian, L., Johansson, M. B., Rensmo, H., and Johansson, E. M. (2017). Highly efficient flexible quantum dot solar cells with improved electron extraction using MgZnO nanocrystals. *ACS. nano.*, 11(8), 8478–8487. https://doi.org/10.1021/acsnano.7b04332

[48] Cappelluti, F., Kim, D., van Eerden, M., Cédola, A. P., Aho, T., Bissels, G., ... and Guina, M. (2018). Light-trapping enhanced thin-film III-V quantum dot solar cells fabricated by epitaxial lift-off. *Sol. Energy. Mater. Sol. Cells.*, 181, 83–92. https://doi.org/10.1016/j.solmat.2017.12.014

[49] Ghosh, P., Sundaram, S., Nixon, T. P., and Krishnamurthy, S. (2022). Influence of nanostructures in perovskite solar cells. *Encycl. Smart. Mater.*, 2, 646–660. https://doi.org/10.1016/B978-0-12-815732-9.00054-1

[50] Li, Z., Klein, T. R., Kim, D. H., Yang, M., Berry, J. J., Van Hest, M. F., and Zhu, K. (2018). Scalable fabrication of perovskite solar cells. *Nat. Rev. Mater.*, 3(4), 1–20. https://doi.org/10.1038/natrevmats.2018.17.

[51] Kumar, N., Rani, J., and Kurchania, R. (2021). Advancement in CsPbBr3 inorganic perovskite solar cells: Fabrication, efficiency and stability. *Sol. Energy.*, 221, 197–205. https://doi.org/10.1016/j.solener.2021.04.042

[52] Yang, D., Yang, R., Priya, S., and Liu, S. (2019). Recent advances in flexible perovskite solar cells: fabrication and applications. *Angew. Chem. Int. Ed.*, 58(14), 4466–4483. https://doi.org/10.1002/anie.201809781.

[53] Wang, J. Y., Huang, C. S., Ou, S. L., Cho, Y. S., and Huang, J. J. (2022). One-step preparation of TiO2 anti-reflection coating and cover layer by liquid phase deposition for monocrystalline Si PERC solar cell. *Sol. Energy. Mater. Sol. Cells.*, 234, 111433. https://doi.org/10.1016/j.solmat.2021.111433

[54] H. Sargent, E. (2005). Infrared quantum dots. *Adv. Mater.*, 17(5), 515–522. https://doi.org/10.1002/adma.200401552

[55] Belghachi, A., and Limam, N. (2017). Effect of the absorber layer band-gap on CIGS solar cell. *Chin. J. Phys.*, 55(4), 1127–1134. https://doi.org/10.1016/j.cjph.2017.01.011

[56] Beaucarne, G. (2007). Silicon thin-film solar cells. *Adv. OptoElectron.*, 2007. https://doi.org/10.1155/2007/36970

[57] Ali, A. M., Rahman, K. S., Ali, L. M., Akhtaruzzaman, M., Sopian, K., Radiman, S., and Amin, N. (2017). A computational study on the energy bandgap engineering in

performance enhancement of CdTe thin film solar cells. *Results. Phys.*, 7, 1066–1072. https://doi.org/10.1016/j.rinp.2017.02.032

[58] Kim, H., Cha, K., Fthenakis, V. M., Sinha, P., and Hur, T. (2014). Life cycle assessment of cadmium telluride photovoltaic (CdTe PV) systems. *Sol. Energy.*, 103, 78–88. https://doi.org/10.1016/j.solener.2014.02.008

[59] Swati, S. I., Matin, R., Bashar, S., and Mahmood, Z. H. (2018, September). Experimental study of the optical properties of Cu2ZnSnS4 thin film absorber layer for solar cell application. *J. Phys. Conf. Ser.*, 1086, 1, 012010. IOP Publishing. https://doi.org/10.1088/1742-6596/1086/1/012010

[60] Emery, K. (2011). Measurement and characterization of solar cells and modules. *Handbook. Photovoltaic Sci. Eng.*, 797–840. https://doi.org/10.1002/9780470974704

[61] Baghel, N. S., and Chander, N. (2022). Performance comparison of mono and polycrystalline silicon solar photovoltaic modules under tropical wet and dry climatic conditions in east-central India. *Clean. Energy.* 6(1), 165–177. https://doi.org/10.1093/ce/zkac001.

[62] Sivaraj, S., Rathanasamy, R., Kaliyannan, G. V., Panchal, H., Jawad Alrubaie, A., Musa Jaber, M., ... and Memon, S. (2022). A comprehensive review on current performance, challenges and progress in thin-film solar cells. *Energies.*, 15(22), 8688. https://doi.org/10.3390/en15228688

[63] Tobbeche, S., Kalache, S., Elbar, M., Kateb, M. N., and Serdouk, M. R. (2019). Improvement of the CIGS solar cell performance: structure based on a ZnS buffer layer. *Opt.l and Quantum Electron.*, 51, 1–13. https://doi.org/10.1007/s11082-019-2000.-z.

[64] Arce-Plaza, A., Sánchez-Rodriguez, F., Courel-Piedrahita, M., Galán, O. V., Hernandez-Calderon, V., Ramirez-Velasco, S., and López, M. O. (2018). CdTe thin films: deposition techniques and applications. *Coat. Thin. Film. Technol.*, 131–148. https://doi.org/10.5772/intechopen.79578.

[65] Zhang, W., You, C., Dan, Z., Wang, W., and Dong, R. (2023, February). Improved performance of Cd-free CZTS thin-film solar cells by using CZTS0. 4Se0. 6 BSF layer. *J. Phy. Conf. Ser* 2418, 1, 012002. IOP Publishing. https://doi.org/10.1088/1742-6596/2418/1/012002

[66] Pinto, A. L., Cruz, L., Gomes, V., Cruz, H., Calogero, G., de Freitas, V., ... and Lima, J. C. (2019). Catechol versus carboxyl linkage impact on DSSC performance of synthetic pyranoflavylium salts. *Dyes. Pigments.*, 170, 107577. https://doi.org/10.1016/j.dyepig.2019.107577

[67] Sannino, G. V., De Maria, A., La Ferrara, V., Rametta, G., Mercaldo, L. V., Addonizio, M. L., ... and Delli Veneri, P. (2021). Development of SnO2 Composites as Electron Transport Layer in Unencapsulated CH3NH3PbI3 Solar Cells. *Solids.*, 2(4), 407–419. https://doi.org/10.3390/solids2040026

[68] Shah, N., Shah, A. A., Leung, P. K., Khan, S., Sun, K., Zhu, X., and Liao, Q. (2023). A Review of Third Generation Solar Cells. *Processes.*, 11(6), 1852. https://doi.org/10.3390/pr11061852

[69] Zhang, F., Lu, H., Larson, B. W., Xiao, C., Dunfield, S. P., Reid, O. G., ... and Zhu, K. (2021). Surface lattice engineering through three-dimensional lead iodide perovskitoid for high-performance perovskite solar cells. *Chem.*, 7(3), 774–785.

36 Optical anisotropies in Co/Fe doped single-walled carbon nanotubes: A theoretical approach

Seema Aggarwal[1,a] and Monica Anand[2,b]

[1]Department of Physics, Ramaiah Institute of Technology, Bengaluru, India
[2]Department of Mathematics, Ramaiah Institute of Technology, Bengaluru, India

Abstract

We have carried out investigations on the optical anisotropies present in a single-walled carbon nanotube (SWCNT) doped with Fe/Co atoms within the framework of density functional theory. The computations are performed with the SIESTA code using the pseudopotential plane wave approach with generalised gradient approximation. The real and imaginary components of dielectric function, refractive index along with reflectivity of pristine, Fe-doped and Co-doped nanotubes have been computed. It is observed that, reported optical properties are highly anisotropic in the low-energy region and the anisotropy shows a remarkable decline with an increase in the energy of electromagnetic radiation. The dielectric constant for light polarised along the tube axis is always greater than that for the perpendicular axis. Our results show a significant improvement in the optical parameters with doping in the near-infrared region (NIR) of radiation as compared to pristine nanotubes for light polarised along the tube axis. The higher values of static dielectric constant and static reflectivity stipulate the metallic behaviour of Co/Fe doped nanotubes. Thus, the transition metal (TM) doped SWCNTs can be a suitable candidate for NIR-optoelectronic devices.

Keywords: Single-walled carbon nanotubes, SIESTA, transition metal, optical properties, density functional theory, anisotropy

1. Introduction

Ever since the advent of carbon nanotubes (CNTs) by Iijima [1], a vast amount of theoretical and experimental research has been done to explore this unique one-dimensional structure. CNTs have drawn massive attention due to their exceptional electronic, mechanical, thermal and optical properties. Various technical applications of CNTs and their future challenges have been discussed by Abdalla et al. [2]. As the Si-based complementary metal–oxide semiconductor (CMOS) technology is on the brink of reaching its absolute limits, researchers have started exploring emerging approaches. One such achievement was the fabrication of

CNT-based field-effect transistors by Zhang et al. [3]. This single-walled carbon nanotubes (SWCNT)-based CMOS technology is much more straightforward and structured as compared to Si-based CMOS technology. It is well known that materials with a thermal interface are necessary for the efficient withdrawal of heat in electronic packaging applications. Fabris et al. [4] have discussed how the introduction of CNTs into thermal interface materials amplifies bulk thermal conductivity. The extraordinary properties of CNTs, such as the high Young's modulus, discussed by Yakobson et al. [5] and the high ductility studied by Gubarev et al. [6] make them suitable for mechanical applications. Additionally, CNTs

[a]smabgl11@gmail.com, [b]amonica@msrit.edu

DOI: 10.1201/9781003545941-36

are also being incorporated in bio-sensing and imaging as optical sensors as discussed by Pan et al. [7] and Hofferber et al. [8].

Further enhancement in the properties of pristine CNTs can be achieved by doping with other metals. Various elements have been used for the doping of carbon nanotubes. Since the use of transition metal-based catalysts is customary in several nanotube fabrication techniques, studies of TM particles doped SWCNTs are of special interest. These studies shed light on the interaction of TM particles with nanotubes, which is necessary to comprehend their possible applications. The hexahedron-metal bond between TM and carbon has a substantial effect on the conductivity of SWCNT films, according to Wang et al. [9] and Shim et al. [10]. Tsukagoshi et al. [11] and Mao et al. [12] examined the interaction between TM particles (Mn, Fe, Co, and Ni) and nanotubes, which could lead to semi-metallic systems of great interest for spintronics devices. Yang et al. [13] and Gulseren et al. [14] contributed significantly to the understanding of titanium-doped SWCNTs for their higher hydrogen storage capacity.

Tabtimsai et al. [15] investigated the interaction of VIIIB transition metal atoms with (5,5) single-walled carbon nanotubes. They discovered that all of these transition metal atoms have a high affinity for SWCNTs. The work of Lafdi et al. [16] further accentuated the magnetic properties of Co-doped SWCNTs. They discerned that Co-doped SWCNTs exhibited superparamagnetic behaviour with lower saturation of magnetisation. Furthermore, Fu et al. [17], Hele et al. [18] and Liu et al. [19] have analyzed the cobalt and nitrogen co-doped CNTs for their exceptional catalytic activity for the oxygen reduction reactions. Meihui et al. [20], Chanda et al. [21] and Vijayan et al. [22] have investigated the optical properties of Co and Fe- doped nanocrystals. The dielectric response of Co-doped reduced graphene oxide was scrutinised by Akhtar et al. [23]. They noticed that

the doped graphene shows a remarkable rise in dielectric permittivity with the increase in frequency. All of the preceding instances highlight the significance of studying TM-doped nanomaterials.

A SWCNT is considered as a rolled-up graphene sheet in the cylindrical form. Avouris et al. [24] have shown that graphene has atomic structural affinities with graphene sheets and is thus known to have identical properties. We, therefore, can expect Co/Fe-doped SWCNTs to show some excellent optical characteristics. To the best of our knowledge, the optical dispersions of TM-doped SWCNTs remain largely unexplored. Researching these characteristics could offer fresh perspectives on how well they function, which is necessary for optoelectronic applications. Our primary goal in this work is to examine how TM-doping affects the optical characteristics of single-walled carbon nanotubes.

A pristine (8,0) SWCNT, consisting of 32 atoms is modelled using Tubegen code. In order to analyse the impact of doping on the optical anisotropies, a single C atom is replaced by Fe /Co atom, resulting in the formation of a Fe-doped/Co-doped SWCNT. This equates to 3.125% of impurity concentration in a pristine nanotube. Further, the density functional theory is employed to determine the optical spectra of each of the simulated. It should be noted that the anisotropic behaviour of optical properties is investigated by considering the polarisation of the electromagnetic field. Here, parallel polarisation has been referred to as z-direction and its perpendicular counterpart has been referred to as x-direction.

2. Computational Method

The optical properties of zigzag (8,0) pristine, Co-doped and, Fe-doped SWCNTs are simulated within the density functional theory approach. The process is executed by applying the generalised gradient approximation (GGA) functional with the Perdew-Burke-Ernzerhof (PBE) subfunctional to solve Kohn-Sham equations. Optimisation of all geometries is conducted using the SIESTA code developed by Soler et al. [25]. To characterise the valence electrons, localised pseudo-atomic orbitals with a double-ζ polarised (DZP) basis set are employed. The computation of electron densities utilises a specific mesh cutoff (300 Ry). The Brillouin zone integration is executed using the Monkhorst-Pack scheme. The k-point sampling of the Brillouin zone is set at 1x1x25 for all the cases. The atoms are fully relaxed until the maximum force acting on the atoms is smaller than 0.04 eV/Å. The smearing broadening in computing the optical properties is kept fixed at 0.5 eV. The optical properties are computed with the polarisation of the electric field vector, along and perpendicular to the tube axis.

3. Results and Discussions

3.1. Electronic Structure

The bandgap is extremely important in understanding the optical characteristics of materials. The optimised band structures of pristine, Co-doped and, Fe-doped SWCNTs are shown in Figure 36.1 (a-c). Fermi levels of all the band structures have been set to zero.

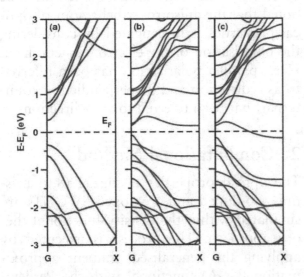

Figure 36.1: The band structures of (a) pristine (b) Fe-doped (c) Co-doped SWCNTs.

Pristine SWCNT demonstrates a direct band gap at G point as 0.648 eV which is in good agreement with the previous calculations by Spataru et al. [26] and Tetik [27]. The band gaps of Co-doped and Fe-doped SWCNTs are reported to be 0.329 eV and 0.230 eV respectively. The existence of dopant atoms creates new levels around the Fermi level, consequently indicate that the impurities form acceptor-like bands. This phenomenon manifests in the form of a metallic property in the nanotubes. The dopant atoms impact the original band structure eminently due to a strong hybridisation with the carbon atoms resulting in a significant reduction in the band gap.

3.2. Optical Properties

In this section, we have explored the optical properties of pristine and TM-doped SWC-NTs, computed for light polarised i) along and ii) perpendicular to the tube axis. The energy of radiation has been taken, in the range 0 to 30 eV to analyze the anisotropies in the optical behaviour.

To evaluate a material's linear response to electromagnetic radiation the complex dielectric function can be written by the following equation:

$$\varepsilon(\omega) = \varepsilon_1(\omega) + \varepsilon_2(\omega) \qquad (1)$$

here $\varepsilon_1(\omega)$ and $\varepsilon_2(\omega)$ denote the real and imaginary part of the dielectric function respectively, for electromagnetic radiation of frequency ω. The values of $\varepsilon_1(\omega)$ and $\varepsilon_2(\omega)$ are determined using Kramers-Kronig relations.

The real part of dielectric constant $\varepsilon_1(\omega)$ for pristine, Fe-doped and Co-doped SWCNTs is shown in Figure 36.2(a-c). The reported values of $\varepsilon_1(\omega)$ are anisotropic, especially, at low energies. It is important to note that as incident photon energy increases, anisotropy diminishes. Additionally, the values of $\varepsilon_1(\omega)$ along the z-direction for Fe-doped and Co-doped nanotubes are found to be remarkably high in the near-infrared energy region. On account of this property, doped nanotubes have substantial potential in NIR photonic applications. The static dielectric constant obtained from the spectra has been reported in Table 1.

It can be noted that the static dielectric constant along the z-direction increases sharply for doped nanotubes. As compared to Co-doped, the increase is more pronounced in Fe-doped nanotubes. A sizeable value of the static dielectric constant further confirms the metallic behaviour of Co-doped and Fe-doped nanotubes. There is no significant change in the static dielectric constant for light polarised along the x-direction.

The spectra of $\varepsilon_2(\omega)$ calculated for pristine, Co-doped and Fe-doped SWCNTs have been shown in Figure 36.3(a-c). In general, optical absorptions are represented by the imaginary component of the dielectric function.

Table 36.1: Static dielectric constant

	x-direction	*z-direction*
Pristine	1.28	2.53
Fe-doped	1.39	10.67
Co-doped	1.37	7.40

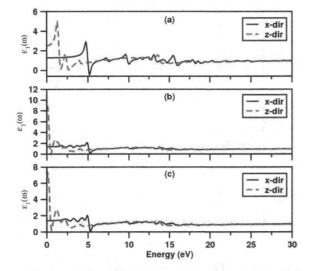

Figure 36.2: The variation of the real part of dielectric function $\varepsilon_1(\omega)$ with energy (eV) for (a) pristine (b) Fe-doped (c) Co-doped SWCNTs.

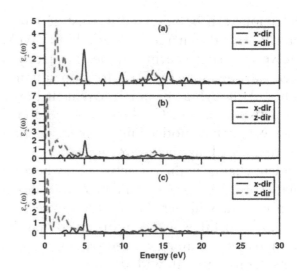

Figure 36.3: The variation of the imaginary part of dielectric function $\varepsilon_2(\omega)$ with energy (eV) for (a) pristine (b) Fe-doped (c) Co-doped SWCNTs.

The peaks corresponding to maximum values of $\varepsilon_2(\omega)$ along the x- and z-directions are reported at 4.97 eV and 1.45 eV for pristine nanotube. For Co-doped maximum absorptions are estimated to be at 5.09 eV and 0.34 eV respectively. The corresponding values for Fe-doped are at 5.07 eV and 0.28 eV. It is observed that the majority of the peaks in both directions are in the energy range of 0 eV to 6 eV. We can, thus, conclude that this energy range is of paramount importance for optical transitions. Furthermore, it is perceived that the positions of peaks corresponding to the maximum value of $\varepsilon_2(\omega)$ along the z-direction for Co-doped and Fe-doped SWCNTs are red-shifted in comparison to that of pristine SWCNT. This can be attributed to the observed band gap of these nanotubes.

Optical refractive index n(ω) is given by

$$n(\omega) = \sqrt{\varepsilon_1(\omega)} \qquad (2)$$

The n) is one of the most important properties to be considered for optical designing and application. The variation of n(ω) for pristine, Co-doped and Fe-doped SWCNTs has been shown in Figure 36.4(a-c). It can be discerned that n(ω) is anisotropic in the low-energy region for all of the three cases. Isotropic behaviour can be observed in pristine nanotubes above the energetic threshold of 20 eV. However, a sharp decline in the anisotropy is seen starting at energy 7 eV for doped nanotubes. Although a sharp improvement in the refractive indices of Co-doped and Fe-doped SWCNTs can be noticed in the energy region up to 1 eV along the z-direction, there is no evident change in the refractive index beyond that. The static refractive index in the z-direction is always higher than in the x-direction as shown in Table 36.2. It is observed that the refractive index along the z-direction records considerable growth with doping.

The optical response of the surface of a material is characterised by its reflectivity, which is another key consideration in the designing of optical devices. The reflectivity $R(\omega)$ is given by

$$R(\omega) = \left(\frac{1 - \sqrt{\varepsilon(\omega)}}{1 + \sqrt{\varepsilon(\omega)}} \right)^2 \quad (3)$$

The $R(\omega)$ of pristine, Co-doped and Fe-doped SWCNTs has been displayed in Figure 36.5(a-c). It can be seen that $R(\omega)$ diminishes significantly after 6 eV. Beyond 20 eV, all the nanotubes are completely transparent which makes them suitable for transparent coating applications. Along the z-direction, a red shift

Table 36.2: Static refractive index

	x-direction	z-direction
Pristine	1.13	1.59
Fe-doped	1.18	3.27
Co-doped	1.17	2.72

Table 36.3: Static reflectivity

	x-direction	z-direction
Pristine	0.00	0.05
Fe-doped	0.01	0.29
Co-doped	0.01	0.21

in the peaks corresponding to the maximum value of $R(\omega)$ can be observed on doping. The static reflectivity for all the samples has been tabulated in Table 36.3. It can be noted that while the static reflectivity along the x-direction is minuscule for all the samples, cobalt and iron doping assists in the achievement of high reflection at zero energy along the z-direction. It has been reported that the static reflectivity of Co-doped and Fe-doped nanotubes are

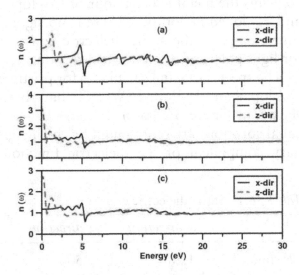

Figure 36.4: The variation of refractive index n(ω) with energy (eV) for (a) pristine (b) Fe-doped (c) Co-doped SWCNTs.

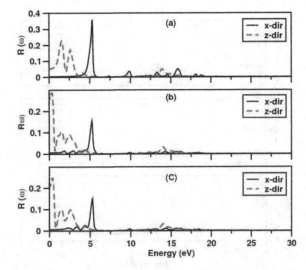

Figure 36.5: The variation of reflectivity R(ω) with energy (eV) for (a) pristine (b) Fe-doped (c) Co-doped SWCNTs.

approximately four and five times greater than that of pristine nanotubes respectively.

We can thus state that the manipulation of the magnitude of reflectivity by the addition of a suitable concentration of dopants can prove to be extensively beneficial for technological applications.

4. Conclusion

To summarise, we have analyzed the influence of Fe and Co doping on the optical anisotropies of SWCNTs. Making use of the Siesta code to employ the first principles approach, we have computed the dielectric constant, refractive index and reflectivity of pristine, Fe-dope and Co-doped SWCNTs. It is to be noted that our results for pristine nanotubes are in good agreement with the earlier studies. This further solidifies the fact that the replacement of carbon atoms with Co/Fe atoms singularly impacts the optical properties drastically. While all the optical properties discussed here are anisotropic in nature, it is found that anisotropy is more pronounced in the near-infrared region. The anisotropy decreases rapidly with an increase in the energy of radiation. It can be observed that the Fe/Co-doped SWCNTs exhibit a significant improvement in the optical properties in the near-infrared region for light-polarised direction in the z-direction. Higher values of static dielectric constant and static reflectivity of TM-doped nanotubes indicate their metallic behaviour which is evident from the band structure diagram as well. It is also to be noted the in perpendicular polarisation, the effect of doping is not as significant. Hence, we can infer that doping with TM atoms can be a simple and effective tool for tuning the optical properties of carbon nanotubes. Thus, it is expected that this analysis may invite further theoretical and experimental studies involving TM-doped functionalised nanotubes for various optoelectronic applications.

References

[1] Iijima, S. (1991). Helical microtubules of graphitic carbon. *Nature*, 354, 56–58.

[2] Abdalla, S., Al-Marzouki, F., Al-Ghamdi, A. A., and Abdel-Daiem, A. (2015). Different technical applications of carbon nanotubes. *Nanoscale Res. Lett.*, 10, 358 (pp10).

[3] Zhang, Z., Liang, X., Wang, S., Yao, K., Hu, Y., Zhu, Y., Chen, Q., Zhou, W., Li, Y., Yao, Y., Zhang, J., and Peng, L. M. (2007). Doping-free fabrication of carbon nanotube-based ballistic CMOS devices and circuits. *Nano Lett.*, 7, 3603–3607.

[4] Fabris, D., Rosshirt, M., Cardenas, C., Wilhite, P., Yamada, T., and Yang, C. Y. (2011). Application of carbon nanotubes to thermal interface materials. *J. Electron. Packag.*, 133: 020902 (pp6).

[5] Yakobson, B. I., Brabec, C. J., and Bernholc, J. (1996). Nanomechanics of carbon tubes: instabilities beyond linear response. *Phys Rev. Lett.*, 76, 2511–2514.

[6] Gubarev, V. M., Yakovlev, V. Y., Sertsu, M. G., Yakushev, O. F., Krivtsun, V. M., Gladush, Y. G., Ostanin, I. A., Sokolov, A., Schafers, F., Medvedev, V. V., Nasibulin, A. G. (2019). Single-walled carbon nanotube membranes for optical applications in the extreme ultraviolet range. *Carbon*, 155, 734-739.

[7] Pan, J., Li, F., and Choi, J. H. (2017). Single-walled carbon nanotubes as optical probes for bio-sensing and imaging. *J. Mater. Chem. B*, 5, 6511–6522.

[8] Hofferber, E. M., Stapleton, J. A., and Iverson, N. M. (2020). Single Walled Carbon Nanotubes as Optical Sensors for Biological Applications. *J. Electrochem. Soc.*, 167, 037530.

[9] Wang, F., Itkis, M. E., Bekyarova, E. B., Tian, X., Sarkar, S., Pekker, A., Kalinina, I., Moser, M. L., and Haddon R. C. (2012). Effect of first row transition metals on the conductivity of semiconducting single-walled carbon nanotube networks. *Appl. Phys. Lett.*, 100, 223111 (pp4).

[10] Shim, D., Jung, S, H., Han, S. Y., Shin, K., Lee, K. H., and Hun. J. (2011). Improvement of SWCNT transparent conductive films via transition metal doping. *Chem. Commun.*, 47, 5202–5204.

[11] Tsukagoshi, K., Alphenaar, B. W., and Ago, H. (1999). Coherent transport of electron spin in a ferromagnetically contacted carbon nanotube. *Nature*, 401, 572–574.

[12] Mao, Y. L., Yan, X. H., and Xiao, Y. (2005). First-principles study of transition-metal-doped single-walled carbon nanotubes. *Nanotechnology*, 16, 3092–3096.

[13] Yang, L., Yu, L. L., Wei, H. W., Li, W. Q., Zhou, X., and Tian, W. Q.(2019). Hydrogen storage of dual-Ti single-walled carbon nanotubes. *Int. J. Hydrogen Energy*, 44, 2960–2975.

[14] Gulseren, O., Yildirim, T., and Ciraci, S. (2001). Tunable adsorption on carbon nanotubes. *Phys. Rev. Lett.*, 87, 116802(pp4).

[15] Tabtimsai, C., Ruangpornvisuti, V., and Wanno, B. (2013). Density functional theory investigation of the VIIIB transition metal atoms deposited on (5,5) single-walled carbon nanotubes. *Physica E*, 49, 61–67.

[16] Lafdi, K., Chin, A., Ali N., and Despres J. F. (1996). Cobalt-doped carbon nanotubes: Preparation, texture, and magnetic properties. *J. Appl. Phys.*, 79, 6007–6009.

[17] Fu, S., Zhu, C., Li, H., Du, D., and Lin, Y. (2015). One-step synthesis of cobalt and nitrogen co-doped carbon nanotubes and their catalytic activity for the oxygen reduction reaction. *J. Mater. Chem. A*, 3, 12718 (pp6).

[18] Hele, G., Qichun, F., Jixin, Z., Jingsan, X., Qianqian, L., Siliang, L., Kaiwen, X., Chao, Z., and Tianxi, L. (2019). Cobalt nanoparticle-embedded nitrogen-doped carbon/carbon nanotube frameworks derived from a metal–organic framework for tri-functional ORR, OER and HER electrocatalysis. *J. Mater. Chem. A*, 7, 3664–3672.

[19] Liu, B., Zhou, H., Jin, H., Zhu, J., Wang, Z., Hu, C., Liang, L., Mu, S., and He, D. (2021).

A new strategy to access Co/N co-doped carbon nanotubes as oxygen reduction reaction catalysts. *Chin. Chem. Lett.*, 32, 535–538.

[20] Meihui, L., Jianping, X., Ximing, C., Xiaosong, Z., Yanyu, W., Ping, L., Xiping, N., Chengyuan, L., and Lan, L. (2012). Structural and optical properties of cobalt doped ZnO nanocrystals. *Supperlatt. Microstruct.*, 52, 824–833.

[21] Chanda, A., Gupta, S., Vasundhara, M., Joshi, S. R., Mutta, G, R., and Singh, J. (2017). Study of structural, optical and magnetic properties of cobalt doped ZnO nanorods. *RSC Adv.*, 7, 50527–50536.

[22] Vijayan, P. P., Thomas, M., and George K. C. (2015). Effect of Fe doping on optical properties of TiO_2 nanotubes. *Adv. Mater. Process. Technol.*, 1, 626–632.

[23] Akhtar, A. J., Gupta, A., Shaw, B. K., and Saha, S. K. (2013). Antiferro quadrupolar ordering in Fe intercalated few layers graphene. *Appl. Phys. Lett.*, 103, 242902 (pp9).

[24] Avouris, P., Freitag, M., and Perebeinos, V. (2008). Carbon-nanotube photonics and optoelectronics. *Nat. Photonics*, 2, 341–350.

[25] Soler, J. M., Artacho, E., Gale, J. D., Gracia, A., Junquera, J., Ordejón, P., and Sánchez-Portal, D. (2002). The SIESTA method for ab initio order-N materials simulation. *J. Phys.: Condens. Matter.*, 14, 2745–2779.

[26] Spataru, C. D., Ismail-Beigi, S., Benedict, L. X., and Louie, S. G. (2004). Excitonic effects and optical spectra of single-walled carbon nanotubes. *Phys. Rev. Lett.*, 92, 077402.

[27] Tetik E. (2014). The electronic properties of doped single walled carbon nanotubes and carbon nanotube sensors. *Condens. Matter Phys.*, 17, 43301 (12pp).

37 Investigating the effects of wire diameter and power input on surface roughness in wire EDM of SAE 4130

G. Narendranath[1,2,a] and J.Udaya Prakash[1]

[1]Department of Mechanical Engineering, Vel Tech Rangarajan Dr. Sangunthala R&D Institute of Science and Technology, Chennai, India
[2]Department of Mechanical Engineering, Siddartha Engineering College (SEAT), Tirupati, India

Abstract

High Strength alloy machining is one of the major concerns for production industries. Special alloys like Stainless steel grades need to be machined for high accuracy with good surface finish is another challenging task. The machining conditions of Wire-EDM can be optimised for high accuracy to produce complex profiles from high strength alloys. This study is focused on the varied wire diameter effect with relative power input for machining. Accuracy of these profiles can be optimised with nozzle gap as another parameter. Selected wire diameters are 0.1, 0.2 and 0.25 mm with power variables of discharge current 20, 25 and 30 amps with a nozzle distance of 15, 25, 35 mm. Machining has been done on 10 mm thickness SAE-4130 alloy. The fixed parameters are gap voltage 50V, pulse on time 10μs and pulse off time 30μs. Better results were found at 0.25 mm diameter with 30 amps and 25 mm nozzle distance in complex profiles after optimisation with Taguchi using Mini-tab software.

Keywords: WEDM, optimisation techniques, surface finish, wire diameter, current

1. Introduction

Wire EDM erodes material by thermo-electric process. In addition to serving as a coolant and clearing away debris, the dielectric fluid submerges the part and wire. For the work piece to have the required high accuracy and three-dimensional shape, the wire is controlled by CNC [1]. Nevertheless, it is not always easy to process super alloys to create high precision with good tolerances and high accuracy; typically, strong production techniques more often non-conventional ones must be used [2]. When removing undesired chips, conventional chip removal operations require a large amount of energy. However, the significant quantity of energy used in machining generates waste heat, which could cause problems with the surface features. [3]. Furthermore, residual stress and burrs also be produced necessitating further post-processing procedures [4]. It is generally important to note that super alloys possessing great strength and resistance to wear are challenging to machine using traditional machining process [5]. Due to this limitation, non-traditional machining techniques have a great deal of promise for producing parts made of super alloy materials that have complex features [6]. NTMT are a group of methods that remove excess material unconventionally [7]. A variety of unconventional machining techniques were created to satisfy the demands of the finished products. In practical terms, non-traditional machining methods have reduced productivity and increased specific energy consumption [8]. Nevertheless, NTMT can produce

[a]udayaprakashj@veltech.edu.in

DOI: 10.1201/9781003545941-37

accurate features with good surface quality if the right processing parameters are strictly followed [9]. Furthermore, it has been demonstrated that certain procedures, like electrical discharge machining (EDM), can create intricate features with good accuracy, tight tolerances, and, to a certain extent, acceptable surface finishes.

2. Literature Review

Gowthaman et al. [10] used a zinc-coated brass wire to study the effect of WEDM process parameters, on the MRR and SR of AISI 4340 alloy steel. The impact of a broad range of WT on recast layer for AISI 304 was investigated by Chaudhary et al. [11]. Van Sy [12] evaluated the influence of input parameters on SR and geometric accuracy during the WEDM process using multi-response optimisation. In order to show how WEDM parameters, affect surface quality and dimensional accuracy, the study looked at these parameters. Using an RSM design known as a face-cantered design, Kumar et al. [13] studied the Wire EDM of steel 304. The investigation focused on the influence of various WEDM machining parameters, SR and MRR. Mathew et al. [14] examined specific machining parameters in their analysis of WEDM on AISI-304.The impact of process parameters on WEDM for machining AISI 304 was examined by Linga durai et al. [15]. The wire cut process parameters were predicted to have the best machining conditions under GV, T_{on}, T_{off}, and WF. Satish Kumar et al.'s study [16] concentrated on how SiC/Al volume fraction and parameter values affected WEDM machining performance. This view point led to the design of a L_9 OA experiment, and the outcomes are analysed by response graphs and extracted variance analysis. The use of a hard facing material with nanostructure in WEDM was investigated by Saha et al. [17]. The experimental parameters were optimised and the different processing properties were characterised. Brass, silver-coated wire and zinc-coated wire were compared for WEDM of Maraging steel by Ruma Sen et al. [18]. The surfaces that underwent

processing using various wire electrodes were examined and evaluated. The zinc-coated wire was found to have superior MRR and surface characteristics. In order to examine the white layer and SR, Radhakrishnan et al. [19] vibrated the wire at various frequencies. ZCWE was added, and this improved both the white layer and the work piece's surface roughness. Keshav Prasad Patel et al.'s study [20] examined the effectiveness of micro wire EDM machining of aluminium material. For precise and tight tolerance machining of conductive materials, WEDM is widely utilised. Using the finite element method (FEM), a straightforward and easily comprehensible 2D model for WEDM has been created. Sivaprakasam et al. [21] investigated the dimensional accuracy in micro-EDM of AMCs and compared the output parameters based on GV, FR and capacitance.

3. Methodology

Taguchi technique is frequently used because of its strong design, which reduces the expense and duration of experimentation. Three input parameters for the experiments are chosen at three different levels: wire diameters of 0.1, 0.2, and 0.25 mm, power variables of 4, 6, and 8 amps, and nozzle distances of 15, 25, and 35 mm. For conducting experiments, standard L_9 OA has been chosen. L_9 OA has nine rows and four columns. Nine rows indicate the number of experiments that must be conducted, and three columns were chosen for three wire EDM input parameters.

3.1. Response Surface Methodology

RSM is helpful for modeling and analysing situations where the goal is to optimise responses that are of interest and are influenced by multiple variables. Because of these qualities, the RSM approach is beneficial for wire EDM modeling and optimisation. The response can be adequately represented by a linear function.

$$Y_U = b_0 + b_1 X_1 + b_2 X_2 + \dots + b_i X_i + e$$

The system is represented by a curved function.

$$Y_U = b_0 + \sum_{i=1}^{K} b_i X_i + \sum_{i=1}^{K} b_{ii} X_i^2$$

Where k is the No. of factors that have been studied and optimised, e is the random error, and i, j, and β are the linear, quadratic, and regression coefficients, respectively. ANOVA is used to estimate the regression model's suitability.

3.2. Brass Wire Electrode

When early EDM users were searching for a more performance-oriented wire, brass EDM wire was the first sensible substitute for copper. Significantly, more tensile strength from the addition of zinc makes up for the relative conductivity losses. Brass emerged as the most popular electrode material for wire EDM.

3.3. Orthogonal Array (OA)

The Taguchi technique is a valuable tool for comprehending and simulating the wire EDM. With little experimentation, a wide

Figure 37.1: WEDM Specimen.

range of parameters can be analysed using the Taguchi method. Because of the many parameters involved in wire EDM, this feature makes the OA approach a valuable one for modelling this process. This method's primary drawback is that the results are only relative.

4. Results and Discussion

The Wire EDM results obtained for SR and Kerf (discharge pulse gap) as per the experimentation conditions in Wire EDM shown in Table 37.4. The work further optimised with response surface method (RSM) using Mini-Tab.

Table 37.1: Process parameters and their levels

Parameters	Level 1	Level 2	Level 3
Diameter (mm)	0.1	0.2	0.25
Current (A)	4	6	8
Distance (mm)	15	25	35

Table 37.2: Chemical composition of AISI 4130 steel alloy

C	Cr	Fe	Mn	Mo	P	Si
0.28 – 0.33	0.80 – 1.1	97.03 – 98.22	0.40 – 0.60	0.15 – 0.25	≤ 0.035	0.15 – 0.30

Table 37.3: Mechanical properties of AISI 4130 steel alloy

Yield Strength (MPa)	Tensile Strength (MPa)	Modulus of Elasticity (GPa)	Poisson's Ratio	Hardness (HRB)	Machinability
435	670	205	197	140	70%

Table 37.4: Experimental results

Exp . No	Current (A)	Wire Diameter (mm)	Nozzle distance (mm)	Surface Roughness (Ra) (μm)	Kerf Width (mm)	Rank
1	20	0.1	15	2.36	0.315	9
2	20	0.2	25	2.29	0.308	7
3	20	0.25	35	2.32	0.305	8
4	25	0.1	25	2.28	0.298	6
5	25	0.2	35	2.05	0.294	2
6	25	0.25	15	2.22	0.297	5
7	30	0.1	35	2.17	0.308	3
8	30	0.2	15	2.21	0.296	4
9	30	0.25	25	2.02	0.293	1

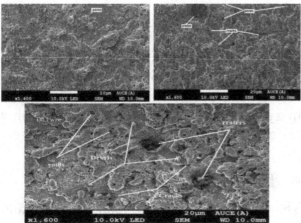

Figure 37.2: Surface roughness texture for different wire diameters.

Table 37.5: Regression equation - Effect of Power Input on SR

$$1 \quad SR \;=\; 2.657 - 0.182\,B - 0.102\,C + 0.045\,B{*}B + 0.025\,C{*}C - 0.020\,B{*}C$$

$$2 \quad SR \;=\; 2.497 - 0.182\,B - 0.102\,C + 0.045\,B{*}B + 0.025\,C{*}C - 0.020\,B{*}C$$

$$3 \quad SR \;=\; 2.447 - 0.182\,B - 0.102\,C + 0.045\,B{*}B + 0.025\,C{*}C - 0.020\,B{*}C$$

Fine texture found after machining with Ø 0.25 mm wire observed while the other diameter machining wires have burn pits lashes and craters on the surface as observed in micro structures shown in Figure 37.2.

The results confirms that a wire diameter of 0.25 with 30 amp power and a nozzle distance of 25 give good results. Based on the wire diameter, the power parameter should be decreased and the nozzle distance should be increased. Much variation found in discharge spark gap. A variation of 0.01 to 0.025 mm with respect to current has been observed.

This may reflect on the surface contact area of the spark to vary SR.

The surface roughness equation for regression with the three factors is as follows

$$SR \;=\; 2.533 - 0.182\,B - 0.102\,C + 0.123\,A\_1 - 0.0367\,A\_2 - 0.0867\,A\_3 + 0.045\,B{*}B + 0.025\,C{*}C - 0.020\,B{*}C$$

Effect of power on the surface finish with respect to factors of diameter and nozzle distance can be defined as given in Table 37.5.

The surface contour plots from the above regression coefficients with 0.25 diameter wire even surface plot observed as shown in Figures 37.3–37.6. The variation in discharge power affecting the surface finish observed in below plots at 30 amp power 0.25 wire diameters and 25 nozzle distance given better results as shown in Figure 37.6.

Individual as well as combined parameters were observed to check the effect of input parameters on SR. The combination of power and wire diameter gives lower surface roughness, then diameter of wire and finally power factor respectively. Individually selected factors have equal importance as shown in Figures 37.7 and 37.8.

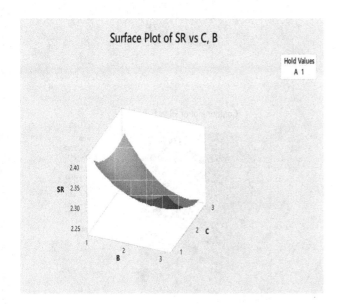

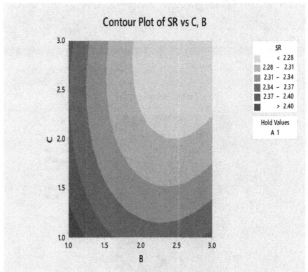

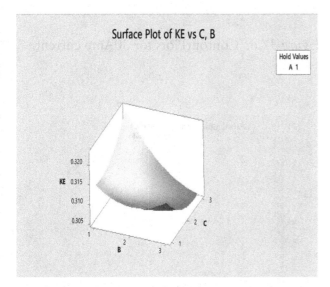

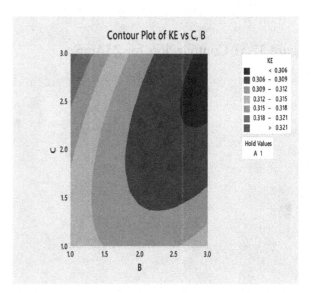

Figure 37.3: Surface Plots at low power input of 20 amps with lower diameter.

Figure 37.4: Contour Plots at 20 amp conditions.

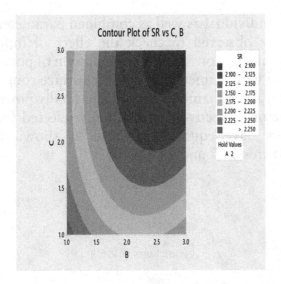

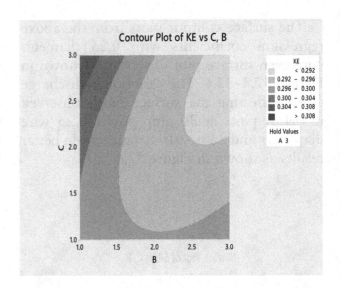

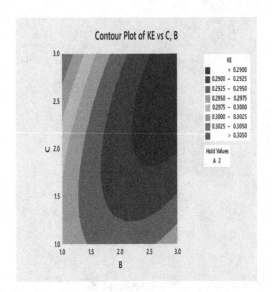

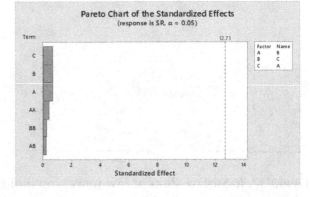

Figure 37.5: ContourPlots for 25Amp.

Figure 37.6: ContourPlots for 30Amp current.

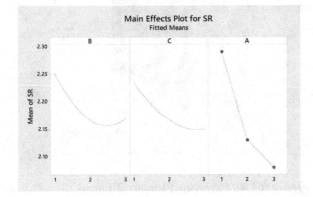

Figure 37.7: Factor analysis of fitted curve Surface Roughness.

Figure 37.8: Factor analysis standard effect on Surface Roughness.

5. Conclusions

Experimentation conducted on SAE 4130 material in wire EDM machining of different profiles. Square shape with corner radius 2R has been selected for testing of surface roughness value and discharge gap estimated with samples obtained. Response surface analysis with selected DoE analyzed with mini-tab software. The following conclusions were drawn after optimisation of results,

1. The effective parameter considered as discharge power and the power relatively vary with the wire diameter.
2. Wire diameter also has a crucial role after verifying micro-structures.
3. Marginal nozzle distance is good to get the surface finish to flush out debris after machining.
4. Overall selected parameters given good response constants may vary in coming research as future enhancement in machining of high strength alloys.

References

[1] Sarala Rubi, C., Prakash, J., Juliyana, S., Čep, R., Salunkhe, S., Kouril, K., and Ramdas Gawade, S. (2024). Comprehensive review on wire electrical discharge machining: a non-traditional material removal process. *Front. Mech. Eng.*, 10.1322605.

[2] Jebarose Juliyana, S., Udaya Prakash, J., Rubi, C. S., Salunkhe, S., Gawade, S. R., Abouel Nasr, E. S., and Kamrani, A. K. (2023). Optimization of wire EDM process parameters for machining hybrid composites using grey relational analysis. *Crystals.*, 13(11), 1549.

[3] Udayaprakash, J., Rajkumar, C., Jayavelu, S., and Sivaprakasam, P. (2023). Effect of wire electrical discharge machining parameters on various tool steels using grey relational analysis. *Int. J. Veh. Struct. Sys.*, 15(2), 203–206.

[4] Farooq, M. U., Ali, M. A., He, Y., Khan, A. M., Pruncu, C. I., Kashif, M., and Asif, N. (2020). Curved profiles machining of Ti6Al4V alloy through WEDM: investigations on geometrical errors. *J. Mater. Res. Technol.*, 9(6), 16186–16201.

[5] Abdulkareem, S., Khan, A. A., and Zain, Z. M. (2011). Experimental investigation of machining parameters on surface roughness in dry and wet wire-electrical discharge machining. *Adv. Mater. Res.*, 264, 831–836.

[6] Datta, S. and Mahapatra, S. (2010). Modeling, simulation and parametric optimization of wire EDM process using response surface methodology coupled with grey-Taguchi technique. *Int. J. Eng., Sci. Technol.*, 2(5), 162–183.

[7] Selvakumar, G., Sarkar, S., and Mitra, S. (2013). An experimental analysis of single pass cutting of aluminium 5083 alloy in different corner angles through WEDM. *Int. J. Mach. Machin. Mater.*, 13(2–3), 262–275.

[8] Narendranath, G. and Prakash, J. U. (2023). Effect of wire EDM process parameters on material removal rate of duplex stainless steel (S31803). *Mater .Today. Proc.*, 92, 424–429.

[9] Gowthaman, P. S., Gowthaman, J., and Nagasundaram, N. J. M. T. P. (2020). A study of machining characteristics of AISI 4340 alloy steel by wire electrical discharge machining process. *Mater. Today. Proc.*, 27, 565–570.

[10] Chaudhary, T., Siddiquee, A. N., and Chanda, A. K. (2019). Effect of wire tension on different output responses during wire electric discharge machining on AISI 304 stainless steel. *Def. Technol.*, 15(4), 541–544.

[11] Van Sy, L. (2021). Multi-Objective Optimization of Processing Parameters in WEDM with Stainless Steel-304 for Die-Angular Cutting. *J. Mech. Eng*, 71, 141–150.

[12] Kalyanakumar, S., Prabhu, L., Saravanan, M., and Imthiyas, A. (2020). Experimental investigation of MRR, RA of 304 stainless steel using WEDM.IOP Conference Series. *Mater. Sci. Eng.*, 993(1), 012035.

[13] Mathew, B. and Babu, J. (2014). Multiple process parameter optimization of WEDM on AISI304 using Taguchi grey relational analysis. *Procedia. Mater. Sci.*, 5, 1613–1622.

[14] Lingadurai, K., Nagasivamuni, B., Muthu Kamatchi, M., and Palavesam, J. (2012).

Selection of wire electrical discharge machining process parameters on stainless steel AISI grade-304 using design of experiments approach. *J. Inst. Eng. (India) Series. C.*, 93, 163170.

[15] Kumar, A., Kumar, V., and Kumar, J. (2012). Prediction of surface roughness in wire electric discharge machining (WEDM) process based on response surface methodology. *Int. J. Eng. Technol.*, 2(4), 708–719.

[16] Satishkumar, D., Kanthababu, M., Vajjiravelu, V., Anburaj, R., Sundarrajan, N. T., and Arul, H. (2011). Investigation of wire electrical discharge machining characteristics of Al6063/SiC p composites. *Int. J. Adv. Manuf. Technol.*, 56, 975–986.

[17] Saha, A., and Mondal, S. C. (2016). Multi-objective optimization in WEDM process of nanostructured hardfacing materials through hybrid techniques. *Measurement.*, 94, 46–59.

[18] Sen, R., Choudhuri, B., Barma, J. D., and Chakraborti, P. (2020). Surface integrity study of WEDM with various wire electrodes: Experiments and analysis. *Mach. Sci. Technol.*, 24(4), 569–591.

[19] Radhakrishnan, P., and Vijayaraghavan, L. (2017). Assessment of material removal capability with vibration-assisted wire electrical discharge machining. *J. Manuf. Process.*, 26, 323–329.

[20] Patel, K. P., Dargar, M. K., and Singh, P. K.(2015). Modeling, Investigation and Analysis of MicroWEDM Process on Aerospace Material, *Int. J. Sci. Eng. Technol.* 662–667. https://doi.org/10.2348/ijset06150662.

[21] Sivaprakasam, P., Udaya Prakash, J., and Hariharan, P. (2022). Enhancement of material removal rate in magnetic field-assisted micro electric discharge machining of Aluminium Matrix Composites. *Int. J. Ambient. Energy.*, 43(1), 584–589.

38 Performance of hot mix asphalt using nano-clay modified binder and polypropylene fiber

Shreyash Sachin shinde[a], Abdulsajid Ganiahamad Gavandi, Vivek Vijay Koli, and Pinki Deb

Civil Engineering, Sharad Institute of Technology College of Engineering, Kolhapur, Maharashtra, India

Abstract

Pavement structures are found to be expansive due to the maintenance cost throughout the life span. To overcome this shortcoming, montmorillonite nano clay is utilized in bitumen as modifier. The initial stage of the study includes the performance in terms of physical and strength characteristics of nano-clay modified bitumen. For preparation of nano-clay modified binder, the nano-clay was blended in bitumen at 2%, 4% and 6% by weight of binder.It was noticed that the addition of nano-clay modifier into bitumen can reduce the penetration value and subsequently enhances the softening point value. Additionally, the performance of mixes made with the unmodified and nano-clay modified bitumen was compared by Marshall stability test and retained Marshall stability ratio.To further improve the cracking resistance of bituminous mixture, polypropylene fiber is introduced into the mix at 0.3%by weight of mixture. It was observed that inclusion of polypropylene fiber has improve the Marshall stability upto to 51.53%, 100.50%, 68.64% and 64.41% respectively for 0%, 2%, 4% and 6% binder content. Additionally, the retained Marshall Stability value of 0%, 2%, 4% and 6% nano-clay modified mix made with 0.3% polypropylene fiber is 82.66%, 93.99%, 87.28% and 83.87% respectively.

Keywords: Hot mix asphalt, nano-clay, polypropylene, Marshall stability, retained Marshall stability

1. Introduction

The ever-increasing demand of the innovative technologies in road infrastructures, several researchers and engineering professionals are trying to improve the long-term performance of existing pavement structure. Flexible pavements are economical compared to rigid pavements in terms of construction cost. However, the maintenance of is too high for flexible pavements which has increased the final cost of flexible pavement. Temperature fluctuation, vehicular load, heavy rainfall results in the premature failure of pavement structure. Because of this, experts have become increasingly interested in modifying the asphalt binder to increase pavement structure strength and decrease early failure.

Montmorillonite nano clay belongs to the smectite family which has been widely utilized in modification of asphalt binder. The composition is based on phyllosilicates that have undergone organic modification;the formula is $(Na, Ca)_{0.33} (Al, Mg)_2 (Si_4O_{10}) (OH)_2 . nH_2O$. Numerous researchers havereported that adding nano-clay rises the softening point and dramatically lowers the viscosity and penetration value [1,2]. The addition of nano-clay as bitumen modifier can improve the performance like stability, resistance to cracking, moisture resistance etc [3–5].

[a]pinkinits93@gmail.com

DOI: 10.1201/9781003545941-38

Cracking of pavement structure is one of the major concern which can be reduced by using fiber in asphalt mix. Various researchers has stated that the inclusion of fiber can reduce the fatigue deformation, increase the tensile strength of mixture. It can also effectively reduce the reflective cracking of pavement structure [6]. Application of different fibers can also increase the viscoelastic properties of binder [7–9].

Recently, several researchers are constantly looking for the new materials and technologies to mitigate the major problems related to asphalt pavement. So, the current study focused on using nano-clay for increasing the engineering properties of neat bitumen. As well as polypropylene fibers are added into the bituminous mix to increase the strength of mix. Thus, it plays a crucial role in prolonging the pavement's useful life.

2. Materials

Locally available aggregate is used in this study. According to MORT&H 2013, the aggregate gradation for bituminous concrete is utilized.[10]. The gradation of aggregate is given in Table 38.1. The properties of aggregate are given in Table 38.2. VG 30 bitumen is used as a binder in mix with specific gravity 0.97. The penetration, ductility and softening point of bitumen is 64 (0.1mm), 45 cm and 48°C. Montmorillonite Nano-clay is used as a bitumen modifier in this study and the physical properties of nano-clay is given in Table 38.3. Additionally polypropylene fiber is used as mixture reinforcement to reduce the failure and deformation.

2.1. Preparation of Nano-Clay Modified Binder

In this laboratory investigation, Marshall Mix design method is used to prepare Marshall samples. Total 1200 gm material including coarse aggregate, fine aggregate and filler were blended together to make the sample. The hand mixing process is adopted by

Table 38.1: Bituminous concrete gradation of aggregate

Sieve Sizes (mm)	Passing by weight (%)	Passing by weight (%)
19	100	100
13.2	90–100	95
9.5	70–88	79
7.75	53–71	62
2.36	42–58	50
1.18	34–48	41
0.6	26–38	32
0.3	18–28	23
0.15	12–20	16
0.075	4–10	7

Table 38.2: Aggregate properties

Properties	Test results
Flakiness	16.8%
Elongation Index	10.5%
Impact value	23.1%
Abrasion value	28.65%

Table 38.3: Properties of nano-clay

Physical Properties of nano clay	Specifications
Water content	Less than 6%
Specific Weight	1.5–1.7 g/cm^3
Particle Size	Less than 10µ´

Table 38.4: Properties of nano-clay modified asphalt binder

Properties	2% nano-clay modified binder	4% nano-clay modified binder	6% nano-clay modified binder
Ductility test, cm	40	38	35
Softening point test (°C)	55°C	57	62
Penetration test, (0.1mm)	55	45	39

blending aggregate and filler at 110°C. The loss of small particles were carefully avoided during mixing. The bitumen was separately heated at 160°C for melting. The bitumen is then added to the mix by mass of aggregate and manually mix for 2–3 minutes to form a uniform texture. Excessive mixing is avoided to control the bleeding of mixture. The resultant mixture then compacted in a standard cylindrical Marshall mould comprising of 101.6 mm dia and 63.5 mm height. Then compaction was done with 75 blows on each side of the cylindrical sample by hammer. The compacted specimen is extracted from the mould by sample extractor and then cured at room temperature for 24 hours. For making fiber reinforced specimens, 0.3% (by weight of 1200 gm) polypropylene was added and mixed thoroughly with aggregate and binder before compaction.

2.2. Preparation of Marshall Sample

In this laboratory investigation, Marshall Mix design method is used to prepare Marshall samples. Total 1200 gm material including coarse aggregate, fine aggregate and filler were blended together to make the sample. The hand mixing process is adopted by blending aggregate and filler at 110°C. The loss of small particles were carefully avoided during mixing. The bitumen was separately heated at 160°C for melting. The bitumen is then added to the mix by mass of aggregate and manually mix for 2–3 minutes to form a uniform texture. Excessive mixing is avoided to control the bleeding of mixture. The resultant mixture then compacted in a standard cylindrical Marshall mould comprising of 101.6 mm dia and 63.5 mm height. Then compaction was done with 75 blows on each side of the cylindrical sample by hammer. The compacted specimen is extracted from the mould by sample extractor and then cured at room temperature for 24 hours. For making fiber reinforced specimens, 0.3% (by weight of 1200 gm) polypropylene was

added and mixed thoroughly with aggregate and binder before compaction.

3. Experimental Programme

3.1. Marshall Stability Test

Marshall test was used for determining resistance to plastic flow of prepared Marshall samples at a constant loading rate of 50mm/min. In order to analyze the optimum bitumen content, Marshall test is used. Marshall test was performed with a variety of percentage of bitumen from 4.5% to 6.5% by weight of aggregate. The volumetric parameters were calculated with the specific formulas.

3.2. Retained Marshall Stability

The test determines the stripping resistance of asphalt mixture. The retained Marshall Stability test is described in IRC: SP 53-2010 on modified asphalt binders and is performed according to ASTM D1075. The standard Marshall specimens with a dimension of 10cm diameter and 6.35 cm height were made for different asphalt mixes. Then asphalt specimens were immersed in water bath for duration of 24 hours at a temperature of 60°C and afterwards the specimens were tested in Marshall testing machine. The results are expressed in percentage with respect to the unconditioned cases. If the Retained Marshall Stability (RMS) is higher it indicates higher the moisture damage resistance of the asphalt mixture. RMS (%) is given by the equation

$$RMS\ (\%) = \frac{Conditioned\ Marshall\ stability}{Unconditioned\ Marshall\ stability} \times 100$$

4. Results and Discussion

4.1. Optimum Binder Content Determination

The optimum binder content (OBC) was determined by adding increasing dosage of bitumen into Marshall mix from 4.5% to 6.5% at 0.5% increment. The Marshall

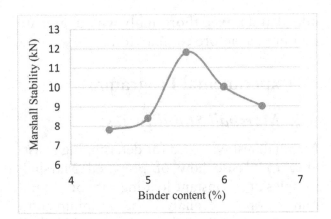

Figure 38.1: Marshall Stability of mixes at different percentage of bitumen.

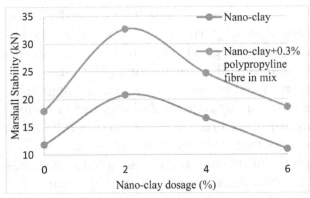

Figure 38.2: Marshall stability of mixes at different dosage of nano-clay.

Table 38.5: Volumetric properties of mixes at optimum binder content

Property	Values
Marshall stability (kN)	11.8
Unit weight (gm/cc)	2.02
Flow (mm)	3.6
Volume of voids (%)	4.3
VMA (%)	15.75
VFB (%)	72.69

stability (MS) is increasing with increasing dosage of binder and the optimum binder content was found at 5.5%. The test results are given inFigure 38.1. The MS was found decreasing due to plastic deformation at higher dosage of binder. The optimum binder content was determined according to highest stability, 4% air void and maximum density. The volumetric properties are given in Table 38.5.

4.3. Influence of Nano-Clay Modified Binder in Asphalt Mix Containing Polypropylene Fiber

The Marshall stability test was conducted for mixes made with nano-clay modified bitumen at optimum dosage of bitumen. The nano-clay modified bitumen is added in mix at 5.5% by weight of mix and tested for Marshall stability. The results are given in Figure 38.2. It is observed that at 2% nano-clay content, the stability value is higher and at further dosages, the stability found decreasing. It is also evident, though, that the stiffening of the binder is the reason why the stability value of the nano-modified binder is always higher than that of the unmodified mix.The addition of 0.3% polypropylene fiber in mix can significantly improve the performance of mix to 51.53%, 100.50%, 68.64% and 64.41% respectively for 0%, 2%, 4% and 6% nano-clay binder mix. It is stated that the mix with 2% nano-clay modified binder and 2% nano clay modified binder with 0.3% polypropylene fiber is the two best combination to use in pavement structure.

4.2. Retained Marshall Stability of Mixes

Retained Marshall stability results of mixes are given in Figures 38.3 and 38.4. The conditioned Marshall stability of mix with nano-clay modified bitumen is 35.59%, 26.92%, 28.91% and 45.45% lower that unconditioned stability. It is stated that the conditioned stability of mix containing nano-clay modified bitumen and polypropylene fiber is 17.33%, 6.0%, 12.71% and 16.12% lower than unconditioned Marshall stability of same mix. It is observed that reduction of stability for mix with polypropylene fiber is comparatively lower as the fiber tried to hold

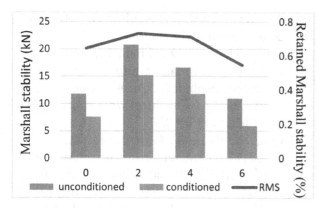

Figure 38.3: Conditioned and unconditioned Marshall stability and Retained Marshall stability of mixes made with nano-clay modified binder.

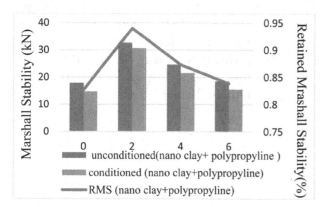

Figure 38.4: Conditioned and unconditioned Marshall stability and Retained Marshall stability of mixes made with nano-clay modified binder and polypropylene fiber.

the mix against deformation. The retained Marshall Stability value of mix made with 0%, 2%, 4% and 6% nano-clay is 64.4%, 73.07%, 71.08% and 54.54% respectively. Additionally, the retained Marshall Stability value of mix made with 0%, 2%, 4% and 6% nano-clay and 0.3% polypropylene fiber is 82.66%, 93.99%, 87.28% and 83.87% respectively.

5. Conclusion

- Adding nano clay in bitumen can reduce the ductility to 12.5% than the VG 30 bitumen.

- Using 2% nano-clay can heighten the softening point of asphalt binder to 14.58%.
- However, the addition of nano-clay into bitumen can also reduce the penetration value to 55mm, 45mm and 39 mm after adding 2%, 4% and 6% nano-clay respectively.
- It can be stated the use of nano-clay can improve the physical properties of asphalt binder.
- Using 2% nano-clay modified binder can enhance the Marshall stability of mix to 76.27% than unmodified mixes. So, optimum dosage of nano-clay is found to be 2% as at higher dosage of nano-clay in bitumen can reduce the mixture properties.
- The addition of 0.3% polypropylene fiber in mix can significantly improve the performance of mix to 51.53%, 100.50%, 68.64% and 64.41% respectively for 0%, 2%, 4% and 6% nano-clay binder mix.
- The retained Marshall Stability value of mix made with 0%, 2%, 4% and 6% nano-clay is 64.4%, 73.07%, 71.08% and 54.54% respectively. the retained Marshall Stability value of mix made with 0%, 2%, 4% and 6% nano-clay and 0.3% polypropylene fiber are 82.66%, 93.99%, 87.28% and 83.87% respectively.
- As per the study, the mix made with 2% nano-clay and 0.3% polypropylene fiber is most robust mix.

Acknowledgement

The authors gratefully acknowledge the staff ofCivil Engineering department and authority of Sharad Institute of Technology, College of Engineering, Yadravfor their cooperation in the research.

References

[1] Ezzat, H., El-Badawy S., Gabr A., et al. (2016). Evaluation of asphalt binders modified with nanoclay and nanosilica. *Procedia*

Eng., 143, 1260–1267, DOI: https://doi. org/10.1016/j.proeng.2016.06.119

[2] Vivek, A. K., Sukhija, M., and Singh, K. L. (2021). Effect of montmorillonitenanoclay and sulphur-modified blends on the properties of bituminous mixes. *Innovative Infrastruct. Solutions,* 6:119 https://doi.org/ 10.1007/s41062-021-00484-2

[3] Jahromi, S. G. andRajaee, S. (2013) Nanoclay-modified asphalt mixtures for eco-efficient construction. In: Nanotechnology in eco-efficient construction: materials, processes and applications. Elsevier Ltd, pp 108–126.

[4] Iskender, E. (2016). Evaluation of mechanical properties of nano- clay modified asphalt mixtures. *Meas. J. Int. Meas. Confed.,* 93, 359–371, DOI: https://doi.org/10.1016/j. measurement.2016.07.045

[5] Ashish, P. K., Singh, D., Bohm, S. (2016). Evaluation of rutting, fatigue and moisture damage performance of nanoclaymodi- fied asphalt binder. *Constr. Build.Mater.,* 113,

341–350, DOI: https://doi. org/10.1016/j. conbuildmat.2016.03.057

[6] Chen, H., Xu, Q., Chen, S., and Zhang, Z. (2009). Evaluation and design of fiber-reinforced asphalt mixtures.*Mater. Des.,* 30, 2595–2603, DOI: https://doi.org/10.1016/ j.matdes.2008.09.030.

[7] Qin, X., Shen, A., Guo, Y., Li, Z., and Lv, Z.(2018). Characterization of asphalt mastics reinforced with basalt fibers.*Constr. Build. Mater.,* 159, 508–516, DOI: https://doi. org/10.1016/j.conbuildmat.2017.11.012.

[8] Qian, S., Ma, H., Feng, J., Yang, R., and Huang, X.(2014).Fiber reinforcing effect on asphalt binder under low temperature.*Constr. Build. Mater.,* 61, 120–124, DOI: https://doi. org/10.1016/j.conbuildmat.2014.02.035.

[9] Khattak, M. J., Khattab, A., Rizvi, H. R., and Zhang, P.(2012). The impact of carbon nano-fiber modification on asphalt binder rheology.*Constr. Build. Mater.,* 30, 257–264, DOI: https://doi.org/10.1016/j. conbuildmat.2011.12.022.

39 Sustainable *Mirabilis jalapa*-derived zinc oxide nanoparticles: synergistic bioactivity in antimicrobial, anticancer, and antioxidant

G. K. Prashanth[1,a], Manoj Gadewar[2], Srilatha Rao[3], H. S. Lalithamba[4], S. H. Prashant[5], M. Mahadevaswamy[6], N. P. Bhagya[7], and B. M. Nagabhushana[8]

[1]Research and Development Centre, Department of Chemistry, Sir M. Visvesvaraya Institute of Technology, Bengaluru, India
[2]Department of Pharmacology, School of Medical and Allied Sciences, KR Mangalam University, Gurgaon, India
[3]Department of Chemistry, Nitte Meenakshi Institute of Technology, Bengaluru, India
[4]Department of Chemistry, Siddaganga Institute of Technology, Tumkur, India
[6]Department of PG Chemistry, JSS College of Arts, Commerce and Science, Mysuru, India
[5]Department of Mechanical Engineering, Sir M. Visvesvaraya Institute of Technology, Bengaluru, India
[7]Department of Chemistry, Sai Vidya Institute of Technology, Bengaluru, India
[8]Department of Chemistry, Ramaiah Institute of Technology, Bengaluru, India

Abstract

In this investigation, sustainable SCS was adopted to produce ZnO NPs, using *Mirabilis jalapa* (*M. jalapa*- MJ) as an unconventional biofuel source. An in-depth phytochemical analysis was carried out to identify the chemicals that help in this combustion process. The resulting ZnO NPs (ZMPs) were characterized by their distinctive wurtzite hexagonal structure, as revealed through PXRD analysis. Further examination of these nanoparticles' surface properties was assessed through the use of advanced techniques, including SEM and TEM. Surface area quantification was conducted using the well-established BET technique, utilizing nitrogen adsorption and desorption. The versatility of these ZMPs extends beyond synthesis and characterization. Their antimicrobial potency against was demonstrated. Additionally, the MTT assay unveiled a dose-dependent anticancer potential against MCF-7 cancer cells, while the DPPH assay showcased their remarkable antioxidant capability. This study not only illuminates the fascinating aspects of sustainable nanomaterial synthesis but also underscores the promising applications of *M. jalapa* as a renewable biofuel source. Furthermore, it highlights the diverse potential of ZMPs in fields ranging from healthcare to environmental conservation, fostering optimism for a sustainable future.

Keywords: Sustainable combustion synthesis, ZnO NPs, *Mirabilis jalapa*, nanoparticle characterization, antimicrobial, anticancer

1. Introduction

Currently, there has been a noticeable increase in interest regarding the synthesis and utilization of nanoparticles across diverse scientific fields [1,2]. Nanoparticles, due to their unique size-dependent properties, have shown promise in fields as diverse as materials

[a]prashanth_chem@sirmvit.edu

DOI: 10.1201/9781003545941-39

science, medicine, and environmental conservation [3,4]. Among the extensive variety of nanoparticles, ZnO NPs have emerged as a focal point of concentrated research interest, primarily owing to their distinct wurtzite hexagonal structure and exceptional properties [5,6].

The significance of sustainable approaches in nanoparticle production has grown substantially. as the world seeks eco-friendly and renewable solutions [7]. In this context, the current work explores sustainable combustion synthesis as a novel approach to producing ZnO NPs. What adds a particularly captivating dimension is the application of this method involves utilizing plant extracts. as an unconventional biofuel source, serving as the precursor material [8].

Solution combustion synthesis utilizing plant extracts is an innovative method in materials science [9–11]. This sustainable approach involves the application of plant source as fuel and metal salts as oxidizers in a solution. The process initiates a combustion reaction, leading to the formation of nanomaterials or nanoparticles. Plant extracts not only act as a green and renewable source but also introduce organic compounds, influencing the synthesis and yielding unique materials [12,13]. This eco-friendly approach offers a promising avenue for the production of various functional materials [14,15], providing information into their structural properties and contributing to the development of sustainable and biocompatible technologies. This novel technique not merely offers an eco-friendly and sustainable method for nanoparticle production but also prompts inquiries about the chemical constituents within *Mirabilis jalapa* in charge of the combustion method.

The research begins by employing qualitative phytochemical screening to elucidate the key chemical components contributing to the combustion process. The resulting ZMPs are subsequently characterized through advanced analytical techniques, such as PXRD, SEM, and TEM. The surface area of these NPs is quantified using the BET method, providing a glimpse into their structural characteristics.

Nevertheless, the goal of this investigation goes beyond the synthesis and characterization of ZMPs. The research delves into the multifaceted applications of these NPs, including their antimicrobial properties. Furthermore, the investigation reveals a dose-dependent anticancer potential against MCF-7 cancer cells, shedding light on their potential in cancer therapy. The study also highlights their remarkable antioxidant capabilities through the DPPH assay, showcasing their potential role in healthcare. This research, therefore, illuminates the intriguing realm of sustainable nanomaterial synthesis and underscores the promising applications of *M. jalapa* as a renewable biofuel source. Moreover, it highlights the diverse potential of ZMPs in various fields, from healthcare to environmental conservation, fostering optimism for a sustainable future. In the subsequent sections, we delve into the experimental methods, results, and discussions that underpin these findings, providing a comprehensive exploration of the potential and implications of sustainable ZMPs synthesis.

2. Materials and Methods

2.1. Chemicals//Reagents

Analytical-grade chemicals and reagents were utilized without additional purification. Leaves were collected from a local garden in Vidyaranyapuara, Bengaluru. underwent thorough washing with ample water and were subsequently left to air-dry for a duration of three weeks.

2.2. Formulation of an Aqueous Extract Derived from the Leaves of M. jalapa

Five grams of dried leaves of *M. jalapa* underwent to soxhlet extraction with 50 mL distilled water as a solvent for 72 hrs.

2.3. Production of ZMPs

The synthesis adheres to the methodology outlined in prior studies [16–20]. 2 grams of [Zn(NO$_3$)$_2$·6H$_2$O] were successfully dissolved in ten mL of distilled water. Subsequent to this step, four mL of infusion of leaf extract was incorporated into the solution, and the amalgamation was stirred comprehensively with the assistance of a magnetic stirrer. The resulting uniform solution was then placed into a muffle furnace, with the furnace's temperature maintained at 375±10°C. Inside the furnace, the solution underwent a dehydration process, eventually igniting and nucleating into finely structured ZMPs.

2.4. Characterization

The structural characteristics of ZMPs were assessed through powder X-ray diffraction, employing a PANalytica IX'pert diffractometer with Cu Kα radiation (λ=1.5418 Å). Surface properties were studies using SEM, utilizing a Carl Zeiss Ultra 55 microscope with a silver coating. Particle shape and size were determined through TEM with a Philips 200 instrument. BET surface area analysis was conducted using a Micromeritics ASAP 2020 instrument. Additionally, a thorough qualitative screening of the aqueous plant leaf extract was performed, following established methods from references [21,22], to identify components related to combustion.

2.5. Evaluation of Antibacterial Potency

The antimicrobial efficacy of ZMPs was evaluated against Gram-positive *C. perfringens* and Gram-negative *S. enterica* using a well diffusion assay, following our previous protocol [23]. Ofloxacin, was used as the positive control.

2.6. Assessment of Anticarcinogenic Activity through the MTT Assay

The anti-cancer effectiveness of ZMPs was investigated using the MTT assay, following the methodology outlined in our earlier works [24]. Cells were cultivated in a 96-well culture plate from Sigma-Aldrich, with specified amounts of NPs (0-300 µg/mL with 2 fold dilution). A cell density of 20,000 cells per well was upheld, and the cells were cultured in the RPMI-1640 medium. They were then exposed to ZMPs for 24 hours. Following the incubation period, MTT was introduced into the wells, and the plates were allowed to incubate for an extra duration of 3 to 4 hours. The mixture was subsequently aspirated, and 100 µL of dimethyl sulfoxide (DMSO) was added to each well, ensuring thorough mixing. To evaluate cell viability, the absorbance at 590 nm was measured using a microplate reader. % inhibition was then calculated as per the following formula:

$$\frac{[A_{test\ sample} - A_{control}] \times 100}{A_{control}}$$

2.7. DPPH Scavenging Activity

To access the antioxidant capabilities of ZMPs, the conventional DPPH assay was adopted with limited modifications [25]. In a nutshell, 3.94 mg of DPPH was dissolved in 10 mL of methanol, creating a one milli molar DPPH solution. Following that, 860 µL of a 50% methanol solution containing ZMPs at concentrations ranging from 25 to 400 mg/mL was combined with 140 µL of the 1 mM DPPH solution. The mixture was then incubated at 37°C for 30 minutes. Subsequently, the absorbance was measured at 520 nm, employing a 50% methanol solution as a blank. A control trial was conducted in the absence of the test sample. The scavenging activity (%) was then determined using the subsequent formula: = [1 − (As/Ac)] × 100.

"As" signifies the peak intensity at 517 nm for the DPPH solvent post-reaction, while "Ac" indicates the peak intensity for the control (without ZMPs).

3. Findings and Interpretations

3.1. PXRD Analysis

Figure 39.1 illustrates the PXRD obtained from ZMPs. Notably, the XRD peaks exhibit clear broadening, a strong indicator that the material prepared consists of nanoscale particles. Upon analyzing these XRD patterns, valuable data was extracted, including peak intensity, position, width, and FWHM. Specifically, the diffraction peaks at 31.84°, 34.48°, 36.32°, 47.61°, 56.58°, 63°, and 67.95°, have been precisely identified as belonging to the hexagonal Wurtzite phase of ZnO. These peaks correspond to the (ICDD) card No. 80-0074. Furthermore, this analysis reaffirms that the NPs are devoid of any impurities, as they exclusively show distinctive XRD peaks associated with ZnO.

To estimate the diameter of the ZMPs, we employed the Debye-Scherrer formula [26]:

$$d = 0.89\lambda/\beta\cos\theta \qquad (1)$$

In the given formula, 0.89 denotes Scherrer's constant, λ stands for the wavelength of X-rays, θ represents the Bragg diffraction angle, and λ signifies the FWHM of the diffraction peak corresponding to the ⟨101⟩ plane. Utilizing these calculations, the crystallite size was determined to be 34 nm

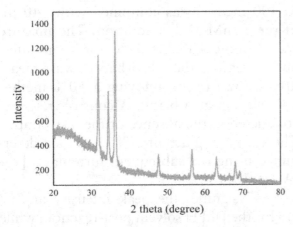

Figure 39.1: PXRD of ZMPs.

Figure 39.2: Plant *M. jalapa.*

3.2. Phyto-Constituents in Leaf

Figure 39.2 displays the leaves of the *M. jalapa* plant. The findings verified the presence of numerous phytochemicals, including flavonoids, carbohydrates, phenolic compounds, alkaloids, and amino acids, aligning closely with previous literature [27].

3.3. Antimicrobial Activity

The results illustrating the antibacterial effects of ZMPs on bacteria are succinctly presented in Table 39.1. These results indicate that ZMPs showcased the highest zone of inhibition (ZOI) at a concentration of 500μg/mL against both bacterial strains. Furthermore, the study of bacterial growth in the presence of these nanoparticles confirmed their bactericidal activity in a concentration-dependent manner, aligning well with existing literature.

3.4. Surface Characteristics

The SEM micrograph, Figure 39.3(a), distinctly depicts the predominantly spherical nature of the samples. Upon closer inspection, it becomes evident that these spherical crystals are interspersed with voids and pores. The formation of these voids and apertures can be attributed to the substantial discharge of gases from the reaction during combustion. It is within these voids of varying sizes and shapes that the individual crystallites intertwine with one another, forming the overall structure. Additionally,

Table 39.1: The antibacterial efficacy of ZMPs

Test organism	Concentration of ZMPs suspensions (µg/mL)				Positive control Ofloxacin
	500	250	125	62.5	(100 µg/mL)
C. perfringens	22.50±0.957	20.75±0.957	16.00±0.125	00.00±0.000	40.75±1.258
S. enterica	18.25±1.500	16.25±1.292	12.50±0.142	00.00±0.000	38.00±0.816

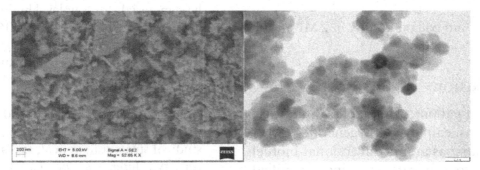

Figure 39.3: (a) SEM, and (b) TEM images of ZMPs.

Figure 39.3(b) showcases the TEM image of the ZMPs. The TEM analysis was conducted with the primary objective of gaining insights into the crystalline characteristics of these NPs. The TEM image reinforces the spherical shape of the particles, but it also reveals that the ZMPs exhibit non-uniform thickness. This data further underscores the diverse structural aspects of these NPs.

The surface area of the ZMPs was assessed through the conventional BET method, which entails the analysis of nitrogen adsorption and desorption isotherms. As a consequence of this examination, this value for ZMPs was calculated to be 21.444 m²/g.

3.5. MTT Assay

Figure 39.4 vividly portrays the cytotoxic impact of ZMPs on the MCF-7 cell line. The results from these investigations unequivocally affirm that the utilization of ZMPs amplifies the sensitivity of cancer cells. Evidently, there is a dosage-dependent decrease in cell viability, with an IC50 value precisely recorded at 76.94 µg/mL.

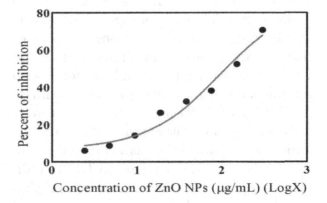

Figure 39.4: Measurement of IC50 in ZMPs.

3.6. Scavenging Activity

The results from the antioxidant investigations and the determination of IC50 values are visually depicted in Figure 39.5. To obtain the IC50 value for the DPPH radical scavenging activity of ZnO nanoparticles, a nonlinear regression analysis, employing a sigmoidal dose-response curve, was conducted utilizing GraphPad Prism 5. The calculated IC50 value was determined to be 109.22 µg/mL.

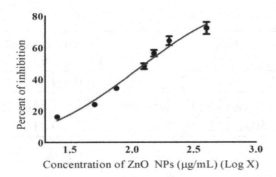

Figure 39.5: Antioxidant activity of ZMPs.

4. Conclusion

In this investigation, sustainable combustion synthesis was employed to generate ZMPs from *M. jalapa* as an unconventional biofuel source. Phytochemical screening identified various chemical constituents in the plant extract. Characterization of the resulting ZMPs revealed their nanoscale nature and a distinctive Wurtzite hexagonal structure. These ZMPs exhibited strong antimicrobial activity against both Gram-positive and Gram-negative bacteria and demonstrated dose-dependent anticancer properties with an IC50 value of 76.94 µg/mL in MCF-7 cancer cells. Additionally, they exhibited remarkable antioxidant capabilities, as evidenced by an IC50 value of 43,338 µg/mL in DPPH radical scavenging. These results not only emphasize the promise of sustainable nanomaterial synthesis but also underscore the significance of *M. jalapa* as a renewable biofuel source, but also underscore the versatile applications of ZMPs across healthcare and environmental conservation, suggesting a promising and sustainable future.

References

[1] Mousavizadegan, M., Firoozbakhtian, A., Hosseini, M., and Ju, H. (2023). Machine learning in analytical chemistry: from synthesis of nanostructures to their applications in luminescence sensing. *TrAC Trends Anal. Chem.*, 167, 117216.

[2] Sadiq, M. U., Shah, A., Haleem, A., Shah, S. M., and Shah, I. (2023). Eucalyptus globulus mediated green synthesis of environmentally benign metal based nanostructures: a review. *Nanomaterials*, 13, 2019. https://doi.org/10.3390/nano13132019.

[3] Nel, A., Xia, T., Mädler, L., and Li, N. (2006). Toxic potential of materials at the nanolevel. *Science*, 311(5761), 622–627.

[4] Yang, Y., Gao, H., and Zhou, W. (2020). Review on the recent progress of photocatalytic materials for low-level hydrogen peroxide detection. *Talanta*, 210, 120665.

[5] Singh, P., Kim, Y. J., and Zhang, D. (2020). Biological synthesis of nanoparticles from plants and microorganisms. *Trends Biotechnol*, 38(5), 588–605.

[6] Singh, R. K., and Singh, D. P. (2018). Hydrothermal synthesis of ZnO nanoparticles: study of structural, optical, and antibacterial activity. *Mater. Sci. Eng.: C*, 93, 813–823.

[7] Yu, S., Fu, X., and Zhang, L. (2021). A green and facile strategy to synthesize ZnO nanoparticles from organic zinc coordination compounds. *J. Alloys Compd.*, 853, 155209.

[8] Vasquez, Y. P., Paes, L. R., and Bernardi, A. C. (2021). Sustainable synthesis of ZnO nanoparticles using an aqueous extract of aloe barbadensis miller and evaluation of their photocatalytic activity. *J. Mol Liq.*, 327, 114671.

[9] Ahsani-Namin, Z., Norouzbeigi, R., and Shayesteh, H. (2022). Green mediated combustion synthesis of copper zinc oxide using Eryngium planum leaf extract as a natural green fuel: excellent adsorption capacity towards Congo red dye. *Ceram. Int.*, 48(14), 20961–20973.

[10] Vinutha, S. A., Meghashree, A. M., Gurudutt, D. M., Kudlur, D. S., Kumar, K. S., Karthik, G., and Mallikarjunaswamy, C. (2023). Facile green synthesis of cerium oxide nanoparticles using Jacaranda mimosifolia leaf extract and evaluation of their antibacterial and photodegradation activity. *Mater. Today: Proc.*, 89(Part 1), 2023, 105-112

[11] Rotti, R. B., et al. (2023). Green synthesis of MgO nanoparticles and its antibacterial properties. *Front. Chem.*, 11, 1143614.

[12] Jeevanandam, J., et al. (2022). Green approaches for the synthesis of metal and metal oxide nanoparticles using microbial and plant extracts. *Nanoscale*, 14(7), 2534–2571.

[13] Kumar, V., et al. (2023). From traditional to greener alternatives: potential of plant resources as a biotransformation tool in organic synthesis. *React. Chem. Eng.*, 8, 2677–2688

[14] Surendra, B.S., et al. (2022). Development of enhanced electrochemical sensor and antimicrobial studies of ZnO NPs synthesized using green plant extract. Sensors International, 3, 100176.

[15] Ahsani-Namin, Z., Norouzbeigi, R., and Shayesteh, H. (2022). Green mediated combustion synthesis of copper zinc oxide using Eryngium planum leaf extract as a natural green fuel: Excellent adsorption capacity towards Congo red dye. Ceramics International, 48(14), 20961–20973.

[16] Patil, K.C., Aruna, S.C., & Mimani, T. (2002). Combustion synthesis: an update. Current Opinion in Solid State and Materials Science, 6(6), 507–512.

[17] Patil, K.C., Hegde, M.S., Tanu, R., et al. (2008). Chemistry of Nanocrystalline Oxide Materials. Singapore: World Scientific.

[18] Prashanth, G.K., Prashanth, P.A., Nagabhushana, B.M., et al. (2018). Comparison of anticancer activity of biocompatible ZnO nanoparticles prepared by solution combustion synthesis using aqueous leaf extracts of Abutilon indicum, Melia azedarach, and Indigofera tinctoria as bio-fuels. Artificial Cells, Nanomedicine, and Biotechnology, 46, 968–979.

[19] Prashanth, G.K., Prashanth, P.A., Bora, U., et al. (2015). In vitro antibacterial and cytotoxicity studies of ZnO nanopowders prepared by combustion-assisted facile green synthesis. Karbala International Journal of Modern Science, 1(2), 67–77.

[20] Prashanth, G.K., Prashanth, P.A., Meghana Ramani, S., et al. (2019). Comparison of antimicrobial, antioxidant, and anticancer activities of ZnO nanoparticles prepared by lemon juice and citric acid fueled solution combustion synthesis. BioNanoScience, 9, 799–812.

[21] Farnsworth, N.R. (1966). Biological and phytochemical screening of plants. Journal of Pharmaceutical Sciences, 55, 225–276.

[22] Harborne, J.B. (1998). Phytochemical Methods (pp. 60–66). London: Chapman and Hall.

[23] Nagarajaiah, S., Nanda, N., Manjappa, P., et al. (2023). Evaluation of apoptosis in human breast cancer cell (MDA-MB-231) induced by ZnO nanoparticles synthesized using Piper betle leaf extract as bio-fuel. Applied Physics A, 129, 461

[24] Prashanth, G.K., Prashanth, P.A., Gadewar, M., et al. (2017). In vitro antibacterial and anticancer studies of ZnO nanoparticles prepared by sugar-fueled combustion synthesis. Advanced Materials Letters, 8, 24–29.

[25] Das, D., Nath, B.C., Phukon, P., Kalita, A., & Dolui, S.K. (2013). Synthesis of ZnO Nanoparticles and Evaluation of Antioxidant and Cytotoxic Activity. Colloids and Surfaces B: Biointerfaces, 111, 556–560. [doi.org/10.1016/j.colsurfb.2013.06.041]

[26] Cullity, B.D. (1967). Elements of X-Ray Diffraction. Addison-Wesley, Reading, Mass, USA, 3rd edition.

[27] Salman, S.M., Ud Din, I., Lutfullah, G., Shahwar, D., Shah, Z., Kamran, A.W., Nawaz, S., & Ali, S. (2015). Antimicrobial activities, essential element analysis, and preliminary phytochemical analysis of ethanolic extract of Mirabilis jalapa. International Journal of Biosciences, 7(4), 186–195.

Abbreviations

ZMPS: ZnO *Mirabilis jalapa* nanoparticles
$[Zn(NO_3)_2 \cdot 6H_2O]$: Zinc nitrate hexahydrate
TEM: Transmission Electron Microscopy
SEM: Scanning Electron Microscopy
BET: Brunauer-Emmett-Teller
PXRD: Powder X-ray Diffraction
C. perfringens: *Clostridium perfringens*
S. enterica: *Salmonella enterica*
MTT: (3-(4,5-Dimethylthiazol-2-yl)-2, 5-Diphenyltetrazolium Bromide).
DPPH: 2,2-diphenyl-1-picrylhydrazyl
SCS: solution combustion synthesis
ICDD: International Centre for Diffraction Data
FWHM: Full width at half-maximum

40 Spectroscopic characteristics and photoluminescence studies of Li+ co-doped Zn_2SiO_4: Eu^{3+} nanophosphors

K. N. Prathibha[1,a,2], B. V. Nagesh[1], and R. Hari Krishna[3,b]*

[1]Department of Physics, M.S. Ramaiah Institute of Technology (Affiliated to Visveswaraya Technological university, Belgaum), Bangalore, India
[2]Department of Physics, M E S College of Arts, Commerce and Science, Bangalore, India
[3]Department of Chemistry, M.S. Ramaiah Institute of Technology, Bangalore, India

Abstract

In this work we have taken the composition of the Zn1.95-x SiO4:Eu0.05Lix (x=0.00 to 0.06) luminescent powders, which were synthesised using the solution combustion process. Various characterisation techniques were employed to determine the crystallographic structure and phase composition, the surface morphological characteristics, including the shape and size of the particles. The techniques utilised include powder X-ray diffraction (PXRD), scanning electron microscopy (SEM), transmission electron microscopy (TEM), Fourier transform infrared spectroscopy (FTIR), and photoluminescence (PL). The photoluminescence excitation spectra of Zn_2SiO_4:Eu^{3+} co-doped Li+ nanophosphors were monitored for the emission wavelength at 614 nm. Interestingly, the photoluminescence emission spectra of Li+ co-doped Zn_2SiO_4:Eu^{3+} upon excitation at 395nm shows a notable increase in PL emission intensity with an increase in Li+ concentration. Excitingly, a stunning orange-red emission was observed at 614 nm, which signifies the effective incorporation of Li+ ions into the Zn_2SiO_4:Eu^{3+} nanophosphors. These findings highlight the significant role of Li+ concentration in influencing the luminescent properties of the nanophosphors. The optimum Li+ concentration for Li+ co-doped Zn_2SiO_4:Eu^{3+} nanophosphors was determined to be x=0.02. This concentration exhibited the highest PL emission intensity, thus indicating its potential for enhanced luminescence in various applications.

Keywords: Photoluminescence, nanophosphors, rare earth metals

1. Introduction

Over the decades, rare earth (RE) ions have been extensively used for the preparation of the luminophore because when rare earth ion doped with suitable inorganic host can act as an efficient luminescent material with high quantum yields, optical stability, energy transfer and emissions. Due to their promising properties of the rare earth doped inorganic nanophosphors, they have been used immensely in various potential and novel applications. Therefore, the selection of luminescent host material plays a major

role in order to modify the emission color of the activator due to their ligand field. Therefore, when nanosised inorganic phosphors incorporated with the trivalent rare earth ions may causes varied luminescence responses. The vacant position caused by the ions and the impurities present in the host matrix may leads a change in luminescence properties. Europium (Eu3+) is indeed a commonly used trivalent rare earth ion in the field of optoelectronics and lighting technology. Its efficient red emission makes it valuable for various applications,

[a]prathibhakn3@gmail.com, belavadinagesh@gmail.com, [b]rhk.chem@msrit.edu

DOI: 10.1201/9781003545941-40

particularly in light-emitting devices. The electronic transitions of Eu3+ ions, such as the 5D0→7F2 transition, lead to red emission, making them suitable for applications where red light is desired. The specific 5D0→7F2 electronic transition at 614nm is significant because it corresponds to the red region of the electromagnetic spectrum. This transition results in the emission of red light, which is essential for achieving vibrant and accurate color representation in display technologies. Overall, the unique optical properties of trivalent Europium, especially its efficient red emission, make it a valuable component in the development of advanced lighting and display technologies.

From the researcher's studies, we also have got to know that luminescence efficiency of phosphors can be enhanced by adding very small quantity of co-dopants to the host matrix. From the recent studies, the incorporation of rare earth co-dopants and non-rare earth co-dopants into phosphors to enhance their efficiency. Rare earth co-dopants are significant due to their involvement in energy transfer mechanisms, up-conversion, and cross-relaxation processes. The effects of non-rare earth co-dopants on luminescence efficiency need more study. However, some non-rare earth co-dopants have been found to increase luminescence efficiency, a topic of recent research interest. In this work, we specifically focuses on the use of alkali element (Li+) as a non-rare earth co-dopant in the synthesis of Zn_2SiO_4:Eu^{3+} phosphors. The selection of Li+ is based on its distinct ionic radius and charge state compared to the host material (Zn2+), causing crystal lattice distortion and disrupting high symmetry around Eu3+.

The distortion of the crystal lattice around Eu3+ leads to improved color parity. So, we have synthesised Li+ co-doped with Zn_2SiO_4:Eu^{3+} by Solution Combustion Synthesis method, a simple and less time-consuming method using oxalyldihydrazide (ODH)

as fuel. This method ensures to achieve a uniform distribution of dopants within the host matrix. The prepared samples with varying Li+ concentrations were studied in detail using various analytical and spectroscopic techniques. In this paper we investigates how co-dopants, particularly Li+, influence and enhance the photoluminescent properties of the Zn_2SiO_4:Eu^{3+} phosphor.

2. Experimental Section

2.1. *Materials and Synthesis*

To preparing Li+ co-doped Zn_2SiO_4:Eu^{3+} samples, we have used high grade purity chemicals which are not required further purification. Zinc Nitrate [Zn(NO3)3], Europium nitrate [Eu(NO3)3], Lithium nitrate [Li(NO3)3], Silicon dioxide(SiO2) and Oxalyhydrazide (C2H6N4O2) is used as an fuel to form products with specific chemical formula Zn1.95-x SiO4:Eu0.05Lix (x=0.00 to 0.06) by solution combustion synthesis method, where reactants need to be combined in appropriate stoichiometric ratios.

Fuel:Oxidizer ratio is calculated by equating F/O to 1 based on the oxidising and reducing valencies of the reactants. Stoichiometric amounts of reactants are dissolved in an aqueous solution. The solution is contained in a cylindrical petri dish with a capacity of nearly 300 ml. The prepared solution is introduced into a muffle furnace.

The furnace is maintained at a temperature of 500°C. The reaction mixture undergoes thermal dehydration, leading to auto-ignition. Gaseous products, including oxides of nitrogen and carbon, are liberated during combustion. The combustion of the fuel, oxalyldihydrazide (ODH), is an exothermic process. The released energy is sufficient for the synthesis process. The combustion propagates throughout the mixture without the need for external heating. The heat generated by the reaction is enough for the redox mixture and the formation of the desired product. The entire process takes approximately 5 to 10 minutes. The result of the combustion process is the production of Li+ co-doped Zn_2SiO_4:Eu^{3+} nano powder. The yield is further calcinated at 1000°C for 3 hours to obtain a crystalline product.

2.2. Characterisation

The Powder X-ray diffraction is being used for the structure characterisation and phase analysis of the samples. The diffraction pattern obtained helps identifying crystal structures and determining the phases present in the samples, which is carried out by Shimadzu 7000s powder X ray diffractometer (with CuKα, 1.541 Å and a nickel filter). The structural and morphological features are being investigated by using Hitachi scanning electron microscope(SEM). During the performance of SEM, a layer of Au is being deposited on the samples; by sputtering deposition method, since insulating samples are being used to get a clear SEM image. Transmission Electron Microscopy (TEM) and high resolution TEM(HRTEM) are examined using Hitachi H-8100 (LaB6 filament, accelerating voltage up to 200 kV) to examines particle size and provides detailed information about the internal structure of the phosphor particles. HRTEM is allowing for a higher resolution imaging . Further, the purity and identification of functional groups in the phosphor samples are being carried out by FTIR from Perkin Elmer spectrometer (Spectrum 1000) by preparing samples with KBr pellet method. And also it is providing information about the chemical composition and bonding in the material. Fluorolog–3 by Jobin Yvon (manufactured by HORIBA) equipped with a 450W xenon lamp as the excitation source is being used to record the photoluminescence (PL) excitation and emission characteristics of the prepared phosphors.

3. Results and Discussions

3.1. *Powder X-ray Diffraction Analysis*

The PXRD pattern of the $Zn_{1.95-x}SiO_4$:$Eu_{0.05}Li_x$ (x=0.00 to 0.06) samples are calcinated at 1000 °C for 3h is as shown in Figure 40.1(a-g) and XRD pattern shows they are well fitted with the orthorhombic crystal structure with Pbca symmetry space group which are readily indexed with the PDF Card -04-020-1855, shows the willemite formation. The dopant material is relatively pure within those concentration ranges, which is often important for ensuring the desired properties and functionality in various applications. This suggests a high level of purity and successful synthesis of the desired Zn_2SiO_4 lattice with Eu3+ and Li+ dopant. This indicates that dopant ions (Eu3+ and Li+) have effectively integrated into the Zn_2SiO_4 lattice. Site substitution occurs when dopant atoms replace atoms in the host lattice structure of a material. So,which helps to predict how well the dopant ions can substitute for the original ions in the Zn_2SiO_4 lattice. XRD peaks are

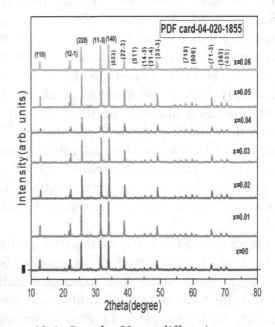

Figure 40.1: Powder X-ray diffraction pattern of $Zn_{1.95-x}SiO_4$:$Eu_{0.05}Li_x$ (x=0.00 to 0.06).

shifted towards higher angle (2θ values) with increase in Li⁺ concentration till (x=0.02), because of the substitution of larger Zn^{2+}ions by smaller Li⁺ ion results shrinking of unit cell also, whereas when x>0.02(with concentration of Li⁺increases) Li⁺ ions into the crystal lattice influences an expansion of the lattice unit cell due to their embedding in interstitial sites. This expansion is reflected in the XRD pattern by a peak shift towards lower 2θ values.

Calculating the crystallite size of the prepared samples using the Debye-Scherrer formula. The Debye-Scherrer formula is expressed as:

$$D = \left[\frac{K\lambda}{\beta \cos\theta}\right]$$ (1)

Where β; the Full Width Half Maximum (FWHM) of the XRD peaks θ in radians; the Bragg angle. Λ; wavelength of the X-rays used (λ = 1.541 Å, K ; the Scherrer constant(usually taken as around 0.9). From the calculations, the crystallite size for all synthesised samples found to be on an average of 38 nm, which confirms the prepared Zn₂SiO₄ crystals are in nanorange (Table 40.1).

3.2. *Fourier Transform Infrared Spectroscopy (FTIR)*

FTIR analysis provided information regarding the chemical bonding and functional

Table 40.1: Crystalline size and strain of $Zn_{1.95-x}SiO_4:Eu_{0.05}Li_x$ (x=0.00 to 0.06) is tabulated below

Concentration (x)	Crystalline size (nm) from Debye sheerer formula	Strain
00	37	0.00137
0.01	39	0.00123
0.02	40	0.00132
0.03	39	0.00127
0.04	34	0.00115
0.05	40	0.00117
0.06	39	0.00105

groups present in the samples. (Figure 40.2) gives the FTIR analysis of the prepared samples $Zn_{1.95}$-xSiO₄:Eu₀.₀₅Liₓ (x=0.03 to 0.06) shows a series of bands are matching with different bonding modes. Band at ~ 864 cm⁻¹ is assigned to the asymmetric stretching mode of SiO_4^{2-} ions. In the context of spectroscopy, the position of an absorption band (in this case, at 864 cm⁻¹) can be indicative of specific vibrational modes associated with the chemical bonds in the material.

The bands at 570–660 cm⁻¹ are attributed to the symmetric stretching modes of ZnO_4^{2-} and Zn-O-Si symmetric stretching modes. The range 570–660 cm⁻¹ suggests multiple vibrational modes within this frequency range associated with the mentioned chemical species. The bands around 1600 cm⁻¹ attributed to H-O-H stretching vibrations, which is due to the absorption of water in the samples, which may be expected during the preparations of pellets in using infrared measurements.

3.3. *SEM Images*

SEM images of Zn₂SiO₄: Eu³⁺ and Zn₂SiO₄: Eu³⁺ with co-dopant Li+ are shown in Figure 40.3(a and b). Cactus like structure can be seen in SEM micrographs, the reason for growth of particles with clusters may be due to samples at high calcination temperature and the sintering temperature. Also it is observed that a large number empty spaces or cavities (within a material or substance) are present in their structures, this may be

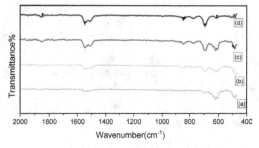

Figure 40.2: Fourier transform infrared spectra of Zn1.95-SiO4:Eu005Lix (x=0.03 to 0.06)

Figure 40.3: Images (a and b) are the SEM micrographs of Zn1.95-xSiO4:Eu0.05Lix (x=0.00 and x= 0.05) nanophosphors.

subjected to a large number of gases escaping during the combustion process of the reaction mixture. Both SEM micrographs shows similar structural features, which indicating (Li+ co-doping with Zn_2SiO_4: Eu^{3+}) that within the studied concentration range, there is no alteration of the shape, size, or arrangement of the phosphor particles. This suggests that varying the concentration of L+ dopants does not lead to noticeable changes in the physical structure or shape of the phosphor particles.

3.4. TEM

Transmission electron microscopy(TEM) images are performed on the Li+ co-doped Zn_2SiO_4 nanophosphor samples. Transmission electron micrographs of selected

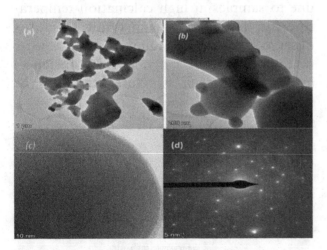

Figure 40.4: TEM images (a,b and c) of $Zn_{1.95-x}SiO_4$:$Eu_{0.05}Li_x$ and (d) SAED pattern of $Zn_{1.95-x}SiO_4$:$Eu_{0.05}Li_x$ (x=0.02).

area electron diffraction (SAED) pattern and high resolution TEM of the Zn1.95-SiO4:Eu0.05Lix (x=0.02)are shown in Figure 40.4. The TEM images(a,b and c) shows the particles are nearly spherical in shape with the diameter 10nm and they are appears to elongated structures formed by the fusion of crystallites or particles. SAED (Figure 40.4 d)image confirms the crystallinity of the prepared samples from clear diffraction spots.

3.5. Photoluminescence

Figure 40.5 shows the excitation spectrum of Zn1.95-x SiO4:Eu0.05Lix (x=0.00 to 0.06), which is having considerable peaks from 300

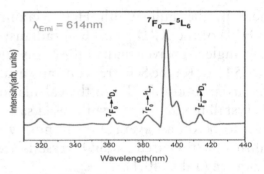

Figure 40.5: Photoluminescence excitation spectra of Zn_2SiO_4:Eu^{3+} co-doped Li^+ nanophosphors for the emission wavelength monitored at 614 nm.

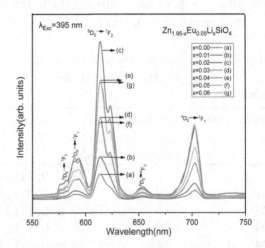

Figure 40.6: Pl emission spectra of $Zn_{1.95-x}SiO_4$:$Eu_{0.05}Li_x$ (x=0.00 to 0.06) nanophosphors monitored at excitation wavelength of 395nm.

to 450nm from f-f transition of Eu3+. The excitation peaks at 362nm ($7F0 \rightarrow 5D4$), 382nm ($7F0 \rightarrow 5L7$), 394nm($7F0 \rightarrow 5L6$), and 416nm($7F0 \rightarrow 5D3$)are observed. Due to the strong absorption band at 394nm, it is confirmed that $Zn1.95-x\ SiO4:Eu0.05Lix$ (x=0.00 to 0.06), the prepared samples can be good red emitting phosphor for WLED applications. Along with these, at 382nm and 416nm we have observed less intense and broad excitation bands that are attributed to the transitions from $7F0 \rightarrow 5G2, 5G3$ (382nm) and $7F0 \rightarrow 5D3$ (416nm) respectively. It is understood that the f-f transitions are forbidden in most of the hosts which results in the excitation spectra of europium (Eu3+) is weak and are not clearly observed in hosts material.

Figure 40.6 shows the photoluminescence emission spectra of $Zn1.95-x\ SiO4:Eu0.05$ for varying Li+ concentration is recorded in below room temperature which is excited at 394nm. A series of emission spectra are observed from 575nm to 700nm and they are related to various concentrations of the Li+ ions results $5D0 \rightarrow 7Fj$ (where j=0 to 4) transitions. Even with the different concentrations of the Li+ ions, the spectra of the phosphors show emission bands centered at 581, 590, 614 doublet 614 and 624nm, 652, and 703nm respectively. They are all associated with $5D0 \rightarrow 7Fj$(j= 0–4) transitions (570–700 nm; the region is considered to be orange–red spectrum) of Eu3+ions. The transitions which are corresponding with the ($5D0 \rightarrow 7F2$) at 590 and 614nm are considered to be most prominent lines. It is noticeable that even with addition of Li+ in $Zn1.95-x\ SiO4:Eu0.05$ will not vary the peak position instead we have seen great increase in the PL emission intensity. The different concentrations of Li+ ions with the $Zn_2SiO_4:Eu^{3+}$ results in increase of PL emission up to x = 0.02 after which it is decreases. This is said to be concentration quenching. Hence, the optimised Li+ concentration in $Zn1.95-x\ SiO4:Eu0.05$ is said to be x = 0.02. Therefore, the enhancement of PL intensity in

co-doping Li+ with $Zn_2SiO_4:Eu^{3+}$ phosphors can be achieved by charge compensation method. When trivalent europium (Eu3+) is replaced with the divalent Zinc(Zn2+), it happens to be charge imbalance in the system. So, two trivalent europium(Eu3+) ions are replaced with the three divalent Zn2+ ions to achieving charge compensation and effective lattice substitution when introducing dopant ions into a crystal lattice. However, a significant difference in the substitution of cation numbers can lead to certain limitations and the increase of cation vacancies. The presence of cation vacancies and point defects in the material acts as a quenching centers. Hence, it is a **challenge in achieving charge neutrality with trivalent to divalent cation replacement. in such case,we can use monovalent cations (alkali metals) for charge compensation.** Therefore, in this study to address the challenge, monovalent cations, specifically one Li+ ions, are employed for charge compensation along with two Zn2+ ions and one Eu3+. The introduction of Li+ co-doping not only aids in charge compensation but also leads to an increase in crystallinity.The improved crystallinity is evidenced by the presence of X-ray diffraction peaks in the material. Also, the charge compensation effect, along with the increased crystallinity, contributes to the enhancement of photoluminescence (PL) emission.

The increased crystallinity of the samples may, results in the incorporation of Li+ ions into the crystal lattice, as part of the charge compensation strategy, is suggested to positively impact the oscillating strengths for optical transitions,which leads to enhancement in the PL emission. However, we have observed increase in the PL emission only up to x = 0.02 beyond which it decreases. This may be due to luminescence quenching, after certain Li+ concentration there may be a creation of oxygen vacancies is going to be increases which causes increase in the probability of non-radiative transitions. The asymmetric ratio of the Eu3+ ion in the host lattice can be calculated by, calculating the ratio of red emission 5D0 → 7F2 (614 nm) to orange emission 5D0 → 7F1 (590 nm) emission of the activator near its site symmetry. The transition corresponds to 5D0 → 7F2 is an electric dipole transition and 5D0 →7F1 is a Magnetic dipole. Therefore, the ratio of electric dipole to the magnetic dipole is called ED/MD ratio. Generally, the crystal field around Eu3+ due to higher symmetry will give a, lower ED/MD value. In my samples the electric dipole transition 5D0 → 7F2 is not allowed when the Eu3+ is located in a centro-symmetric environment whereas the magnetic dipole transition 5D0–7F1 is insensitive to local symmetry.

4. Conclusion

We have successfully synthesised the Zn1.95-x SiO4:Eu0.05Lix (x=0.00 to 0.06) orange-red light emitting nanophosphor by the solution combustion method using ODH as fuel. The powder x-ray diffraction analysis shows the formation of crystalline with a willemite phase. The SEM micrographs reveal that particles are porous and contain voids due to high temperature presence. TEM analysis indicates that most particles are nearly spherical in shape with a diameter of 10nm, and they appear fused together, leading to elongated structures. PL excitation and emission spectra of Li+ co-doped Zn_2SiO_4:Eu^{3+} nanophosphors were recorded. As the concentration of Eu3+ ions increased in the samples (up to a concentration of x = 0.02), the photoluminescence emission also increases; beyond this concentration, quenching is observed. This suggests that Li+ co-doping enhances the PL emission intensity and the crystallinity of the phosphors. We conclude that luminescence improvement can be achieved through a charge compensation method in systems with Li+ co-activators.

References

[1] Monika, D. L., Nagabhushana, H., Hari Krishna, R., Nagabhushana, B. M., Sharma, S. C., and Thomas, T. (2014). Synthesis and photoluminescence properties of a novel Sr2CeO4:Dy nanophosphor with enhanced brightness by Li co-doping. *RSC Ad.*, 4, 38655–38662.

[2] Prathibha, K. N., Hari Krishna, R., Nagesh, B. V., Prakashbabu, D., Panigrahi, B. S., and Ananthanarayanan, R. (2021). Investigation of luminescence spectroscopic characteristics in Eu3+-doped Zn_2SiO_4 by Judd–Ofelt parameters. *J. Mater. Sci*0197–20210.

[3] Shivakumara, J., Chikkahanumantharayappa, Hari Krishna, V, Ashoka, S., and Nagaraju, G. (2018). CdSiO3:Eu3+ nanophosphor: one pot synthesis and enhancement of orange–red emission through Li+ co-doping. *J. Mater. Sci.*12986–12992.

[4] Dhananjaya, N. (2012). Enhanced photoluminescence of Gd”2O”3:Eu^3^+ nanophosphors with alkali (M=Li^+, Na^+, K^+) metal ion co-doping. *Spectrochim. Acta, Part A*. 86 (2012) 8–14.

[5] Prakashbabu, D., Ramalingam, H. B., Hari Krishna, R., Nagabhushana, B. M., et al. (2016). Charge compensation assisted enhancement of photoluminescence in combustion derived Li co-doped cubic ZrO :Eu nanophosphors. *Phys. Chem. Chem. Phys.* 18 (42), 29447–29457.

[6] Dillip, G. R., Dhoble, S. J., Manoj, L., Madhukar Reddy, C., and Deva Prasad Raju, B. (2012). A potential red emitting K4Ca(PO4)2: Eu3+ phosphor for white light emitting diodes. *J. Lumin.*132(11) , 3072–3076.

[7] Li, J., Wang, Y., and Liu, B. (2010). Influence of alkali metal ions doping content on photoluminescence of (Y, Gd) BO3:Eu red phosphors under VUV excitation. *J. Lumin.* 130(6), 981–985.

[8] Liu, Z., Feng, Y., Jiao, D., Zhang, N., and Jiao, H. (2008). Morphology and luminescent properties of Al3+/Mg2+-doped Y2O3:Eu3+ phosphor. *J. Mater. Sci.* 43, 1619–1623.

[9] Dhananjaya, N., Nagabhushana, H., Nagabhushana, B. M., Rudraswamy, B., Shivakumara, C., and Chakradhar, R. P. S. (2011). Effect of Li+- ion on enhancement of photoluminescence in Gd2O3:Eu3+ nanophosphors prepared by combustion technique. *J. Alloys Compd* 509(5), 2368–23741.

[10] Varma, A., Mukasyan, A. S., Rogachev, A. S., Manukyan, K. V. (2016). Solution combustion synthesis of nanoscale materials. *Chem. Rev.*, 116(23), 14493–14586.

[11] Dhananjaya, N., H. Nagabhushana, H., Nagabhushana, B. M., Rudraswamy, B., Shivakumara, C., Narahari, K., and Chakradhar, R. P. S. (2012). Enhanced photoluminescence of Gd2O3:Eu3+ nanophosphors with alkali (M=Li+, Na+, K+) metal ion co-doping. *Spectrochim. Acta Part A.* 86, 8–14.

[12] Bommareddy, P., Kumar, A., Rathore, A., Negi, D., and Saha, S. (2022). Engineering crystal structure and spin- phonon coupling in Ba1-xSrxMnO3. *J. Magn. Magn. Mater.* 541, 168539.

[13] Chen. (2011). New method for preparation of luminescent lanthanide materials BaMoO4:Eu3+. *Mater. Technol. Adv. Perform. Mater.* 26(2), 67–70.

[14] [14] Liang, H., Jia, L., Chen, F., Jing, S., and Tsiakaras, P. (2022). A novel efficient electrocatalyst for oxygen reduction and oxygen evolution reaction in Li-O2 batteries: Co/CoSe embedded N, Se co-doped carbon. *Appl. Catal.*, B.317, 121698.

[15] Monika, D. L., Nagabhushana, H., Sharma, S. C., Nagabhushana, B. M., and Hari Krishna, R. (2014). Synthesis of multicolor emitting Sr2–xSmxCeO4 nanophosphor with compositionally tuneable photo and thermoluminescence. *Chem. Eng. J.* 253, 155–164.

[16] Mamatha, G. R., Radha Krushna, B. R., Malleshappa, J., Sharma, S. C., et al. (2024). Investigating the influence of mono-, di-, and trivalent co- dopants (Li+, Na+, K+, Ca2+, Bi3+) on the photoluminescent properties and their prospective role in data security applications for SrAl2O4:Tb3+ nanophosphors synthesized via an eco-friendly combustion method. *Mater. Sci. Eng.*, B 299, 117008.

41 Investigation of CNT-enhanced aluminum matrix composites for high-wear resistance and self-lubrication

P. Gurusamy[1], G. Boopathy[2], G. Balaji[3], and N. Ramanan[4,a]

[1]Centre for Additive Manufacturing, Chennai Institute of Technology, Chennai, India
[2]Depatment of Aeronautical Engineering, Vel Tech Rangarajan Dr. Sagunthala R&D Institute of Science and Technology, Chennai, India
[3]Department of Aeronautical Engineering, Hindustan Institute of Technology and Science, Chennai, India
[4]Department of Mechanical Engineering, Sri Jayaram Institute of Engineering and Technology, Chennai, India

Abstract

In modern industrial applications, there is a growing need for lightweight components with exceptional wear resistance properties. These parts find application in numerous industrial domains. Additionally, materials with intrinsic self-lubricating qualities are becoming more and more necessary. In this study, we employed carbon nanotubes (CNTs) as nanoscale additions to improve an aluminum matrix composite. This particular material was selected due to its remarkable resistance to wear and its capacity to perform an efficient self-lubricating agent. In order to do this, the composite was mixed with varying quantities of CNTs, and stir casting was used during the casting process. The castings were then wire-cutted in order to prepare them for testing using electrical discharge machining (EDM). These samples underwent extensive testing and analysis, and data was obtained and displayed graphically. Wear testing carried out through the pin-on-disc method, and the corresponding output were highlighted through images. Additionally, machining tests using EDM wire cutting technology were conducted, leading to an analysis of the white coating that was formed on the material. Finally, this study tackles the urgent need for lightweight, self-lubricating materials in industrial applications, with a specific emphasis on the incorporation of carbon nanotubes into an aluminum matrix composite to improve wear resistance and self-lubrication qualities.

Keywords: Aluminum matrix composite, Carbon nanotubes (CNTs), self-lubrication, wear resistance, Electrical discharge machining (EDM)

1. Introduction

The current state of industrial production is marked by a growing need for materials that combine strong wear-resistant qualities with minimal weight. The necessity for high-performance components across a range of industries is what motivates this pursuit. The vital function that these materials play in enhancing the effectiveness, robustness, and general dependability of machinery and structures serves as further evidence of their significance. The essential relationship between wear resistance and self-lubrication lies at the core of this criterion. The search for robust, long-lasting industrial components begins with an understanding of a material's abrasion resistance, which quantifies its capacity to tolerate degradation brought on by friction, abrasion, or other mechanical forces. Simultaneously, the new attribute of self-lubricating qualities in materials is gaining significance.

[a]ramananinjs@gmail.com

DOI: 10.1201/9781003545941-41

The integration of wear resistance and autonomous lubrication is critical in high-performance settings where efficiency and dependability are paramount. Materials that can endure difficult working conditions while lowering energy dissipation through friction are becoming more and more important to various industries, including aerospace, automobile, heavy equipment, and precision equipment. Depending on this background, we explore the incorporation of carbon nanotubes (CNTs) with aluminum matrix composite. This composite material is purposefully chosen because of its remarkable wear resistance and natural capacity to act as a self-lubricating agent. Our goal is to contribute meaningful insights to the development of lightweight, high-performance materials as we navigate the complex interactions between various material properties. By conducting this research, we hope to address the broader issues raised by wear and friction and advance the development of materials that are resilient and provide a self-sustaining mechanism to improve performance.

Interest in the primary application of aluminum-metal matrix composite material has increased significantly in view of their widespread use in several fields such as automotive, aerospace, and defense. Most of the research activities in this field have mainly focused on metal matrix composites with a single reinforcement. In this research, we used stir casting technology to produce aluminum-metal matrix composite material reinforced with multiwalled carbon nanotubes (MWCNTs) and silicon carbide (SiC). It is important to note that the stir casting process often results in non-uniform distribution of MWCNTs. By applying sonication, we were able to achieve a more uniform gain distribution. It is noteworthy that there are limited previous studies documenting non-conventional machining of aluminum-SiC-MWCNT metal matrix composites.

CNT-reinforced aluminum matrix composites, featuring the integration of carbon nanotubes into an aluminum matrix, show great potential in terms of their ability to withstand high levels of wear and their inherent ability to behave independently. [1–4] Reinforcement with reduced graphene oxide (rGO) significantly increases the hardness of the composite, and the maximum hardness is observed when the weight percent (wt%) is 0.4 rGO. Resistance against wear of hybrid CNT-rGO/aluminum (Al) composites is pointedly better than that of rGO/Al composites [5]. Stir casting technology is used to synthesise Al-CNT composites. Various ratios of multi-walled carbon nanotubes have been used to strengthen the aluminum matrix [6]. A two-step approach was used to synthesise the aluminum-based metal matrix composite. Experimental procedures include performing potentiodynamic tests, electrochemical tests, linear polarisation tests, and evaluation of electrochemical impedance spectra [7].

Composites reinforced CNTs coated with titanium carbide (TiC) exhibit an increased yield stress when subjected to high-temperature deformation, in contrast to composites reinforced with uncoated CNTs. The reinforcing effect of these composites increases by 14% to 37% with increasing temperature [8] utilising a multi-step strengthening method with microchannels has significantly improved grain refinement. Incorporation of his CNTs into the Al matrix improved the tensile strength and decreased the specific wear rate. Studies have been conducted to investigate the effectiveness of approaches to introduce reinforcement into Al-CNT nanocomposites. The microstructural, mechanical, and tribological properties of different manufacturing processes are compared [9]. Al-CNT-Sn composites are created by pressing and sintering. Different weights of MWCNTs are used. CNT increases hardness and wear resistance. Friction and wear increase with temperature, but decrease with increasing oil [10].

The presence of MWCNT clusters advances the tribological stuffs of the composite. As the volume fraction increases, the wear losses and friction coefficient decrease

[11]. Techniques such as vacuum sintering, cold compaction, and solution ball milling were used. The investigation's main goal was to investigate strengthening mechanisms by using theoretical models and microstructure analysis [12]. Addition of 1 wt % of 50 wt % CNT and 50 wt % MgO reduced the wear weight loss of the Al-Cu-Mg alloy by about 49%. The wear weight loss of hybrid composite increased by 73% when the load increased from 2 to 5 N. [13] Low loading rates and high temperatures cause CNT/Al composites to lose strength. At temperatures as high as 603 K, the presence of CNTs enhances microstructural stability and restricts the migration of dislocations, resulting in improved strength [14]. CNT-reinforced aluminum composites exhibit increased strength and strain rate sensitivity. Innovative models based on microstructural analysis improve prediction accuracy of mechanical properties [15].

The addition of the SiC transition layer improves the interfacial fusion and increases the load-bearing capacity. Using CNT-reinforced AMC significantly improves mechanical performance [16]. CNTs incorporated into SLM-fabricated aluminum matrix nanocomposites exhibit excellent wear resistance. The inclusion of CNTs as reinforcement increases the hardness of the material surface and forms a well-defined hierarchical microstructure [17]. Mechanical properties of CNT/Al composites are improved by CNT reinforcement, matrix particle refinement, and the presence of a layered structure. The properties of model size, interfacial behavior, volume fraction of CNTs, and the presence of layered construction on the mechanical properties of CNT/Al composites are also investigated [18]. The process exposes a mixture of aluminum powder and ceramic reinforcement to high temperatures in a nitrogen-rich environment to produce aluminum matrix composites (AMCs) without relying on external influences or catalysts [19]. Aluminum/CNT composites can be used in weight-sensitive applications in the aerospace industry. A process known as Pressure Assisted Fast Electric Sintering (PAFES) has remarkable ability to effectively sinter the aforementioned aluminum/CNT composites [20].

The principal goal of this research is to achieve several goals in the exploration of composites consisting of multiwalled carbon nanotubes (Al-SiC-MWCNT) and aluminum-silicon carbide. The production of these composite materials is first carried out as precisely as possible, ensuring a systematic and controlled process. Next, a thorough analysis is carried out to determine the mechanical characteristics of the Al-SiC-MWCNT composites that have been created. This includes critical elements like tensile strength, fracture toughness and hardness. The goal of this work is to advance the mechanical properties of composite materials by optimising WEDM parameters. Processing conditions must be carefully considered. This study also addresses composite material wear behavior. It offers information about their resilience and longevity in various settings. The understanding and utilisation of Al-SiC-MWCNT composites in high-performance industrial domains are enhanced by this research.

1.1. Materials and Methods

In the course of this research, the manufacturing process of metal matrix composites requires the use of AA 6061 aluminum alloy as the base material. Silicon carbide and MWCNT reinforcements are incorporated into the powder structure to improve mechanical properties. The silicon carbide used in this study is characterised by a mesh size of 220. The manufacturing process also requires aluminum oxide and aluminum alloy ingots. Carefully cut the aluminum alloy ingot into small dimensions of 1cm x 1cm x 3mm. This size optimisation facilitates seamless placement of the aluminum alloy pieces into the graphite crucible during the subsequent melting process, ensuring uniform and controlled melting of the components.

1.2. Fabrication Process

Preheating the material to be cast is the first stage in the stir casting procedure [21,22]. A crucible made of graphite is filled with an aluminum rod that weighs about 1.5 kg. The reinforcement is then mixed proportionately in another crucible and heated similarly. For three to four hours, the reinforcements are each heated to 400°C while the aluminum rods are heated to 900°C. Following this preheating phase, reinforcement is added to the aluminum melt and both crucibles are removed. To eliminate gases and contaminants, deaerators and dedusters are introduced to the mixture. In order to completely fuse the aluminum alloy with the CNT and silicon carbide powders, the composite material in the crucible is heated further in a furnace. Place the crucible on a stirrer after it has been heated for two hours. Use the blade to rotate at 550 rpm to mix forcefully, and stir continuously for around 6 to 10 min. Mixture starts to harden and a matrix structure forms throughout this phase. The well-mixed composite material is put into a heated mold at 550°C to achieve the required mold form, ensuring total homogeneity. The composition particulars of the four samples are shown in Table 41.1.

Table 41.1: Matrix and reinforcement composition

Samples	AA6061 (%)	SiC (%)	CNT (%)
1	99.5	0	0.5
2	98	2	0
3	97.5	2	0.5
4	95.5	4	0.5
5	93.5	6	0.5

1.3. Evaluation of Mechanical Performance

The study of mechanical behavior has proven to be an important step in the evaluation of composite materials, providing essential insight into their structural strength and performance properties. Several mechanical tests are used to study the response of composite materials under different loading conditions. Tensile testing evaluates a material's performance and elasticity, while compression testing evaluates its ability to withstand compressive forces. Impact testing simulates sudden loading conditions and evaluates a material's resistance to dynamic forces. Additionally, hardness testing quantifies the penetration resistance of the material [23–25]. Taken together, these diverse tests contribute to a comprehensive considerate of the mechanical performance of composite materials then help optimise their designs and applications in various industries.

1.4. Tensile Test

Following machining, tensile testing is carried out in compliance with ASTM E8 guidelines. The specimen is securely fastened between the testing machine's two jaws. While the other end experiences a load that increases progressively, one end stays in place. A significant piece of information that indicates tensile strength of the composite is the recorded load at which the material breaks. This uniform testing process adds to a thorough understanding of mechanical nature of materials and ensures the accuracy and dependability of the measured tensile strength values.

1.5. Compression Test

According to the ASTM guidelines, the compression tests entail clamping the ends of the sample between two plates and then applying a weight that causes compression. Carefully measuring and comparing the resulting deformations to the applied loads is done. With this specific testing process, important mechanical properties can be determined, such as the material's yield strength and its yield strength under compression. Compression testing provides important insights into the structural behaviour and performance characteristics of a material by methodically evaluating its response to compressive stresses. This helps in the development and application of materials in technical and industrial environments.

1.6. Impact Test

Impact tests are carried out by applying sudden and dynamic loads to composite samples in compliance with ASTM standards. Measuring the energy absorbed by the sample during its fracture is the main objective of this test. Impact testing yields useful insights into the resilience and toughness of materials by measuring the results in joules. This allows for an assessment of the materials' capacity to survive rapid and severe loading circumstances. As a result, the information gathered using this testing methodology is essential for fully describing how materials behave in dynamic environments and providing a thorough picture of how well they function in practical applications where impact forces are common.

1.7. Hardness Test

Hardness is a fundamental property that has an inverse connection with toughness. It is a amount of a material's capability to withstand scratching. According to the guidelines provided by the ASTM, the Brinell hardness test is used to estimate the hardness of various sample materials. This specific evaluation method is applied in a methodical manner to samples that span a wide range of material compositions. The test indicated above uses a ball penetrator with a 10 mm diameter that is standardised and carefully made of hardened steel. A precisely applied stress of 500 kg/mm² allows the material's hardness to be

determined during test execution based on the impressions that are generated. The utilisation of this standardised test approach ensures the collection of reliable and consistent results. As such, this makes it easier to fully understand the characteristics linked to material hardness and provides essential data for a wide range of industrial uses.

1.8. Machining of Composites

Wire Electrical Discharge Machining (WEDM) is utilised to shape and frame composites accurately and successfully. The method includes utilising an electrically conductive wire as an anode to disintegrate the fabric through electrical releases. WEDM is particularly useful for machining composites since it can make complex shapes and keep up tight resiliences [26–28]. It's critical to evaluate the composite material's performance following machining, especially in relation to wear resistance. To find out how well a material resists abrasion and friction under particular circumstances, wear testing is done. When precise machining using Wire EDM is combined with wear testing afterward, composite materials are better understood and used in more industrial settings.

1.9. Wire Electrical Discharge Machining

The YCM W500 model of WEDM equipment is used for the successive machining steps in stir casting process used to produce aluminum-silicon carbide through MWCNT composite. Brass wire with a 0.25mm diameter is the electrode used in this process. Four input parameters, namely current (A), pulse-on time (µs), pulse-off time (µs), and tension in wire (N), are systematically changed at three distinct levels in order to maximise the machining conditions. Concurrently, distilled water is used as dielectric medium, the feed rate of wire is kept at 3 mm/min, and the gap voltage is kept at 40V. The machining process does not start until the system is completely submerged in water. The specimen's depth of cut is set at 15 mm to guarantee an extensive and regulated machining procedure for the aluminum-silicon carbide with MWCNT composite.

1.10. *Wear Test*

Through a Pin-on-Disc device, wear experiments were performed on the aluminum-silicon carbide composite including Multi-Wall Carbon Nanotube (MWCNT) in compliance with ASTM G99 standard. In the experimental setup, 8 mm diameter and 25 mm length cylindrical specimens were subjected to dry sliding wear against a revolving disk composed of EN-32 steel hardened to 65 RC. The Shimadzu electronic weighing apparatus, which has precision of 0.0001 g, was used to measure the specimens' initial and ending weights. This accurate measurement made it easier to calculate wear rates and assess how well the material performed under particular wear conditions, which gave important information about its tribological properties [29]. The utilisation of a standardised approach guarantees consistency and dependability in the evaluation of the composite material's wear resistance.

2. Results and Discussion

2.1. *Mechanical Behaviour of Composites*

The experimental results Table 41.2 provides a thorough overview of the mechanical properties of aluminum-silicon carbide with Multi-Wall Carbon Nanotube (MWCNT) composite. Tensile strength values range from 154.9 MPa to 186.14 MPa, indicating the samples' ability to withstand axial stress. Compression load values range from 75.70 kN to 87.93 kN, showing varying resistance to compressive forces. Hardness values range from 47.5 to 55.4, giving insights into resistance to indentation. Impact resistance ranges from 2 J to 4 J, reflecting the samples' ability to withstand sudden dynamic loads.

2.2. *Results of WEDM Process*

The outcomes of the experiment Table 41.3 offers an in-depth evaluation of the WEDM settings and associated results for several tests. Specific variables for each experiment include Tension in Wire (N), pulse on time (µs), pulse off time (µs), and current (A). Material Removal Rate (MRR) is measured in cm³/min, showcasing the rate of material removal during machining. For both MRR and surface roughness (µm), signal-to-noise (S/N) ratios are computed to show how each parameter affects the responses [30–33].

2.2.1. Parameter Effects on Material Removal Rate

When it comes to MRR, negative S/N ratios indicate a preference for lower values to improve machining efficiency. In the same way, lower surface roughness values are preferred when surface roughness S/N ratios are

Table 41.2: Results of mechanical testing

Sl.No.	Test samples	Compressive strength (kN)	Tensile strength (MPa)	Hardness (BHN)	Impact strength (J)
1	Sample1	81.21	165.22	55.2	5
2	Sample2	81.24	189.21	56.2	3
3	Sample3	85.52	159.23	54.2	4
4	Sample4	83.24	154.21	55.5	3
5	Sample5	84.21	153.33	55.2	5

Table 41.3: Results for WEDM process with L-16 Taguchi array

Exp.No.	Current A	P_{on} (μs)	P_{off} (μs)	Stress in wire (N)	MRR (cm³/min)	Ratio of S/N for MRR
1	35	10	12	2	0.006	−42.651
2	35	9	13	3	0.005	−43.144
3	35	15	18	5	0.006	−42.685
4	35	13	16	2	0.009	−38.772
5	37	10	15	4	0.007	−47.337
6	37	12	16	6	0.009	−38.505
7	37	14	18	5	0.008	−38.899
8	37	16	19	3	0.010	−38.024
9	42	10	14	5	0.006	−40.337
10	42	12	16	4	0.008	−40.074
11	42	14	18	3	0.011	−36.701
12	42	16	18	5	0.013	−35.259
13	45	10	15	3	0.012	−37.798
14	45	12	16	3	0.014	−36.961
15	45	14	12	5	0.018	−34.626
16	45	16	10	2	0.019	−38.126

negative. With the help of this large dataset, it is possible to analyze each machining parameter's effect on WEDM's performance in a methodical manner, which makes it easier to optimise the machining conditions and produce better surface quality and increased efficiency.

The plot graph that shows the average effect makes it clear that a significant MRR is seen when the machining parameters are changed to 42A for the current, 14 μs for the pulse on time (P_{on}), 10 μs for the pulse off time (P_{off}), and 4N for the wire tension. Notably, the optimal values derived from the experimental outcomes closely correspond to the average effect values, with the exception of the pulse-on time. A thorough confirmation test was performed to verify the mixture of variables. The confirmation test results verify the accuracy of the parameters produced from the average effect plot in terms of obtaining the best rates of material removal during the WEDM process. Tables 41.4 and 41.5 present the reaction tables for S/N ratios

and the ANOVA results related to MRR, respectively. Likewise, the mean effects are shown in Figure 41.1.

2.3. Results of Wear Test

The experimental results Table 41.5 provides a detailed insight into the wear parameters and corresponding outcomes for various experiments.

The comprehensive overview of wear rates observed under different test conditions

Table 41.4: Response table for MRR S/N ratios

Levels	Current A	P_{on} μs	P_{off} μs	Tension in Wire
1	−40.8	−42.8	−38.3	−35.2
2	−38.9	−42.2	−38.1	−36.7
3	−36.3	−36.7	−37.8	−48.1
4	−38.1	−34.3	−36.9	−37.1
Delta	4.7	4.5	0.8	1.0
Rank	1	2	4	3

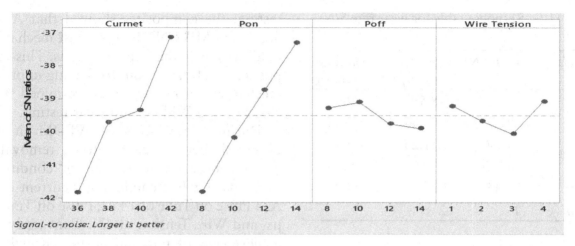

Figure 41.1: MRR mean effect Plot for S/N ratio.

Table 41.5: Results for wear test with L_{16} Taguchi array

ExperimentNo.	Load (N)	Speed of Disc(rpm)	Sliding velocity (m/s)	Wear rate (10^{-7} cm^3/min-m)
1	6	600	2	1.56
2	6	700	3	1.82
3	6	1300	4	1.85
4	6	1500	5	1.85
5	13	600	3	2.37
6	13	600	2	3.49
7	13	1100	3	3.99
8	13	1400	4	3.95
9	18	600	2	3.85
10	18	700	3	4.54
11	18	1100	2	4.75
12	18	1400	3	5.85
13	23	500	5	4.70
14	23	600	2	5.03
15	23	1100	3	5.55
16	23	1400	2	5.75

for the aluminum-silicon carbide with Multi-Wall Carbon Nanotube (MWCNT) composite is provided by the experimental results Table 41.6 The table encompasses various parameters including load (N), disc speed (rpm), sliding velocity (m/s), wear rate (10^{-7} cm^3/min-m), and the Signal-to-Noise (S/N) ratio for wear rate. Wear rate has an apparent pattern that is dependent upon the specific combinations of load, disc speed, and sliding velocity. For instance, a higher load often results in a higher wear rate, and varying sliding velocities and disc speeds influence various wear performances [34–37].

The response tables for S/N ratios and ANOVA results pertaining to wear rate are displayed in Tables 41.6 and 41.12, respectively. Likewise, the mean effects are shown

Table 41.6: Response table for wear rate S/N ratios

Levels	Load (N)	Speed of Disc (rpm)	Sliding velocity
1	–4.82	–8.59	–9.65
2	–11.25	–10.45	–9.85
3	–13.85	–11.25	–9.95
4	–15.56	–11.06	–9.65
Delta	10.15	3.57	0.47
Rank	1	2	3

in Figure 41.2 and equation of regression displayed in Table 41.6.

The mean effect plot indicates that the sliding velocity of 1 m/s, 400 rpm disc speed, and 5 N load result in the lowest wear rate. Given how closely these ideal values match the actual outcomes, it's possible that a confirmation test is not required. Strong comprehension of the factors affecting wear rate in the aluminum-silicon carbide with MWCNT composite is indicated by the convergence between the experimental data and the optimal values in the mean effect graph.

4. Conclusions

Stir Casting method was successfully used to fabricate Aluminum–SiC–MWCNT Composite, laying the groundwork for the rest of the investigation. Following mechanical testing, the composites showed that Al-4% SiC-0.5% MWCNT had the best mechanical qualities out of the four variations. This composite, which stands out for having improved mechanical properties, was examined more closely using WEDM and wear testing.

During the process of WEDM, it was observed that a higher rate of removal of material occurred under specific conditions, particularly when employing a current of 42 A, a Pulse on Time of 14 μs, Pulse off Time 10 μs, and Wire Tension of 4N. Conversely, the achievement of minimal Surface Roughness in Wire EDM required a different combination of parameters: a current of 36 A, Pulse on Time of 8 μs, Pulse off Time of 10 μs, and Wire Tension 1N. Wear testing additionally brought details on the best parameters for reducing wear rate, which included providing a 5N force, rotating the disc at 400 rpm, and sliding at a speed of 1 m/s. These thorough results greatly advance our understanding of the manufacturing, mechanical properties, and machining properties of Al-SiC-MWCNT composites.

References

[1] Abdeltawab, N., Esawi, M. A. M., and Wifi, A. (2023). Investigation of the Wear Behavior of Dual-Matrix Aluminum–(Aluminum–Carbon Nanotube) Composites. *Metals.*, 13(7), 1167.

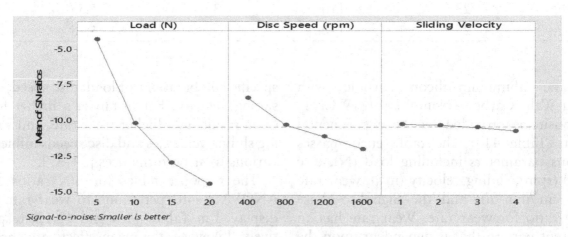

Figure 41.2: Wear Rate Mean Effects Plot for S/N Ratios.

[2] Van Trinh, P., Lee, J., Kang, B., Minh, P. N., Phuong, D. D., and Hong, S. H. (2022). Mechanical and wear properties of SiCp/CNT/Al6061 hybrid metal matrix composites. *Diam. Related. Mater.*, 124, 108952.

[3] Yu, T., Liu, J., He, Y., Tian, J., Chen, M., and Wang, Y. (2021). Microstructure and wear characterization of carbon nanotubes (CNTs) reinforced aluminum matrix nanocomposites manufactured using selective laser melting. *Wear.*, 476, 203581.

[4] Maniraj, S., Anand, K., Anbarasu, R., Aravindan, A. A., Gokul, G., and Logendran, R. (2022). Impacts of carbon nano tubes (CNT) and boron carbide (B_4C) particles on material properties of al 6061. *Mater. Today. Proc.*, 66, 738–742.

[5] Nyanor, P., El-Kady, O., Yehia, H. M., Hamada, A. S., Nakamura, K., and Hassan, M. A. (2021). Effect of carbon nanotube (CNT) content on the hardness, wear resistance and thermal expansion of in-situ reduced graphene oxide (rGO)-Reinforced aluminum matrix composites. *Met. Mater. Int.*, 27, 1315–1326.

[6] Tiwari, S. K., Dasgotra, A., Singh, V. K., Umamaheswararao, A., and Pandey, J. K. (2021). Wear Characteristics of Aluminum Composite Reinforced by Multiwall Carbon Nanotubes. In Recent Advances in Sustainable Technologies: Select Proceedings of ICAST 2020. 137–144. Springer Singapore.

[7] Popov, V. V., Pismenny, A., Larianovsky, N., Lapteva, A., and Safranchik, D. (2021). Corrosion resistance of Al–CNT metal matrix composites. *Mater.*, 14(13), 3530.

[8] Aborkin, A. V., Bokaryov, D. V., Pankratov, S. A., and Elkin, A. I. (2023). Increasing the Flow Stress during High-Temperature Deformation of Aluminum Matrix Composites Reinforced with TiC-Coated CNTs. *Ceram.*, 6(1), 231–240.

[9] Sharma, A., Fujii, H., and Paul, J. (2020). Influence of reinforcement incorporation approach on mechanical and tribological properties of AA6061-CNT nanocomposite fabricated via FSP. *J. Manuf. Processe.*, 59, 604–620.

[10] Dhore, V., Rathod, W., and Patil, K. (2021). Mechanical and tribological attributes of Al-CNT- Sn composites prepared by press and sintering. *J. Compos. Sci.*, 5(8), 215.

[11] Babu, J. S. S., Srinivasan, A., and Kang, C. G. (2021). Tribological and nano-scratch properties of aluminum (A356) based hybrid composites reinforced with MWCNTs/alumina fiber. *Met. Mater. Int.*, 27, 3666–3680.

[12] Nyanor, P., El-Kady, O., Yehia, H. M., Hamada, A. S., and Hassan, M. A. (2021). Effect of bimodal-sized hybrid TiC–CNT reinforcement on the mechanical properties and coefficient of thermal expansion of aluminium matrix composites. *Met. Mater Int.*, 27, 753–766.

[13] Ergül, E., Kurt, H. İ., Oduncuoğlu, M., and Can, Ç. İ. V. İ. (2020). Wear properties of Al-Cu-Mg composites reinforced with MGO and MWCNT under different loads. *Int. J. Mater.Eng. Technol.*, 2(2), 70--5.

[14] Wang, M., Shen, J., Chen, B., Wang, Y., Umeda, J., Kondoh, K., and Li, Y. (2022). Compressive behavior of CNT-reinforced aluminum matrix composites under various strain rates and temperatures. *Ceram. Int.*, 48(7), 10299–10310.

[15] Wang, M., Li, Y., Chen, B., Shi, D., Umeda, J., Kondoh, K., and Shen, J. (2021). The rate- dependent mechanical behavior of CNT-reinforced aluminum matrix composites under tensile loading. *Mater. Sci.Eng,.* A, 808, 140893.

[16] Zhang, X., Hou, X., Pan, D., Pan, B., Liu, L., Chen, B., and Li, S. (2020). Designable interfacial structure and its influence on interface reaction and performance of MWCNTs reinforced aluminum matrix composites. *Mater. Sci. Eng* , A, 793, 139783.

[17] Yu, T., Liu, J., He, Y., Tian, J., Chen, M., and Wang, Y. (2021). Microstructure and wear characterization of carbon nanotubes (CNTs) reinforced aluminum matrix nanocomposites manufactured using selective laser melting. *Wear.*, 476, 203581.

[18] Su, Y., Li, Z., Jiang, L., Gong, X., Fan, G., and Zhang, D. (2014). Computational structural modeling and mechanical behavior of carbon nanotube reinforced aluminum matrix composites. *Mater. Sci. Eng.*, A, 614, 273–283.

[19] Lee, K. B., Kim, S. H., Kim, D. Y., Cha, P. R., Kim, H. S., Choi, H. J., and Ahn, J. P. (2019). Aluminum matrix composites manufactured using nitridation-induced self-forming process. *Sci. Rep.*, 9(1), 20389.

[20] Marchisio, S., Manfredi, D., Deorsola, F. A., Biamino, S., Fino, P., and Pavese, M. (2012, June). Carbon nanotubes-reinforced aluminium with improved yield strength and toughness. In Proceedings of the 15th European Congress on Composite Materials, ECCM15, Venezia, Italy (pp. 24–28).

[21] Hooda, B., Thakur, S., Khurana, S., and Khokher, V. Metal Matrix Alloy AA 6061 Produced by Stir Casting Method. *IJFMR-Int. J. Multidiscip. Res.*, 5(2), 1–12.

[22] Gupta, M. K. (2023). Mechanical behaviors of Al 6063/TiB2 composites fabricated by stir casting process. *Mater. Today., Proc.*, 82, 222–226.

[23] Boopathy, G., Vijayakumar, K. R., Chinnapandian, M., and Gurusami, K. (2019). Development and experimental characterization of fibre metal laminates to predict the fatigue life. *Int J Innov Technol Explor Eng*, 8(10), 2815–2819.

[24] Kumaar, R. K., Kannan, G., Boopathy, G., and Surendar, G. (2017). Fabrication and Computational Analysis of Cenosphere Reinforced Aluminum Metal Matrix Composite Disc Brakes. *Technol*, 8(6), 553–563

[25] Boopathy, G., Vijayakumar, K. R., and Chinnapandian, M. (2017). Fabrication and Fatigue Analysis of Laminated Composite Plates. *Int. J. Mech. Eng. Technol.*, 8(7), 388–396.[26] Samson, R. M., Nirmal, R., and Ranjith, R. (2021). Parametric optimization of wire-EDM machining of nimonic 80a using response surface methodology. *IOP Conf. Ser. Mater. Sci. Eng.*, 1130(1), 012078. IOP Publishing.

[26] Boopathy, G., Prakash, J. U., Gurusami, K., and Kumar, J. S. P. (2022). Investigation on process parameters for injection moulding of nylon 6/SiC and nylon 6/B$_4$C composites. *Mater. Today, Proc.*, 52, 1676–1681.

[27] Arunadevi, M., and Prakash, C. P. S. (2021). Predictive analysis and multi objective optimization of wire-EDM process using ANN. *Mater. Today. Proc.*, 46, 6012–6016.

[28] Kaushik, J., Khan, H. A., Tiwari, A., Nafees, K., Varshney, S., and Singh, S. P. (2022). A review on application and optimization processes used for wear testing machine (pin on disc apparatus). *Mater. Today. Proc.*, 64, 1440–1444.

[29] Boopathy, G., Prakash, J. U., Gurusami, K., and Kumar, J. S. P. (2022). Investigation on process parameters for injection moulding of nylon 6/SiC and nylon 6/B$_4$C composites. *Mater. Today. Proc.*, 52, 1676–1681.

[30] Doreswamy, D., Bongale, A. M., Piekarski, M., Bongale, A., Kumar, S., Pimenov, D. Y., ... and Nadolny, K. (2021). Optimization and modeling of material removal rate in wire-EDM of silicon particle reinforced Al6061 composite. *Mater.*, 14(21), 6420.

[31] Arunadevi, M., and Prakash, C. P. S. (2021). Predictive analysis and multi objective optimization of wire-EDM process using ANN. *Materials Today. Proceedin Mater. Today. Proc* , 46, 6012–6016.

[32] Boopathy, G., Gurusami, K., Chinnapandian, M., and Vijayakumar, K. R. (2022). Optimization of process parameters for injection moulding of nylon 6/SiC and nylon 6/B$_4$C polymer matrix composites. *Fluid. Dyn. Mater. Process.*, 18(2), 223–232.

[33] Singh, K., Singh, A. K., Chattopadhyay, K. D., and Juyal, A. (2022). Optimization of tool wear rate during EDM of HSLA steel. *Mater. Today. Proc.*, 69, 361–364.

[34] Boopathy, G., Vanitha, V., Karthiga, K., Gugulothu, B., Pradeep, A., Pydi, H. P., and Vijayakumar, S. (2022). Optimization of tensile and impact strength for injection moulded nylon 66/SiC/B$_4$C composites. *J. Nanomater.*, 1, 1–9.

[35] Reddy, V. V., Mandava, R. K., Rao, V. R., and Mandava, S. (2022). Optimization of dry sliding wear parameters of Al 7075 MMC's using Taguchi method. *Materials Today Proc.*, 62, 6684–6688.

[36] Khan, M. M., Dey, A., and Hajam, M. I. (2022). Experimental investigation and optimization of dry sliding wear test parameters of aluminum based composites. *Silicon.*, 14(8), 4009–4026.

42 Next generation industrial automation and cloud systems – a review

M. S. Rishikesh[a], Praveen Sai, and Prashanth S. Humnabad

Department of Mechanical Engineering, Sir M Visvesvaraya Institute of Technology, Bengaluru, India

Abstract

The manufacturing industry is undergoing an immense change as a result of the implementation of cutting-edge technologies that include AI, Industrial IoT (IOT), Cloud computing, Big data analytics and Automation. These developments are transforming the manufacturing processes by enabling automation and optimisation inside fully digital smart factories. Traditionally, purpose control in manufacturing has been dominated by purpose-built, proprietary gear and software. Modern breakthroughs in software, development, automation and security, on the other have succeeded in the commoditisation of industrial process control. This shift to commodity hardware and software brings along many new advantages. Edge computing has been enabled through the combination of cloud technologies along with process control systems, enhancing data availability for Industrial IoT (IOT). Convergence of data from a variety of sources has been supported by cloud environments to facilitate analytical decision making and application interoperability. The future of distributed control systems is being transformed by such collective developments and the possibilities for modern manufacturing are expanding. They deliver the modern, network base experience equivalent to today's automatic cloud deployment. The fundamental architectural concept involves Advanced Computer Platforms (ACP's) managing control tasks and dedicated Distributed Control Nodes (DCNs) responsible for carrying out control processes.

Keywords: AI, cloud system, industrial automation, industrial revolution

1. Introduction

Industrial Revolution 4.O refers to the trend towards automation and data exchange in manufacturing technologies and processes. It encompasses various technologies such as cyber physical systems (CPS), Internet of Things (IoT), cloud computing, cognitive computing, and artificial intelligence. This revolution has led to development of the advanced systems and they are capable of self-awareness, self-learning, autonomous decision-making, self-execution, and adaption for production.

Automation plays a crucial role in this revolution by enabling machines to complete missions with small human intervention. Free people from related and repetitive tasks, allowing them to focus on more complex creative challenges. For instance, automation is essential for operating cutting-edge

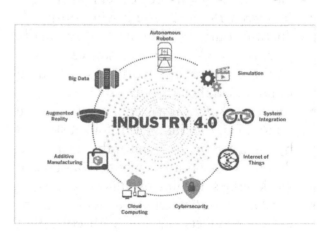

Figure 42.1: Industrial revolution 4.0.
Source: Google

[a]rishikesh091101@gmail.com

DOI: 10.1201/9781003545941-42

technologies like drones, self-driving cars, farm tools, robots, cloud robotic arms, Internet of Things components, and network security systems.

The rise of the automation has transformed multiple industries and it is expected to create new job opportunities while also raising concerns about job displacement. Telecom providers are leveraging automation to deliver reliable services and support the widescale rollout of 5G networks. The efficient operation and scale of automation are key factors for success in this era of the industrial revolution 4.0.

2. Objectives

The next generation industrial automation and cloud systems are driven by the variety of specific objectives aimed to enhance the product quality and production quantity. Details of the both the objective is described below:

2.1. Objective of Industrial Automation 4.0

- *Enhanced Product Quality:* Implementing the quality control measures and real-time monitoring to make sure the consistently high product quality. Automation and data analytics can help in identify and rectify defects in real-time.
- *Increased Productivity:* it will boost production output without proportionally increase in labour or in resources. Automation systems can work 24x7, leading to increased throughput.
- *Improved Safety:* it will enhance workplace safety by automating hazardous tasks and by using sensors to monitor the safety conditions. Collaborative robots (cobots) can work alongside humans safely.
- *Customer Satisfaction:* it will able to deliver products on time with consistent quality and to meet the customer expectations. Enhanced flexibility in production allows it for quicker response to customer demands.

- *Long-term Sustainability:* It develops a long-term strategy for the technology adoption and integration, by considering evolving industry trends and future-proofing systems
- *Competitive Advantage:* It gains a competitive edge by adopting the advanced automation and the cloud systems, allowing it for faster response to the market changes, customisation of products, and shorter time-to-market.
- *Talent Retention and Training:* It attract and retain skilled employees by providing opportunities for training and upskilling in the context of automation and advanced technologies

2.2. Objective of Cloud System

- *Real-time Data Insights:* It Leverage data analytics and AI/ML algorithms to gain actionable insights from the vast amount of data generated by IoT devices. This can lead to better decision-making and process optimisation.
- *Environmental Sustainability:* It reduces the energy consumption, waste, and carbon emissions by optimising resource usage and implementing eco-friendly manufacturing practices.
- *Data Security:* Implement robust cybersecurity measures to protect the sensitive data and ensure the integrity of industrial automation systems.
- *Remote Monitoring and Control:* It enable remote access to the industrial processes and equipment and allowing it for off-site monitoring and control. This can improve operational flexibility and reduce the need for physical presence on the factory floor.

3. Work Related

- *Industrial Automation and Industry Revolution 4.0:* Research related to the industrial revolution 4.0, it focuses on the integration of IoT, cloud computing, and automation in the manufacturing.

The studies on the use of cyber-physical systems (CPS) and the smart factories to improve the industrial processes and supply chain management.

- *IoT and Sensor Technologies:* Research on the deployment of IoT sensors and devices in industrial settings for the data collection, process monitoring, and asset tracking.
- *Cloud Computing in Manufacturing:* Research on the adoption of the cloud computing platforms in the manufacturing is for data storage, analysis, and remote monitoring.

  Case studies of companies that have successfully implemented cloud-based manufacturing solutions.
- *AI&ML in Industrial Automation:* Research on the application of the AI and ML algorithms for predictive maintenance, quality control, and process optimisation.

  Studies on computer vision and natural language processing (NLP) for the automation in industrial settings.
- **Edge Computing and Fog Computing:** Research on edge and fog computing technologies to process the data closer to the data source, reducing latency and bandwidth requirements. Case studies of edge computing applications in robotics and autonomous systems.
- *5G Connectivity and Low-Latency Networking:* Research on the impact of 5G networks on industrial automation, enabling low-latency communication and real-time control.

  Studies on the use of private 5G networks in the industrial environments.
- *Robotics and Collaborative Robots (Cobots):* Research on the integration of robotics, including the cobots, in the industrial automation and their role in enhancing the productivity and safety.

  Case studies of successful cobot deployments in manufacturing.

4. Advantage

- **Flexibility and scalability:** The cloud-based technologies enable a greater scalability and flexibility. It can possibly be set up more quickly by scaling existing systems to meet the client demands, reducing the cost of ownership through economies of scale.
- **Improved data availability:** The cloud-based environments facilitate data and convergence across multiple sources, improving data availability for the better decision-making.
- **Edge computing:** Edge computing can be performed using process control systems paired with the cloud technologies, enabling real-time data processing at the network's edge.

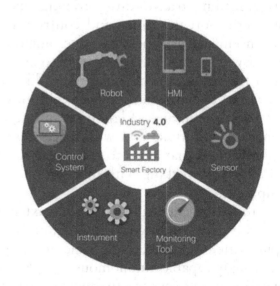

Figure 42.2: Advantage of industrial automation and cloud system.

Source: https://labhgroup.com/insights/6-benefits-of-industry-4-0/

- **Enhanced security:** Next-generation industrial automation systems emphasise security measures to safeguard the vital infrastructure from the cyberattacks.
- **Digital transformation:** Industrial automation4.0 and cloud solutions enable a digital transformation program by allowing the process makers to gather, process and act on data from the factory floor.
- **Enhanced Productivity:** Automation systems can operate around the clock and increasing productivity without the need for extra human labour.
- **Predictive Maintenance:** IoT sensors and data analytics provide predictive maintenance, minimising the equipment downtime and the prolonging machine lifespan.

5. Effects

The next generation of industrial automation and cloud technologies is predicted to have a substantial influence on numerous parts of the industry. While I don't have access to the specifics of the situation you stated, I can provide some broad perspectives.

Automation and Artificial Intelligence (AI), according to a McKinsey briefing report are altering organisations and contributing to economic development by increasing productivity. These technologies are also assisting in the resolution of social issues like as health and climate change. They will, however affect the nature of labour, resulting in changes in vocations and workforce transitions. Workers may need to learn new skills and adjust to working with more proficient equipment.

Automation and AI's significance extends beyond typical industrial automation and sophisticated robotics. New generations of increasingly capable autonomous systems are appearing in a variety of situations, from self-driving cars on the road to automatic checkouts in supermarkets. These accomplishments have been aided by recent advances in AI, machine learning algorithms, processing power and data availability.

6. Overview

The next generation of industrial automation and cloud technologies is predicted to have a substantial influence on numerous parts of the industry. While I don't have access to the specifics of the situation you stated, I can provide some broad perspectives.

Automation and Artificial Intelligence (AI), according to a McKinsey briefing report are altering organisations and contributing to economic development by increasing productivity. These technologies are also assisting in the resolution of social issues like as health and climate change. They will, however affect the nature of labour, resulting in changes in vocations and workforce transitions. Workers may need to learn new skills and adjust to working with more proficient equipment.

Automation and AI's significance extends beyond typical industrial automation and sophisticated robotics. New generations of increasingly capable autonomous systems are appearing in a variety of situations, from self-driving cars on the road to automatic checkouts in supermarkets. These accomplishments have been aided by recent advances in AI, machine learning algorithms, processing power and data availability.

7. Conclusion

Cloud platforms are crucial for the next generation of industrial automation. They provide scalable computing power and storage, enabling data storage, analysis, and remote access. Advanced analytics and AI algorithms are used to gain insights and make data-driven decisions.

Cloud-connected systems allow remote access and control of industrial processes and equipment, reducing the need for physical presence on the factory floor. Cloud-based platforms are highly scalable, allowing organisations to adapt to changing production demands and easily expand or downsize their operations. Robust cybersecurity measures are essential to protect sensitive data and maintain the integrity of industrial automation systems. Ensuring that different automation and control systems can work together seamlessly is essential. Industry standards like OPC UA and MQTT facilitate interoperability.

Automation streamlines processes, reduces manual labour, and optimises resource utilisation, resulting in increased operational efficiency and higher productivity. More creative and research-focused tasks, driving innovation within organisations.

Automation ensures consistent product quality by minimising variations and errors in manufacturing processes, enhancing customer satisfaction. Cloud systems provide access to real-time data analytics, enabling data-driven decision-making, process optimisation, and proactive issue resolution.

Allowing businesses to respond to changing market conditions and production demands swiftly. Cloud-connected systems enable remote access and control of industrial processes, reducing the need for on-site presence and enhancing operational flexibility.

Automation and cloud systems optimise resource usage, reduce waste, and contribute to environmental sustainability by minimising energy consumption and emissions. Which also in return deliver high-quality goods and services.

Automation technology necessitates a workforce with new skills, necessitating investment in training and upskilling to ensure that employees can work effectively with these systems.

References

[1] Lu, B., Bateman, N., & Cheng, K. (2016). Cloud manufacturing: from concept to practice. *Enterprise Information Systems*, 10(2), 135–154.

[2] Tao, F., Cheng, J., Qi, Q., Zhang, M., Zhang, H., & Sui, F. (2018). Digital twin-driven product design, manufacturing and service with big data. *The International Journal of Advanced Manufacturing Technology*, 94(9–12), 3563–3576.

[3] Xu, X. (2012). From cloud computing to cloud manufacturing. *Robotics and Computer-Integrated Manufacturing*, 28(1), 75–86.

[4] Wang, L., Törngren, M., & Onori, M. (2015). Current status and advancement of cyber-physical systems in manufacturing. *Journal of Manufacturing Systems*, 37, 517–527.

[5] Lee, J., Bagheri, B., & Kao, H. A. (2015). A cyber-physical systems architecture for industry 4.0-based manufacturing systems. *Manufacturing Letters*, 3, 18–23.

[6] Zhong, R. Y., Xu, X., Klotz, E., & Newman, S. T. (2017). Intelligent manufacturing in the context of industry 4.0: a review. *Engineering*, 3(5), 616–630.

[7] Jeschke, S., Brecher, C., Meisen, T., Özdemir, D., & Eschert, T. (2017). Industrial Internet of Things and Cyber Manufacturing Systems. In *Industrial Internet of Things* (pp. 3–19). Springer, Cham.

43 Design analysis of electric vehicle for E-commerce application

S. M. Sanjay Kumar[1,a], M. R. Praveen Kumar[2], A. Harshith[3], B. T. Naveen[3], R. Pavan[3], and M. Prasanna[3]

[1]Department of Mechanical Engineering, SJB Institute of Technology, Bengaluru, Karnataka, India
[2]Department of Mechanical Engineering, Bangalore Institute of Technology, Bengaluru, Karnataka, India
[3]Department of Mechanical Engineering, Department of Mechanical Engg., SJB Institute of Technology, Bengaluru, Karnataka, India

Abstract

Over numerous years, the automobile industry has faced enormous engineering hurdles in order to fulfil the needs caused by fuel shortages, fuel costs, pollution control, raw material costs, and safety. In order to tackle these problems, a broad range of new models have arisen or joined the market in quest of greater strength, safety, and fuel efficiency. As a result, three-wheeled motor vehicles have been created for public and private transportation purposes all throughout the world, particularly in developing nations. The goal of this project is to enhance the body and structure design of three-wheeled vehicles. The exact design will be created with CAD software [Unigraphics NX], and the analysis will be performed with ANSYS. The body design will be specified in accordance with the corporate specifications. The project's goal is to create a new electric vehicle with a load capacity of 500 kg. To examine the divergence of the acquired findings, the computed theoretical results were near to the ANSYS results.

Keywords: Electric vehicle (EV), chassis, three-wheeled vehicles (TWV), load capacity, E-commerce

1. Introduction

In today's world, transportation is vital. It has played a crucial role in the majority of historical societies. A fair indicator of a nation's economic development is the quality of its transportation infrastructure. The movement of people, products, and the facilities needed to support them can all be categorized as transportation. Different kinds of three-wheeled vehicles (TWV) are a crucial component of India's transportation network.[6].

Automotive manufacturers are now designing cars with higher energy efficiency and lower dimensions to maximise the utilisation of existing streets and roads. The three-wheel platform appears to be introduced organically by the concept of smaller, more energy-efficient automobiles for personal transportation. Generally, opinions are either highly opposed to the three-wheel configuration or firmly in favour of it. Supporters cite improved handling qualities, reduced production costs, and a mechanically simpler chassis. The three-wheeler's tendency to topple is criticised by its detractors. Both viewpoints are valid. Three-wheelers require less weight and are less expensive to construct. Three-wheeled cars can be configured with two wheels for power and steering in the back, one wheel for steering in the front, two wheels for power and steering in the front, two wheels for steering and two wheels for

[a]sanjay20376@gmail.com

DOI: 10.1201/9781003545941-43

power in the rear, or any other arrangement. The driver and passengers each have a bench seat, and the chassis and body are composed of pressed steel.

In Ethiopian cities, Bajaj three-wheelers have generally gained a lot of notoriety as a means of transportation. These cars are becoming more and more in demand every day. Due to a dearth of domestic auto businesses, Ethiopia is importing three-wheelers and their replacement parts, which is causing a significant loss of foreign cash and impeding the nation's economic development. Along with the driver, it has three passengers. Three-wheelers will always be a vital component of the city's transit network. They are numerous and rapidly increasing in number, so it is imperative that the advancement of technology proceed more quickly. Three-wheelers account for a sizable portion of urban mobility as intermediate public transportation.

2. Literature Survey

Kaushalendra Kumar Dubey et al. [1] The mechanical strength, durability, fatigue, and other material behaviour studies of ORV chassis are included in this research project. The goal of the proposed study is to determine the best methods for creating the safest roll cages (RCs) possible while taking into account many factors such as ergonomics, manufacturability, material selection, and design that must be considered while constructing off-road vehicles

M. Palanivendhan et al., [2] discussed about the development and design of automobiles The vehicle's self-weight and the entire external load, including passenger weight, are built upon the chassis frame, making the chassis an essential part of the automobile production sector. One important and essential component of the car is the chassis. Off-road vehicle (ORV) development and design will be impacted by design standards, material choices, and component convenience. The goal of the proposed study is to determine the best methods for

creating the safest roll cages (RCs) possible while taking into account many factors such as ergonomics, manufacturability, material selection, and design that must be considered while constructing off-road vehicles.

R.V. Patil et al. [3]. Through stress analysis, the three-wheeler chassis is optimised in this study to save weight and expenses. The front portion of the chassis was the only one to receive this optimisation; the specifications for the rear portion of the chassis remained the same. The optimisation resulted in stress and displacement charts. When the calculated outcomes for the optimised chassis are contrasted with those of the current chassis, it is discovered that the optimised chassis is safe.

Mohammad Waseem et al. [4] The three-wheeled "electric" vehicle's 3D chassis is made with the SolidWorks modeling environment. Using inertia relief analysis, the directional deformation, equivalent Von Misses stress, and force response under 2G vertical loading are found for the current and modified chassis of the three-wheeled vehicle.

Michal Janulin et al., [5] explored a number of options for building an electric vehicle for urban use that has a low-voltage DC motor power supply system. A simulation model was created and validated using a built-in city car with a low voltage electric motor system. The constructed simulation model was detailed and an examination of the energy flow was carried out (Figures 43.1–43.3).

3. Objective and Inputs for the Body Design

The following are the main objective of EV body design:

- To reduce the weight of the body to improve the vehicle performance. Lighter the vehicle, the lesser amount of energy it will necessitate to drive.
- To design the electric vehicle for commercial purpose like E-commercial vehicle and vegetable cart.

- To assemble the body in such a way that it distributes the weight of the body equally on chassis.
- To design vehicle for overall load carrying capacity of 500 Kg.

3.1. *Chassis Dimensions*

- Length: 2350 mm
- Width: 900 mm
- Height: 290 mm
- Thickness: 30*50mm & 40*60mm (hollow pipe)

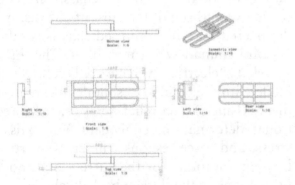

Figure 43.1: Chassis [6]dimensions and drafting.

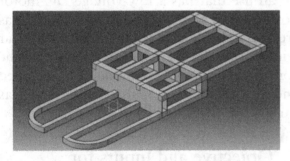

Figure 43.2: 3D Model of Chassis [6].

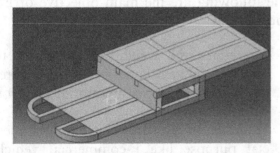

Figure 43.3: 3D Model of Chassis [6] with Sheet Metal Assembly.

4. Ansys Analysis

This type of vehicle is used to deliver goods in logistics, online shopping in cities & town. An E-cargo rickshaw is provided with a tilted carriage system. It ensures easy adjustment in carrying of goods in unfavorable street conditions.

4.1. *Data*

The vehicle body weight -150 kg
Chassis weight -100 kg
Load on the body -500 kg

4.2. *Static Analysis on E-Commerce Vehicle*

The step-by-step procedure for static structural problem can be stated as follows (Figures 43.4 and 43.5):

Step 1:- Geometry
Step 2:- Modelling
Step 3:- Meshing
Step 4:- Fixed support
Step 5:- Applying loading conditions
Step 6:- Solver
Step 7:- Computation of element strains and stresses.

4.3. *Static Analysis of the E-Commerce Vehicle*

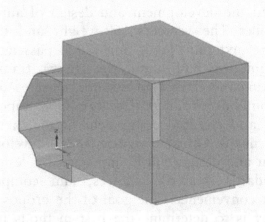

Figure 43.4: 3D model of E-commerce vehicle.

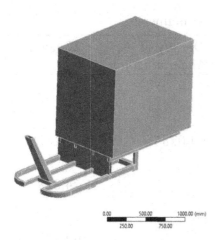

Figure 43.5: Modelling.

4.4. Container Weight Calculation for Different Material & Gauges

Formula:
Weight=Length *Width * Thickness * Density

For,
CRC (Cold Rolled Carbon Steel)
 Density=0.00785 g/mm^3
 Base-Gauge 18,
  T -1.31 mm
  L-1500 mm
  W-1100 mm
  Weight = 16.967 Kg
 Side-Gauge 23,
  T-0.78 mm
  L-1500 mm
  W-1500 mm
  Weight = 13.776 Kg
 Top & Back-Gauge 23,
  T-0.78 mm
  L-1500 mm
  W-1100 mm
  Weight = 10.102 Kg
Total Weight of The Container,
Weight = Base +(side*2) + Top + (Behind*2)
 = 16.967 + (13.776*2) + 10.102 +
 (10.102*2)
Weight = 74.828 Kg.

CRC (Cold Rolled Carbon Steel)
 Density=0.00785 g/mm^3
 Base-Gauge 14,
  T -1.99 mm
  L-1500 mm
  W-1100 mm
  Weight = 25.775 Kg
 Side-Gauge 17,
  T-1.46 mm
  L-1500 mm
  W-1500 mm
  Weight = 25.787 Kg
 Top & Back-Gauge 23,
  T-1.46 mm
  L-1500 mm
  W-1100 mm
  Weight = 18.91 Kg
Total Weight of The Container,
Weight = Base +(side*2) + Top + (Behind*2)
 = 25.775 + (25.787*2) + 18.91 +
 (18.91*2)
Weight = 134.079 Kg.

Aluminum Alloy Sheet
 Density=0.00281 g/mm^3
 Base-Gauge 9,
  T -2.91 mm
  L-1500 mm
  W-1100 mm
  Weight = 13.492 Kg
 Side-Gauge 10,
  T-2.59 mm
  L-1500 mm
  W-1500 mm
  Weight = 16.375 Kg
 Top & Back-Gauge 23,
  T-1.46 mm
  L-1500 mm
  W-1100 mm
  Weight = 12.008 Kg
Total Weight of The Container,
Weight = Base +(side*2) + Top + (Behind*2)
 = 13.492 + (16.375 *2) + 12.008 +
 (12.008 *2)
Weight = 82.266 Kg.

Table 43.1: Material characteristics [6] to calculate weight of container

Material	Gauge	Thidkness (mm)	Density of Material (g/mm³)	Weight (kg)
CRC (cold rolled steel)	12 (for base)	2.75	0.00785	211.43
	13 (rest of all)	2.37		
	14(for base)	1.99		134.079
	17(rest all)	1.46		
	18(for base)	1.31		74.828
	23 (rest of all)	0.78		
CRC & Aluminum	18(crc for base)	1.31	0.00785	62.134
	23 (crc for 3 side)	0.78		
	20 (rest other)	0.81	0.00281	
GI Sheet	G17 (for base)	1.46	0.0078	93.234
	G2O(rest of all)	1.01		
Aluminum Alloy	G9 (for base)	2.91	0.00281	60.442
	G10(rest of all)	2.59		
	GIO(for base)	2.59		82.266
	G12 (rastof of fill)	2.05		

5. Results and Discussion

The following are the results obtained in the Tables 43.2 and 43.3 indicated below. The overall chassis design is within the safe limits as checked with the ANSYS results (Figures 43.6–43.11).

Table 43.2: Mechanical properties of chassis material [6]

Structural Steel		
S.No	Properties	Values
1	Young's Modulus	2e+05MPa
2	Poison's Ratio	0.3
3	Density	7.85e–06kg/mm³
4	Yield strength	250MPa

Table 43.3: Static structural analysis results chassis

Results	Mininum	Maximum	Units
Equivalent Stress	2.7907e-010	45.503	MPa
Max Principal Stress	-9.0929	43.295	MPa
Min Principal Stress	-39.685	4.9504	MPa
Total Deformation	0	1.0464	mm

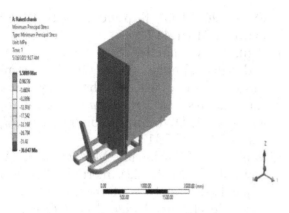

Figure 43.6: Meshing.

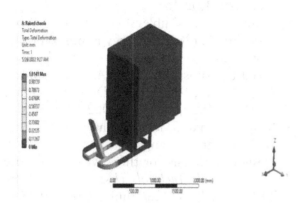

Figure 43.7: Static Structural Support.

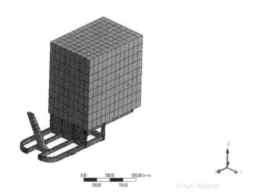

Figure 43.8: Equivalent stress.

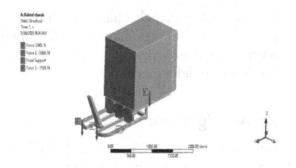

Figure 43.9: Maximum principal stress

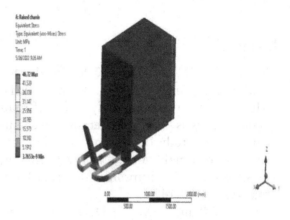

Figure 43.10: Minimum principal stress.

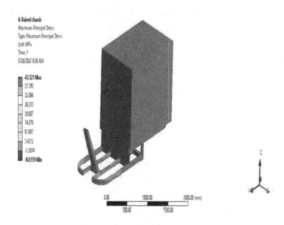

Figure 43.11: Total deformation.

6. Conclusions

These are the probable outcome of three-wheeler electric vehicle chassis design:

• Three-wheeled vehicle body design of Chassis for commercial need was successfully designed, as per the inputs given by the company.

• The design was modified by taking feedback from the customers.

• As per the survey, compared to previous design of the chassis we have made a compactable & more reliant design modification of existing design.

• The overall structural, theoretical and ANSYS results for the displacement of structure within the expected limit.

- A new electric vehicle chassis with a load capacity of 500 kgs has been designed with equivalent stress of 45.503 N/mm², 1.0464 mm deformation.

Acknowledgement

I would like to thank the team of OKTRA Electric Vehicle company for giving us an opportunity to design chassis for E-commerce vehicle.

References

[1] Dubey, K. K., Pathak, B., Singh, B. K., Rathore, P., and Yadav, S. R. S. (2021). Mechanical strength study of off-road vehicle chassis body materials mechanical strength study of off-road vehicle chassis body materials. *Mater. Today Proc.*, 46, 6682–6687, DOI: https://doi.org/10.1016/j.matpr.2021.04.147

[2] Palanivendhan, M., Devanand, S., Chandradass, J., Philip, J., and Reddy, S. S. (2021). Design and analysis of 3-wheeler chassis. *Mater. Today Proc.*, 45, 6958–6968, DOI: https://doi.org/10.1016/j.matpr.2021.01.417

[3] Patil, R. V., Lande, P. R., Reddy, Y. P., and Sahasrabudhe, A. V. (2017) Optimization of three wheeler chassis by linear static analysis. *Mater. Today Proc.*, 4 (8), 8806–8815, DOI: https://doi.org/10.1016/j.matpr.2017.07.231

[4] Waseem, M., Ahmad, M., Parveen, A., and Suhaib, M. (2022). Inertial relief technique based analysis of the three-wheeler E-vehicle chassis. *Mater. Today Proc.*, 49, 354–358, DOI: https://doi.org/10.1016/j.matpr.2021.02.158

[5] Janulin, M., Vrublevskyi, O. and Prokhorenko, A. (2022). Energy minimization in city electric vehicle using optimized multi-speed transmission. *Int. J. Automot. Mech. Eng.*, 19 (2), 9721–9733, DOI: https://doi.org/10.15282/ijame.19.2.2022.08.0750

[6] Sanjay Kumar, S. M., Harshith, A. Naveen, B. T., Pavan, R., and Prasanna, M. (2023). Design of chassis for three wheeled electric vehicle. *Bull. Technol. Hist.*, 23 (7), 109–114, DOI:10.37326/bthnlv22.1/1239(UGC care journal)

44 An engineering comparison for suitability of PLA material for light engineering activity using FEA simulation techniques

V. N. Shailaja[1,a], M. Rajesh[2,b], M. Appaiah[1,c], M. R. Deepthee[3,d], and B. Ravishankar[1,e]

[1]Department of Industrial Engineering and Management, B.M.S. College of Engineering, Bangalore, India

[2]Department of Industrial Engineering and Management, Ramaiah Institute of Technology, Bangalore, India

[3]Department of Mathematics, B.M.S. College of Engineering, Bangalore, India

Abstract

The use of Poly Lactic acid (PLA) material infused deposit Modelling (FDM) process is widespread. It is convenient due to factors such as i) cost, ii) environmental impact such as low fume generation during 3D Printing iii) easy availability from many supplier's and iv) percentage filling of material can be altered, making it a versatile material supportive of frugal engineering activity for development purpose. Studies are carried out to compare 2D beam structure models prepared to same dimensions using different filling densities of PLA's - 30%, 60%, 90% and 100%. The above models are tested using 250 kN capacity Servo Hydraulic Universal Testing Machine for parameters – Normal and Buckling stress. FEA studies are also considered to compare with actual test results. If appropriate PLA material spec is available/defined in the FEA database, we conclude that close correlation is observed between simulation and actual test results. However, it is required that FEA boundary conditions must be clearly defined. Simulations of PLA material can be used in lieu of test results for light engineering / prototyping development.

Keywords: PLA, FEA, 3D-printing, material testing, in-fill percentage

1. Introduction

One of the most commonly used natural raw material used in 3D PrintingisPolylactic Acid and it is commonly known as PLA plastic. It is popular as a base or filler material for FDM process [1–5]. It was discovered by Wallace Carothers in the 1920s. This does not require any heated bed as it can be printed at lowest temperature set between 190 and 220°C. The properties of PLA plastics have a

Density of 1.24 g/cm³, Tensile Strength of 50 MPa and Flexural Strength 80 MPa. Compared to other materials PLA is environmentally friendly when disposed of correctly, easy to 3D print, comes in different colour options enabling to provide various properties and appearances as it has wide

[a]vnshailaja.iem@bmsce.ac.in, [b]mrajeshiem@msrit.edu, [c]mayurappaiah.iem@bmsce.ac.in, [d]deeptheemr.maths@bmsce.ac.in, [e]drravi.iem@bmsce.ac.in

DOI: 10.1201/9781003545941-44

range of composites[6]. Thus, it can be used in applications like food containers and medical devices. In our study different filling densities of PLA's i.e 30%, 60%, 90% and 100% 3D printing models are prepared for testing.Higher the filling density and higher the layer height, stronger would be the yielded 3D printed part [7]. PLA material properties are known for its brittleness and not for strength. It has very low flexural strength compared to other plastics, but a stiff plastic. The outer fibres of the material are subjected to tension and the inner fibres are under pressure. Hence the bending stress i.e the flexural stress is maximum at the outer most fibres (both tensile and compression) and minimises as they approach towards the neutral fibre which exhibits inhomogeneous stress distribution. In bending test, the bending stress is greatest at the centre of the specimen with highest deflection. In this paper studies are carried out to compare 2D beam structure models prepared to same dimensions. These models are tested using 250 kN capacityServo Hydraulic Universal Testing Machine [8]. The test parameters include – Normal and Buckling stress. FEA studies are also considered to compare with actual test results.

Various tests like normal (tensile / compressive), impact and bending are considered to assess the mechanical properties to understand the composites with indicators. From these tests the mechanical characterisation reveals the behaviour of composites suitability for the intended purpose [9,10]. Multiple properties can be compared to find out the specific application of composites and the design criteria [11].

1.1. Tensile Properties

One of the most common methods of testing is the tensile testing, as is very simple and one of the most common type of mechanical testing method. This test is used to determine the σ-Є (stress-strain) behaviour of a polymer material composite [9]. In this test a pull type of force is exerted and the required stress is quantified. These samples are tested with different fillings till the specimen fails. σ-Є curves are then analysed for the test results like "E" (Young's Modulus) and the percentage of elongation [12]. The stretch can also be determined [13].

1.2. Flexural Properties

The bending/flexural/transverse test is typically used to measure the bending strength and modulus. Especially for structural applications it's one of the best and basic parameters used to find the feasibility of the composite materials [10]. These flexural properties are the outcome of both normal (tensile/compressive) and tangential stresses. Rectangular c/s(cross-section) component can be loaded either in 3 or 4 point bending modes[9].Maximum breaking stress of the component at the failure is indicated by the bending/flexural strength[9].The flexural modulus indicates the ratio of applied stress by deflection from the stress-strain deflection curve[9,13].These two test values help us to identify the components resistance to bending force applied.

The Figure 44.1 above depicts the steps in the 3D part fabrication process [14].

The mechanical properties of PLA are as shown in Table 44.1. These values were considered for our test purpose.

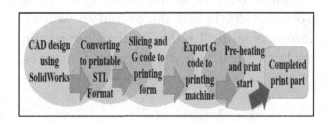

Figure 44.1: Steps in 3D Parts Fabrication.

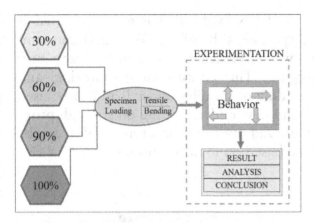

Figure 44.2: Test Process Adopted with PLA Material.

2. Methodologyand Comparision Study

2.1. *Polylactic Acid - Material Selection and Properties*

The material under study, Polylactic Acid (PLA) serves as the base component. The raw material was sourced from M/s. WoL (World of Lilliput) based in Mumbai, India. The properties used in this study are shown in Table 44.1 below.

2.2. *Testing Standards and Dimensional Preparation of Specimen*

The specimens for testing (tensile / bending) were created with CAD package from Solid-Works. This was necessary for 3D Printing process. Specimen preparation was as per the following standards [15,16]:

Tensile Specimen   ASTM D638

Bending Specimen: ASTM D790

2.3. *Specimen Creation with 3D FDM Machine*

In this study, Ultimaker 2 make 3D Printer, as shown in Figure 44.4, is used. It consists of single extrusion Bowden

Tube and head, automatic height adjusting system, nozzle system, etc. It is optimised for

a 2.85 mm dia. filament (1.75 mm filament can also be used) [17]. The schematic of the system is shown in Figure 44.5 below.

Table 44.1: Important properties–PLA

Property	Value	Unit
Tensile Strength	250	N/mm²
Yield Strength	10	N/mm²
Shear Modulus	318.9	N/mm²
Poisson's Ratio	0.35	-
Density	3410	Kg/m³

Figure 44.4: 3D printer – ultimaker 2.

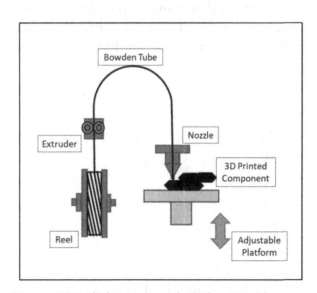

Figure 44.5: Schematic – ultimaker 2.

The printer utilises the Fused Deposition Modelling (FDM) method to carry out printing. It is versatile, capable of movement in the XYZ Axes, and has the requisite degrees of freedom to fabricate parts in a variety of different forms. Each Specimen takes about 35 minutes for printing, and some additional time is required for curing, cutting and other post processing activities.

For our study, 8 specimens were fabricated (4 tensile/ 4 bending), each with varying amounts of material in-fill. All required specimens were obtained using FDM technique, and were printed to dimensions specified (above mentioned ASTM Standards).

2.4. Tensile Test and Bending Test

Testing was done to understand the specimen strength w.r.t. specific load conditions (Tensile / Bending) and in-fill percentages (30%, 60%, 90% and 100%). The specimen loading is shown in Figure 44.6 below.

The specimen properties were revealed during testing, and results were tabulated and analysed. Tensile experiments were done with a 250 KN Universal Testing Machine (UTM) from M/s. MTS, USA. The Test set up is as shown in Figure 44.7 below, and testing was as per ASTM D638 Standard, with dimensions as shown in Figure 44.8 below. The speed was maintained at 1mm / min, the specimen clamped onto the UTM and a controlled force applied whilst being continuously monitored

by the 1 KN load cell. The test terminated at the specimens' breaking point, and one sample was stretched till fracture to glean new insights. The continuous monitoring and recording of samples under load can be used to gain new understanding on strength properties, and optimisation of assemblies can be done based on load conditions.

Figure 44.7: Tensile test setup - 250 kilo newton UTM.

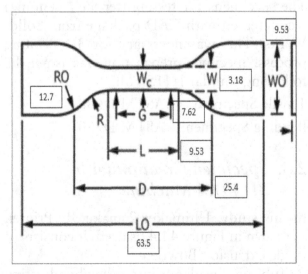

Figure 44.8: TensileSpecimen Dimensions (ASTM D638).

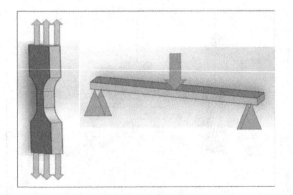

Figure 44.6: Specimen loading for tensile/ bending test.

Bending Test was carried out with Test Equipment from M/s. Edutech, UK. The test set up is as shown in Figure 44.9 below. Specimens were prepared as per ASTM D790 Standard, with dimensions as shown in Figure 44.10 below. Testing was done for maximum load capacity of test bench (i.e. 720 gram), and the maximum deflection was recorded in each case. The test was terminated at the maximum possible loading of the specimen. The results were tabulated and graphs plotted for further analysis and study.

2.5. FE Analysis and Creation of Library Models

FE Analysis is a tool for optimisation to produce product enhancements – lightweight, durable, easier to manufacture, alternate material, etc. Large digital structures are broken down into standard elements, whose nodes act as a pathway for loading and displacement. An understanding is obtained on likely material behaviour and typical mode of failure by using this method. In the present study, we use CosmosXpress FEA Suite from Dassault Systems for studying fatigue on the material due to loading conditions. The PLA material properties are not available in the FEA Model Library, and hence PLA material and its properties were added. This is shown in Figure 44.11 below. Four Assemblies were prepared using SolidWorks (Safety Valve, Plummer Block, Screw Jack and Knuckle Joint – Figure 44.12 below) and some critical components of these assemblies were taken up for FE Analysis.

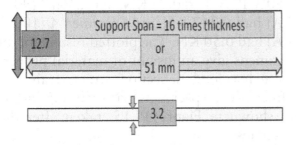

Figure 44.9: Bending test setup – edutech test bench

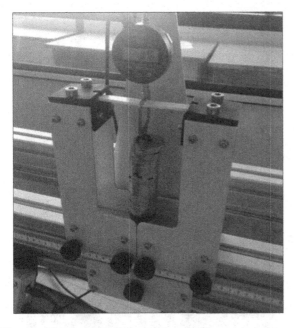

Figure 44.10: Bendingspecimen dimensions (ASTM D790)

Category: Plastic
Name: Custom Plastic - PLA 3D Printing

Property	Value	Units
Elastic Modulus	3500	N/mm^2
Poisson's Ratio	0.35	N/A
Shear Modulus	318.9	N/mm^2
Mass Density	3410	kg/m^3
Tensile Strength	250	N/mm^2
Compressive Strength		N/mm^2
Yield Strength	10	N/mm^2
Thermal Expansion Coefficient		/K
Thermal Conductivity	0.1	W/(m-K)

Apply Close Save Config... Help

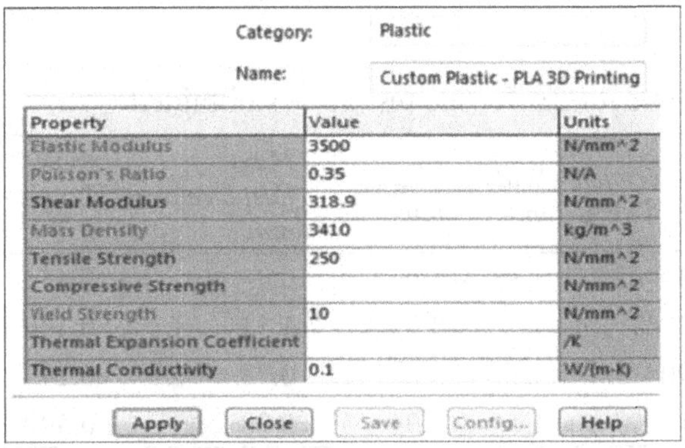

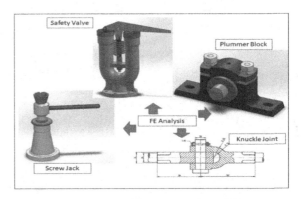

Figure 44.11: PLA properties added to material library

Figure 44.12: Assemblies considered for FE analysis

3. Results and Discussion

The Specimens taken for test had differences which were acceptable. Factors such as environment, temperature, molecular weight and grade of material used, loading region, etc. play a role in the obtained results.

3.1. *Tensile test Results - Specimens with Different In-fill*

Tensile testing was done for 4 specimens. The specimen details and experimental results of the test are shown in Table 44.2 below.

The values in Table 44.2 include inputs such as In-Fill Percentages. Dimensionally, all specimens are prepared to the same ASTM D638 Standard Outputs were obtained such as Axial Load, Deflection, etc. from which Stress and Strain values were derived, and the Stress-Strain graph plotted, as shown in Figure 44.13 above. The graph of tensile stiffness

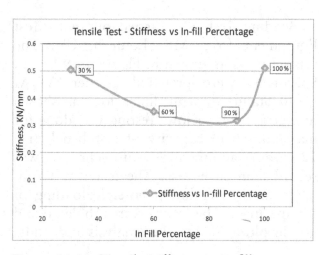

Figure 44.14: Tensile stiffness vs. in-fill percentage

was also plotted, as shown in Figure 44.14 above. Specimen 1 with PLA in-fill of 30% had a maximum load of 0.453 KN, Specimen 2 (60% PLA) had 0.53 KN, Specimen 3 (90% PLA) had 0.247 KN and Specimen 4 (100% PLA) had 0.50 KN. The plotted results for all specimens give an insight of material properties such as load, deflection, stress, strain and stiffness. The specimens with varying in-fills are shown in Figure 44.15, taken after the tensile test.

Table 44.2: Tensile test values obtained

Specimen In-Fill %	Width mm	Thickness mm	Max. Displacement mm	Max. Load KN
30%	3.39	3.05	0.9	0.453
60%	3.47	3.05	1.51	0.530
90%	3.55	3.08	0.78	0.247
100%	3.46	3.04	0.98	0.500

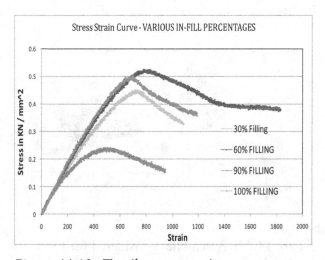

Figure 44.13: Tensile stress-strain curve

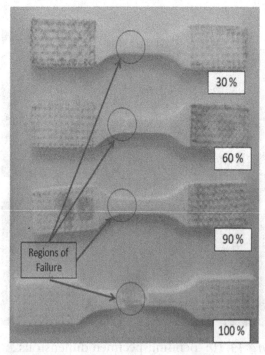

Figure 44.15: Specimens after tensile test

From the results, we see that highest stiffness is when the in-fill percentage is least, i.e at 30%, and then stiffness drops down, before again rising at around 100% in-fill, i.e. pure solid material (zero or no void).All failures occurred in the narrowest region of the specimen, with the appearance of a physical crack. The load cell on the Test Bench is able to sense the failure of the specimen and terminates the test at this point. One of the Specimens, Specimen 4, at 100% in-fill, continued testing till fracture, but as no new in-sight was gleaned, it was deemed sufficient to test the other specimens only till failure, stop the test and then carry out the tabulation and analysis activity. While conducting this experiment, there exists a possibility of experimental errors taking place, and factors which could influence test results include molecular weight of samples used, loading regions having full or no material (voids), the characteristics of PLA material, etc.

3.2. Bending Test Results for PLA Specimens

Bending Test was carried out for specimens with different in-fill percentages, details and experimental results of the test are shown in Table 44.3 below.

As shown in Table 44.3, in-fill percentages are the inputs and load and deflection are the outputs. From this, bending stiffness is calculated and tabulated. All specimens were fabricated with dimensions as per ASTM D790 Standard. The bending stiffness graph is shown in Figure 44.16 above. The maximum possible load of test bench (720

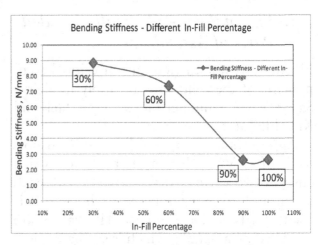

Figure 44.16: Bending stiffness vs. in-fill percentage

gram or 7.06 Newton) was used for testing all specimens, and the corresponding deflection was recorded and tabulated. We see that Specimen 1 with least in-fill (30%) has highest stiffness, and lowest deflection of 0.80 mm, Specimen 2 (60% in-fill) has 0.96 mm, Specimen 3 (with 90% in-fill) has deflection of 2.72 mm, and Specimen 4 (100% In-fill) has 2.68 mm deflection.

The results show that the highest bending stiffness occurs at lowest in-fill of 30%, and then falls with increasing in-fill percentages, with least stiffness seen at 100% in-fill (i.e. pure material). None of the specimens experienced any failure during the test. Bending Test samples are shown in Figure 44.17, taken after the test.

Table 44.3: Bending Test Values obtained

Specimen \In-Fill %	Load (gram)	Load (Newton)	Max. Displacement (mm)	Max. Stiffness (N/mm)
30%	720	7.06	0.80	8.83
60%	720	7.06	0.96	7.35
90%	720	7.06	2.72	2.60
100%	720	7.06	2.68	2.63

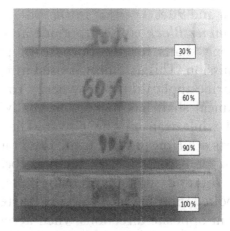

Figure 44.17: Specimens after bending test

3.3. FE Analysis – Critically Loaded Parts of Assemblies

FE Analysis was carried out for critical components of the following assembly models – Safety Valve, Plummer Block, Screw Jack and Knuckle Joint.

Safety valve is also called as Ramsbottom Safety Valve. It is a safety device which ensures two valves work in unison with each other, and are not tampered / adjusted. It is used widely in locomotives. The valves are operated by a handle, which is spring loaded. The spring is held in position by a *shackle*, a critically loaded part subjected to tension. This component was taken up for FE Analysis, with material being PLA and considering in-fill percentage as 100% (pure material). Under an axial load of 25 N , simulating tensile loading, the material deforms as shown in Figure 44.18. The maximum deformation occurs in the region of the upper hole, which also tends to be the region of least Factor of Safety. This results are interpreted as follows: the material tends to deform in the upper hole region , which is the weakest, having least FoS. So, at higher loading, failures will occur in this region. As per the tensile test results, an in-fill percentage of 100% will result in highest stiffness and ability to take highest possible loads. A 30% in-fill component can also be favourably used; such a component under tensile loading will give comparable performance, while resulting in 70% material saving. This is preferred to in-fill percentages of 60% and 90%.

Plummer Block is used to guide and support long shafts for the purpose of power transmission. This assembly is used in Industry, marine applications, etc. where a rotating mechanical shaft has to be continually supported while transmitting power. The critical parts are *Housing* and *Cap*. The Housing supports all components such as bearing shells, rotating shaft, cap, bolts, nuts, etc. The bolted housing is subjected to tensile forces in opposite directions when accepting the bearing shell during assembly. This is seen

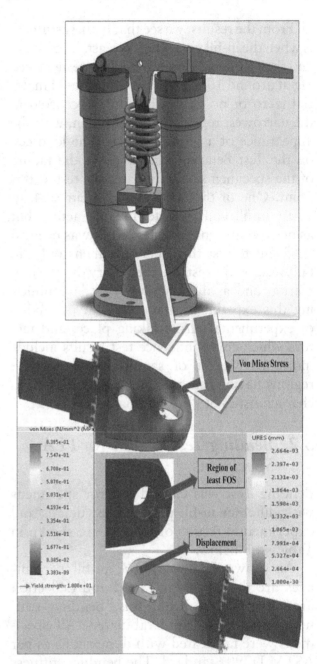

Figure 44.18: FE analysis of shackle (safety valve assembly)

in Figure 44.19 above. The cap experiences a bending stress when bolted at both ends and while accommodating the shaft and upper bearing shell. This results in a bending load with both ends fixed and being loaded with a 25 N force in the middle. The cap experiences displacement at the ends and at the central hole located in the centre of the cap. The region of least FOS is this central hole itself. The body housing under tensile

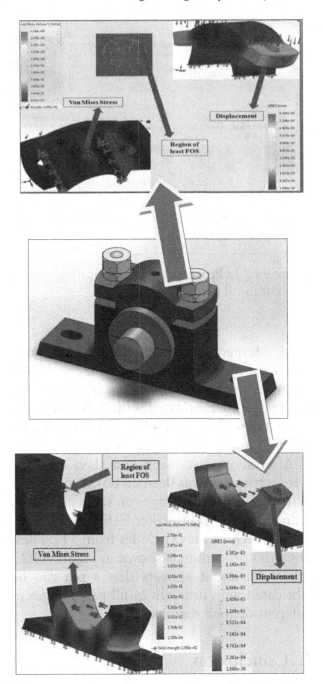

Figure 44.19: FE analysis of cap and housing (plummer block assembly)

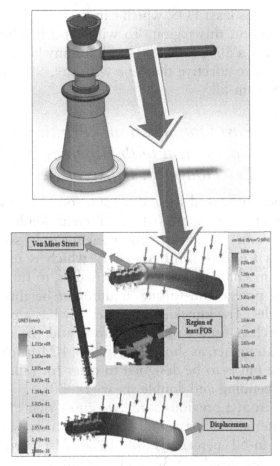

Figure 44.20: FE analysis of tommy bar (screw jack assembly)

action of 25 N, in opposite directions tends to deform, with least FOS being seen at the loading point. From bending test results, the cap can be made with 30% in-fill, having best stiffness. Higher in-fill percentages (90% and 100%) are not recommended for this part. The Housing can be with 100% in-fill or 30 % in-fill, especially in the region of tensile

load application / Von Mises stress. This is preferred over 60% in-fill, while 90% in-fill can also be used.

Screw Jack is used to lift weights with less effort using principle of screw threads. The *tommy bar* is used to turn the screw spindle and raise its height, thereby lifting a weight. One section of the tommy bar is inserted into the spindle, while the other protrudes out. A weight acting on the tommy bar can be idealised as a bending load on a cantilever beam, and is thus analysed for a bending load. In Figure 44.20 above, under action of a 25 N load at one end, we see that while maximum displacement is at the free end of the tommy bar, the maximum stress occurs in the portion inside the spindle, in the region where there is a junction of two different diameters, and this also has least FOS, which can lead to a likely failure in this region. To withstand bending forces, a 30% (or 60%) in-fill tommy bar will be more effective that one with say, 90% or 100% in-fill.

3.4. *Part Optimisation – Printing a Component with Different In-fill Percentages in Different Regions*

The authors discussed about some critical components subjected to different loads at the same time. Such a component can be seen in a **Knuckle Joint Assembly** which is used for mechanical power transmission, to transmit power and torque at different heights / angles. The knuckle joint assembly has a pin-which is used to hold the fork and the eye rod together, as shown in Figure 44.21 below. The pin is to be designed to cater for failure by bending and double shear [18–20].

From a previous study,we know that PLA material has maximum shear resistance with an in-fill percentage of 40%. Also, from our test results, a 30% PLA in-fill gives better results against bending loads. To obtain the best of both worlds, a brief study was undertaken to understand if 3D fabrication

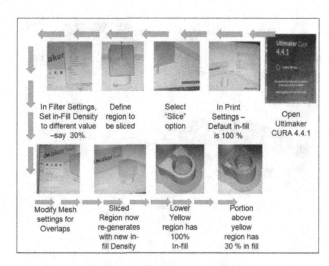

Figure 44.22: Procedure -part fabrication with different in-fill percentage

/ G-Code generation can be done on same part with different in-fill percentages. If, for example, the knuckle joint pin is considered, can the pin befabricated with different in-fill percentages (30% and 40%) at regions under different loads (bending and shear)?

After some discussion and checking software settings, it was deemed possible to print different in-fill percentages on same part. This is done by using Ultimaker CURA 4.4.1 Software, which creates G Codes from STL Files. The brief procedure is shown in Figure 44.22 below. Hence, it appears that a pin can be fabricated with different in-fill percentages in different defined regions.

4. Conclusion

A combination of material testing and FE Analysis is useful in not only understanding material properties under different loads, but also in quickly simulating assembly models to better understand their behaviour and likely failure under different load conditions (bending, tensile, shear, etc.). The ability to fabricate PLA components with different in-fill percentages allows not only material (and hence cost) saving, but also optimise the component for the load acting on it. The conclusions of the study undertaken are as follows:

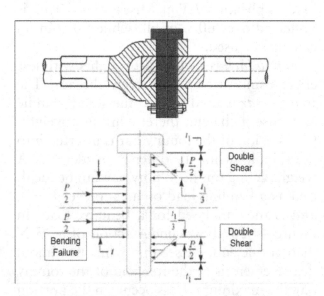

Figure 44.21: Pin subjected to both bending and shear load (Knuckle Joint Assembly)

1. While PLA material with pure material (100% in-fill percentage) has the best performance under tensile loads, that with 30% in-fill is almost as good, and markedly better than samples with higher infill percentages of 60% and 90%.
2. PLA material with lower in-fill percentage (30% and 60%) is better suited for moderate bending loads (720 grams/7 Newton) than that with higher in-fill percentage (90% and 100%).
3. FE Analysis is a quick and useful technique to simulate PLA and other Materials; however, non-linear plastic behaviour is better understood when combined with actual sample testing.
4. It is possible to fabricate a critical component with different in-fill percentages to better withstand different load conditions.
5. PLA is an excellent and easy to fabricate prototyping material, which can be further developed with greater understanding and testing of its properties.

Acknowledgements

The Authors would like to thank the following people for their support and encouragement:
The Principals of MSRIT and BMSCE for their constant support and encouragement. The Head of the Department, IEM, MSRIT. Head of the following Departments of BMSCE - IEM, ME, Aerospace, Mathematics and R&D. Dr. Anil Chandra and Mr. Mahesh (B.S. Narayana Centre for Structural Integrity Studies, BMSCE), Dr. Chethana (Aerospace Engineering, BMSCE), and Shri Praveen (3DPLM Lab, BMSCE).

References

[1] Cano-Vicent, A., Tambuwala, M.M., Hassan, S.S., Barh, D., Aljabali, A.A.A., Birkett, M., Arjunan, A., and Serrano-Aroca, Á. (2021). Fused deposition modelling: Current status, methodology, applications and future prospects. *Addit. Manuf.*, 47, 102378.

[2] Radzuan, N.A.M., Khalid, N.N., Foudzi, F.M., Royan, N.R.R., and Sulong, A.B. (2023). Mechanical analysis of 3D printed polyamide composites under different filler loadings. *Polymers*, 15, 1846.

[3] Ahmad, M.N., Ishak, M.R., Taha, M.M., Mustapha, F., and Leman, Z. (2023). A review of natural fiber-based filaments for 3D printing: filament fabrication and characterization. *Materials*, 16, 4052.

[4] Mazzanti, V., Malagutti, L., and Mollica, F. FDM 3D Printing of Polymers Containing Natural Fillers: A Review of their Mechanical Properties. Polymers **2019**, 11, 1094.

[5] Rajeshkumar, G., Arvindh Seshadri, S., Devnani, G.L., Sanjay, M.R., Siengchin, S., Maran, J. P., Al-Dhabi, N. A., Karuppiah, P., Mariadhas, V. A., Sivarajasekar, N., Ronaldo Anuf, A. (2021). Environment friendly, renewable and sustainable poly lactic acid (PLA) based natural fiber reinforced composites – A comprehensive review, *J. Cleaner Prod.*, 310, 127483. https://doi.org/10.1016/j.jclepro.2021.127483

[6] Moradi, M., Rezayat, M., Rozhbiany, F.A.R., Meiabadi, S., Casalino, G., Shamsborhan, M., Bijoy, A., Chakkingal, S., Lawrence, M., Mohammed, N., et al. (2023). Correlation between infill percentages, layer width, and mechanical properties in fused deposition modelling of poly-lactic acid 3D printing. *Machines*, 11,950. https://doi.org/10.3390/machines11100950

[7] Ranakoti, L, Gangil, B, Mishra, S. K, Singh, T, Sharma, S, Ilyas, R. A., El-Khatib, S. (2022). Critical review on polylactic acid: properties, structure, processing, biocomposites, and nanocomposites. *Materials (Basel)*, 15(12), 4312. doi: 10.3390/ma15124312. PMID: 35744371, PMCID: PMC9228835.

[8] Lusiak, T. and Knec, M. (2018). Use of ARAMIS for fatigue process control in the accelerated test for composites, *Trans. Res. Proc.*, 35, 250–258. https://doi.org/10.1016/j.trpro.2018.12.023. (https://www.sciencedirect.com/science/article/pii/S2352146518303636)

[9] Shahzad, A. (2017). 11-Mechanical Properties of Lignocellulosic Fiber Composites. In *Lignocellulosic Fibre and Biomass-Based Composite Materials*, Jawaid, M.,

Md Tahir, P., Saba, N., Eds., Woodhead Publishing: Sawston, UK, 193–223. ISBN 978-0-08-100959-8.

[10] Shesan, O.J., Stephen, A.C., Chioma, A.G., Neerish, R., and Rotimi, S.E. (2019). Improving the mechanical properties of natural fiber composites for structural and biomedical applications. In *Renewable and Sustainable Composites*, Agwuncha, C.S., Ed., IntechOpen: Rijeka, Croatia, ISBN 978-1-78984-216-6.

[11] Sadasivuni, K.K., Saha, P., Adhikari, J., Deshmukh, K., Ahamed, M.B., Cabibihan, J.J. (2019). Recent advances in mechanical properties of biopolymer composites: a review. *Polym. Compos.*, 41, 32–59.

[12] Long, S., Zhong, L., Lin, X., Chang, X., Wu, F., Wu, R., and Xie, F. (2021). Preparation of formyl cellulose and its enhancement effect on the mechanical and barrier properties of polylactic acid films. *Int. J. Biol. Macromol.*, 172, 82–92.

[13] Zhang, K., Chen, Z., Smith, L.M., Hong, G., Song, W., and Zhang, S. (2021). Polypyrrole-modified bamboo fiber/polylactic acid with enhanced mechanical, the antistatic properties and thermal stability. *Ind. Crops Prod.*, 162, 113227.

[14] Moradi, M., Rezayat, M., Rozhbiany, F.A.R., Meiabadi, S., Casalino, G., Shamsborhan, M., Bijoy, A., Chakkingal, S., Lawrence, M., Mohammed, N., et al. (2023). Correlation between infill percentages, layer width, and mechanical properties in fused deposition modelling of poly-lactic acid 3D printing. *Machines*, 11, 950. https://doi.org/10.3390/machines 11100950

[15] ASTM Standards, Standard Test Method for Tensile Properties of Plastics, ASTM International, PA, USA. 2012. Available online: https://www.astm.org/d0638-03.html

[16] ASTM Standards, Standard Test Method for Tensile Properties of Plastics, ASTM International, PA, USA. 2012. Available online: https://www.astm.org/d0790-17.html

[17] Ultimaker 2 User Manual, https://support.makerbot.com/s/article/1667337895718, (accessed on 12-11-2023)

[18] *Design of Machine Elements: For V Semester B.E. Me/Ip/Au/Im/Ma As Per The New Syllabus Of Vtu- 1* (1st ed., Vol. 1). (2009). [1]. Sapna Book House – Bangalore, ISBN10-8128002376

[19] Ehrmann, G. and Ehrmann, A.(2021). Investigation of the shape-memory properties of 3D Printed PLA structures with different infills. *Polymers*, 13,(1),164. https://doi.org/10.3390/polym13010164

[20] Sandanamsamy, L., Harun, W.S.W., Ishak, I. et al. (2023). A comprehensive review on fused deposition modelling of polylactic acid. *Prog AdditManuf*, 8, 775–799. https://doi.org/10.1007/s40964-022-00356-w

45 Effect of alloying elements on mechanical properties of LM6 alloy on addition of magnesium, copper and zinc

V. Ashith Raj[1,a], M. R. Shivakumar[1,a], M. Shilpa[1,a], Murali Krishna Panchagam[2,b], and B. Sanath Kumar[1,a]

[1]Industrial Engineering and Management, Ramaiah Institute of Technology, Bangalore, India
[2]Department of Chemistry, Ramaiah Institute of Technology, Bangalore, India

Abstract

For last several decades, aluminium and its alloys are used in automotive industries because of their properties such as low density, good malleability, formability, and high corrosion resistance, high electrical and thermal conductivity. Pure aluminium and its alloys, however, suffer from low strength and hardness. By incorporating alloying elements in aluminium alloys, as well as by heat treatment, these properties may be altered. In this regard, the present work is carried out by developing new alloy to improve some of the properties of LM6 by the addition of alloying elements: magnesium, copper and zinc. Alloys are produced through stir casting process and Design of Experiments (DOE) is adopted. Mechanical properties of developed alloys are investigated; X-ray Diffraction (XRD) and microstructural examination is carried out to check the quality of castings that are produced. Trend of the properties of the developed alloys are analysed. Also, Analysis of Variance (ANOVA) is carried out to confirm the significance of the alloying elements. Confirmation test is carried out at optimised levels of alloying elements and the tensile strength and hardness properties are noted. The identified optimum levels are Cu - 6%, Zn - 3% and Mg - 6% respectively. With this experimentation, it is noted that the tensile strength has increased from 166.3Mpa to 213.33Mpa and Vickers's hardness has increased from 95HV to 143.443HV when compared with base LM6 alloy.

Keywords: Analysis of variance, design of experiments, LM6 alloy, tensile strength, hardness

1. Introduction

Aluminium alloys are not only popular ecause of their unique strength relative to weight and material modulus, but they also have low wear resistance and poor hardness [1]. To improve the hardness and wear resistance, new aluminium alloys are being developed through research, with appropriate composition of alloying elements [2]. To improve their properties, alloys are processed with appropriate heat treatment and mechanical treatment methods, and composite materials are developed by reinforcing the hard ceramic materials [3].

LM6 is a popular aluminium alloy used in aerospace, marine and automobile applications [4]. LM6 aluminium alloy is a eutectic alloy; it has good impact strength, moderate durability, and also it is adaptable to both sand and permanent mould casting. It is relatively easy to machine and shape, which allows its use for precise and consistent manufacturing. It has good resistance to corrosion, and is able to retain its strength and

[a]vashithraj123@gmail.com, mrshivakumar@msrit.edu, mallikashilpa@gmail.com, sanathkumarb03@gmail.com, [b]muralikp@msrit.edu

DOI: 10.1201/9781003545941-45

other properties even after being exposed to higher temperatures [5]. These characteristics of LM6 makes it an ideal candidate for use in high-temperature applications [6]. It is an ideal material for use in aerospace applications where weight is a critical factor. LM6 suffers from low hardness and tensile strength and this work aims to improve these properties [7].

Literature shows the study of the effect of copper content on the microstructure and properties of LM6 aluminum alloy. The results showed that increasing the copper content resulted in a finer and more uniform distribution of silicon particles, which led to improved mechanical properties. Literature shows the influence of zinc on the solidification characteristics and mechanical properties of LM6 aluminum alloy [8]. The results demonstrated the effect of zinc on solidification characteristics and resulted in refined grain structure, along with improved mechanical properties. Literature presents the examination towards the study of influence of magnesium element on the microstructure and mechanical properties of LM6. The results showed that increased magnesium content led to refined microstructure and better mechanical properties, such as tensile strength and hardness [9].

From the literature, it is observed that the desirable properties of LM6 aluminium alloy may be improved by the addition of alloy elements. Literature shows that, alloying elements such as, magnesium, copper and zinc are the contributing to the enhancement of strength and hardness on LM6.

Present work aims to develop a new aluminium alloy to improve the properties such as tensile strength and hardness of the most popularly used LM6, by the addition of alloying elements: magnesium, copper and zinc. Stir casting method of alloy development is adopted, which is economical and suitable method. Statistical approach of experimentation is used so as to reduce the experimentation efforts, to optimise the composition of

alloying elements and to identify the significance of alloying elements in the new alloy.

2. Materials and Casting Preparation

LM6 is used as base material which is procured for the purpose of experimentation. The composition of LM6 is 87% aluminium, 11.5% silicon, 0.1% magnesium, 0.1% copper, 0.1% zinc, 0.6% iron, 0.5% manganese, 0.2% titanium and smaller percentages of other elements. Magnesium, copper and zinc are the alloying elements used in this experimentation, and are procured in chip forms. Stir casting methodology is used to produce the castings having different compositions [10].

3. Experimentation and Analysis

Taguchi's orthogonal array approach is used for the experimentation [11]. Magnesium, copper and zinc are considered as factors in the experimentation. Literature reveals the maximum amount of alloying elements that can be added to the base aluminium, which is shown in Table 45.1.

The factors and their levels considered for the present experimentation are shown in the Table 45.2. These levels are selected based on

Table 45.1: Maximum % of alloying elements in Al

Alloying element	Max. % in Al
Copper	6.5%
Zinc	5%
Magnesium	10%

Table 45.2: Percent allying elements and levels

Factors	Copper (%)	Zinc (%)	Magnesium (%)
Level 1	2	1	3
Level 2	4	2	6
Level 3	6	3	9

acceptance of elements by the aluminium for alloy development. The uncontrollable factors such as maximum melt temperature, pouring temperature of melt, the stirring speed, and pre-heat temperature of die are maintained at the same conditions during all the castings' preparation [12]. This is to nullify the effect of uncontrolled factors on the properties of the newly developed castings.

Taguchi's L9 orthogonal array layout and experimentation layout for three factors each at three levels is shown in Table 45.3.

Nine castings are developed using stir casting method as per the experimental layout shown in Table 45.3. LM6 alloy was melted in the crucible at 875°C in an electric resistance furnace and calculated amounts of copper, zinc and magnesium are added to the melt [13]. After the addition of alloying elements, the melt was stirred for two minutes at constant speed of 500 rpm to ensure proper mixing of alloying elements in the melt. Hexachloroethane tablets were added to the melt as degassing agent, to remove trapped gases and avoid blow holes in the castings. The slag was removed. Pre-heated (300°C) steel mould having runner and riser was used to make the castings. Engine oil of 40 grade was applied to the inner surface of the mould to avoid sticking of castings to the surface of the mould [14]. The molten metal was poured into the mould at 650°C and allowed to solidify for a period of 30 minutes. Test specimens were prepared as per the standards using Computer Numerical Control (CNC) turning machine.

To perform the hardness test, the surface of the specimen was polished and Vickers's hardness testing machine was used. To conduct the tensile test, the specimens were prepared as per ASTME8 standards. The microstructural examination of the specimens was made using the inverted metallurgical microscope. X-ray diffraction test was used to determine the crystal structure of the specimen material.

4. Results and Discussion

Microstructural examination was carried out on all the specimens; it revealed the clear grain boundaries and there were no impurities or inclusions in the prepared castings. XRD test was carried out at 20° using Nifilter and CuKa radiation on Rigaku desktop Miniflex II X-ray diffract meter and the spectra graph is shown in Figure 45.1. Each and every peak of the spectra has been analysed. The peaks in XRD spectra reveal that Al, Cu, Zn and Mg have been successfully incorporated in LM6 alloy. Figure 45.1 shows the XRD graph of experimental run number 3.

Table 45.3: L9 Orthogonal Array Layout and Experimental Layout

Runs	L9 Orthogonal Array			Physical Experimental Layout		
	Column 1	Column 2	Column 3	Copper (%)	Zinc (%)	Magnesium (%)
1	1	1	1	2	1	3
2	1	2	2	2	2	6
3	1	3	3	2	3	9
4	2	1	2	4	1	6
5	2	2	3	4	2	9
6	2	3	1	4	3	3
7	3	1	3	6	1	9
8	3	2	1	6	2	3
9	3	3	2	6	3	6

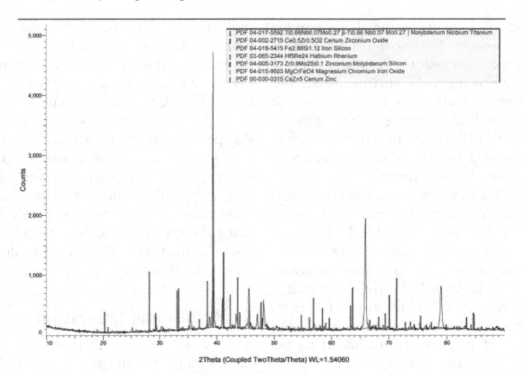

Figure 45.1: XRD graph for one experimental trial

4.1. Tensile Test

Tensile test was carried out on all the specimens and the Ultimate Tensile Strength (UTS) was noted. These test results were converted into Taguchi's Signal-to-Noise (S/N) ratios. S/N ratio for larger-the-better characteristic was used, as one of the objectives of the work is to maximise the tensile property of the developed castings. Table 45.4 shows the tensile test results and S/N ratio values.

The optimum combination of elements and their level to maximise tensile strength is Cu 6%, Zinc-3% and Mg-6% is identified by the response Table 45.5. ANOVA was performed on S/N ratio (UTS) and p-values are shown in Table 45.6. The p-value is less

Table 45.4: Tensile test result and S/N ratio (UTS)

Run No.	Cu (%)	Zn (%)	Mg (%)	UTS (MPa)			Mean UTS	S/N ratio (UTS)
				T1	T2	T3		
1	2	1	3	195	197	198	196.667	45.8741
2	2	2	6	197	198	199	198.000	45.9331
3	2	3	9	198	199	201	199.333	45.9911
4	4	1	6	198	202	199	199.667	46.0052
5	4	2	9	199	203	201	201.000	46.0631
6	4	3	3	202	204	201	202.333	46.1209
7	6	1	9	208	207	208	207.667	46.3473
8	6	2	3	208	207	211	208.667	46.3882
9	6	3	6	214	210	214	212.667	46.5530

than 0.05 for cupper. This shows that copper is significant factor. The p-values of zinc and magnesium are 0.061 and 0.491 respectively which are greater than 0.05 and hence, are not significant.

Confirmation test was carried out at the optimum level combination of the alloying elements: Cu 6%, Zn-3% and Mg-6%. UTS for the same is obtained as 213.33MPa and is shown in the Table 45.7.

4.2. Hardness Test

Hardness test was carried using micro-Vickers' hardness testing machine. The hardness results are shown in Table 45.8. These test results were transferred into S/N ratios. S/N for larger-the-better characteristic was used, as one of the objectives of the work is to maximise the hardness of the developed castings. Optimum levels of alloying elements are

Table 45.5: Response table for S/N ratio (UTS)

Level	Cu	Zn	Mg
1	45.93	46.08	46.13
2	46.06	46.13	**46.16**
3	**46.43**	**46.22**	46.13
Delta	0.50	0.15	0.04
Rank	1	2	3

Table 45.6: ANOVA test results for S/N ratio (UTS)

Source	DF	Seq SS	Adj SS	Adj MS	F	p-values
Cu	2	0.397986	0.397986	0.198993	185.17	0.005
Zn	2	0.032865	0.032865	0.016433	15.29	0.061
Mg	2	0.002230	0.002230	0.001115	1.04	0.491
Error	2	0.002149	0.002149	0.001075		
Total	8	0.435231				

Table 45.7: Confirmation test results (UTS)

Run No.	Alloying elements (%)			Ultimate tensile strength (MPa)			
	Cu	Zn	Mg	T1	T2	T3	Mean
10	6	3	6	215	213	212	213.33

Table 45.8: Hardness test result and S/N ratio (VHN)

Run No.	Cu (%)	Zn (%)	Mg (%)	Vickers hardness number (VHN)				S/N Ratio (VHN)
				T1	T2	T3	Mean	
1	2	1	3	121.00	119.33	121.00	120.443	41.6151
2	2	2	6	125.10	122.70	125.33	124.377	41.8936
3	2	3	9	128.00	128.33	127.66	127.997	42.1438
4	4	1	6	131.66	129.33	133.33	131.440	42.3725
5	4	2	9	133.00	133.66	129.33	131.997	42.4085
6	4	3	3	131.00	135.33	128.66	131.663	42.3836
7	6	1	9	139.13	140.66	137.33	139.040	42.8615
8	6	2	3	135.33	135.33	141.00	137.220	42.7435
9	6	3	6	144.00	143.33	143.00	143.443	43.1335

384 Advanced Materials for Engineering Sciences

Table 45.9: Response table for S/N Ratio of hardness

Level	Cu	Zn	Mg
1	41.88	42.28	42.25
2	42.39	42.35	42.47
3	42.91	42.55	42.47
Delta	1.03	0.27	0.22
Rank	1	2	3

selected from the response Table 45.9. The optimum combination of elements and their levels to maximise the hardness are Cu 6%, Zinc-3% and Mg-6%.

ANOVA was performed on S/N ratio and p-values are shown in Table 45.10. The p-value of Cu, Zn and Mg is less than 0.05. This shows that all the alloying elements are significant factors.

Confirmation test was carried out with optimum combination of alloying elements- Cu 6%, Zn-3% and Mg-6%. Vickers hardness of the optimum combination of alloying elements casting is 143.443 HV and is shown in the Table 45.11.

5. Conclusions

In this paper, efforts are made to study the effect of major alloying elements Cu, Zn and Mg with LM6 aluminium alloy on the tensile strength and hardness. Castings are developed by varying Cu, Zn and Mg in the range 2%–6%, 1%–3% and 3%–9%, with identified levels. XRD test was conducted to confirm the incorporation of the alloying elements in the developed castings. Microstructure examination conducted using optical microscope to confirm the quality in terms of free from blow holes, cracks and also know the grain boundaries.

S/N ratio analysis and ANOVA are carried out to find the optimum combination of the alloying elements. LM6 after being added with Copper, Zinc and Magnesium tensile strength, showed significant improvement in its hardness. The optimum alloy combinations, identified for tensile strength and hardness are Cu - 6%, Zn - 3% and Mg - 6%. Confirmation tests are carried out and the results are validated. The tensile strength of the developed alloy increased from 166.3Mpa to 213.33Mpa, Vickers' hardness increased from 95HV to 143.443HV when compared with the base LM6 alloy. The newly developed aluminium alloy is useful structural materials for applications in the aerospace, marine and automotive industries. Developed alloy is very much ssuitable for motor car and road transport fittings, water-cooled manifold and jackets, thin section and intricate castings such as housing, meter cases and switchboxes.

Table 45.10: ANOVA Test result for hardness test

Source	DF	Seq SS	Adj SS	Adj MS	F	p
Cu	2	1.58740	1.58740	0.79370	558.16	0.002
Zn	2	0.11961	0.11961	0.05980	42.06	0.023
Mg	2	0.09818	0.09818	0.04909	34.52	0.020
Error	2	0.00284	0.00284	0.00142		
Total	2	1.80803				

Table 45.11: Confirmation test results of Vickers hardness

Run No.	Alloying elements (%)			Vickers hardness			
	Cu	Zn	Mg	T1	T2	T3	Mean
10	6	3	6	144.00	143.33	143.00	143.443

References

[1] Agarwal, M., Singh, A., and Srivastava, R. (2018). Influence of powder-chip based reinforcement on tensile properties and fracture behaviors of LM6 aluminum alloy. *Trans. Indian Inst. Met.*, 71, 1091–1098.

[2] Srinivas, D., MC, G., Hiremath, P., Sharma, S., Shettar, M., and Jayashree, P. K. (2022). Influence of various trace metallic additions and reinforcements on A319 and A356 alloys—a review. *Cogent Eng.*, 9(1), 2007746.

[3] Callegari, B., Lima, T. N., and Coelho, R. S. (2023). The influence of alloying elements on the microstructure and properties of Al-Si-based casting alloys: a review. *Metals*, 13(7), 1174.

[4] Varshney, D. and Kumar, K. (2021). Application and use of different aluminium alloys with respect to workability, strength and welding parameter optimization. *Ain Shams Eng. J.*, 12(1), 1143–1152.

[5] Hamritha, S., Shilpa, M., Shivakumar, M. R., Madhoo, G., Harshini, Y. P., and Harshith, H. (2021). Study of mechanical and tribological behavior of aluminium Metal Matrix Composite reinforced with alumina. *Mater. Sci. Forum*, 1019, 44–50.

[6] [6] Janamatti, S. V., Ganesh Rao, I. N., Rakesh, H., Manasa, T., and Arul Mary, A. (2017). Experimental study on mechanical properties of LM6 metal matrix composite with Ti-boron reinforcement. *Int. Res. J. Eng. Technol.*, 2395(0072), 2461–2468.

[7] Ahmed, S. and Arora, R. (2017). Optimization of turning parameters of Aluminum 6351 T6 using Taguchi decision making technique. *Int. J. Data Network Sci.*, 1(2), 27–38.

[8] Shivakumar, M. R., Hamritha, S., Shilpa, M., and Gouda, K. S. (2020). Optimization of milling process parameters of aluminium alloy LM6 using response surface methodology. *J. Phys.: Conf. Ser.*, 1706(1), 012217.

[9] Avazkonandeh Gharavol, M. H., Haddad-Sabzevar, M., and Haerian, A. (2009). Effect of copper content on the microstructure and mechanical properties of multipass MMA, low alloy steel weld metal deposits. *Mater. Des.*, 30(6), 1902–1912.

[10] Singh, G. and Kumar, A. (2019). A review paper on stir casting of reinforced aluminium Metal Matrix Composite (MMC). *Int. J. Curr. Eng. Technol.*, 9(3), 432–439.

[11] Shivakumar, M. R., Hamritha, S., Shilpa, M., Sobarad, P., and Madhosh Gowda, S. (2022). Optimization of heat treatment parameters to improve hardness of high carbon steel using Taguchi's orthogonal array approach. *Key Eng. Mater.*, 933, 129–136.

[12] Chelladurai, S. J. S., Arthanari, R., Krishnamoorthy, K., Selvaraj, K. S., and Govindan, P. (2018). Effect of copper coating and reinforcement orientation on mechanical properties of LM6 aluminium alloy composites reinforced with steel mesh by squeeze casting. *Trans. Indian Inst. Met.*, 71, 1041–1048.

[13] Nallusamy, S. (2016). A review on the effects of casting quality, microstructure and mechanical properties of cast Al-Si-0.3 Mg alloy. *Int. J. Perform. Eng.*, 12(2), 143.

[14] Sivaprakash, A. and Sathish, S. (2013). Investigation of microstructure and mechanical properties of squeeze cast LM6 alloy with varying contents of Al2O3 and Si3N4-a review. *Int. J. Curr. Eng. Technol.*, 2(2), 207–212.

46 Process automation and smart manufacturing

K. Srujan[1], D. Tejesh[1], and Prashant S. Humnabad[2,a]

[1]Department of Mechanical Engineering, Sir M Visvesvaraya Institute of Technology, Bengaluru, India
[2]Mechanical Engineering, Sir M Visvesvaraya Institute of Technology, Bengaluru India

Abstract

Process automation is revolutionising industries by streamlining operations and enhancing efficiency. Through the integration of advanced technologies like robotics, artificial intelligence, and the Industrial Internet of Things (IIoT), repetitive and manual tasks are being automated. Furthermore, real-time data collection and analysis empower decision-makers with insights to optimise processes, reduce downtime, and improve resource utilisation. Process automation is at the forefront of improving manufacturing, supply chain management, and many other industries, resulting in cost savings and a more agile response to market demands. Smart manufacturing is a transformative paradigm leveraging cutting-edge technologies to revolutionise industrial operations. This approach integrates the Industrial Internet of Things (IIoT), Artificial Intelligence, Automation, and Data Analytics to enhance efficiency and flexibility. Through real-time data collection and analysis, smart manufacturing empowers organisations with actionable insights to optimise processes, reduce production downtime, and improve resource allocation. It reduces the human error and increases productivity; which not only minimises the cost associated with manufacturing but also enables a more agile and responsive approach to customer demands. Smart manufacturing is redefining industries, unlocking new levels of productivity and competitiveness while paving the way for a more intelligent and interconnected future in the industrial landscape.

Keywords: Automation, smart manufacturing, IIOT, artificial intelligence

1. Introduction

Process automation has become an essential driver of efficiency and competitiveness. With the rapid advancements in technology, businesses are increasingly turning to automation to enhance their operations. Process automation leverages tools like robotic process automation, artificial intelligence, and machine learning to streamline and optimise workflows.

These technologies allow organisations to minimise human involvement in routine, repetitive tasks, reducing errors and increasing productivity. In various industries such as manufacturing, finance, healthcare, and customer service, process automation is revolutionising how businesses operate. It not only leads to cost savings but also allows employees to focus on more creative and strategic tasks, ultimately improving the overall performance and adaptability of companies in the ever-evolving industrial landscape. Process automation is thus a critical element for modern businesses to stay competitive and agile.

Smart manufacturing is at the forefront of a technological revolution that is transforming the way products are designed, produced, and delivered. Smart manufacturing, often referred to as Industry 4.0, is a holistic approach that integrates cutting-edge technologies such as the Internet of Things (IoT), artificial intelligence (AI), data analytics, and automation into the manufacturing process.

This approach allows manufacturers to create highly efficient and agile production

asrujankarisiddaiah@gmail.com

DOI: 10.1201/9781003545941-46

systems. It enables real-time data collection and analysis, predictive maintenance, and the seamless communication of machines and systems. Smart manufacturing not only increases productivity and quality but also enhances customisation and flexibility in responding to changing market demands. Furthermore, smart manufacturing fosters sustainability and reduces waste by optimising resource usage. It's a key driver in the current industrial landscape, promoting competitiveness, innovation, and the ability to adapt to rapidly changing market conditions. In essence, smart manufacturing is revolutionising the industry by creating more connected, data-driven, and efficient manufacturing ecosystems.

2. Mathematical Model

2.1. *Mathematical model of Process Automation and Smart Manufacturing*

Comprehensive mathematical model that combines both process automation and smart manufacturing in the present industrial context is a complex task. However, we can provide a simplified mathematical representation that integrates some of the key components of both concepts. Let's consider a generic manufacturing process:

Let:
P: represent the production output.
Q: represent product quality.
T: represent production time.
C: represent production cost.
U: represent resource utilisation.
D: represent data and connectivity factors.
S: represent sustainability metrics.
A: represent the level of automation.
M: represent maintenance costs.

A simplified mathematical model that incorporates both process automation and smart manufacturing could be expressed as:

$$P = f(Q, T, C, U, D, S, A, M)$$

Where:
f is a complex function that relates production output (P) to product quality (Q), production time (T), production cost (C), resource utilisation (U), data and connectivity factors (D), sustainability metrics (S), the level of automation (A), and maintenance costs (M).

The actual form of the function f and the relationships between these variables would depend on the specific manufacturing process, industry, and the technologies being utilised. These relationships can be highly nonlinear, and advanced mathematical techniques, such as optimisation, simulation, and machine learning, may be employed to model and optimise the system effectively.

The parameters, data sources, and algorithms used in this model would be customised to align with the specific goals and requirements of the manufacturing process, which might include maximising efficiency, minimising costs, improving quality, reducing maintenance efforts, enhancing sustainability, and optimising the degree of automation.

In practice, developing and implementing mathematical models for process automation and smart manufacturing often require a multidisciplinary approach, involving experts in mathematics, engineering, data science, and industry-specific knowledge to design and optimise the manufacturing processes effectively.

3. Objectives

Process automation and smart manufacturing are driven by a range of specific objectives aimed at enhancing efficiency, productivity, and competitiveness. Below are detailed objectives for both concepts:

3.1. *Objectives for Process Automation*

- *Minimise Human Intervention*: One of the primary objectives of process automation is to reduce human involvement in repetitive, manual tasks, allowing employees to focus on more strategic and creative responsibilities.

- **Increase Efficiency:** Automation aims to optimise processes, reduce cycle times, and improve resource utilisation, ultimately increasing the overall efficiency of operations.
- **Reduce Errors:** Automation helps eliminate human errors, ensuring consistent and accurate results, which is particularly crucial in industries with strict quality standards.
- **Enhance Cost Savings:** By streamlining processes and reducing labor costs, process automation leads to significant savings, making operations more cost-effective.
- **Improve Scalability:** Automated systems are often designed to be easily scalable, accommodating changes in production volume and demand with minimal disruption.
- **Enable Data-Driven Decision-Making:** Automation generates large amounts of data, supporting informed decision-making and continuous process improvement through data analytics.
- **Enhance Customer Satisfaction:** Efficient and error-free processes lead to improved product quality and on-time deliveries, which in turn enhances customer satisfaction.
- **Comply with Regulations:** In regulated industries, automation helps ensure compliance with industry-specific standards and government regulations.

3.2. Objectives for Smart Manufacturing

- **Real-Time Monitoring and Control:** Smart manufacturing systems provide real-time data on various aspects of production, enabling immediate adjustments and improvements.
- **Predictive Maintenance:** One of the key objectives is to predict when machinery and equipment require maintenance, reducing downtime and maintenance costs.

- **Quality Optimisation:** Smart manufacturing focuses on consistently producing high-quality products by identifying and addressing quality issues in real time.
- **Resource Efficiency:** It aims to optimise resource utilisation, including energy, materials, and labor, to minimise waste and environmental impact.
- **Customisation and Flexibility:** Smart manufacturing allows for more personalised and customised production, responding to varying customer demands efficiently.
- **Data-Driven Decision-Making:** It leverages advanced data analytics and artificial intelligence to make data-driven decisions, optimising processes and product designs.
- **Supply Chain Integration:** Integrating supply chain processes into smart manufacturing ensures better coordination and synchronisation of production with the supply and demand chain.
- **Cybersecurity and Data Protection:** As smart manufacturing relies heavily on data and connectivity, one of the key objectives is to ensure the security and protection of sensitive information from cyber threats.
- **Sustainability and Environmental Impact:** Reducing waste, energy consumption, and emissions are essential objectives, aligning with global efforts to minimise the environmental impact of manufacturing.
- **Enhance Collaboration:** Smart manufacturing fosters collaboration between humans and machines, allowing for more efficient and responsive teamwork.
- **Competitiveness and Innovation:** Ultimately, the objective is to maintain competitiveness and foster innovation by staying at the forefront of technological advancements and industry best practices.

These objectives are not exhaustive and may vary depending on the industry, organisation, and specific goals. In practice, businesses implement process automation and smart manufacturing with a combination of these objectives to achieve a competitive edge in today's industrial landscape.

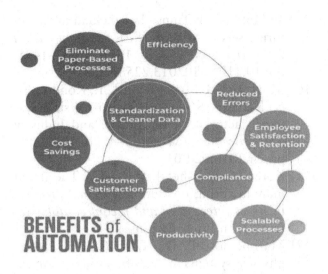

Figure 46.1: Benefits of automation in industrial 4.0

4. Advantages

In the present industrial landscape, process automation and smart manufacturing offer substantial advantages that are transforming the way businesses operate. Process automation streamlines workflows, reducing manual labor and increasing operational efficiency, resulting in significant cost savings and improved product quality. Automated systems consistently perform tasks, allowing for easy scalability and generating valuable data for data-driven decision-making. On the other hand, smart manufacturing provides real-time monitoring and predictive maintenance, minimising downtime and ensuring high product quality. It optimises resource utilisation, enables customisation, and fosters sustainability while promoting flexibility and responsiveness to changing customer demands. These advantages make both process automation and smart manufacturing pivotal components of modern industrial strategies, enhancing productivity, competitiveness, and environmental sustainability.

5. Overview

In the contemporary industrial landscape, process automation and smart manufacturing are paramount. Process automation utilises technology to streamline routine tasks, significantly enhancing efficiency and cost-effectiveness while reducing human errors. It's instrumental in standardising processes, optimising resource utilisation, and ensuring regulatory compliance. Smart manufacturing, on the other hand, leverages cutting-edge technologies like the Internet of Things and artificial intelligence to create agile, data-driven production systems. This approach enables real-time monitoring, predictive maintenance, quality optimisation, and resource efficiency. It also fosters customisation, flexibility, and sustainability, making it a pivotal strategy for modern businesses. In the present industry context, the synergy between process automation and smart manufacturing is revolutionising operations, increasing competitiveness, and propelling organisations towards excellence.

6. Conclusion

Process automation and smart manufacturing stand at the forefront of the present industrial landscape, ushering in a new era of efficiency, competitiveness, and innovation. Process automation's ability to reduce manual labor, minimise errors, and improve productivity is driving cost savings and enhancing product quality. Simultaneously, smart manufacturing, with its integration of advanced technologies, empowers real-time monitoring, predictive maintenance, and data-driven decision-making. This combination leads to remarkable gains in operational efficiency, quality, and sustainability. As industries evolve to meet changing market demands, process automation and smart manufacturing provide the critical tools and strategies necessary to stay ahead. The synergy between these approaches represents a pivotal shift toward a more agile and connected industrial future, where businesses can thrive by optimising their operations and addressing the challenges of the modern world. It is clear that in the present aspect of the industry, embracing both process automation and smart manufacturing is not merely a choice but a necessity for organisations seeking to remain competitive and adaptable in a rapidly changing world.

Acknowledgement

The authors gratefully acknowledge the Professor, Head of the department of Mechanical Engineering for their cooperation in the research.

References

[1] Li, X., Tao, F., Zhang, L., & Sui, F. (2018). Predictive maintenance in manufacturing. *IEEE Transactions on Industrial Informatics*, 14(4), 1835–1845.

[2] Ivanov, D., & Dolgui, A. (2019). Supply chain optimization using IIoT. *International Journal of Production Research*, 57(15–16), 5108–5124.

[3] Shekhar, S. S. (2019). Artificial intelligence in automation. *Research Review International Journal of Multidisciplinary*, 4(6), 14–17.

[4] Rai, R., Tiwari, M. K., Ivanov, D., & Dolgui, A. (2021). Machine learning in manufacturing and industry 4.0 applications. *International Journal of Production Research*, 59(16), 4773–4778.

[5] Al-Fuqaha, A., Guizani, M., Mohammadi, M. et al. (2015). Internet of Things: A Survey on Enabling Technologies, Protocols, and Applications. *IEEE Communications Surveys & Tutorials*, 17, 2347–2376

[6] Lee, J., Bagheri, B., & Kao, H. A. (2015). A cyber-physical systems architecture for industry 4.0-based manufacturing systems. *Manufacturing Letters*, 3, 18–23. doi:10.1016/j.mfglet.2014.12.001

[7] Monostori, L., Kádár, B., Bauernhansl, T., Kondoh, S., & Kumara, S. (2016). Cyber-physical systems in manufacturing. *CIRP Annals*, 65(2), 621–641. doi:10.1016/j.cirp.2016.06.002

[8] Thoben, K.-D., Wiesner, S., & Wuest, T. (2017). Data-driven business models: Challenges and opportunities in the era of big data. *Journal of Business Models*, 5(1), 39–53.

[9] Tao, F., Cheng, Y., Da Xu, L., Zhang, L., & Li, B. H. (2018). CCIoT-CMfg: Cloud computing and Internet of Things-based cloud manufacturing service system. *IEEE Transactions on Industrial Informatics*, 10(2), 1437–1445. doi:10.1109/TII.2013.2258010

[10] Monostori, L., Bauernhansl, T., Kondoh, S., & Kumara, S. (2016). Cyber-physical systems, factories of the future, and Industry 4.0. *Procedia CIRP*, 44, 1–4. doi:10.1016/j.procir.2016.02.001

[11] Koren, Y., & Usher, J. M. (2017). The global manufacturing revolution: *Product-process-business integration and reconfigurable systems*. John Wiley & Sons.

[12] Maropoulos, P. G. (Ed.). (2015). Cyber-physical systems in manufacturing. Springer.

[13] Rüßmann, M., Lorenz, M., Gerbert, P., Waldner, M., Justus, J., Engel, P., & Harnisch, M. (2015). Industry 4.0: The future of productivity and growth in manufacturing industries. Boston Consulting Group.

[14] Tao, F., Zhang, L., Venkatesh, V. C., Luo, Y., & Cheng, Y. (2019). Cloud manufacturing: A computing and service-oriented manufacturing model. *IEEE Transactions on Industrial Informatics*, 10(2), 1437–1445. doi:10.1109/TII.2013.2258010

[15] Lasi, H., Fettke, P., Kemper, H.-G., Feld, T., & Hoffmann, M. (2014). *Industry 4.0. Business & Information Systems Engineering*, 6(4), 239–242. doi:10.1007/s12599-014-0334-4

[16] Xu, L. D., He, W., & Li, S. (2014). Internet of Things in industries: A survey. *IEEE Transactions on Industrial Informatics*, 10(4), 2233–2243. doi:10.1109/TII.2014.2300753

[17] Lu, Y., Xu, X., Wu, D., & Liu, Y. (2017). Smart manufacturing systems for Industry 4.0: Conceptual framework, scenarios, and future perspectives. *Frontiers of Mechanical Engineering*, 12(1), 137–150. doi:10.1007/s11465-017-0447-4

[18] Wang, S., Wan, J., Li, D., & Zhang, C. (2016). Implementing smart factory of Industrie 4.0: An outlook. *International Journal of Distributed Sensor Networks*, 12(1), 3159805. doi:10.1155/2016/3159805

[19] Wei, J., Luo, Y., Xiong, H., & Zhang, L. (2016). Enabling cyber-physical systems for intelligent manufacturing in cloud manufacturing environments. *Journal of Intelligent Manufacturing*, 27(3), 493–510. doi:10.1007/s10845-014-0932-4

47 Strength and microstructural behaviour of areca nut fiber and laterite sand-based sustainable concrete

R. Vijayasarathy[a], Asha Waliitagi, and K. Chandana

Department of Civil Engineering, Atria Institute of Technology, Bangalore, India

Abstract

In concrete, aggregate plays a vital role. The strength and other parameters mainly depend upon the quality of aggregates that we use in the concrete. The aggregates used in concrete are granite stones as coarse aggregates and river sand as fine aggregate, but this leads to severe depletion of natural resources and also sand mining. Sand mining affects the environment and also the ecology system near mining areas. A novel attempt is made by replacing river sand with laterite sand (0% to 100% with an interval of 20%). If the usage of laterite as building material results in good for building construction to some extent sand mining can be avoided. Additional to this to improve the performance of concrete agro waste areca nut fiber (0.5%) was utilised. Since areca nut fiber is one of the waste materials obtained in areca nut farming, which is incorporated in concrete, it would be beneficial for farmers by making areca nut fiber an economical construction material. This novel research aims to investigate the workability, strength properties and microstructural behavior of laterite sand and areca nut fiber-based concrete. The results were found to be in a decrement trend.

Keywords: Areca nut fibre, laterite sand, carbonation and workability

1. Introduction

Introducing natural fibers in concrete leads to eco-friendly and green construction. This research also aims in utilising the economical material that is laterite sand as fine aggregate and also agricultural waste, and areca nut fiber as strength enhancement material. The usage of natural fibers has also been found to be beneficial [1,2]. Natural fibers such as coconut coir, flax, hemp, bamboo, jute, etc. are researched to check their efficiency as fiber reinforcement in concrete. The main parameters to be considered are the length and diameter of the fiber. Jute fiber with optimum length and diameter gave better performance and justified that jute can be used in concrete for making FRC composites [3]. The research on the replacement of fine aggregates is essential to reduce the burden on the requirement of river sand since it is obtained from the mining of rivers. The laterite sand present in the coastal part of Karnataka has a distinct particle size distribution making it a prominent alternative for river sand in structural concrete production.

As the population grows and rates of urbanisation increase, the demand for sand needed in construction also increases. The perennial problem that India is facing is sand mining. Excessive use of sand and gravel causes depletion of river sand. It affects vegetation, causes erosion of soil, pollutes water sources and affects the aquatic species thereby coastal ecosystem is damaged.

[a]rvsarathycivil@gmail.com

DOI: 10.1201/9781003545941-47

Areca nut coir is used for the strength enhancement of the concrete; the aspect ratio of the fiber used is the length and diameter of the fiber [4]. It could be very economical if used and resulted better helps the local areca nut farming farmers also financially. Through testing the properties of the laterite sand and also the laterite sand's characteristics it was classified according to the standards [5]. Industries, marine structures, automobile industries, and aerospace industries have a wide range of applications of low-cost and lightweight concrete in construction. By the process of alkali treatment using a solution of NaOH (6%), the fiber is used in concrete [6]. The areca nut fiber is replaced with cement with varying proportions from 0 to 3% of the weight of the concrete. The compressive strength and flexural strength after 28 days of curing for 0.5% areca nut were found to be optimum. Very limited studies are available on a combination of areca nut fiber and laterite sand. So this research mainly focuses on a combination of areca nut fiber and laterite sand-based concrete.

2. Materials and their Properties

2.1. Materials

In this research areca nut fiber and laterite sand were utilised in concrete. Areca nut coir [7] is obtained from the surrounding places of Chickkaballapura. The physical properties of areca nut fiber are depicted in Table 47.1. laterite sand with specific gravity 2.6 and fineness modulus 3.2 is collected from the coastal areas. Ordinary Portland Cement of 53 grade as specified to IS-4031:2017 [8] was utilised in our experimental work.

Natural coarse aggregates (crushed granite stones) of particle size 20 mm were utilised in our experimental work. The aggregates were tested for their physical properties as per IS: 383:2019 [9]. Crushed stone sand was used as fine aggregate. water free from deleterious materials is used for making concrete.

Table 47.1: Physical properties of areca nut fiber

Physical properties	Values
Diameter	0.36
Density	1.2 g/cm^3
Tensile strength	112 N/mm^2
Water absorption	48%
Aspect ratio	40-70 (L/d)

Laterite (Figure.47.1) is one of the highly formed sediment materials that is formed due to weathering process. The chemical composition of laterite sand varies widely based on the genesis, climate conditions, and age of lateralisation. Lateritic contains more than 60% Fe_2O_3 and little Al_2O_3. The chemical analysis of Indian soils shows that soils are rich in oxides of iron and aluminum but poor in nitrogen, potassium, lime and organic matter. India produces and also consumes areca nut fiber in the large amount. Major States producers Karnataka-40% Kerala-25%, Assam-10% Climate. Areca nut (Figure 47.2) can be grown in different areas which receive an annual rainfall of 750 mm in major parts of Karnataka and 4500 mm malnad of Karnataka. areca nut fiber can be grown in red soil, and also grow in clay fertile loamy soil. Sandy, alluvial, and Sticky clay are not suitable for areca nut cultivation.

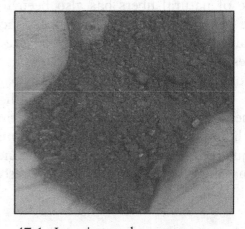

Figure 47.1: Laterite sand.

Figure 47.2: Areca nut fiber.

3. Experimental Investigations

The concrete specimen of size (150*150*150) mm were cast without arena nut fiber and laterite sand as conventional concrete and various percentages of laterite sand along with the areca nut coir of (0.5%) was also cast. A slump test was performed for all the proportions taken. The cubes of size as mentioned were prepared. Machine-mixed concrete was poured into the mould and specimens

Table 47.2: Mix ratio

Cement (kg/m³)	Fine Aggregate (kg/m³)	Coarse Aggregate (kg/m³)	Water Cement ratio (kg/m³)
394.32	854.01	1503.52	197.16
1	2.16	3.81	0.45

Table 47.3: Mix proportion of laterite sand and areca Nut fiber

MIX ID	Areca nut fiber (%)	Laterite sand Replacement (%)
ML0	0	0
ML20	0.5	20
ML40	0.5	40
ML60	0.5	60
ML80	0.5	80
ML100	0.5	100

were prepared simultaneously. After casting the very next day the cube was demoulded and placed in the curing water tank. After 7 days and 28days of curing the specimen were tested for compressive and split tensile strength. The mix ratio is depicted in Table 47.2. The various mix proportion of areca nut fiber and laterite sand is shown in Table 47.3 in which areca nut is used at a uniform percentage of 0.5% and laterite sand varies from 0 to 100% with an interval of 20%.

3.1. Properties of Fresh Concrete

Fresh concrete is normally a wet mix of concrete before it starts to set that is the concrete that is in the plastic state. The slump cone test (Figure 47.3) is the commonly used method for measuring the consistency or workability of concrete. The mould of 300mm frustum cone having a bottom diameter of 200mm and top diameter of 100mm was placed on the smooth leveled surface and the prepared fresh concrete was poured into it and tamped 25 times with 16mm diameter steel rod and top surface was struck with a trowel. The mould was lifted slowly and a decrease in height was measured for each trail mix. A slump cone test has been carried out to check the workability of the mixes. A slump value of around 70mm was observed. Due to the non-addition of superplasticiser, the workability of concrete was

Figure 47.3: Slump cone test.

affected. If superplasticisers were added high slump value might be achieved.

3.2. Compressive Strength Test

As the desirable characteristics of concrete mainly depend qualitatively on the strength of concrete. The compressive strength of the cube test was conducted through a compressive Testing Machine (CTM) (Figure 47.4), the cube cast was cured for 7days and 28days. After the completion of the curing period, the compressive strength of casted cubes was checked through CTM (hardened stage properties). The size and weight of the casted cubes were noted down. The compressive strength was performed according to the IS-516:2021 [10].

The compressive strength test for the casted specimen which was cured for 7days and 28days was conducted.The values are depicted in Figure 47.5. It was found that the compressive strength of all mix that is ML0, ML20, ML40,ML60,ML80 And ML100 were found that 18.96, 16.63, 12.64, 11.90, 11.64 and 10.32 N/mm² respectively for 7days curing, And fo 28days it was found that compressive strength was 35.55, 28.93, 26.15, 24.00, 23.87 and 22.10 N/mm² respectively. The decrease in compressive strength was observed as an increase in laterite sand. This could be because of the high water absorption property of laterite sand due to excess water content the strength may

be poor and no bond between the cement and materials used in the concrete. And also may be due to natural fiber. The addition of some super plasticisers would have improved the strength [11]. Improvement of strength with areca nut and tile powder as an additive has been reported [12]. The control concrete gave higher strength than other mixes.

3.3. Split Tensile Test

The concrete cylinders were casted of size (150mm dia and 300mm height) and cured for 28days, after the completion of the curing period the specimens were subjected to the split tensile test (Figure 47.6). The cylinders cast were placed in the compressive strength machine and subjected to a constant load. The load to the deformation was noted down and split tensile strength is calculated.

Below Figure 47.7 represents the split tensile strength test for the casted specimen which was 28 days aged. As the graph indicates the strength decreased as the % of Laterite increased, that is ML0, ML20, ML40, ML60, ML80 and ML100 were found to be 3.25, 2.54, 1.98, 1.55, 1.27 and 1.19 N/mm² respectively. From the graph, it is observed that as the % of laterite sand increased the split tensile strength decreased [13,14]. Since the casted cylinders contain a constant % of areca nut fiber the split tensile strength was expected more but the results were too low this could be because the natural fiber could not perform better on its strength [15]. We

Figure 47.4: Compressive testing machine.

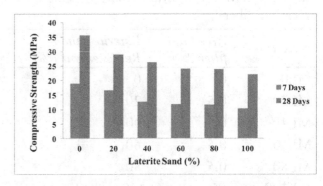

Figure 47.5: Values of Compressive strength test.

Figure 47.6: Split Tensile Test.

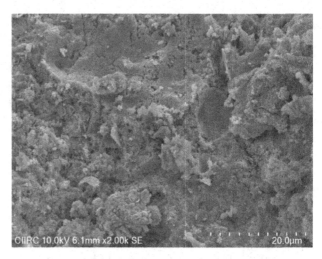

Figure 47.8: SEM image of 0.5% areca nut fiber and 20% laterite sand.

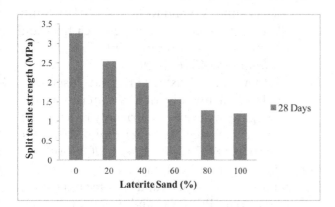

Figure 47.7: Graph of Split Tensile Test for 28days curing.

can overcome this by adding some admixture which usually reduces water content and enhances the strength.

3.4. Microstructural Behavior

From the SEM image of concrete made up of 0.5% areca nut fiber and 20% laterite sand (Figure.47.8) it can be inferred that the addition of laterite sand and areca nut fiber makes the mix porous thereby its strength gets reduced. Microstructural cracks were observed which in turn affects the strength of the concrete. Similar to these results, microstructure of concrete varies for different type of fine aggregate and that influences

the strength property of concrete [16]. The inclusion of laterite sand and areca nut fiber bonding between the concrete materials reduces which results in the development of a weak interfacial zone. The utilisation of 25% laterite sand improved the strength properties of concrete [17]. The formation of white patches indicates the chemical reaction between materials. The hindrance in mixing may also be the reason for the formation of non-clustered concrete.

Conclusions

Based on the tests performed and tests that are going to be performed on various materials to be used in the production of laterite – areca nut fiber-based concrete

- Based on the initial tests performed, it can be concluded that the materials are suitable to experiment on laterite–areca nut fiber-based concrete.
- After testing the specimen for 7 days and 28days curing, the compressive strength of areca nut fiber and laterite sand-based concrete was reduced compared to the conventional concrete.
- The split tensile test results of the areca nut fiber and laterite sand-based concrete

specimen were minimal when compared to control concrete.

- 20% replacement of laterite sand and 0.5% areca nut mix showed better strength than other laterite sand-based concrete
- A combination of areca nut fiber and laterite sand-based concrete without admixtures showed lesser strength

References

[1] Aziz, M. A., Paramasivam, P. and Lee, S. L. (1981). Prospects for natural fibre reinforced concretes in construction. *Int. J. Cem. Compos. Lightweight Concr.* 3 (2) 123–132.

[2] Tioua, T., Kriker, A., Bali, A., Barluenga, G. and Behim, M. (2017). Properties of self-compacting concrete with natural and synthetic fibers. Spec Publ 240:24.1–24.10.

[3] Aftab, M. D. S. Gupta, S.D., Zakaria, H.M., and Karmakar, C.(2020). Scope of improving mechanical characteristics of concrete using natural fibre a reinforcing material. *Malays. J. Civ. Eng.*, 32(2),49–57.

[4] Padmaraj, N.H., Keni, L.G., Chetan, K.N. and Mayur S. (2018). Mechanical characterization of areca husk coir fiber reinforced hybrid composites. *Mater. Today. Proc.*, 5, 1292–1297.

[5] Chandrasasi., Marsudi, S. and Suhartanto, E.(2021). Determination of Types and Characteristics of Laterite Soil as Basic Land for Building Construction. *IOP Conf. Series: Earth and Environmental Science*,930012041 doi:10.1088/1755-1315/930/1/012041

[6] Divakar, L., Babu, A.P.V., Chethan Gowda, R.K., Nithin Kumar, S. (2020). Experimental Study on Mechanical Properties of Areca Nut Fibre-Reinforced Self-compacting Concrete. In: Drück, H., Mathur, J., Panthalookaran, V., Sreekumar, V. (eds) Green Buildings and Sustainable Engineering. Springer Transactions in Civil and Environmental Engineering. Springer, Singapore. https://doi.org/10.1007/978-981-15-1063-2_25

[7] Muralidhar, N., Kaliveeran, V., Arumugam, V., and Srinivasula Reddy, I. (2019). A study on areca nut husk fiber extraction, composite panel preparation and Mechanical characteristics of the composites. *J. Instit. Eng.*

[8] IS: 4031 Part-11 (1988). Methods of physical tests for hydraulic cement. Bureau of Indian Standards, New Delhi, India.

[9] IS: 383 (2016). Specification for Coarse and Fine Aggregates from Natural Sources for Concrete. Bureau of Indian Standards, New Delhi, India.

[10] IS: 516 (1959). Methods of tests for strength of concrete. Bureau of Indian Standards, New Delhi, India.

[11] Itagi, M., and Annapurna, B.P. (2019). Evaluation of strength properties of hybrid fibre (plastic +coir +areca nut husk) reinforced concrete. *Int. J. Innov. Technol. Explor. Eng,* 8(8),3098-3101

[12] Aishwarya, G.I., Dinesh, T., Neha Andrews., Sreelekshmi, V.S and Resmi, V. (2019). Study of strength of concrete using areca fiber and tile powder as additives. *Int. Res. J. Eng. Technol.*6(4),4016-4019

[13] Mathew, B., Christy, C.F., and Joseph, V. (2018). An investigation of laterite as fine aggregate to develop ecofriendly mortar. IOP conference Series. *Mater. Sci. Eng*,431(8)2006 doi:10.1088/1757-899X/431/8/082006

[14] Temitope, F. A., Adebayo, O. S. and Olasehinde A. (2017). SDA and laterite applications in concrete: Prospects and effects of elevated temperature. Department of civil engineering, Ekiti State university, Ekiti state, Nigeria.

[15] Ashok, R.B., Srinivas, C.V., and Basavaraju, B.(2018) A review on the mechanical properties of areca fiber reinforced composites. *science and technology materials*.30(2),120-130

[16] Waliitagi, A., Rathanasalam. V.S., Kishore, H.R.B., and Chithambaram, S.J.(2023). The implications of sustainable fine aggregate on self-compacting concrete: a review, IOP Conf. Ser.: *Mater. Sci. Eng.* 1273, 012007.

[17] Raja, R., and Vijayan, P.(2019). Strength and microstructural behaviour of concrete incorporating laterite sand in binary blended cement. Department of Civil Engineering, Guindy, Anna university, Chennai, Tamil Nadu.

Printed in the United States
by Baker & Taylor Publisher Services